Environmental History

Volume 17

Series Editor

Mauro Agnoletti, Florence, Italy

The series intends to act as a link for ongoing researches concerning the historical interrelationships between man and the natural world, with special regard to the modern and contemporary ages. The main commitment should be to bring together different areas of expertise in both the natural and the social sciences to help them find a common language and a common perspective. Interdisciplinarity and trans-disciplinarity are needed for more and better understanding of the environment and its history, with new epistemological frameworks and methodological practices. The links between human activities and flora, fauna, water, soil, are examples of the most debated topics in EH, while established discipline like forest history, agricultural history and urban history are also dealing with it. The human impacts on ecosystems and landscapes over time, the preservation of cultural heritage, studies of historical trajectories in pattern and processes, as well as applied research on historical use and management of landscapes and ecosystems, are also taken into account. Other important topics relate to the history of environmental ideas and movements, policies, laws, regulations, conservation, the history of immaterial heritage, such as traditional knowledge related to the environment.

Borna Fuerst-Bjeliš · Jelena Mrgić · Hrvoje Petrić ·
Matija Zorn · Žiga Zwitter
Editors

Environmental Histories of the Dinaric Karst

Springer

Editors
Borna Fuerst-Bjeliš
Department of Geography
Faculty of Science
University of Zagreb
Zagreb, Croatia

Jelena Mrgić
Department of History
Faculty of Philosophy
University of Belgrade
Belgrade, Serbia

Hrvoje Petrić
Department of History
Faculty of Humanities and Social Sciences
University of Zagreb
Zagreb, Croatia

Matija Zorn
ZRC SAZU
Anton Melik Geographical Institute
Ljubljana, Slovenia

Žiga Zwitter
Department of History
Faculty of Arts
University of Ljubljana
Ljubljana, Slovenia

ISSN 2211-9019 ISSN 2211-9027 (electronic)
Environmental History
ISBN 978-3-031-56088-0 ISBN 978-3-031-56089-7 (eBook)
https://doi.org/10.1007/978-3-031-56089-7

This Springer imprint is published by the registered company Springer Nature Switzerland AG
The registered company address is: Gewerbestrasse 11, 6330 Cham, Switzerland

Paper in this product is recyclable.

Preface

The cooperation in the field of environmental history in the Dinaric Karst countries[1] started in the early twenty-first century. The journal *Economic- and Ecohistory: Journal for Economic History and Environmental History* (*Ekonomska i ekohistorija: Časopis za gospodarsku povijest i povijest okoliša*), published in Zagreb since 2005, attracted researchers from Croatia, its neighbouring Dinaric Karst countries, and environmental historians from elsewhere in Europe and beyond. In addition, a small but intensive international environmental history conference titled *Man, Nature and Environment between the Northern Adriatic and the Eastern Alps in Premodern Times*, with 23 presentations from Austria, Croatia, Hungary, Italy, Serbia, Slovenia, and Switzerland, was held in Ljubljana in 2014. The results were published in an edited volume[2]—including Jelena Mrgić's study *Rocks, Waters and Bushes—What Did the Ragusan Commune Acquire from the Bosnian King in 1399?*[3] and Dušan Mlacović's paper *The Communes of the Northern Istria in the Late Middle Ages and Wood*,[4] contributions to the mediaeval (and early modern) environmental history of the Dinaric Karst. The environmental history cooperation in the Dinaric Karst countries was further strengthened by the edited volume *Environmentalism in Central and Southeastern Europe: Historical Perspectives* (2017),[5] which included two Dinaric Karst environmental history papers: *The Origins and Developmental Course of the Plitvice Lakes National Park: From "The Devil's Garden" to the UNESCO World Natural Heritage Register* by Ivan Brlić and Anita Bušljeta Tonković,[6] and *Contrasting the "Sunny Side": Goli Otok and the Islandness of a Political Prison in the Croatian Adriatic Sea* by Milica Prokić.[7]

[1] Kranjc (2004).

[2] Štih and Zwitter (2014).

[3] Mrgić (2014).

[4] Mlacović (2014).

[5] Petrić and Žebec Šilj (2017).

[6] Brlić and Bušljeta Tonković (2017).

[7] Prokić (2017).

In 2017, one of the particularly significant gatherings of environmental historians in the world, the large multi- and interdisciplinary biennial conference titled *Natures in between. Environments in areas of contact among states, economic systems, cultures and religions,* under the auspices of the *European Society for Environmental History* (ESEH), was hosted by the Department of History at the Faculty of Humanities and Social Sciences and the Department of Geography at the Faculty of Science, University of Zagreb. The conference opened with the guest plenary lecture *From Determinism to Complexity: Historical Disaffiliation in Climate Change and Migration* by Andrew W. A. Baldwin. The programme included the plenary roundtable *Trespassing: Environmental History and the Challenges of Migrations*, 291 further conference lectures, a plenary poster session with twenty presentations, eight more roundtables, and two extensive discussions, all held in four days. The conference was attended by participants from several continents.[8] One of the key motivations behind the organisers' decision to host the conference was to bring more researchers from south-east Europe and its surroundings closer to the European and worldwide mainstream environmental history research. At the ESEH General Assembly held during the Zagreb conference, Prof. Borna Fuerst-Bjeliš—an environmental historian and historical geographer, a member of the local organising committee, and one of the main conference organisers—successfully ran for the post of the ESEH vice-president. In her plenary talk, she emphasised the need to dedicate more energy to the research of the Dinaric Karst's environmental history in the near future. The initiation of an edited volume on the environmental histories of the Dinaric Karst, aimed at increasing the visibility of environmental history research from south-east Europe and its vicinity, was a part of her ESEH vice-presidency programme for the 2017–2019 period.

The existing ESEH regions of Croatia and Serbia, and Slovenia—where some environmental history, historical ecology, environmental archaeology, and historical geography research had been carried out even before the country was included in any of the ESEH regions—should form a Croatian–Serbian–Slovenian region, following the model of the successful ESEH regions spanning across several countries, as proposed by the environmental historian Prof. Hrvoje Petrić. At that time, Petrić was the ESEH regional representative for Croatia, the editor-in-chief of the journal *Economic- and Ecohistory*, and, together with Fuerst-Bjeliš, a member of the local organising committee and one of the main 2017 Zagreb ESEH conference organisers. His proposal aimed to foster the cosmopolitan environmental history in the region and beyond. The ESEH members from Croatia, Serbia, and Slovenia fully supported this idea. At the 10th biennial ESEH conference, held in Tallinn in 2019, the ESEH General Assembly endorsed this initiative as well,[9] and thus, the ESEH region Croatia–Serbia–Slovenia was established.

The events and ideas presented above paved the way towards the volume titled *Environmental Histories of the Dinaric Karst* in Springer's book series *Environmental History*. An international volume focusing on this topic was one of the main

[8] European Society for Environmental History (2017).

[9] Zwitter (2017); Zwitter (2019).

References

Brlić I, Bušljeta Tonković A (2017) The origins and developmental course of the Plitvice Lakes National Park: from 'the devil's garden' to the UNESCO World natural heritage register. In: Petrić H, Žebec Šilj I (eds) Environmentalism in Central and Southeastern Europe: historical perspectives. Lexington Books, Lanham, Boulder, New York, London, pp 175–196

European Society for Environmental History: biennial conference (2017). European Society for Environmental History, Zagreb. http://eseh.org/wp-content/uploads/ESEH2017PrintedProgramme.pdf

Kranjc A (2004) Dinaric Karst. In: Gunn J (ed) Encyclopedia of caves and karst science. Fitzroy Dearborn, New York and London, pp 287–288. https://doi.org/10.4324/9780203483855

Mihevc A, Prelovšek M, Zupan Hajna N (eds) (2010) Introduction to the Dinaric Karst. Karst Research Institute at ZRC SAZU, Postojna

Mlacović D (2014) The communes of the northern Istria in the Late Middle Ages and wood. In: Štih P, Zwitter Ž (eds). Man, nature and environment between the Northern Adriatic and the Eastern Alps in premodern times. Ljubljana University Press, Faculty of Arts, and Historical Association of Slovenia, Ljubljana, pp 86–101

Mrgić J (2013) Zemlja i ljudi: Iz istorije životne sredine Zapadnog Balkana [Land and people: from the environmental history of the Western Balkans]. Equilibrium, Beograd

Mrgić J (2014) Rocks, waters and bushes—what did the Ragusan commune acquire from the Bosnian king in 1399? In: Štih P, Zwitter Ž (eds) Man, nature and environment between the Northern Adriatic and the Eastern Alps in premodern times. Ljubljana University Press, Faculty of Arts, and Historical Association of Slovenia, Ljubljana, pp 52–73

Petrić H, Žebec Šilj I (eds) (2017) Environmentalism in Central and Southeastern Europe: historical perspectives. Lexington Books, Lanham, Boulder, New York, London

Prokić M (2017) Contrasting the 'sunny side': Goli Otok and the islandness of a political prison in the Croatian Adriatic Sea. In: Petrić H, Žebec Šilj I (eds) Environmentalism in Central and Southeastern Europe: historical perspective. Lexington Books, Lanham, Boulder, New York, London, pp 197–222

Štih P, Zwitter Ž (eds) (2014) Man, nature and environment between the Northern Adriatic and the Eastern Alps in premodern times. Ljubljana University Press, Faculty of Arts, and Historical Association of Slovenia, Ljubljana

Zwitter Ž (2017) 9. bienalna konferenca Evropskega združenja okoljskih zgodovinarjev: Natures in between: environments in areas of conflict among states, economic systems, cultures, and religions [9th European society for environmental history biennial conference: Natures in between: environments in areas of conflict among states, economic systems, cultures, and religions]. Econ and Ecohist: J Econ Hist Environ Hist 13:167–168

Zwitter Ž (2019) 10. bienalna konferenca European society for environmental history: boundaries in/of environmental history, Talin, 21–24. August 2019 [10th European society for environmental history biennial conference: boundaries in/of environmental history, Tallinn, 21–24 August 2019]. Econ and Ecohist: J Econ Hist Environ Hist 15:159–160 planned joint activities on the Croatia–Serbia–Slovenia region's agenda for the 2019–2024 period. Namely, the Dinaric Karst is one of the two environmental macroregions that connects all countries in this ESEH region and beyond. Consequently, fostering environmental history research in the Dinaric Karst is one of the possible strategies for raising the research interest in environmental history in the broader region and strengthening the position of south-east Europe and its surroundings in the European research network.

This inter- and multidisciplinary volume, linking humanities and natural sciences, consists of sixteen chapters written by thirty-one authors and editors, researchers from twenty institutes, departments, as well as cultural and natural heritage protection institutions in seven European countries. While most authors are geographers and historians, four archaeologists, an archaeobotanist, a forestry scientist, a palynologist, and a researcher at the interface of biology and climatology also participate.

The volume's editors have research and academic teaching and outreach experience in environmental history and historical geography. The editorial board includes Borna Fuerst-Bjeliš and Hrvoje Petrić as the ESEH members from Croatia, Jelena Mrgić, the former ESEH regional representative for Serbia and author of the environmental history university textbook *Land and People: From the Environmental History of the Western Balkans* (2013),[10] as well as Matija Zorn and Žiga Zwitter, the ESEH members from Slovenia (the latter has also been the regional representative for the Croatia–Serbia–Slovenia region since 2019). The editors work at three leading universities with the longest academic tradition at the state level—the University of Belgrade (the Department of History at the Faculty of Philosophy), the University of Ljubljana (the Department of History at the Faculty of Arts, i.e. the humanities), and the University of Zagreb (the Department of Geography at the Faculty of Science and the Department of History at the Faculty of Humanities and Social Sciences)—as well as at the ZRC SAZU (Research Centre of the Slovenian Academy of Sciences and Arts) Anton Melik Geographical Institute in Ljubljana.

The volume editors wish to thank the researchers for their contributions, the authors' and editors' institutions and projects for their support, the anonymous reviewers for their thorough reviews, and Springer's editors Doris Bleier, Rajan Muthu, and all the Springer production team for their outstanding support.

Zagreb, Croatia — Borna Fuerst-Bjeliš
Belgrade, Serbia — Jelena Mrgić
Zagreb, Croatia — Hrvoje Petrić
Ljubljana, Slovenia — Matija Zorn
Ljubljana, Slovenia — Žiga Zwitter
July 2024

[10] Mrgić (2013).

Contents

Editors and Contributors

About the Editors

Borna Fuerst-Bjeliš University of Zagreb, Faculty of Science, Department of Geography, Zagreb, Croatia
Senior researcher and full professor. Her research interests include environmental history and landscape change, focusing spatially mainly on the Mediterranean karst region, including the development of research methods (historical cartography and GIS technology), borderlands, spatial perception, and spatial (regional) identities. She is the (co-)author/(co-)editor of books on topics of *Historical Geography of Croatia*, *Mediterranean Identities*, and *Globalization, Marginalization and Conflict*. She served as a vice-president of the *European Society for Environmental History* (2017–19), and she is currently the chair of the *Commission on Marginalization, Globalization and Regional and Local Responses of the International Geographical Union* (2020–present).

Jelena Mrgić University of Belgrade, Faculty of Philosophy, Department of History, Belgrade, Serbia
Associate professor. She teaches *Historical geography* and *Introduction to environmental history*. Her focus in research changed from historical geography of the mediaeval and early Ottoman Bosnia to preindustrial environmental history of the Western Balkans, changes of settlement patterns and cultural landscapes, wine and brandy production history in lieu of climate changes during the LIA, to emotional history as an aspect of studying environment, past and recent crises. She has been a representative of Serbia in the *European Society for Environmental History* (ESEH) since 2010, now jointly represented in the Region of Slovenia-Croatia-Serbia. She has written three books, several book chapters in edited volumes, and research articles in peer-reviewed journals and is active in reviewing papers for research journals in environmental history.

Hrvoje Petrić Department of History at the Faculty of Humanities and Social Sciences, University of Zagreb, Zagreb, Croatia
Professor of history and head of the Chair of Mediterranean, Central and Southeast European History. His research deals with environmental history, economic history, and Croatian early modern history. He has written several books and a series of articles (more than 20). He is the editor of "Ekonomska i ekohistorija" [*Economic and Environmental History*] Journal. He served as a representative of Croatia (2011–2019) and board member (2015–2019) of the European Society for Environmental History. He is currently vice-president of the *Historical Association of Croatia* and president of the *Society for Croatian Economic History and Environmental History.*

Matija Zorn ZRC SAZU Anton Melik Geographical Institute, Ljubljana, Slovenia
Head of the Anton Melik Geographical Institute of the Research Centre of the Slovenian Academy of Sciences and Arts (Ljubljana, Slovenia) and a lecturer at the Faculty of Humanities of the University of Primorska (Koper, Slovenia). He is a geographer and a historian, Ph.D. His main research interest includes physical geography (especially geomorphology), natural hazards, land degradation, environmental history, historical geography, and geographic information systems. Most recently he co-edited a book titled *The Geography of Slovenia: Small But Diverse* (Springer Nature). He is the chairman of the *Commission on Land Degradation and Desertification of the International Geographical Union.*

Žiga Zwitter Department of History, Faculty of Arts, University of Ljubljana, Ljubljana, Slovenia
Assistant professor of early modern history. He graduated in history and geography. In 2015 he obtained his Ph.D. in history. In 2016–2017 he was a postdoctoral research fellow at the Vienna's Institute of Social Ecology. His research has focused on environmental history. He has been leading the Croatia–Serbia–Slovenia region in the *European Society for Environmental History* and has served as a vice-president of the *Historical Association of Slovenia.* He is the main author of the forthcoming monograph *Historical Biodiversity in the Alps: Grassland Agroecosystems in the Last Millennium.*

Contributors

Maja Andrič ZRC SAZU Institute of Archaeology, Ljubljana, Slovenia

Ante Blaće Department of Geography, University of Zadar, Zadar, Croatia

Anica Čuka Department of Geography, University of Zadar, Zadar, Croatia

Marin Cvitanović Department of Life and Environmental Sciences, Faculty of Science and Technology, Bournemouth University, Poole, UK

Anamarija Durbešić Department of Forestry, Sector for Planning, Analysis, Forestry and Informatics, Hrvatske šume d.o.o., Zagreb, Croatia

Tomaž Fabec Institute for the Protection of the Cultural Heritage of Slovenia, Ljubljana, Slovenia

Josip Faričić Department of Geography, University of Zadar, Zadar, Croatia

Borna Fuerst-Bjeliš Department of Geography, University of Zagreb Faculty of Science, Zagreb, Croatia

Mauro Hrvatin ZRC SAZU Anton Melik Geographical Institute, Ljubljana, Slovenia

Dirk Nikolaus Karger Swiss Federal Research Institute WSL, Birmensdorf, Switzerland

Klemen Kocjančič Defence Research Centre, University of Ljubljana's Faculty of Social Sciences, Ljubljana, Slovenia

Bine Kramberger Institute for the Protection of Cultural Heritage of Slovenia (ZVKDS), Centre for Preventive Archaeology (CPA), Ljubljana, Slovenia

Ivan Laković Historical Institute of the University of Montenegro, Podgorica, Montenegro

Jelena Lazarević Biotechnical Faculty of the University of Montenegro, Podgorica, Montenegro

Philip Mason Institute for the Protection of Cultural Heritage of Slovenia (ZVKDS), Centre for Preventive Archaeology (CPA), Ljubljana, Slovenia

Martin Meiske Research Institute for the History of Science and Technology at the Deutsches Museum, Munich, Germany

Dimitrij Mlekuž Vrhovnik Department of Archaeology, University of Ljubljana's Faculty of Arts, Ljubljana, Slovenia;
Institute for the Protection of the Cultural Heritage of Slovenia, Ljubljana, Slovenia

Jelena Mrgić Department of History, University of Belgrade Faculty of Philosophy, Belgrade, Serbia

Filip Pavelić Ph.D. Student, Catholic University of Croatia, Zagreb, Croatia

Hrvoje Petrić Department of History, University of Zagreb Faculty of Humanities and Social Sciences, Zagreb, Croatia

Petra Radeljak Kaufmann Department of Geography, University of Zagreb Faculty of Science, Zagreb, Croatia

Mislav Radošević Institute of Social Ecology (SEC), University of Natural Resources and Life Sciences (BOKU), Vienna, Austria

Meta Remec Institute of Contemporary History, Ljubljana, Slovenia

Daniela Ribeiro ZRC SAZU Anton Melik Geographical Institute, Ljubljana, Slovenia

Aneja Rože Kravanja Municipality of Ilirska Bistrica, Ilirska Bistrica, Slovenia

Matej Simčič Institute of the Republic of Slovenia for Nature Conservation, Novo Mesto Regional Unit, Novo Mesto, Slovenia

Andrej Studen Institute of Contemporary History, Ljubljana, Slovenia

Ivan Tekić Oikon Ltd. - Institute of Applied Ecology, Zagreb, Croatia

Tjaša Tolar ZRC SAZU Institute of Archaeology, Ljubljana, Slovenia

Matija Zorn ZRC SAZU Anton Melik Geographical Institute, Ljubljana, Slovenia

Žiga Zwitter Department of History, University of Ljubljana's Faculty of Arts, Ljubljana, Slovenia

Chapter 1
Some Insights into the Environmental History of the Dinaric Karst

Žiga Zwitter, Borna Fuerst-Bjeliš, Jelena Mrgić, Hrvoje Petrić, and Matija Zorn

Abstract The explanations of the terms *environmental history* and *karst* are followed by an introduction to the Dinaric Karst and a selection of groundbreaking examples of the early intellectual history of karst research in this environmental macroregion, placing the Dinaric Karst on the European intellectual history map. Then, the scope and contents of the volume *Environmental Histories of the Dinaric Karst* are outlined. The research presented in this monograph is based on several classic research topics of the interactions between human society and the rest of nature in karst, but it implements fresh, modern environmental history approaches.

Keywords Environmental history · Dinaric Karst · Karst science · Intellectual history · Landscape change

Ž. Zwitter (✉)
Department of History, University of Ljubljana's Faculty of Arts, Ljubljana, Slovenia
e-mail: ziga.zwitter@ff.uni-lj.si

B. Fuerst-Bjeliš (✉)
Department of Geography, University of Zagreb Faculty of Science, Zagreb, Croatia
e-mail: bornafb@geog.pmf.unizg.hr

J. Mrgić
Department of History, University of Belgrade Faculty of Philosophy, Belgrade, Serbia
e-mail: jmrgic@f.bg.ac.rs

H. Petrić
Department of History, University of Zagreb Faculty of Humanities and Social Sciences, Zagreb, Croatia
e-mail: hrvoje.petric@ffzg.unizg.hr

M. Zorn
ZRC SAZU Anton Melik Geographical Institute, Ljubljana, Slovenia
e-mail: matija.zorn@zrc-sazu.si

B. Fuerst-Bjeliš et al. (eds.), *Environmental Histories of the Dinaric Karst*, Environmental History 17, https://doi.org/10.1007/978-3-031-56089-7_1

1.1 Introduction

Forsaking and transcending the contaminated notions of the (Western) Balkans, this volume bears a title that rises above divergences and particularities. It explores the "*history of the mutual relations between humankind and the rest of nature*,"[1] "*socioecological formations through time*," as the overarching research topic in environmental history, which sees nature and human society as intertwined and transcends the usual disciplinary boundaries of history and natural sciences.[2] The concept of *environment* can be defined in various ways. This volume follows Ian Gordon Simmons's definition of the environment as the elements of nature in relation to human beings, where there is a transfer of energy between them.[3]

The research focuses on the Dinaric Karst (Fig. 1.1), an environmental and transpolitical region, in accordance with John Robert McNeill's observation that political boundaries rarely matter in the history of ecological processes and interactions between human society and nature. Thus, environmental regions usually represent more relevant study areas than political regions in environmental history.[4] The Dinaric Karst is the origin of the term "karst", which is used globally in the sense of "*rock dissolutional processes and the landforms and groundwater systems derived from them*".[5] In the context of global karst regions, the Dinaric Karst is an extensive area possessing a high degree of landscape evolution complexity.[6]

Geographically, the Dinaric Karst is the carbonate part of the Dinaric Alps,[7] which form the western part of the Balkan Peninsula between north-eastern most Italy and southern Slovenia in the northwest and northern Albania or even northwestern Greece[8] in the southeast. It covers an area of about 70,000 km^2 and extends over 650 km in eight or nine countries depending on the definition[9]: Italy, Slovenia, Croatia, Bosnia and Herzegovina, Serbia, Montenegro, Kosovo, Albania, and, according to some authors, also Greece. If we define it as a region in eight countries (Fig. 1.1), its boundaries within the Dinaric Alps more or less follow the boundary of the carbonate rocks, although some non-carbonate rocks (e.g. flysch) are also present. Its northern limit is the boundary between the Alps and the Dinaric Alps. The western end is located at the Adriatic Sea, so the Dinaric Karst also includes the islands in eastern Adriatic and several submarine environments. The eastern

---

[1] McNeill (2003) (quote from page 6); see also Winiwarter and Knoll (2007).

[2] Armiero (2024).

[3] Simmons (2008), 3 *et passim.*

[4] McNeill (2003), pp. 18, 35.

[5] Ford (2004); Ford (2007), pp. 677–678 (quote from page 677), 683; Gams (1973), pp. 45–50; Gams (1993); Gams (2004), pp. 11–12, 20–22; Ferk and Zorn (2015), pp. 368–374.

[6] Ferk and Zorn (2015), pp. 372–373; Ford (2007), pp. 677–678, 683.

[7] Gams (1974), p. 270; Gams (1993), p. 110; Gams (2004), p. 21; Herak (1965); Lučić (2019), pp. 20–21; Roglić (1965); Roglić (1969); Zupan Hajna (2019), p. 354.

[8] Gams (1993), p. 110.

[9] See Footnote 7. The territory measuring about 70,000 km^2 includes the area on Fig. 1.1 as well as some submarine karst environments.

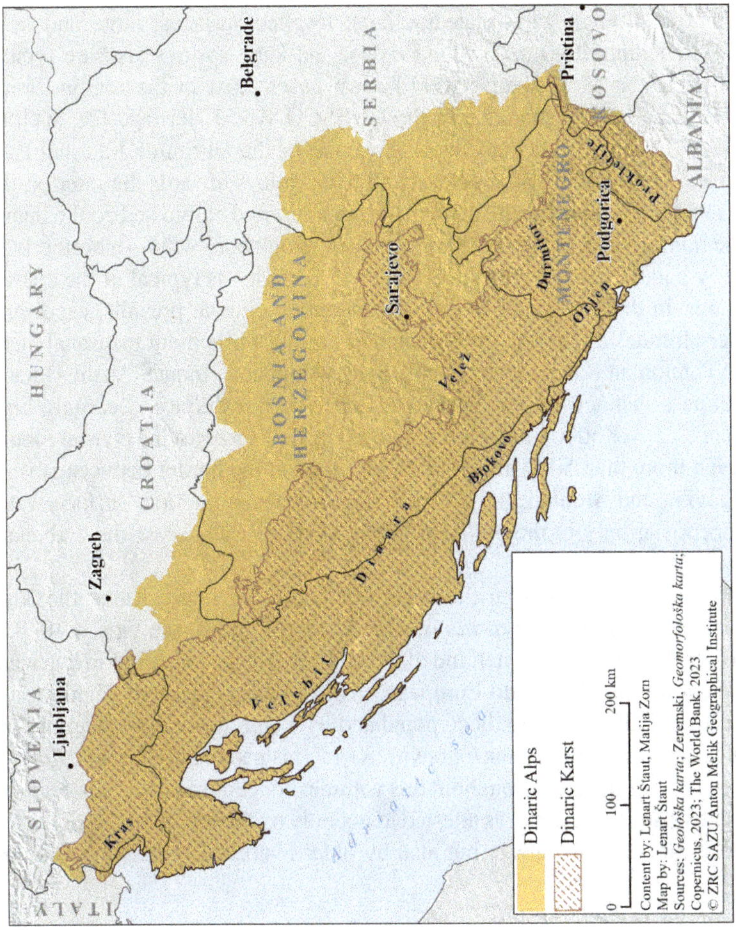

Fig. 1.1 Map of the Dinaric Karst. *Source* Own elaboration

boundary is not entirely clear, as there is much alternation between carbonate and non-carbonate rocks within the Dinaric Alps. The south-eastern boundary is also not very distinct but roughly follows the boundary of the carbonate rocks in the Prokletije Mountains.[10]

The most typical karst features in the Dinaric Karst are karst depressions of various sizes: *dolines*, collapse *dolines*, *uvalas*, and *poljes*[11] (over 40–see Fig. 12.4f): "*nowhere else in the world is there such a pattern of large, complexly linked, tectonic* poljes."[12] The distribution of *dolines* can be very dense, up to 200 *dolines*/km^2 (Figs. 4.3 and 4.7). High karst plateaus, large levelled surfaces, large and deep caves (even exceeding 1000 m), sinking rivers, and karst springs also are typical of the Dinaric Karst. More than 20,000 known caves exist in the region. Some geomorphosites have been included in the UNESCO World Heritage list, such as Plitvice Lakes (Croatia), Škocjan Caves (Slovenia) or the Durmitor National Park (Montenegro).[13] Along the Adriatic coast and river valleys towards the interior, the climate is (sub)Mediterranean with dry and hot summers and humid and cool winters. The average rainfall there is about 800 mm/year. A strong bora wind—a northerly to north-easterly katabatic wind with speeds up to 200 km/h—is typical of the coastal zone in winter. In the interior of the region, mountain climate prevails, especially at the higher altitudes and in the inter-mountain basins, while the transitional areas towards the Pannonian Basin feature a temperate continental climate. As the Dinaric Alps constitute a distinct orographic barrier, some parts exhibit exceedingly high precipitation, e.g. over 3000 mm/year in the Bosnia and Herzegovina coastal mountains and even more than 5000 mm/year on Mt Orjen at the border between Bosnia and Herzegovina and Montenegro. Despite such high precipitation, surface water is sparse, occurring only in the *poljes* or larger rivers that can cross the carbonate rocks.[14]

The word "karst" comes from the north-westernmost Dinaric Karst, the Kras Plateau (It. *Carso*, from Lat. *Carusadius*, *Carsus*; Germ. *Karst*, see Figs. 4.10, 8.1, 11.1, and 16.2), located in the hinterland of the Gulf of Trieste, today in Slovenia and Italy. The word's origin is pre-Indo-European.[15] Since during the last millennium and beyond, the Kras Plateau has not been populated by a significant percentage of the German-speaking population,[16] the toponym "*Kras*" (instead of the German "*Karst*") is used to refer to this plateau throughout this volume. For centuries, this has been the most frequently used word form, as attested to not only by the toponym "*Kras*" in the contemporary Slovenian language but also by older historical sources mentioning

[10] Ford (2007), p. 677; Gams (1969), p. 565; Gams (1993), p. 110; Zupan Hajna (2019), p. 353.

[11] Gams (1969), pp. 567, 569; Zupan Hajna (2019), p. 358.

[12] Ford (2007), p. 683.

[13] Gams (1969), p. 564; Zupan Hajna (2019), pp. 353, 358, 360.

[14] Gams (1969), p. 567; Zupan Hajna (2019), pp. 353–354, 356.

[15] Ford (2004); Ford (2007), p. 677; Gams (1973), pp. 45–50; Gams (1993); Gams (2004), pp. 11–12, 20–22; Ferk and Zorn (2015), pp. 368–374.

[16] Bajt and Vidic (2011), pp. 79, 136, 149, 159.

the vernacular toponym "*Grast*" in 1177,[17] "*Krass*" in a sixteenth-century Slovenian manuscript, as well as in several Slovenian quotes recorded in the seventeenth and eighteenth centuries.[18]

Expressions describing diverse karst phenomena have long existed in regional languages and dialects. In the eyes of today's geomorphology, some of these terms refer to a single geomorphic form, while others are names for a broader group of geomorphic phenomena.[19] The word "*pećina*"—which refers to a cave in some parts of Croatia, throughout Serbia and Montenegro, and in most of Bosnia and Herzegovina, as reported in the 1980s[20]—was recorded in the toponym "*Pechice*" in the eleventh century, "*Pechie*" in the twelfth century (not every letter always fully reflects the contemporaneous use in the records of mediaeval toponyms), "*Pechina*" in 1652, and "*Pechine osia Grotta*" on an eighteenth-century map. All these eleventh-to-eighteenth-century toponyms referred to the cave Pećina, a mediaeval possession boundary located in a prominent position on the Dinaric Karst Adriatic island of Ugljan near Zadar.[21] Some forty kilometres north of the northern Dinaric Karst, in the Southern Limestone Alps of Dleskovec Plateau (Sln. *Dleskovška planota*), the Slovenian word "*zijalka*" ("*auf zwo oder dreÿ stainen grẅben, so … sÿaukha genant werrden*") was recorded in a manuscript containing court testimonies about the boundaries of a compact holding of an isolated mountain farm (the Planinšek farm), collected in the year 1588.[22] In the regional dialects, "*zijalka*" means a cave, often useable as a shelter.[23] The spelling mistakes contained in the quote indicate that the translator of the Slovenian testimonies to the official German language was not in perfect command of the target language. Hence, his broader translation of "*zijalka*" is awkward: in the regional dialects, the term "*zijalka*" excludes *dolines* as well as rock shelters.[24] Well aware of the imprecise German translation, in 1588, the author of the text added the Slovenian word for the sake of clarity. Some regional Slavic terms

[17] Gams (1973), p. 46.

[18] Golec (2015a), pp. 446–448, 451; Ferk and Zorn (2015), pp. 369–370; Panjek (2015), p. 15; see also Gams (1973), pp. 45–46.

[19] Compare Gams (1973), pp. 51–53; Malez (1984), pp. 76, 91.

[20] Božić (2020), pp. 1–2, 5–6.

[21] Malez (1984), pp. 74–80. Erroneously, Malez claims that a map dating back to 1166 has been preserved, where the spelling of this toponym largely corresponds to the present one for the first time ("*Pechine*"). On page 78, Malez published a reproduction of the document indicating that that map had been drawn in the eighteenth century (this dating was confirmed by Josip Faričić, University of Zadar's Department of Geography, based on the drawing technique). Malez's dating mistake was reproduced in subsequent literature (e.g. Božić (2020), p. 1; Kranjc (2010), p. 9).

[22] SI NŠAL 20, box 97, Verschiedene Untertanenangelegenheiten [Various cases concerning the subjects of the seigneury of Gornji Grad (Germ. Oberburg)], sixteenth–nineteenth century, 1524–1596: 1588, Am S: Andreas tag … The word "*sÿaukha*" was underlined by Žiga Zwitter.

[23] Badjura (1953), pp. 294–297; Cevc (1977).

[24] Compare the general use of "*zijalka*" in Badjura (1953), pp. 294–297 and in Cevc (1977).

for karst phenomena in the Dinaric Karst, such as *doline,*[25] *uvala,*[26] *polje,*[27] *ponor,*[28] and *kamenitza,*[29] became part of the international karst science terminology[30] and are therefore also used in this volume. This is closely related to the intellectual history of karst research in the Dinaric Karst, briefly outlined below.

The "*Dinaric karst is the landscape where the* [western] *sciences of karstology and speleology took their origin and started to develop.*"[31] In the seventeenth century, when modern science was born in Europe[32]—and almost concurrently with the karst research in China by Xu Xiake, the "*father of karst studies in China*" (1586–1641)[33]—the northern Dinaric Karst was explored by a local intellectual, the Carniolan nobleman and polymath Johann Weichard Valvasor (1641–1693), active in the fields of natural sciences, geography, history, ethnology, and engineering.[34] His opus, rooted in previous explorations but including extensive original fieldwork results, placed regional research on the intellectual map of Europe. As a graduate of the Jesuit Grammar School in Ljubljana,[35] Valvasor undertook his Grand Tour, which was exceptionally long in his case: it lasted more than ten years and significantly broadened his intellectual horizons. He visited places from Denmark in the north to Tunis in the south and France in the west.[36] Valvasor's Grand Tour included visits to several karst caves, like the then-famous Baumannshöhle in Harz (today in Germany).[37] Equipped with such knowledge, Valvasor carried out his own investigations in the northern Dinaric Karst. The world history of karst science acknowledges him as the first true speleologist in Europe.[38] He is also the first known author of a tentative estimate of the dynamics of speleothem growth, at least in the West, based on his observations of carbonate deposited in dated cave flowstone incisions, while also being aware of the diversity of speleothem growth at different sites.[39] In 1687, Valvasor was elected a member of the prestigious scientific Royal Society in London owing to his description of the functioning of the intermittent karst lake in

---

[25] Ford (2007), pp. 677–680; Williams (2004); Gams (1973), pp. 51–53; see also Panjek (2015), pp. 130–131.

[26] E.g. Ćalić (2011).

[27] E.g. Bonacci (2004a).

[28] E.g. Bonacci (2004b).

[29] E.g. Cucchi (2009).

[30] Ferk and Zorn (2015), pp. 373–374; Kranjc (2010), p. 9; compare U.S. Environmental Protection Agency (2002); Gunn (2004).

[31] Kranjc (2010), quote from page 9; see also Gams (2004), especially 20–22. But see also LaMoreaux and Stevanović (2015), pp. 8–9.

[32] Rossi (2011).

[33] LaMoreaux and Stevanović (2015), p. 9.

[34] Golec (2015b), pp. 91–92, 95–96; Reisp (1983), pp. 260–262; Reisp (2001), p. 3.

[35] Golec (2014a).

[36] Golec (2014a), pp. 28–29; Golec (2014b), pp. 50–63.

[37] Golec (2014b), pp. 57, 60–61.

[38] Shaw (2007), pp. 14, 16, 22, 255, 282; see also Kranjc (1990), p. 213 and LaMoreaux and Stevanović (2015), p. 9.

[39] Zwitter (2014), p. 654; LaMoreaux and Stevanović (2015), p. 9.

the Cerknica Polje. At a Royal Society meeting, the renowned British astronomer Edmond Halley presented the mechanisms behind the filling and disappearance of Lake Cerknica, using Valvasor's findings as a foundation.[40] The "*fame of the Dinaric Karst phenomena*" in the north-westernmost Dinaric Karst in the following centuries was largely due to Valvasor's publications and subsequent historical researchers.[41] The beginnings of (geo)tourism in the area preceded Valvasor: as early as at the beginning of the seventeenth century, the access to the Vilenica Cave on the Kras Plateau, situated close to an important commercial route, near the Habsburg port of Trieste (today in Italy) and the princely stud farm at Lipica (see Chap. 16), was subject to an entrance fee.[42]

Based on the significant yet gradual progress in scientific karst knowledge since the seventeenth century, the extensive nineteenth-century geological explorations paved the way for the evolution of systematic karst research in the Dinaric Karst in the second half of the century.[43] One of the incentives for the Dinaric Karst research of the time was to provide more plumbing water to the quickly growing city of Trieste, the Habsburg Monarchy's main port.[44] The construction works on the railway connection between Vienna and Trieste, which became operational in 1857 and crossed the Kras Plateau, represented another major stimulation for karst research.[45] Severe floods in some *poljes* also encouraged karst explorations to mitigate the flood hazard.[46] The second half of the nineteenth century was characterised by rapid general developments in physical geography in Europe, especially geomorphology, which became the leading physical geography branch in the late nineteenth century.[47] At the University of Vienna, a Department of Geography was established in 1884, building on the tradition of earlier geography studies there. At the time, it was one of only three Departments of Geography that existed at Central European universities.[48] Unsurprisingly, the late nineteenth century was a period of swift geography research developments at the University of Vienna, which evolved into a karst research centre. The railway connections between the capital of the Habsburg

---

[40] Bidovec and Golec (2015). About Valvasor's Lake Cerknica research, see also Habič (1990); Shaw (2004); Zwitter (2014), pp. 625–626, 629–630, 632, 634–635, 639–640, 647, 652, 655–658, 660, 664–665.

[41] Kranjc (2010), quote from page 11; see also Ferk and Zorn (2015), pp. 371–374; Gams (2004), pp. 15–22. Rarely, Jovan Cvijić still referred to Valvasor: e.g. Cvijić (1925), pp. 4, 7.

[42] Kavrečič (2017), pp. 261–277; Zorn et al. (2009), p. 3; compare Zwitter (2014), p. 655.

[43] Ferk and Zorn (2015), p. 371 (literature in footnotes 24–26), 373–374; Gams (2004), pp. 15–22; Kranjc (2010), pp. 11–12; LaMoreaux and Stevanović (2015), pp. 10, 12–13; Malez (1984), pp. 77, 80, 82–87; Trimmel (2004), p. 335.

[44] Mattes (2016), p. 15; Panjek (2015), pp. 115–122; Moritsch (1969), pp. 136, 146, *et passim*; compare Kalc (2010).

[45] Kranjc (2010), p. 11; Kavrečič (2017), pp. 262, 275–276, *et passim*; compare Panjek (2015), pp. 113–115.

[46] Mattes (2016), p. 15.

[47] Cigale and Ogrin (2016), pp. 139–144.

[48] Mattes (2016), p. 19.

Monarchy and a few Dinaric Karst regions were an advantage in this respect.[49] From 1885 to 1906, the leading European geomorphologist Albrecht Penck (1858–1945), who had studied in Leipzig and Munich, worked as a physical geography professor at the University of Vienna, giving Viennese physical geography a far-reaching international reputation.[50] "*Penck was the most prominent protagonist of the new discipline, karstology.*"[51] Jovan Cvijić (1865–1927), a Serbian geographer who came from a karst region in Western Serbia and was strongly inspired by his doctoral studies with Penck in Vienna—which included geology courses by Eduard Suess, "*one of the foremost tectonic geologists of the time*"—became "*the foremost karstologist of his generation and, arguably, the greatest karst 'generalist' of the nineteenth and early twentieth centuries.*" *Das Karstphänomen* (1893), his most influential publication,[52] represented a turning point, "*the beginning of the intensive study of karst morphology, including the use of the term karstification.*"[53] "Das Karstphänomen *provided the first major overview of karst terrains, clearly describing, with case study examples,* [the] *complex aspects of karstification.*"[54] Apart from his own research, Cvijić cited many leading specialists in this synthesis. Later, some of Cvijić's explanation attempts were found to be wrong, and the scope of his definition of karst was narrower than today. "*Nevertheless, his contributions to karst science have been fundamental.*"[55] Cvijić substantially developed his understanding of karst further in his ensuing publications, and his final and highly significant work was published posthumously in 1960.[56] For instance, Cvijić "*described karst surface lowering*" and "*argued that, with so many variables involved, no single, general rule covers the progression of all aspects of karstification, and stressed that, contrary to common expectation, knowledge of underground drainage is the key to understanding surface karst.*" Karst hydrology research progressed as well, and Cvijić's contribution was significant.[57]

This short overview of the history of karst explorations in the Dinaric Karst highlights the major intellectual history contribution of the karst studies in the macroregion this volume focuses on to the global development of karst science. Its further development is marked by the dispersion of ideas and investigations conducted worldwide rather than concentrated in only a few regions, especially since the second half of the twentieth century.[58]

---

[49] Gams (2004), pp. 21–22; Kranjc (2010), pp. 11–12; Panjek (2015), p. 32; see also Mattes (2016), pp. 14–15 and Mrgić (2017), p. 85; a railroad history map: Bruckmüller and Stekl (1998).

[50] Bernleithner (1965), pp. 94, 100, 102; Ford (2007), p. 675.

[51] Kranjc (2010), p. 12; see also Gams (1973), p. 46.

[52] Cvijić (1893); see Ford (2007), (first quote from page 675); LaMoreaux and Stevanović (2015), pp. 11–13; Lowe (2004), p. 380 (second quote); Mattes (2016), pp. 13, 18–23; Mrgić (2017), p. 78.

[53] Ford (2007), pp. 676–680, 683; Kranjc (2010), p. 12 (quote).

[54] Lowe (2004), p. 380.

[55] Ford (2007), (quotes from pages 676–677); Ferk and Zorn (2015), pp. 374–375.

[56] Cvijić (1960).

[57] Ford (2007), pp. 676–677, 680–683; LaMoreaux and Stevanović (2015), pp. 12–13; Lowe (2004), p. 380 (quotes); Kranjc (2010), p. 12.

[58] Kranjc (2010), pp. 12–13; Ferk and Zorn (2015), pp. 373–375; LaMoreaux and Stevanović (2015), pp. 13–17. Among the many recent karstology publications focused on the Dinaric Karst, see, e.g.

Few karstologists limited their research to geomorphology and hydrology during the earlier stage of the karst science development. As a polymath, Valvasor also focused on human activities in karst, such as the challenges and opportunities in agriculture, fishing, or water provisioning.[59] Cvijić was not only a physical geographer, but also an ethnographer and anthropogeographer, recognised by the French anthropogeography school. Thus, he lectured at the Sorbonne and was cited by Paul Vidal de la Blache and the *Annales* historian Lucien Febvre.[60] Late in his career, Cvijić published a paper titled *Karst and Man* (*Karst i čovek*, 1925), briefly but systematically presenting several main topics of the interactions between human society and nature in the Dinaric Karst, mainly in the rural areas.[61] The study synthesises results based primarily on extensive fieldwork in the Dinaric Karst in the late nineteenth and early twentieth centuries. It highlights the diversity of karst environments, the main environmental challenges for people (farmers) in karst: the widespread scarcity of arable land, water supply challenges often faced by the local inhabitants and livestock, floods in the flood-prone *poljes* in the context of agriculture and fishing, and the absence of forest in large sections of the (sub)Mediterranean karst, which he interpreted as a consequence of society-nature interactions. In line with the prevalent theory of possibilism in anthropogeography, Cvijić presented a selection of human adaptations to these challenges in the Dinaric Karst.[62] A significant part of his observations and data can be used in modern environmental history research,[63] while some interpretations require further evidence and better contextualisation.[64]

Historical historiographies and geographies themselves are historical sources as well, inspired and influenced by the aspirations and viewpoints of the author(s) or those who had commissioned these works.[65] Valvasor and Erasmus Francisci, the co-author of Valvasor's most extensive though not always most accurate publication, *The Glory of the Duchy of Carniola* (1689), aimed primarily at describing, documenting features and events, as well as attracting, educating, and amusing the readers.[66] Cvijić's anthropogeography, written in times of modern nation-building

---

Gams (2004); Garasic (2021); Knez et al. (2020); Parise et al. (2018); Mihevc et al. (2010); Milanović (2023); and papers in the journal *Acta carsologica*, one of the foremost karstology journals in the world, published by the ZRC SAZU Karst Research Institute.

[59] Zwitter (2014).

[60] Ford (2007), p. 676; Grčić (2015), p. 180; Mrgić (2017), pp. 78–79; compare Mattes (2016), pp. 21–24, 26. About the Annales School, see Burke (1990).

[61] Grčić (2015).

[62] Cvijić (1925).

[63] Compare Zwitter (2019), pp. 225–229; for a landscape history approach to the Kras Plateau, see, e.g. Panjek (2015); an environmental history study based on the sixteenth- and seventeenth-century archival records: Zwitter (2015).

[64] See, for instance, a critique of Cvijić's historical migration studies due to their excessive reliance on oral traditions and lack of information from contemporaneous sources in Grčić (2015), p. 178, and Grafenauer (1991).

[65] Compare Winiwarter (2016).

[66] Reisp (1983), pp. 257–263; Reisp (2001), p. 9; Zwitter (2014); Žvanut (2014), pp. 173–193, especially pages 184–186; see also Shaw (2004).

and ever-increasing conflicts, was strongly inspired by "*searching for geographic, ethnic or cultural data to scientifically* [prove] *national unity*."[67] Late in his career, Cvijić perceived the Dinaric Karst as "*geomorphic phenomena and a source of national consciousness, which should unify all people of the same language*."[68] Since Cvijić, the epistemological background, concepts, methods, and state of research have evolved significantly. This book considers several topics he focused on in his paper *Karst and Man* from a modern, environmental history perspective.

1.2 Contents and Scope of This Volume

This volume is the first multiauthored inter- and multidisciplinary monograph on the history of Dinaric Karst environments and society-nature interactions therein. It is written by thirty-one authors: geographers, historians, archaeologists, an archaeobotanist, an environmental engineer, a forestry scientist, a palynologist, and a researcher at the interface of biology and climatology. The authors work in Croatia, Germany, Montenegro, Serbia, Slovenia, Switzerland, and the United Kingdom.

The title of this publication, with environmental histories in the plural, emphasises the diversity of the research questions and approaches in this volume on the Dinaric Karst. They all add their pieces to the mosaic of environmental history. The present publication also includes a few historical ecology and historical geography studies.

The chapters address several key research topics of the society-nature interactions in the karst landscapes, forming separate book sections or parts; within each section, chapters are ordered chronologically and from more general to more specific. Part I approaches some of the main aspects of the holistic transformations of cultural landscapes[69] in the northern and western Dinaric Karst through time; Part II focuses on the histories of woodland and forests; Part III on the histories of water and climate; while Part IV presents the histories of environmental degradation, as the low water self-purification capacity or the challenges that rocky soils pose for agriculture in several types of karst environments make these environments particularly sensitive or challenging from the economic perspective. Authors explore several types of socio-natural sites, i.e. nexuses between practices and their material precipitates, i.e. arrangements as hybrids connecting the environment and the people.[70] The book also deals with the transformations of socio-natural sites over time, including, for instance, the black pine and other forests, waterscapes, and caves. Some chapters focus on diverse karst environments, from mountains and plateaus to plains, *dolines*, *poljes*, and the Adriatic coast and islands.

Chronologically, the volume deals with periods from prehistory to contemporary history—those characterised by all three major sociometabolic regimes throughout

[67] Mattes (2016), quote from page 26.

[68] Mattes (2016), p. 24.

[69] See Baker (1992).

[70] Winiwarter et al. (2013), pp. 105–108.

human history. It begins with the uncontrolled solar energy system of the foraging societies, though most of the monograph is dedicated to the periods of the controlled solar energy system of the agrarian regime and the energy system of industrial societies.[71] The Bronze Age and the times between the Early Modern Period to contemporary history receive the most research attention in this monograph. This makes sense because these periods were characterised by particularly substantial human transformations of the Dinaric Karst environments, as explained in this book. Depending on the study subjects and methods, this publication combines long-term research (in some cases spanning over millennia) and insights into the selected (short series of) events and processes relevant in the broader historical context.

Spatially, most contributions deal with the Dinaric Karst of modern Croatia and Slovenia, though Bosnia and Herzegovina, and Montenegro are also represented in the volume. With various educational backgrounds, training, experience, individual publications, and personal research affinities, we have composed this book in a relatively long period between 2020, marred and marked by the covid-19 pandemic, and 2023, during the record-breaking summer heat. As the present work aimed to include as much of the Dinaric Karst terrain as possible and a greater diversity of "stories" about it, the omitted texts called for profound theoretical and practical improvements. Thus, we encouraged the authors to keep perfecting their work. The process of preparing a scientific book and the circumstances involved in such an endeavour demand that the knowledge produced be placed in a specific spatial, temporal, and intellectual context.[72] Thus, this book includes only a selection from a wide variety of potential research subjects and methodologies.

1.2.1 The Structure of This Volume

Part I—Cultural Landscape

The Holocene Vegetation and Land Use History in the Northern Dinaric Karst

Chapter 2, written by the palynologist Maja Andrič and Dirk Nikolaus Karger, a researcher at the intersection of biology and climatology, provides a rich regional synthesis of palynological research at nine sampling sites located throughout a large area of the northern Dinaric Karst. Pollen diagrams are compared with model-based climate reconstructions (CHELSA-TraCE21k), and the results indicate that humans have significantly influenced vegetation in the northern Dinaric Karst since the introduction of agriculture and pastoralism. The first traces of human influence (small-scale forest clearing, agriculture, and grazing) can be found in the Neolithic period from about 6000 cal BP. More intensive agropastoral activities were identified in the Bronze Age, especially in the western part of the region, e.g. in the Mediterranean

[71] See Krausmann et al. (2016).

[72] Haraway (1988).

coastal areas and on islands, where deciduous taxa were replaced by evergreen vegetation. Between the end of the Bronze Age (3200 cal BP) and the Iron Age (2800 cal BP), a more open landscape emerged, resembling the present conditions. During the Roman period, clearings probably occurred near important trade and military routes. The area under consideration lacks detailed studies of the Holocene climate, making it difficult to distinguish between anthropogenic and climate-induced vegetation change. Nonetheless, the results show that over the past 4000 years, human activities have altered vegetation more considerably than the climate. The last 2000 years have been poorly covered by the existing palynology analyses but are dealt with in Chaps. 3, 4, 5, 6, 7, 8, and 9.

Prehistoric to Mediaeval Woodland Management and Agriculture in the Westernmost Carpathian Basin and the Surrounding Northern Dinaric Karst: A Case Study of Archaeobotanical Analyses from the Multiperiod Site at the Obrežje International Border Crossing

In Chap. 3, the archaeobotanist Tjaša Tolar and archaeologists Philip Mason and Bine Kramberger present a detailed novel archaeobotanical analysis of wood (charcoal) and fruit/seed remains from three chronological cross sections—the Late Bronze Age, the Early Roman period, and the Late Middle Ages—in the contact zone between the Gorjanci Hills Dinaric Karst and the Sava River plain at the Obrežje international border crossing between Slovenia and Croatia. These remains, accumulated in multi-layered archaeological deposits, represent a valuable tool for reconstructing the past vegetation and its use by the (pre)historical populations. The data interpretation focuses on land use and especially forest use during the respective chronological periods. Agriculture in the prehistoric settlement is evidenced by the culturally autochthonous taxa, while some fruits imported from the Mediterranean indicate trade and exchange. The finds from numerous archaeological sites show repeated (and preferential) settlement in certain environmental zones, such as fertile terraces and hills in the contact zone between the Sava Valley and the Gorjanci Hills, in contrast to the sparse evidence from the Lower Sava plain.

With more than 500 archaeobotanically studied and chronologically and spatially defined samples of charcoal and fruit or seed remains, the archaeobotanical research at Obrežje is among the most extensive in Slovenia. Some of the results from the charcoal remains indicate differences in wood supply areas through time, suggesting that, from the viewpoint of the local population, the role of the Gorjanci Dinaric Karst forests might have changed in different periods.

The Formation of the Kras Landscape from Prehistory to the Early Modern Period

In Chap. 4, the archaeologists Dimitrij Mlekuž Vrhovnik and Tomaž Fabec introduce novel approaches to the long-term history of the Kras Plateau. Instead of applying a traditional archaeology approach focusing on the prehistorical finds from the isolated cave sites or hilltop settlements, this chapter focuses on the space between these sites and provides a groundbreaking *longue durée* examination of the Kras Plateau landscape transformations from prehistory to the Early Modern Period. The authors meticulously combine the data collected through remote sensing techniques and environmental archaeology fieldwork in the relevant *dolines* as well as geochemical and soil magnetic susceptibility analyses. The results show two periods of increased human influence on the Kras Plateau: the Bronze Age and the Early Modern Period. During the rest of the time, human landscape transformation was almost negligible. The Bronze Age was the period of large-scale forest clearing and formation of the cultural landscape, focused on livestock farming. The period after 1500 AD is associated with the most intensive deforestation and labour-intensive farming, including overgrazing, which gradually turned some parts of the Kras Plateau into a stone desert. The increased human influence during the Bronze Age and especially the Early Modern Period is visible in the sedimentation intensity. During the last millennium, an average of 0.9 m of soil was deposited in the studied *dolines*, 0.6 m only in the three centuries of the Early Modern Period. The Bronze Age layers are half as thin as those deposited during the last millennium, with an average thickness of about 0.45 m, even though the Bronze Age lasted much longer. Thus, the Early Modern Period was the era of the strongest farming impacts in the Kras environments.

Landscape Change and Fire Risk in the Croatian Dinaric Karst—Looking Back and Moving Forward and

Land Use/Land Cover Changes on Croatian Islands Since the Beginning of the Twentieth Century—Drivers and Consequences

The last two chapters of the cultural landscape book section deal with long-term land use/land cover change from the early nineteenth and early twentieth century onwards. The time span of these studies is mainly determined by the availability of cadastral documents, the source type providing the oldest data these chapters are based on.

The research conducted by the geographers Ivan Tekić, Borna Fuerst-Bjeliš, Anamarija Durbešić, Petra Radeljak Kaufmann, and Marin Cvitanović (Chap. 5) addresses the complex relationship between landscape change and fire risk. It contributes to the understanding of the historical relationships between the socio-economic and landscape changes taking place over two centuries in three main Dinaric socio-environmental zones—the islands, the littoral, and the hinterland—manifested in depopulation, land abandonment, and ecological succession. Each zone underwent a specific cultural landscape change, as landscapes are socio-natural hybrids. This chapter also examines how these changes affected the transformation of forest fire risk between the nineteenth and twenty-first centuries using a model based on land use/vegetation cover, climate parameters (temperature, precipitation, and humidity) as dynamic variables, and lithology, soils, orography, and exposure as locally specific constant factors. This way, the chapter shows a significant increase in fire risk in all three zones: high-risk areas strongly prevail in all zones. Contrary to the usual findings, this research indicated that the rising temperatures and decreased precipitation are not the most important variables that increase the fire risk in the Croatian Dinaric Karst. Instead, the increased fire risk stems mainly from the homogenisation of land cover, i.e. the predominance of the highly flammable maquis and scrubland at the expense of pastures. Based on a model of development scenarios, the authors indicate that tourism monoculture will most likely increase the forest fire risk in the future.

In Chap. 6, the geographers Ante Blaće, Marin Cvitanović, Anica Čuka, and Josip Faričić focus on three eastern Adriatic Dinaric Karst islands, namely Krk, Žirje, and Korčula, and examine the main factors influencing land use changes and landscape development in the twentieth century, as revealed mainly by the archival cadastral data and the Corine Land Cover. The authors conduct an in-depth analysis of the social processes in the relevant timeframe and their land use impact, meticulously analysing the underlying drivers. The results indicate that the peak of agricultural activities and land use on the Adriatic islands was reached at the beginning of the twentieth century. In the 1920s, the island population started to decline, leading to the abandonment of agricultural land and ecological succession. Similarly to other socio-environmental zones of the Croatian Dinaric Karst, as shown in Chap. 5, only a small part of the land is still cultivated, mainly by cash crops such as olive trees and grapes, while Mediterranean maquis and forests cover most of the island surface, increasing the risk of summer fires and threatening the traditional agricultural landscape. Although most islands, especially the smaller ones, face a continuous decline in agricultural land, some larger islands have seen an increase in agricultural land, especially olive groves and vineyards, partly due to the subsidies after Croatia's accession to the EU. The research confirms that the most significant change in land cover, especially after 2012, is related to forest degradation, directly linked to forest fires, as also shown in Chap. 5.

Part II—Woodland and Forests

Deforestation and Reforestation Attempts of the Military Frontier's Coastal Dinaric Karst (1762–1881), Croatia

The complex interplay between the natural environment of the Dinaric Karst, the local population, and the Habsburg Military Frontier's administration regarding forest management is the subject of a thorough investigation by the historians Filip Pavelić, Hrvoje Petrić, and Mislav Radošević in Chap. 7. The authors meticulously research the documentary and legislative records, extensive reports of professional forest administrators, and narrative evidence to reconstruct the many layers of the environmental transformation processes. The latter fit in several groups of factors, affecting the soil depletion and erosion due to deforestation and climate regime (strong bora wind, torrents, and flash floods), and were accelerated by the unsustainable logging practices of the local inhabitants. Their traditional agropastoral economy provided meagre economic substance, therefore encouraging them to engage in illegal timber cutting and sell the wood for additional income. Meanwhile, sheep and goat herding further contributed to the deforestation of extensive areas. From the first attempt to reverse the forest degradation process—the Theresian Forest Order (1762)—to the establishment of the Karst Reforestation Inspectorate (1878), all efforts of the central government in the territory of the Military Frontier have been closely examined and outlined. They consisted of appointing general foresters and their deputies in the field, drawing up detailed forest reports, and implementing strict measures for the protection of forest resources. Additionally, there were attempts at afforestation with various species, with various degrees of success. The state was mainly interested in higher quantities of timber for shipbuilding and trade. The representatives of the forestry profession and those of the central authorities did not share the same vision of the environment and the way of exploiting its resources. Namely, the implementation of the German model of permanent forest management contrasted with the traditional logging and animal husbandry practices and the local population's way of life.

Afforestation with Non-native *Pinus nigra* in the Karstic Areas of Southwestern Slovenia Since the Mid-Nineteenth Century: Environmental Implications

In Chap. 8, Meta Remec, a historian interdisciplinarily cooperating with forestry scientists, adopts a different view and addresses the issue of afforestation projects carried out since the nineteenth century on the Kras Plateau with the non-native black pine (*Pinus nigra*), with the primary aim of increasing the forest area on Kras Plateau. The black pine ("pioneer of the bare karst") stands were considered and planned as a transitional phase of afforestation. In reality, this type of afforestation resulted in non-native monocultural forest stands as a profound and extensive long-term human intervention in the environment, leading to rapid and drastic changes in land use patterns with far-reaching consequences for the entire ecosystems, society, and the economy in general. Some of these impacts were not foreseen at all. One of them was the infestation of the non-native pine processionary moth (*Thaumetopoea pityocampa*), which followed the introduction of the non-native black pine forests and led to aggressive pest control, first mechanical and later chemical, resulting in the contamination of the groundwater in the karst areas and the entire local environment. Black pine forests have survived. In recent years, the removal of black pine has become an essential part of the strategy to save endangered species, preserve the cultural landscape, and perhaps even revive agriculture in these areas. Nevertheless, black pine eventually became integral to the karst landscape and its identity.

Remec's research is based on rich archival and published sources. The institutional archival sources analysed were produced in the local, regional, and state institutions of three different states, and they are kept in the state and regional archives. Chapter 8 also uses newspaper articles, published legislative sources, and forestry and Provincial Committee reports.

Traditional Management of Forest Resources in Mountain Sites of the Dinaric Transhumance in Montenegro

The last chapter in the woodland and forests section explores transhumance—one of the most widespread and long-lived practices in the Mediterranean coastal areas, the Mediterranean mountains, and their continental hinterland. Transhumance is a form of pastoralism in which livestock are herded seasonally—in mountain pastures during the summer, lowland pastures during the cold parts of the year, and along the migration routes linking both areas during the transitional periods. Transhumance co-shaped the local and regional economies, societies, livestock, and ecosystems. In 2019, it was inscribed on UNESCO's Representative List of the Intangible Cultural Heritage of Humanity. Transhumance involves specific and dynamic traditional ecological knowledge, not limited to pastures. Chapter 9 is a result of interdisciplinary cooperation between the historian Ivan Laković and forestry scientist Jelena Lazarević. They focused on the traditional management of forest resources in the seasonal mountain settlements in the Dinaric Karst of Montenegro. The authors used a mixed-methods approach, including surveys, interviews, and fieldwork. The research can be seen as the first attempt to analyse transhumance throughout Montenegrin history through the lens of forests.

Part III—Water and Climate

Scarcity Versus Abundance: A Case Study of the Hardship and Imminent Famine in the District of Senožeče in 1864 and 1865

In Chap. 10, the historian Andrej Studen explores the issue of the vulnerability of farm economies in the northern Dinaric Karst to natural hazards during a specific period that rarely receives this kind of environmental history attention. Namely, this chapter focuses on the time when the railway connection became operational, while environmental history subsistence crises studies in Central Europe usually explore either the great subsistence crises that had caused famine earlier[73] or the specific contexts of the interplay between disastrous harvest failures and the war economy—a topic that has recently attracted the research attention focusing on the First World War.[74] Chapter 10, based on rich archival and newspaper sources, puts the exceptionally rainy and cold weather of 1864 and the extraordinary winter of 1864/1865—a Little Ice Age-type weather anomaly during the waning Little Ice Age—in the broader context of the integrated farm economy in the area under consideration and the general economic crisis in the rural areas and its specific local dimensions. Under such circumstances, the impacts of the exceptional weather on the local farm economy were disastrous. The study emphasises the highly significant micro-environmental diversity in the northern Dinaric Karst and the diversity of the local farm economies, all of which affected the scale of the subsistence crisis in question. The issue of crisis mitigation is also discussed. The study identifies an ambiguous role of the railway in the local economy during the 1860s regarding vulnerability to natural hazards.

Water Infrastructure in Kras in the Nineteenth Century Between Continuity and Change

In Chap. 11, the historian Aneja Rože Kravanja engaged in multiperspective research, contrasting the traditional water management and changes in the water supply infrastructure during the modern period in two cadastral municipalities: Dutovlje and Lokev. These two microregions in the Kras Plateau shared some common features

[73] Compare White et al. (2018).

[74] Pfister et al. (2018), p. 126.

but had also experienced different economic development throughout history. The author used a vast diversity of sources for the highly detailed reconstruction of the types of water reservoirs (*kals*, open-type *štirnas*, and *lokevs*), traditionally built and modernised, if possible. She determined their exact number and locations in situ during fieldwork and interviews with the locals, gathering their valuable testimonies on traditional knowledge and water management. The Josephine land survey maps from the late eighteenth century, the Franciscan cadastre from the beginning of the nineteenth century, and the censuses of 1902, 1905, and 1908 represent solid historical bases upon which the author built a micro-historical portrait of the Kras Plateau region. The administrative and political changes during the nineteenth century resulted in the engagement of state engineers and sanitary experts, whose work improved the water supply infrastructure in those communities and thus contributed to the well-being of people and livestock as well as improved their economic standard.

Climate and Discharge Trends, and Flood Hazard in Slovenia's Dinaric Karst Region Since the Mid-Twentieth Century

In Chap. 12, the geographers Mauro Hrvatin and Matija Zorn adopt a historical hydrology approach and focus on the socio-natural sites of rivers in the Slovenian Dinaric Karst from 1961 to 2020. Based on extensive meteorological and hydrological measurement data, the chapter provides new insights into the trends in average annual temperature, annual precipitation, days with precipitation over 0.1 mm, days with snow cover, minimum annual discharges, mean annual discharges, maximum annual discharges, and absolute maximum annual discharges, as well as discharge regimes and flood hazard. To determine the trend of change in the selected climate and hydrological variables, the Mann–Kendall test and Theil–Sen estimator were used at the chosen temperature, precipitation, and gauging stations. During the 1961–2020 period, temperatures increased by 2.5 °C on average, precipitation decreased by 50–400 mm on average, and the number of days with snow cover declined by 30–60%, or by an average of thirty-five days a year. All of this had a substantial impact on the water conditions. The most significant changes in the discharge regime, as seen by comparing the 1991–2020 and the 1961–1990 discharge regime data sets, can be observed in the rivers formerly characterised by highly pronounced snow retention in late autumn and winter. The decrease in the average minimum annual and mean annual discharges is alarming. A downward trend can also be observed for many average maximum annual discharges. In contrast, higher flood risk is possible due to more significant deviations in the absolute maximum annual discharge of the Dinaric Karst rivers in recent years. Environmental and economic historians should meticulously consider these and other historical hydrology results, including the history of river bed transformations when interpreting earlier historical climate change and its impacts on ecosystems and people.

Part IV—Environmental Degradation

Empire, Extraction, and Externalisation. Wood Impregnation in the Early Twentieth-Century Bosnia and Herzegovina and Its Precarious Legacy

In Chap. 13, the historian Martin Meiske gives an example of how a single company within the Austro-Hungarian Empire—Guido Rütgers KG—could cause many human health and environmental problems by impregnating railway sleepers with creosote, a highly toxic material. Even though five locations in today's Austria have been designated as hazardous, those in Bosnia and Herzegovina have not been registered even today. Furthermore, the author contextualises a plethora of issues from the current debates concerning environmental justice and post-colonial hazardous legacies. Also in Bosnia and Herzegovina as the Austro-Hungarian "colony," constructing and maintaining a dense railway network was essential for the Habsburg imperial plans—political, military, and economic (natural resource extraction). Like other imperial states at the end of the nineteenth and the beginning of the twentieth century, Austria-Hungary paid little or no regard to the local population and the environment. Meiske conducted meticulous research in the company's archive, interpreting the findings within the framework of the Dinaric Karst's environmental history and the current protective measures, demonstrating the importance of detecting many layers of entanglement between the socio-political structures and the natural environment.

Examples of Human Disturbances in the Dinaric Karst Landscape Since the Early Twentieth Century: A Case Study of Bela Krajina

Chapter 14, written by the environmental engineer Daniela Ribeiro and geographer Matej Simčič, analyses the disturbances in the karst landscape of the Bela Krajina region in southeast Slovenia since the beginning of the twentieth century. The following major landscape disturbance types were identified: management of litter-raking forests; massive illegal dumping of waste condensers in the sensitive karst landscape; landscape transformations through intensive agriculture, such as the removal of large quantities of protruding stones; and excessive application of fertilisers and agricultural plant protection products, such as pesticides. At the beginning of the twenty-first century, excessive application of liquid manure wreaked havoc on the grassland ecosystems,[75] affecting groundwater quality and karst cave ecosystems. Together with the levelling of *dolines*, all of these practices were identified as causing the most severe degradation. However, there is a high rate of diversity within the degradation types discussed in this chapter. As litter raking caused the depletion of soil nutrients in the forests, a forest usually considered degraded developed (see Chap. 16). This degradation type, characteristic of extensive (labour-intensive)[76] historical agriculture of the agrarian sociometabolic regime's controlled solar energy system, also had some ecosystem effects considered desired from the viewpoint of environmental protection. The other degradation types presented in

[75] Dietl and Lehmann (2004).

[76] See Netting (1993).

this chapter belong to the sociometabolic regime of industrial societies, whose detrimental impacts on the ecosystems are far more severe.

From Karstology to War Crimes. Third Reich's Military and Scientific Exploitation of the Dinaric Karst

In Chap. 15, the historian Klemen Kocjančič presents his pioneering research into a specific karst aspect of the Second World War: the Third Reich's exploration of the northern Dinaric Karst in the Operational Zone of the Adriatic Littoral as an example of the "misuses of science for war crimes". One of the topics discussed in this chapter is the special Nazi military unit for karst warfare, which was active in the northern Dinaric Karst for almost two years and "*committed the largest number of war crimes in the Slovenian territory of all the German units that operated against the Slovenian Partisans*". Military Dinaric Karst uses and warfare on both sides resulted in the region's environmental degradation, the most notorious example being the vast fuel explosion and fire in the famous Postojna Cave. The German Army planned to put the caves in the relevant area to even more extensive military use, but fortunately, this was never realised. After the end of the war, several shafts and caves were used as mass graves during the mass killings in Slovenia; this was first and foremost a large-scale disaster of human ethics, while in terms of environmental history, this action also had locally significant ecological impacts.

Kocjančič's study is based on the documents kept in the German Federal Archives, the National Archives and Records Administration in the USA, and the archives of the Institute of Contemporary History in Slovenia. It narrows the general gap in the historical military geography and military environmental history research into the north-western Dinaric Karst, where most existing studies focus on the First World War.

Part V contains the conclusions, further contextualisations of the book contents, a critical reflection on the narratives of degradation and improvement in the environmental history of the Dinaric Karst, and the outlook.

Acknowledgements The work on this chapter was funded by (1) the Ministry of Science, technological development, and innovations of the Republic of Serbia, University of Belgrade—Faculty of Philosophy (no. 451-03-47/2023-01/200163), (2) the Slovenian Research and Innovation Agency, research programmes No. P6-0101 (Geography of Slovenia) and No. P6-0235 (Slovene History), (3) the University of Zagreb's Grant for research on the influence of paleo and recent climate changes on processes in marine and continental environments in 2020–2023 (Nos. 20285705; 20286205; 20286474; 20286585). The authors wish to thank Borut Praper, who copy-edited the text, and Lenart Štaut, who drew the map.

References

Published Sources

Copernicus (2023) European digital elevation model (EU DEM) v 1.1. European Environment Agency, Copenhagen

The World Bank (2023) World country polygons. The World Bank, Washington

Archival Source

SI NŠAL [Archdiocesan Archives in Ljubljana, Slovenia]: NŠAL 20, Gornji Grad Avstrija: box 97

Literature

Armiero M (2024) Breaking the green cage: environmental history everywhere! Econ Ecohist 18

Badjura R (1953) Ljudska geografija: terensko izrazoslovje [Vernacular geography: terminology collected through field-work]. Državna založba Slovenije, Ljubljana

Bajt D, Vidic M (eds) (2011) Slovenski zgodovinski atlas [Historical atlas of the Slovenian territory]. Nova revija, Ljubljana

Baker ARH (1992) Introduction. In: Baker ARH, Bigger G (eds) Ideology and landscape in historical perspective: essays of the meaning of some places in the past. Cambridge University Press, Cambridge, pp 1–13

Bernleithner E (1965) Sechshundert Jahre Geographie an der Wiener Universität. Studien zur Geschichte der Universität Wien 3:55–125

Bidovec M, Golec B (2015) Valvasor in Edmond Halley: o vprašanju polihistorijevega članstva v angleški Kraljevi družbi (Royal Society) in posledicah prekinjenih stikov z njo [Valvasor and Edmond Halley: On the question of the polymath's membership in the Royal Society and the consequences of his severed contacts with it]. Zgodovinski časopis 69(3–4):324–335

Bonacci O (2004a) Poljes. In: Gunn J (ed) Encyclopedia of caves and karst science. Fitzroy Dearborn, New York, pp 599–600. https://doi.org/10.4324/9780203483855

Bonacci O (2004b) Ponors. In: Gunn J (ed) Encyclopedia of caves and karst science. Fitzroy Dearborn, New York, pp 600–601. https://doi.org/10.4324/9780203483855

Božić V (2020) O izrazima spilja /špilja/ i pećina [About the terms of *spilja/špilja/*and *pećina*]. Hrvatski planinski savez. https://www.hps.hr/files/data/170/1%20Vladimir%20Bozic.pdf

Brückmüller E, Stekl H (1998) Das Eisenbahnnetz der Habsburgermonarchie bis 1914. In: Putzger FW, Brückmüller E (eds) Historischer Weltatlas zur allgemeinen und österreichischen Geschichte. Verlag Hölder-Pichler-Tempsky and ÖBV Pädagogischer Verlag, Wien, p 67

Burke P (1990) The French historical revolution: the annales school, 1929–1989. Polity Press, Cambridge

Cevc T (1977) Vpliv zemljišča in družbeno-zgodovinskih razmer na življenje pastirjev v naravnih zavetiščih na planinah v Kamniških Alpah [The influence of environment and social history conditions on the life of shepherds in the natural shelters in the alps of the Kamnik Alps]. Slovenski etnograf 30:93–112, 121–122

Cigale D, Ogrin D (2016) Alexander Georg Supan: v Sloveniji spregledan geograf slovenskega rodu [Alexander Georg Supan: a geographer of Slovenian origin, overlooked in Slovenia]. Dela 45:135–163. https://doi.org/10.4312/dela.45.135-163

Cucchi F (2009) Kamenitzas. In: Ginés A, Knez M, Slabe T, Dreybrodt W (eds) Karst rock features: Karren sculpturing. Založba ZRC, Ljubljana, pp 139–149. https://doi.org/10.3986/9789610502968.
Cvijić J (1893) Das Karstphänomen: Versuch einer morphologischen Monographie. Ed. Hölzel, Wien
Cvijić J (1925) Karst i čovek [Karst and man]. Glasnik geografskog društva 11:1–11
Cvijić J (1960) La géographie des terrains calcaires. Académie Serbe des Sciences et des Arts, Beograd
Ćalić J (2011) Karstic uvala revisited: toward a redefinition of the term. Geomorphology 134(1–2):32–42. https://doi.org/10.1016/j.geomorph.2011.06.029
Dietl W, Lehmann J (2004) Ökologischer Wiesenbau: Nachhaltige Bewirtschaftung von Wiesen und Weiden. Österreichischer Agrarverlag, Leopoldsdorf
Ferk M, Zorn M (2015) Kras—večpomenski izraz z mednarodno veljavo [Kras—a polysemous term with international recognition]. Kronika: časopis za slovensko krajevno zgodovino 63(3):367–378
Ford D (2004) Karst. In: Gunn J (ed) Encyclopedia of caves and karst science. Fitzroy Dearborn, New York, pp 473–475. https://doi.org/10.4324/9780203483855
Ford D (2007) Jovan Cvijić and the founding of karst geomorphology. Environ Geol 51:675–684. https://doi.org/10.1007/s00254-006-0379-x
Gams I (1969) Some morphological characteristics of the Dinaric Karst. Geograph J 135(4):563–572. https://doi.org/10.2307/1795104
Gams I (1973) The evolution of the Slovene words 'kras' and 'dolina' into international terms until the end of [the] 19th century. In: Gams I, Kunaver J, Radinja D (eds) Slovene karst terminology. Katedra za fizično geografijo Oddelka za geografijo FF, Univerza v Ljubljani, Ljubljana, pp 45–54
Gams I (1974) Kras: zgodovinski, naravoslovni in geografski oris [Karst: A historical, natural-sciences and geographical outline]. Slovenska matica, Ljubljana
Gams I (1993) Origin of the term 'karst,' and the transformation of the Classical Karst (kras). Environ Geol 21:110–114. https://doi.org/10.1007/BF00775293
Gams I (2004) Kras v Sloveniji v prostoru in času [Karst in Slovenia in space and time]. Založba ZRC, Ljubljana
Golec B (2014a) Polihistor brez univerzitetne izobrazbe: regularno šolanje Janeza Vajkarda Valvasorja [A polymath without a university education: regular schooling of Johann Weichard Valvasor]. Šolska kronika 47(1–2):7–35
Golec B (2014b) Valvasorjev izvor, družina in mladost: stare neznanke v novi luči (3. del) [Valvasor's origin, family and youth – the old unknowns in a new light (part 3)]. Kronika: časopis za slovensko krajevno zgodovino 62(1):15–66
Golec B (2015a) Kras, Kraševci in kraška zemljepisna imena v zgodnjih slovenskih besedilih [Kras, its people and geographic Kras names in early Slovenian texts]. Kronika: časopis za slovensko krajevno zgodovino 63(3):445–456
Golec B (2015b) Valvasorji: med vzponom, Slavo in zatonom [Die Valvasors: Zwischen Aufstieg, Ehre und Untergang]. Založba ZRC, Ljubljana
Grafenauer B (1991) Migracije v jugoslovanskih deželah od 15. do 18. stoletja in njihov pomen za nadaljnji narodnostni razvoj [Die Migration in den jugoslawischen Ländern vom XV. bis zum XVIII. Jahrhundert und ihre Bedeutung für die fernere Nationale Entwicklung]. Zgodovinski časopis 45(2):203–209
Garasic M (2021) The Dinaric Karst System of Croatia: Speleology and Cave Exploration. Springer, Cham. https://doi.org/10.1007/978-3-030-80587-6
Grčić M (2015) The human-geographical school of Jovan Cvijić. In: Jović V, Kostić A (eds) Jovan Cvijić: life, work, times: on the occasion of 150th anniversary of Jovan Cvijić's birth. Serbian Academy of Sciences and Arts, Belgrade, pp 177–182
Gunn J (ed) (2004) Encyclopedia of caves and karst science. Fitzroy Dearborn, New York. https://doi.org/10.4324/9780203483855

Habič P (1990) Valvasorjev in današnji pogled na presihanje Cerkniškega jezera [Valvasor's and present views on the intermittent Lake Cerknica]. In: Vovko A (ed) Valvasorjev zbornik: ob 300 letnici izida Slave vojvodine Kranjske. SAZU, Ljubljana, pp 226–231

Haraway D (1988) Situated knowledges: the science question in feminism and the privilege of partial perspective. Femin Stud 14(3):575–599

Herak M (1965) Geologische Übersicht des Dinarischen Karstes. Naše jame 7(1–2):5–11

Kalc A (2010) Problematika popisa stanovništva slobodnolučkog Trsta od početka sveaustrijske statistike stanovništva 1754. godine do kraja vojnog konskripcijskog sustava sredinom 19. stoljeća [Issues regarding the population census in the free port of Trieste since the beginning of all-Austrian population statistics in 1754 until the end of the military conscription system in the mid-19th century]. Historijski zbornik 63(1):149–157

Kavrečič P (2017) Tourism as a source of non-agricultural rural income: the case of the Karst (16th–19th centuries). In: Panjek A, Larsson J, Mocarelli L (eds) Integrated peasant economy in a comparative perspective: Alps, Scandinavia and beyond. University of Primorska Press, Koper, pp 205–227

Knez M, Otoničar B, Petrič M, Pipan T, Slabe T (eds) (2020) Karstology in the Classical Karst. Springer, Cham. https://doi.org/10.1007/978-3-030-26827-5

Kranjc A (1990) Valvasor in kraško podzemlje [Valvasor and underground karst]. In: Vovko A (ed) Valvasorjev zbornik: ob 300 letnici izida Slave vojvodine Kranjske. SAZU, Ljubljana, pp 211–219

Kranjc A (2010) Short history of research. In: Mihevc A, Prelovšek M, Zupan Hajna N (eds) Introduction to the Dinaric Karst. Karst Research Institute at ZRC SAZU, Postojna

Krausmann F, Weisz H, Eisenmenger N (2016) Transitions in sociometabolic regimes throughout human history. In: Haberl H, Fischer-Kowalski M, Krausmann F, Winiwarter V (eds) Social ecology: society-nature relations across time and space. Springer, Cham, pp 63–92. https://doi.org/10.1007/978-3-319-33326-7_3

LaMoreaux J, Stevanović Z (2015) Historical overview on karst research. In: Stevanović Z (ed) Karst aquifers: characterization and engineering. Springer, Cham, pp 3–18. https://doi.org/10.1007/978-3-319-12850-4_1

Lowe DJ (2004) Geoscientists. In: Gunn J (ed) Encyclopedia of caves and karst science. Fitzroy Dearborn, New York, pp 380–383. https://doi.org/10.4324/9780203483855

Lučić I (2019) Presvlačenje krša: povijest poznavanja Dinarskog krša na primjeru Popova polja [Karst molting: the history of knowledge of the Dinaric Karst on the example of Popovo Polje]. Synopsis, Zagreb

Malez M (1984) Povijest speleoloških istraživanja u Hrvatskoj [Geschichte der Höhlenforschungen in Kroatien]. In: Deveti jugoslavenski speleološki kongres. Savez speleologa Jugoslavije and Speleološko društvo Hrvatske, Zagreb, pp 73–102

Mattes J (2016) Coming from abroad: the discourse on scientific centralism and Cvijić's studies in Vienna. In: Jović V, Petrović AM (eds) Proceedings of the 150th anniversary of Jovan Cvijić's birth. Serbian Academy of Sciences and Arts, Belgrade, pp 13–28

McNeill JR (2003) Observations on the nature and culture of environmental history. Hist Theory 42:5–43. https://doi.org/10.1046/j.1468-2303.2003.00255.x

Mihevc A, Prelovšek M, Zupan Hajna N (eds) (2010) Introduction to the Dinaric Karst. Inštitut za raziskovanje krasa ZRC SAZU, Postojna

Milanović P (2023) Karst of east Herzegovina and Dubrovnik littoral. Springer, Cham. https://doi.org/10.1007/978-3-031-28120-4

Moritsch A (1969) Das nahe Triester Hinterland: Zur wirtschaftlichen und sozialen Entwicklung vom Beginn des 19. Jahrhunderts bis zur Gegenwart. Hermann Böhlau's Nachf., Wien

Mrgić J (2017) Something old, something new: historical geography and environmental history discourse in Serbia. In: Petrić H, Žebec Šilj I (eds) Environmentalism in Central and Southeastern Europe: historical perspectives. Lexington Books, Lanham, pp 77–90

Netting RMC (1993) Smallholders, householders: farm families and the ecology of intensive, sustainable agriculture. Stanford University Press, Stanford

Panjek A (2015) Paesaggio culturale e ambiente del Carso: L'uso delle risorse naturali in età moderna. Založba Univerze na Primorskem, Koper

Parise M, Gabrovsek F, Kaufmann G, Ravbar N (eds) (2018) Advances in karst research: theory, fieldwork and applications. Geol Soc London Spec Publ 466. https://doi.org/10.1144/SP466

Pfister C, Camenisch C, Dobrovolný P (2018) Analysis and interpretation: Temperature and precipitation indices. In: White S, Pfister C, Mauelshagen F (eds) The Palgrave handbook of climate history. Palgrave Macmillan, London, pp 115–129. https://doi.org/10.1057/978-1-137-43020-5_11

Reisp B (1983) Kranjski polihistor Janez Vajkard Valvasor [The Carniolan polymath Johann Weichard Valvasor]. Mladinska knjiga, Ljubljana

Reisp B (2001) Valvasorjeva skicna knjiga za topografijo Kranjske 1678–1679 [Valvasor's sketchbook for the topography of Carniola 1678–1679]. In: Gostiša L (ed) Janez Vajkard Valvasor: Topografija Kranjske 1678–1679: Skicna knjiga: Faksimiliran natis originala iz Metropolitanske knjižnice v Zagrebu. Mladinska knjiga, Ljubljana, pp 3–17

Roglić J (1965) The delimitations and morphological types of the Dinaric Karst. Naše jame 7(1–2):12–20

Roglić J (1969) Geografski aspekti Dinarskog krša [The Dinaric Karst from the geographic aspect]. Krš Jugoslavije 6:19–38

Rossi P (2011) La nascita della scienza moderna in Europa. Laterza, Roma

Savezni geološki zavod (1970) Geološka karta SFR Jugoslavija 1:500.000 [Geological map SFR Yugoslavia 1:500,000]. Savezni geološki zavod, Beograd

Shaw TR (2004) Valvasor: a common error about his publications on Cerkniško jezero, Slovenia. Acta Carsol 33(2):313–317. https://doi.org/10.3986/ac.v33i2.307

Shaw TR (2007) History of cave science. Sydney Speleological Society, Sydney

Simmons IG (2008) Global environmental history: 10,000 BC to 2000 AD. Edinburgh University Press, Edinburgh

Trimmel H (2004) Europe, central: history. In: Gunn J (ed) Encyclopedia of caves and karst science. Fitzroy Dearborn, New York, pp 335–336. https://doi.org/10.4324/9780203483855

U.S. Environmental Protection Agency (2002) A lexicon of cave and karst terminology with special reference to environmental karst hydrology. U.S. Environmental Protection Agency, Washington, DC

White S, Brook J, Pfister C (2018) Climate, weather, agriculture, and food. In: White S, Pfister C, Mauelshagen F (eds) The Palgrave handbook of climate history. Palgrave Macmillan, London, pp 331–353. https://doi.org/10.1057/978-1-137-43020-5_27

Williams P (2004) Dolines. In: Gunn J (ed) Encyclopedia of caves and karst science. Fitzroy Dearborn, New York, pp 304–310. https://doi.org/10.4324/9780203483855

Winiwarter V (2016) Method précis: working with historical material. In: Haberl H, Fischer-Kowalski M, Krausmann F, Winiwarter V (eds) Social ecology: society-nature relations across time and space. Springer, Cham, pp 411–415

Winiwarter V, Knoll M (2007) Umweltgeschichte: Eine Einführung. Böhlau, Köln

Winiwarter V, Schmid M, Hohensinner S, Haidvogl G (2013) The environmental history of the Danube River Basin as an issue of long-term socio-ecological research. In: Singh SJ, Haberl H, Chertow M, Mirtl M, Schmid M (eds) Long-term socio-ecological research: studies in society-nature interactions across spatial and temporal scales. Springer, Dordrecht, pp 103–122. https://doi.org/10.1007/978-94-007-1177-8_5

Zeremski M (ed) (1992) Geomorfološka karta Jugoslavije 1:500.000 [Geomorphological map of Yugoslavia 1:500,00]. Geografski institut "Jovan Cvijić", Odbor za geodinamiku Srpske akademije nauka i umetnosti, Beograd

Zorn M, Erhartič B, Komac B (2009) La Slovénie, berceau du géotourisme karstique. Karstologia 54:1–10. https://doi.org/10.3406/karst.2009.2655

Zupan Hajna N (2019) Dinaric karst–geography and geology. In: White WB, Culver DC, Pipan T (eds) Encyclopedia of caves. Academic Press, London, pp 353–362. https://doi.org/10.1016/B978-0-12-814124-3.00039-X

Zwitter Ž (2014) Okolje na Kranjskem v 17. stoletju po *Slavi vojvodine Kranjske* [Die Umwelt in Krain im 17. Jahrhundert in der *Die Ehre Deß Hertzogthums Crain*]. In: Weiss J (ed) Studia Valvasoriana: zbornik spremnih študij ob prvem integralnem prevodu Die Ehre Deß Hertzogthums Crain v slovenski jezik. Zavod Dežela Kranjska, Ljubljana, pp 611–702

Zwitter Ž (2015) Izbrani vidiki okoljske zgodovine kobilarne Lipica od ustanovitve do začetka tridesetih let 17. stoletja [Selected environmental history aspects of the Lipica stud farm from its foundation to the early 1630s]. Kronika: časopis za slovensko krajevno zgodovino 63(3):457–474

Zwitter Ž (2019) Doktorati slovenskih študentov geografije na dunajski univerzi od leta 1900 do konca prve svetovne vojne [Doctoral theses by Slovenian students of geography at the University of Vienna between 1900 and 1918]. In: Smolej T (ed) Zgodovina doktorskih disertacij slovenskih kandidatov na dunajski Filozofski fakulteti (1872–1918). Znanstvena založba Filozofske fakultete, Ljubljana, pp 223–422

Žvanut M (2014) Veliki mali svet Erasma Franciscija [Die große kleine Welt Erasmus Franciscis]. In: Weiss J (ed) Studia Valvasoriana: zbornik spremnih študij ob prvem integralnem prevodu Die Ehre Deß Hertzogthums Crain v slovenski jezik. Zavod Dežela Kranjska, Ljubljana, pp 173–193

Žiga Zwitter Assistant professor of early modern history. He graduated in history and geography. In 2015 he obtained his PhD in history. In 2016–2017 he was a postdoctoral research fellow at the Vienna's Institute of Social Ecology. His research has focused on environmental history. He has been leading the Croatia–Serbia–Slovenia region in the *European Society for Environmental History*, and has served as a vice-president of the *Historical Association of Slovenia*. He is the main author of the forthcoming monograph *Historical Biodiversity in the Alps: Species-Rich Grassland Agroecosystems in the Last Millennium*.

Borna Fuerst-Bjeliš Senior researcher and full professor. Her research interests include environmental history and landscape change, focusing spatially mainly on the Mediterranean karst region, including the development of research methods (historical cartography and GIS technology), borderlands, spatial perception and spatial (regional) identities. She is (co-)author/(co-)editor of books on topics of *Historical Geography of Croatia, Mediterranean Identities*, and *Globalization, Marginalization and Conflict*. She served as Vice-President of the *European Society for Environmental History* (2017–2019) and she is currently Chair of the *Commission on Marginalization, Globalization and Regional and Local Responses of the International Geographical Union* (2020-present).

Jelena Mrgić Associate professor. She teaches *Historical geography* and *Introduction to environmental history*. Her focus in research changed from historical geography of the mediaeval and early Ottoman Bosnia to preindustrial environmental history of the Western Balkans, changes of settlement patterns and cultural landscapes, wine and brandy production history in lieu of climate changes during the LIA, to emotional history as an aspect of studying environment, past and recent crises. She has been a representative of Serbia in the *European Society for Environmental History* (ESEH) since 2010, now jointly represented in the Region of Slovenia-Croatia-Serbia. She has written three books, several book chapters in edited volumes and research articles in peer reviewed journals and is active in reviewing papers for research journals in environmental history.

Hrvoje Petrić Professor of history. His research deals with environmental history, economic history and Croatian early modern history. He has written several books and a series of articles (more than 20). He is the editor of "Ekonomska i ekohistorija" [*Economic and Environmental History*] Journal.

Matija Zorn Head of the Anton Melik Geographical Institute of the Research Centre of the Slovenian Academy of Sciences and Arts (Ljubljana, Slovenia), and a lecturer at the Faculty of Humanities of the University of Primorska (Koper, Slovenia). He is a geographer and a historian, PhD. His main research interest includes physical geography (especially geomorphology), natural hazards, land degradation, environmental history, historical geography and geographic information systems. Most recently he co-edited a book titled *The Geography of Slovenia: Small But Diverse* (Springer Nature). He is Chairman of the *Commission on Land Degradation and Desertification of the International Geographical Union.*

Part I
Cultural Landscape

Chapter 2
The Holocene Vegetation and Land-Use History in the Northern Dinaric Karst

Maja Andrič and Dirk Nikolaus Karger

Abstract This study compiles pollen records from four study sites in southern Slovenia (Cerkniško jezero and Mlaka) and north western Croatia (Prapoče and Čepićko polje) to examine how and when climatic variations or human activities influenced the vegetation of the Dinaric Karst during the last 10,000 years. Analysis of the existing palynological data shows significant differences in forest composition over time. Late-successional and drought-intolerant tree taxa (fir, beech) were prevalent in mountainous hinterland during periods of wetter and/or colder climate (ca. 6800, 6000–5600, 4000 cal. BP), and lower human impact. The first traces of human impact (small-scale forest clearing, agriculture, grazing) appear in the Neolithic period from about 6000 cal. BP. More intensive agro-pastoral activities were recorded in the Bronze Age, especially in the western part of the region. In some areas, beech and fir forests were partially replaced by mixed oak forests; between the end of the Bronze Age (3200 cal. BP) and the Iron Age (2800 cal. BP), a more open landscape similar to today appeared. The area around Cerkniško jezero was more forested (mainly fir and beech) it was not until the Roman period (1600 cal. BP) that forests were cleared on a large-scale. In the east (Bela Krajina, Mlaka), the main species was beech, whose presence in the landscape has been greatly reduced by permanent small-scale deforestation and burning. At Mlaka, vegetation recovered rapidly after clearing during most of the Holocene; the present-day open landscape only appeared at about 1000 years ago.

Keywords Pollen · Vegetation history · Archaeology · Land-use · Holocene

---

M. Andrič (✉)
ZRC SAZU Institute of Archaeology, Ljubljana, Slovenia
e-mail: maja.andric@zrc-sazu.si

D. N. Karger
Swiss Federal Research Institute WSL, Birmensdorf, Switzerland
e-mail: dirk.karger@wsl.ch

B. Fuerst-Bjeliš et al. (eds.), *Environmental Histories of the Dinaric Karst*, Environmental History 17, https://doi.org/10.1007/978-3-031-56089-7_2

2.1 Introduction

Long-term (millennial) changes in vegetation can be reconstructed by studying fossil pollen preserved in lake and marsh sediments. In the Dinaric Karst, Holocene vegetation was strongly influenced by climate and humans. Therefore, it is important to study the long-term vegetation history to better understand the reasons for environmental changes. However, in karst environments, wetlands suitable for palynological studies are very rare due to the lack of surface water. At many study sites, pollen is not preserved or only partly preserved, resulting in fragmented sediment sequences with insufficient radiocarbon dates and low resolution of analysed pollen samples. Nevertheless, several pollen diagrams have been produced in the last decades (Table 2.1, Fig. 2.1). This article aims to provide an overview of previous palynological research in order to identify the similarities and differences between the regions of the northern Dinaric Karst, to better understand the mechanisms of environmental change and to plan future research. In particular, four recently analysed sites (Prapoče, Čepićko polje, Gorenje jezero and Mlaka; Figs. 2.2, 2.3, 2.4, 2.5 and 2.6) with high-temporal resolution and relatively good chronological control, covering most of the Holocene, will be compared with model based climate reconstructions (CHELSA-TraCE21k)[1] to better understand how vegetation was shaped by climate variability and human influence.

2.2 Short Review of to Date Palynological Research

2.2.1 Western Coast of Istria

The first palynological studies in this region were conducted in the 1960s when Beug[2] analysed several sediment cores, on the west coast of Istria, in marine, freshwater and brackish environments. The results suggest that in the 5th millennium cal. PB, sub-Mediterranean deciduous oak forests predominated in Istria, with a belt of evergreen trees, mainly evergreen oaks (*Quercus ilex*) in a natural zone 20–30 m wide along the Adriatic coast. Beug[3] assumed that with the beginning of the Roman colonization (2200 cal. BP) the forest consisting of deciduous oaks was cleared in the area, and that evergreen species (mainly *Phillyrea* and *Quercus ilex*) formed a wider belt of scrub. At the same time, *Juglans* and *Castanea* expanded indicated by an increase in pollen from these genera. This vegetation change however, has not been dated by radiocarbon. Similarly, a significant human influence on the environment was also found further north, at Škocjanski zatok[4] (indicated by Cerealia, and increased

[1] Karger et al. (2023).

[2] Beug (1967); Beug (1977).

[3] Beug (1977).

[4] Culiberg (1995).

Table 2.1 Palynological research in the northern Dinaric Karst

Study site	Altitude/ coordinates	Type of study site (size, diameter)	Timeframe	References
Škocjanski zatok (Koper, Slovenia)	1 m a.s.l. 45°21′N, 13°45′E	Marsh at the mouth of the Rižana river	ca. 7000–500 cal. BP?	Culiberg (1995)
Palu (Istria, Croatia)	0 m a.s.l. 45°01′N, 13°42′E	Brackish marsh (200 m^2)	ca. 5000–2000 cal. BP?	Beug (1977); Beug (1967)
Limski kanal (Istria, Croatia)	0 m.a.s.l	Marine core	ca. 5000–2000 cal. BP?	Beug (1977); Beug (1967)
Busuja Bay (Istria, Croatia)	– 1 to 2 m a.s.l. 45°15′N, 13°34′E	Marine core	5200–3200 cal. BP	Kaniewski et al. (2018)
Lake Vrana (Island of Cres, Croatia)	10–15 m a.s.l. 43°53′N, 15°34′E	Freshwater lake (5.75 km^2, 5.5 × 1.5 km)	17.500–0 cal. BP	Schmidt et al. (2000)
Prapoče (Istria, Croatia)	480 m a.s.l. 45°25′N, 14°04′E	Marsh (3 km^2, 600 m)	9400–0 cal. BP	Andrič and Willis (2003); Andrič (2004, 2006)
Čepićko polje (Istria, Croatia)	27 m a.s.l. 45°10′N, 14°9′E	Dried-out lake (5.4–8.6 km^2, 2.5 × 4 km)	6800–1400 cal. BP	Balbo et al. (2006)
Gorenje jezero (Cerkniško jezero, Slovenia)	550 m a.s.l. 45°43′N, 14°24′E	Marsh (intermittent lake) (26–38 km^2, 10.5 × 4.7 km)	9800–0 cal. BP	Andrič and Willis (2003); Andrič (2004)
Mlaka (Bela krajina, Slovenia)	140 m a.s.l. 45°30′N, 15°12′E	Marsh (148 m^2, 30 m)	9500–0 cal. BP	Andrič and Willis (2003); Andrič (2004, 2007)

Castanea, Juglans, Olea and *Vitis*). Higher values for *Abies*, *Fagus* and *Carpinus betulus* were found there, which is due to the less Mediterranean climatic conditions and the larger pollen catchment area of the brackish water basin.

Recently, the influence of humans on the Middle/Late Holocene vegetation in western Istria has been studied in greater depth.[5] Palynological studies in the Busuja Bay (Table 2.1, Fig. 2.1) revealed agro-pastoral activities (grazing, agriculture, fires) and a transition from deciduous to evergreen vegetation around 5000 cal. BP. There was a gradual increase in coastal pine and oak forests (*Pinus, Quercus ilex, Sanguisorba, Mercurialis*), which peaked in the Bronze Age around 4300 cal. BP, and shrub forests (Ericaceae, *Juniperus, Helianthemum* and *Cistus* family, *Phillyrea,*

[5] Kaniewski et al. (2018).

Fig. 2.1 Map of palynological study sites mentioned in the text. Altitude: light grey > 600 m a.s.l., dark grey > 1000 m a.s.l

Pistacia) which dominated from 3850 cal. BP. This is much earlier than suggested by Beug.[6] An increase of olive (*Olea*) pollen was also detected at 5050 cal. BP (similarly as at Lake Vrana, together with *Vitis*, Cerealia type and *Plantago*, see below and reference[7]) when a wild olive variety was probably planted around Busuja bay.[8] A greater influence of seawater between 4200–3850 cal. BP could be linked to the last phase of postglacial marine transgression and drier climatic conditions.[9]

[6] Beug (1977).

[7] Schmidt et al. (2000).

[8] Kaniewski et al. (2018).

[9] Ibid.

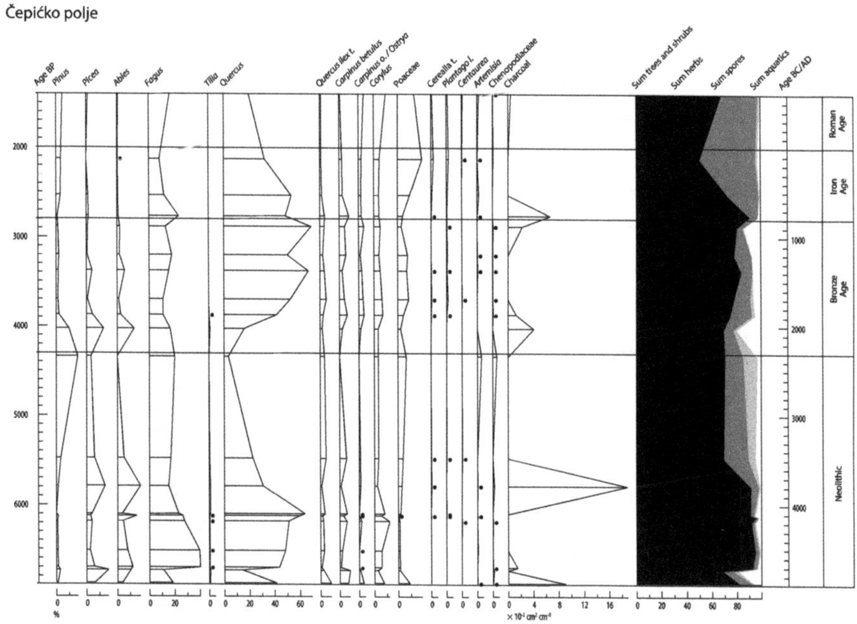

Fig. 2.2 Čepićko polje. Short pollen diagram of selected taxa. Values < 0.5% are marked with a solid dot. See the English translations of the Latin plant names in Appendix 2.1

2.2.2 *Island Cres and the Hinterland of Istria*

The longest sequence in this region (about. 17,500 years) was analysed in Lake Vrana (island of Cres) (Table 2.1, Fig. 2.1). Using multi-proxy research (cladocerans, diatoms, geochemistry, plant macrofossils, pollen), changes in lake level and vegetation were reconstructed.[10] The shallow lake and marsh phases occurred at the end of the late glacial and beginning of the Holocene; the lake was surrounded by mixed forests (*Pinus, Quercus, Betula, Fraxinus*). The present deep lake formed in the pluvial period at ca. 9100–9700 cal. BP (i.e. 8.5 ^{14}C ky BP), when beech (*Fagus*) started to spread. The first traces of human influence appeared as early as in the Neolithic period (at 120 cm core depth, i.e. ca. 6200 cal. BP): deforestation (degradation of deciduous oak forest, increased concentration of evergreen oaks and microcharcoal) and increased pollen of *Juniperus,* Cerealia, *Plantago, Olea* and *Carpinus betulus.*[11] In the Bronze Age, after about 5000 cal. BP human impact increased, as in Busuja Bay.[12] The later appearance of the present vegetation (*Quercus ilex, Ostrya/Carpinus orientalis, Carpinus betulus, Fraxinus ornus*) and the replacement of deciduous oaks by evergreen oaks is probably a consequence

[10] Schmidt et al. (2000).

[11] Ibid.

[12] Kaniewski et al. (2018).

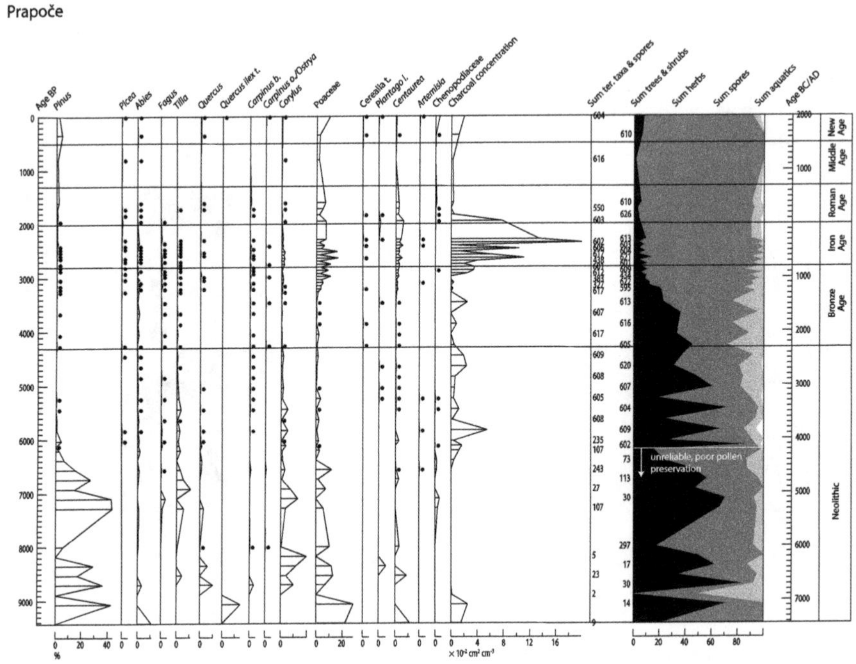

Fig. 2.3 Prapoče. Short pollen diagram of selected taxa. Values < 0.5% are marked with a solid dot. See the English translations of the Latin plant names in Appendix 2.1

of forest clearing (evidenced by an increased influx of allochtonous sediment into the lake), associated with an increase in summer drought (4.2 ky BP event).[13] The most recent part of the core has however not been accurately dated by radiocarbon. Abundance of *Juglans* and *Castenea* increased during the transition between the Iron Age and to the Roman period. In the mediaeval part of the core at 40 cm, i.e. around 850 cal. BP, if linear interpolation is used for age-depth modelling, another *Juniperus* peak indicates that forests were cleared for pastures, while *Olea* cultivation was intensified.

The two sequences in Istrian hinterland cover much shorter periods than at Lake Vrana, but show similar traces of human influence on vegetation. At Čepićko Polje[14] (Table 2.1, Fig. 2.1), pollen preserved only in a part of the sequence (6800–1400 cal. BP) suggests that around 6600 cal. BP, Čepićko Polje was surrounded by a mixed forest (*Fagus, Quercus*–mostly deciduous type of oak, *Abies, Corylus*), while after 6500 cal. BP, the composition of the forest changed towards a more open forest type (decline of *Fagus*) (Fig. 2.2). Geochemical analyses of sediments indicate a decrease in the water level of the lake (visible by a carbonate peak and increase in organic matter) and gradually drier hydrological conditions. The increased concentration of

[13] Schmidt et al. (2000).

[14] Balbo et al. (2006).

Gorenje jezero

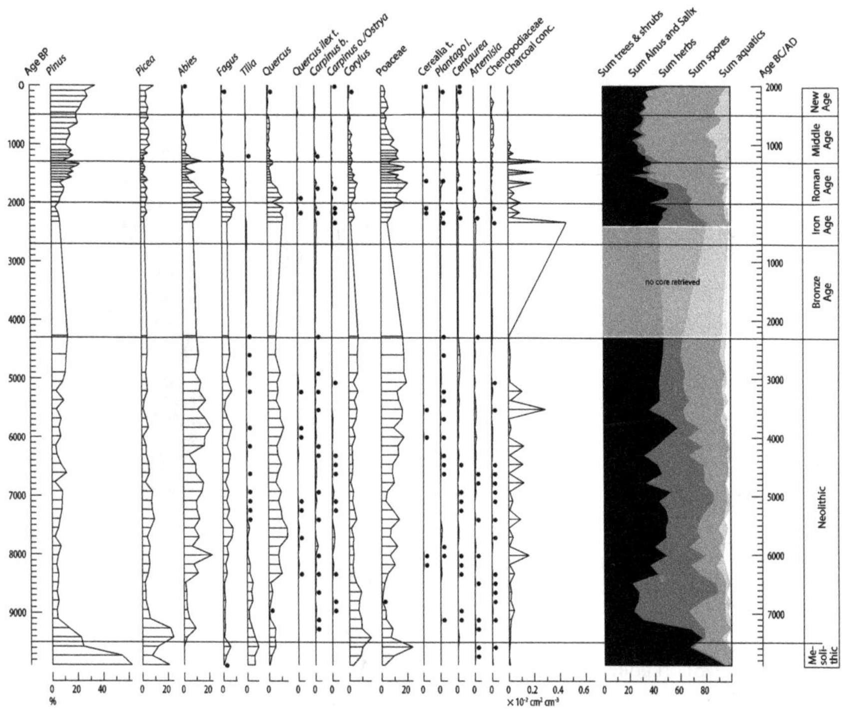

Fig. 2.4 Gorenje jezero. Short pollen diagram of selected taxa. Values < 0.5% are marked with a solid dot. See the English translations of the Latin plant names in Appendix 2.1

microscopic charcoal and the first Cerealia-type pollen grains at around 6150 cal. BP indicates small-scale Neolithic agricultural practices around the Polje, where numerous archaeological sites have been discovered.[15] In the middle section of the core, dated to around 4300–4100 cal. BP, pollen is poorly or not at all preserved, indicating drier hydrological conditions. During the Bronze and Iron Ages, human influence increased: around 3900 cal. BP the proportion of tree and shrub taxa decreased and the number of herbs indicates a more open landscape with agriculture (Cerealia type pollen) and grazing (Poaceae, *Plantago lanceloata*). Sedimentological analyses indicate wetter hydrological conditions and faster sedimentation rates.[16]

The Prapoče (Table 2.1, Fig. 2.1) sediment core covers the last 9500 years (Fig. 2.3), but the pollen of the early Holocene part of the core is poorly preserved, so that the high proportion of *Pinus* pollen (up to ca. 40%) between 9400 and 6800 cal. BP might indicate a relatively open landscape or be a consequence of selective pollen degradation. Therefore, only the vegetation evolution of the last 6300 years

[15] Miracle and Forenbaher (2005); Miracle and Forenbaher (2006).

[16] Balbo et al. (2006).

Mlaka

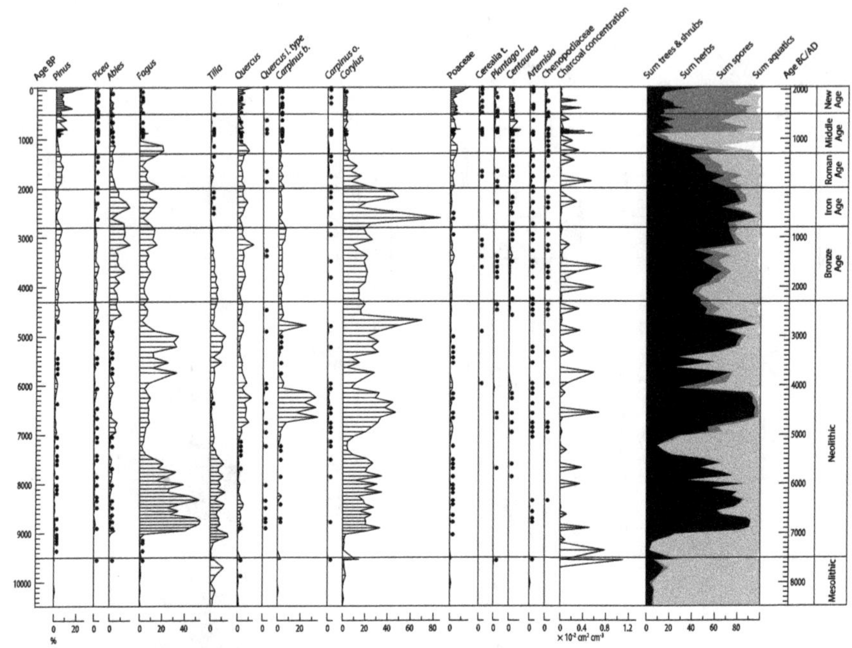

Fig. 2.5 Mlaka. Short pollen diagram of selected taxa. Values < 0.5% are marked with a solid dot. See the English translations of the Latin plant names in Appendix 2.1

or so can be reliably reconstructed[17] (Fig. 2.3). Between about 6300 and 4000 cal. BP, a mixed forest of conifers and deciduous trees (dominated by *Tilia*, but also *Quercus, Carpinus betulus, Corylus, Abies* and *Fagus*) grew in the region. Towards the end of the 5th millennium cal. BP, the landscape gradually became more open, and around 4100 cal. BP, Cerealia-type pollen appeared. Between 4000 and 3500 cal. BP, the proportion of *Abies* also increased. The Prapoče sequence shows a significant human influence on the vegetation in the Bronze Age, when, due to forest deforestation/burning, agriculture and grazing, present-day vegetation developed from about 3100 cal. BP.

2.2.3 Cerkniško Jezero and Bela Krajina Regions

The palynological sequences of Gorenje jezero (located in the Cerkniško jezero region) and Mlaka (located in the Bela Krajina region) (Table 2.1, Fig. 2.1) are further away from the Adriatic Sea, so, due to the more humid and less Mediterranean climate, *Fagus* and *Abies* were more widespread than in other locations in Istria and

[17] Andrič and Willis (2003); Andrič (2004); Andrič (2006).

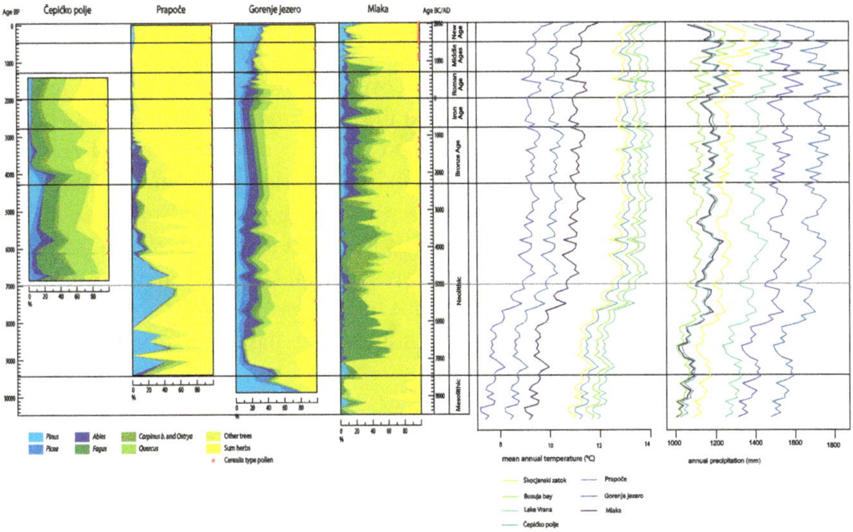

Fig. 2.6 Comparison of four study sites (Čepićko polje, Prapoče, Gorenje jezero and Mlaka) showing the main characteristics of the vegetation (forest composition and landscape openness) with mean annual temperature and annual precipitation according to CHELSA-TraCE21k v1.0 model. See the English translations of the Latin plant names in Appendix 2.1. *Source* Karget et al. (2021)

Cres. At the beginning of the Holocene (10,000–9000 cal. BP), Gorenje jezero was surrounded by *Pinus* and *Picea* forests, which were gradually replaced by broadleaf taxa (*Fagus, Tilia, Quercus, Corylus*) and later by *Alnus* and *Abies* (Fig. 2.4).[18] In 8000 cal. BP *Abies* was probably the most common tree in the region. Major deforestation of the forest does not appear until the Roman Period (1700 cal. BP), when the current open landscape emerged.

In the early Holocene (10,400–9000 cal. BP), the Mlaka site was surrounded by a forest composed mainly of *Tilia* coupled with *Corylus, Ulmus, Quercus, Pinus* and *Betula* (Fig. 2.5).[19] Micro-charcoal records indicate that forest fires were quite frequent in this landscape. Around 8900 cal. BP, the forest composition suddenly changed, probably due to a wetter climate, and a dense forest dominated by *Fagus*, started to grow around the Mlaka site. After about 7800 cal. BP, *Fagus* began to decline, the landscape became more open, and by 7300 cal. BP, the proportion of tree pollen had fallen to early Holocene values. It is not known whether this change in vegetation is a consequence of climate variations (drier climate) and/or human influence (see reference[20]). Around 7000 cal. BP, the forest regenerated with *Corylus* and *Quercus* forests, followed by *Carpinus betulus* (6700–6100 cal. BP). During this period, the impact of Neolithic populations on the landscape is clearly visible in the

[18] Andrič and Willis (2003).

[19] Andrič and Willis (2003); Andrič (2007).

[20] Andrič (2007).

form of deforestation and forest burning. Pollen of *Plantago lanceolata* and Cerealia type during this time indicates the presence of livestock and agriculture when also numerous archaeological sites have been discovered in the area.[21] Human influence reached its peak around 5800 cal. BP, when the *Carpinus betulus* forest was burnt and anthropogenic indicator pollen taxa increased. Vegetation recovered and the *Fagus* forest expanded again around 5700 cal. BP, but brief phases of forest burning and clearing with rapid vegetation recovery[22] ($\leq$ 50 years) lasted for most of the Holocene. From about 4500 cal. BP onwards human pressure on the environment increased, resulting in a change in forest composition and considerable expansion of *Abies*. After about 2000 cal. BP, the landscape became progressively more open and around 900 cal. BP, when intensive human influence on the environment began, the present open landscape emerged.

2.3 Vegetation and Human Impact

The vegetation in the northern Dinaric Karst has been significantly influenced by humans since the introduction of agriculture and pastoralism. The first traces of Neolithic human influence on the vegetation were discovered at Lake Vrana and at Čepićko polje. At both study sites, the first small-scale deforestation and burning (decrease in *Quercus, Fagus, Abies*, increase in the concentration of microscopic charcoal) and the first pollen grains of Cerealia type appeared around 6200 cal. BP.[23] Similarly, the first clear traces of local Neolithic agriculture and pastoralism (deforestation/burning of the beech forest) were recorded in Bela Krajina (Mlaka) around 5800 cal. BP.[24] In the Cerkinško jezero (Gorenje jezero), a larger intermittent lake with larger catchment area (and analysed with a lower resolution), traces of regional human activities (e.g. deforestation/burning of beech and fir forest) are less evident.[25]

In the Bronze Age, human pressure on the environment intensified. In Istria (Busuja Bay)[26] and on the island of Cres (Lake Vrana),[27] coastal pines (*Pinus nigra*) and evergreen oaks (*Quercus ilex*) began to spread gradually after 5000 cal. BP, together with the appearance of pollen from cultivated species (such as Cerealia, *Olea* and *Vitis*). The greatest expansion of shrubs (*Juniperus, Phillyrea, Pistacia*) and evergreen oaks is dated after 4000 cal. BP, which may be due to human influence and a drier climate.[28] The pollen of the cultivated taxa reached its peak around

---

[21] Budja (1992); Mason (1994); Mason (2001); Mason and Andrič (2009).

[22] Šilc and Andrič (2012).

[23] Schmidt et al. (2000); Balbo et al. (2006).

[24] Andrič (2007).

[25] Andrič and Willis (2003).

[26] Kaniewski et al. (2018).

[27] Schmidt et al. (2000).

[28] Kaniewski et al. (2018).

3800 cal. BP which is consistent with archaeological findings. In the Bronze Age, Istria was densely populated with large proto-urban settlements located about 5–10 km apart. For example, the settlement Monkodonija near Rovinj (about 20 km south of Busuja Bay) was occupied in the Early/Middle Bronze Age (ca. 3750–3400 cal. BP). It was estimated that ca. 1000 people lived in the settlement.[29] The main crops cultivated around this settlement were wheat (*Triticum monococcum* and *T. dicoccum*), barley (*Hordeum vulgare*), broomcorn millet (*Panicum miliaceum*) and legumes (*Pisum sativum, Vicia ervilia, Vicia faba*).[30] The main animal species in Mokodonija were sheep, goats and cattle, and hobut bones of domestic pigs and horses were also found.[31] The impact of these communities on the vegetation in Istria was considerable. A decline in late successional arboreal taxa (*Abies, Fagus*), but also *Pinus, Picea* and an increase in anthropogenic indicator taxa (*Plantago lanceloata* and Cerealia) after about 4000 cal. BP were also found at Čepićko Polje. Numerous hilltop prehistoric settlements have also been found further north, in Čičarija[32] and Komenski Kras,[33] but for many of these sites, there is no precise chronology (they were probably founded around 4000 cal. BP) and no archaeobotanical/archaeozoological studies have been conducted.[34] As in Busuja Bay, human influence on vegetation (deforestation/burning, cereal pollen) began to increase at Prapoče (Čičarija) around 3800 cal. BP and the open landscape, similar to today's appeared as early as 3100 cal. BP.

In Bela Krajina (Mlaka), deforestation and/or fire clearing of the beech forest intensified in the Early Bronze Age, when there were archaeological sites in the area, but the landscape was still forested (about 40–70% of the tree pollen).

In the Iron Age, the vegetation of the northern Dinaric Karst remained similar to that of the Bronze Age, while in the Roman period, the landscape became more open at Čepićko polje, Gorenje jezero and Mlaka. Archaeobotanical research conducted at the "Obrežje International Border Crossing" site, located approximately 60 km northeast of Mlaka, indicates that humans used thermophilic communities of mixed deciduous oaks with wide diversity of species as the main source of firewood and harvested fruit/hazelnuts.[35] Similar vegetation still grows today in the Obrežje region,[36] while the remaining beech forests at Mlaka were cleared later, in the Middle Ages, around 900 cal. BP.[37]

In summary, it can be said that the first human influence on vegetation was in the Neolithic period. With the beginning of agriculture and grazing, at least from about 6200–5800 cal. BP, humans burned/cut small areas of mixed forests, mainly

---

[29] Hänsel et al. (2015).

[30] Kroll (2015).

[31] Toškan et al. (2020).

[32] Buršić-Matijašić (2021).

[33] Bratina (2021).

[34] Kuzmanović and Mihovilić (2021).

[35] See also Chap. 3, in this volume: Tolar et al. (2024).

[36] Ibid.

[37] Andrič (2007).

consisting of deciduous oak and, inland, beech and/or fir, in order to create areas for crops and pasture. In the Bronze Age, human pressure on the vegetation intensified. The landscape opened up considerably, especially in Istria, probably due to the increase in human population (evident by highland fortified settlements) and metallurgical activities. In the Mediterranean coastal areas and on the islands, deciduous taxa have been replaced by evergreen vegetation, while inland *Fagus* and *Abies* have declined. These changes in forest composition may be associated with both, human influences and climatic variation (drier Mediterranean climate after 4200 cal. BP,[38] and references cited there; see below). Another phase of deforestation and spread of Cerealia, *Vitis, Olea, Juglans, Castanea* can be traced back to Roman period, when clearings probably took place in the vicinity of important trade and military roads (e.g. at Cerkniško jezero). In the hinterland areas characterised by a wetter climate, a lower prehistoric population density and a higher presence of late successional tree species (*Fagus, Abies*), the forest was quickly re-established and the open landscape, which resembles today's appeared only later, e.g. at Mlaka only in the Middle Ages.

2.4 Vegetation and Climate

It is difficult to address the effects of climate variability on vegetation because detailed palaeoclimatic studies of the fossil record have not yet been conducted in the region, and vegetation has been significantly shaped by humans. However, pollen results can be compared with paleoclimate studies in other regions and with data from global climate models (e.g. reference[39]) (Fig. 2.6).

Only at Lake Vrana, a complete transition between the late glacial and the Holocene has been preserved and researched.[40] The data show that at the beginning of the Holocene, when the climate became warmer and probably somewhat wetter,[41] there were fluctuations and a rise in the water level of Lake Vrana and an increase in the pollen of *Pinus nigra, Quercus* and *Fraxinus ornus*.[42] At some sites (Mlaka, Prapoče), the pollen in the early Holocene sediment was not preserved, indicating still rather dry conditions due to seasonal contrasts (higher summer solar radiation and higher summer temperatures).[43]

Changes in vegetation over the last 10,000 years have been at least partly preserved at several sites. The results for selected tree taxa (showing past forest composition) vs. herbaceous (reflecting landscape openness) and Cerealia-type (agriculture) pollen were plotted on the same timescale for ease of comparison (Fig. 2.6). These pollen

[38] Bini et al. (2019).

[39] Karger et al. (2023); Karger et al. (2020).

[40] Schmidt et al. (2000).

[41] Kutzbach and Guetter (1986); Feurdean et al. (2008); Lotter et al. (2000); von Grafenstein et al. (2000); Peyron et al. (2005).

[42] Schmidt et al. (2000).

[43] Feurdean et al. (2008).

diagrams were compared to the mean annual temperature and annual precipitation data for each study area, extracted from CHELSA-TraCE21k (Fig. 2.6). These data show that at the three hinterland study sites (Prapoče, Gorenje jezero and Mlaka), simulated mean annual temperatures were about 1.5–4.5 °C lower and annual precipitation about 200–500 mm higher than in Istria and the Mediterranean coast (except for Mlaka where only about 1000–1200 mm was measured) throughout the Holocene period. After 10,000 cal. BP, *Fagus* (beech) and *Abies* (fir) advanced at Gorenje jezero and Mlaka. At Gorenje jezero, *Abies* was more widespread (especially after about 8300 cal. BP), while at Mlaka, where forest fires were more frequent and the climate was drier (~ 400 mm annual precipitation) today and in the past (Fig. 2.6),[44] *Fagus* prevailed since 8900 cal. BP. No significant human impact on the vegetation was found, and it was assumed that the expansion of *Fagus* and *Abies* was due to a wetter climate (reference[45] and references cited therein), although no significant increase in mean annual precipitation was found (Fig. 2.6). Both trees are sensitive to drought, late frosts, and frequent disturbances[46] such as grazing (especially fir[47]) and early Holocene forest fire regimes with frequent natural fires.[48] It is assumed that both taxa could only have expanded later in the Holocene, under human influence (after disturbances)[49] and when the climate became less seasonal (i.e. less continental), and colder/wetter due to changes in the amount of solar radiation that reaches the surface.[50] The proportion of *Fagus* pollen also increased at Lake Vrana and a deep water lake was formed, which can be correlated with the pluvial period on the Dalmatian coast.[51]

Between about 6700 and 6000 cal. BP, a major change in forest vegetation occurred at Mlaka: *Fagus* decreased, while *Corylus, Quercus* and *Carpinus betulus* increased. At other study sites in the region (Fig. 2.6) and in Ljubljansko barje (central Slovenia),[52] only a less pronounced decline of *Fagus* and/or *Abies* was observed. Possibly, the decline of *Fagus* at Mlaka was a consequence of the drier climate (see reference[53]), which was more pronounced than in other study sites located in near mountainous hinterland areas that received more orographic precipitation (Fig. 2.6). The palaeoclimate model data (Fig. 2.6) indicate a decrease in annual precipitation by 100 mm and an increase of average temperature by 0.5 °C between 7100 and 6300 cal. BP. The decline in lake level was observed at study sites in Switzerland, Germany,[54]

44 Ogrin (1996).

45 Andrič and Willis (2003).

46 Ellenberg (1988).

47 Tinner et al. (1999).

48 Feurdean et al. (2008).

49 Giesecke et al. (2017).

50 Tinner and Lotter (2006).

51 Wunsam et al. (1999); Schmidt et al. (2000).

52 Andrič et al. (2008).

53 Andrič (2007); Andrič et al. (2008).

54 Haas et al. (1998); Kalis et al. (2003).

and also in Čepićko polje.[55] The diatom and geochemical record of alpine lakes in central Austria suggests climatic warming between 7300 and 6000 cal. BP.[56] Similarly, the $\delta^{18}O$ record at Lake Geneva (6350–6000 cal. BP)[57] and Grotta di Ernesto (6800–6000 cal. BP)[58] suggests a warmer and drier climate.

After 6000 cal. BP, the proportion of *Fagus* and *Abies* pollen increased at all study sites (Fig. 2.6). This change in forest composition can be related to the global cold and wet period,[59] the rise in lake levels,[60] the advance of glaciers after 5800 cal. BP,[61] and ice rafting[62] and peat formation between 5100 and 4400 cal. BP in northern Europe.[63] Data from palaeoclimatic models indicate that at around 6000 cal. BP, annual precipitation was about 200 mm higher than in the early Holocene at all sites studied (Fig. 2.6).

After 4500 cal BP, the world climate became wetter. Due to minimal solar activity, summer temperatures in the Northern Hemisphere probably decline.[64] However, the controversial question of whether the climate actually cooled or warmed in the late Holocene is subject of heated debate (see reference[65]). Alpine lakes recorded high water levels and an increased frequency of flooding between 4200 and 2400 cal. BP.[66] The wetter climate of the late Holocene is clearly reflected in an increase of *Abies* at Mlaka (Fig. 2.6). At Prapoče and Čepićko polje *Abies* started to increase a few centuries later, but declined in the Middle/Late Bronze Age, probably due to clearing/burning of forests and grazing.

Further south, in the coastal areas of Istria, the situation is less clear due to the lack of study sites covering the early/middle Holocene. In the late Holocene (after 4200 cal. BP), the influence of climate on the vegetation of Istria was significant, as was the human influence in the Bronze Age, which masked the climate signal. It is assumed that "mediterraneanization", the appearance of Mediterranean maquis, i.e. evergreen and sclerophyllous shrubs or heaths vegetation, developed in response to the drier climate in the Late Holocene and/or the intense human influence on the environment since the Bronze Age.[67] Based on of lake level data, it has been suggested that in the Late Holocene the climate south of about 40° N latitude became drier

[55] Balbo et al. (2006).

[56] Schmidt et al. (2006).

[57] Anadón et al. (2006).

[58] McDermott et al. (1999).

[59] Mayewski et al. (2004); Haas et al. (1998).

[60] Magny (2004); Magny and Haas (2004).

[61] Denton and Karlén (1973); Seppä et al. (2001).

[62] O'Brien et al. (1995); Bond et al. (1997).

[63] Seppä et al. (2001).

[64] Denton and Karlén (1973); Wanner et al. (2011).

[65] Wanner (2021).

[66] Vanniere et al. (2013); Wirth et al. (2013); Sabatier et al. (2017); Rapuc et al. (2018); Andrič et al. (2020).

[67] Jalut et al. (2000); Reille and Pons (1992); Roberts et al. (2004); Roberts et al. (2011); Sadori et al. (2011).

than in the Middle Holocene, while north of 40° N latitude it became wetter.[68] The northern Dinaric Karst is located in north central Mediterranean, (46–45°N), north of the 40°N. The wetter climate of the Late Holocene is clearly evident at the study sites in the northern part of our study area, while further south, in Istria, a short dry phase/lake minima was noted in Busuyja Bay and Čepićko Polje, dated to about 4200 cal. BP.[69] A review of 4200 cal. BP Mediterranean palaeoclimate data suggests that there was a climatic dry period around 4200 cal. BP, but with significant interregional heterogeneity in vegetation and predictions from climate models.[70] In the northern Dinaric Karst, differences in forest composition between the different study sites were also significant. In the northern part of the region, *Abies* and *Fagus* were more abundant and no traces of the 4200 cal. BP drought were detected. Paleoclimate model data showed variations in temperature and precipitation, but no major differences were detected between the study sites. In the central Mediterranean, stronger storm periods occurred during the last 4500 years, leading to coastal flooding.[71]

Over the last 4000 years, human activities have altered vegetation more than climate, so that it is difficult to distinguish between climate change and human influence. Climate change has probably not been strong enough to cause significant changes in forest composition (at the lowland study sites) and human activities have changed forest composition more than climate. In addition, the sampling resolution at many study sites was not detailed enough to detect changes in the short-term. So far, research has focused on vegetation changes during archaeological periods, while the last few centuries (for which we have historical data) are underrepresented.

2.5 Conclusions

The vegetation of the North Dinaric Karst has been considerably influenced by climate fluctuations and human activities. The first traces of human influence (small-scale clearing/burning of forests, agriculture, grazing) can be identified in the Neolithic period from at least ca. 6000 cal. BP, while more intensive agro-pastoral activities have been recorded in the Bronze Age, especially in the western part of the region. The tree taxa at the end of succession (*Fagus, Abies*) were replaced by light-demanding vegetation at the beginning of succession (e.g. *Quercus*). In the late Holocene, when the climate in the southern Mediterranean became drier,[72] evergreen Mediterranean vegetation spread to the coastal areas. Human influence was more intense, and led to soil erosion, changes in microclimate and irreversible changes in

---

[68] Magny et al. (2013).

[69] Balbo et al. (2006); Kaniewski et al. (2018).

[70] Sadori et al. (2011); Bini et al. (2019).

[71] Kaniewski et al. (2016).

[72] Jalut et al. (2000); Reille and Pons (1992); Roberts et al. (2004); Roberts et al. (2011); Sadori et al. (2011); Magny et al. (2013).

vegetation.[73] In the hilly areas with a wetter climate and deeper soils, the vegetation recovered more quickly, and the mediterraneanization that began later was less intense.[74] In hinterland of the North Dinaric Karst, there were very open landscapes, similar to those of today, only in Roman (Cerkniško jezero) and/or mediaeval times (Bela Krajina).

The vegetation history of the last 2000–1000 years has hardly been researched. Since no detailed studies of the Holocene climate have been carried out in the region, it is difficult to distinguish between anthropogenic and climate-induced vegetation. Climate variability in the late Holocene was probably not strong enough to cause significant changes in forest composition, and human activities have changed forest composition more than climate. In addition, the sampling resolution at many study sites was not detailed enough to detect these short-term changes. Many studies focused on Holocene environmental changes during different archaeological periods, neglecting the more recent centuries when more historical written sources were available. At many sites studied, the upper sedimentary layers are too dry or have been drained, so pollen is not preserved. Therefore, further studies of the past environment, vegetation and climate are needed to better understand the effects of climate variability on Holocene vegetation in the northern Dinaric Karst.

Acknowledgements This research was funded by Slovenian Research Agency, programmes P6-0064 and P6-0064-22, "Archaeological research." Figures were prepared by Tamara Korošec. We are grateful to Nina Caf for corrections and comments on the first draft of the paper. Comments and corrections of an anonymous reviewer are gratefully acknowledged.

Appendix 2.1: Latin Names of Plant Taxa Mentioned in the Text

Abies	fir
Alnus	alder
Artemisia	mugwort/wormwood/sagebrush—family Asteraceae
Betula	birch
Carpinus betulus	hornbeam
Carpinus orientalis/Ostrya	oriental/hop hornbeam
Castanea	chestnut tree
Centaurea	knapweeds
Cerealia	cereal type pollen
Chenopodiaceae	goosefoot

(continued)

[73] Kaligarič et al. (2006).

[74] Sadori et al. (2011).

(continued)

Cistus	rock rose—family Cistaceae
Corylus	hazel
Ericaceae	heather
Fagus	beech
Fraxinus excelsior type	ash
Fraxinus ornus	manna ash
Helianthemum	rock rose
Hordeum vulgare	barley
Juglans	walnut
Juniperus	juniper
Mercurialis	mercuries
Olea	olive
Panicum miliaceum	broomcorn millet
Phillyrea	mock privet
Picea	spruce
Pinus	pine
Pinus nigra	black pine
Pisum sativum	peas
Pistacia	pistachio
Plantago sp.	plantain
Plantago lanceolata	ribwort plantain
Poaceae	grasses
Quercus robur type	deciduous oak
Quercus ilex	evergreen oak
Sanguisorba	burnet
Tilia	lime
Triticum	wheat
Ulmus	elm
Vicia ervilia	ervil
Vicia faba	broad bean
Vitis	grapevines

References

Anadón P, Moscariello A, Rodríguez-Lázaro J, Filippi ML (2006) Holocene environmental changes of Lake Geneva (Lac Léman) from stable isotopes ($\delta^{13}C$, $\delta^{18}O$) and trace element records of ostracod and gastropod carbonates. J Palaeolimnol 35:593–616. https://doi.org/10.1007/s10933-005-4009-5

Andrič M (2004) Paleookolje v Sloveniji in severnem delu hrvaške Istre v pozni prazgodovini [The vegetation of Slovenia and northern Istria in late prehistory]. Arheološki vestnik 55:509–525

Andrič M (2006) Prapoče pollen core and Holocene vegetation change in northern Istria. In: Miracle PT, Forenbaher S (eds) Prehistoric herders of northern Istria: The archaeology of Pupićina cave, vol 1. Arheološki muzej Istre, Pula, pp 31–62

Andrič M (2007) The Holocene vegetation development in Bela krajina (Slovenia) and the impact of first farmers on the landscape. Holocene 17(6):763–776. https://doi.org/10.1177/0959683607080516

Andrič M, Willis KJ (2003) The phytogeographical regions of Slovenia: a consequence of natural environmental variation or prehistoric human activity? J Ecol 91(5):807–821. https://doi.org/10.1046/j.1365-2745.2003.00808.x

Andrič M, Kroflič B, Toman MJ, Ogrinc N, Dolenec T, Dobnikar M, Čermelj B (2008) Late quaternary vegetation and hydrological change at Ljubljansko barje (Slovenia). Palaeogeogr Palaeoclimatol Palaeoecol 270:150–165. https://doi.org/10.1016/j.palaeo.2008.09.008

Andrič M, Sabatier P, Rapuc W, Ogrinc N, Dolenec M, Arnaud F, von Grafenstein U, Šmuc A (2020) 6600 years of human and climate impacts on lake-catchment and vegetation in the Julian Alps (Lake Bohinj, Slovenia). Quat Sci Rev 227:106043. https://doi.org/10.1016/j.quascirev.2019.106043

Balbo AL, Andrič M, Rubinić J, Moscariello A, Miracle PT (2006) Palaeoenvironmental and archaeological implications of a sediment core from Polje Čepić, Istria. Croatia. Geol Croat 59(2):109–124. https://doi.org/10.4154/GC.2006.08

Beug H-J (1967) On the forest history of the Dalmatian coast. Rev Palaeobot Palynol 2:271–279. https://doi.org/10.1016/0034-6667(67)90156-X

Beug H-J (1977) Vegetationsgeschichte Untersuchungen im Küstenbereich von Istrien (Jugoslawien). Studies on the vegetation history of the coastal area of Istria (Yugoslavia). Flora 166(4):357–381. https://doi.org/10.1016/S0367-2530(17)32153-9

Bini M, Zanchetta G, Perşoiu A, Cartier R, Català A, Cacho I, Dean JR, Rita FD, Drysdale RN, Finnè M, Isola I, Jalali B, Lirer F, Magri D et al (2019) The 4.2 ky BP event in the Mediterranean region: an overview. Clim past 15:555–577. https://doi.org/10.5194/cp-15-555-2019

Bond G, Showers W, Chesbey M, Lotti R, Almasi P, deMenocal P, Priore P, Cullen H, Hajdas I, Bonani G (1997) A pervasive millennial-scale cycle in north Atlantic Holocene and Glacial climates. Science 278(5341):1257–1266. https://doi.org/10.1126/science.278.5341.1257

Bratina P (2021) Najstarejša zgodovina Komenskega Krasa. Štanjel od prazgodovine do poznega Rima [The oldest history of Komenski Kras. Štanjel from Prehistory to late Rome]. In: Friš F, Matjašič Friš M (eds) Kaštelir: Prazgodovinska gradišča in etnobotanika za trajnostni turizem. Univerza v Mariboru, Maribor, pp 53–65. https://doi.org/10.18690/978-961-286-492-7

Budja M (1992) Neolithic and Eneolithic settlement patterns in Bela krajina region of Slovenia. Memoire del Museo Civico di Storia Naturali di Verone (Iia Serie). Sezione Scienze Dell'uomo 4:119–127

Buršić-Matijašić K (2021) Gradine od Krasa do Kvarnera, dio kulturnog i krajobraznog identiteta sjeverne Istre [Hillforts from Kras to Kvarner, part of the cultural and landscape identities of northern Istria]. In: Friš D, Matjašič Friš M (eds) Kaštelir: Prazgodovinska gradišča in etnobotanika za trajnostni turizem. Univerza v Mariboru, Maribor, pp 81–93

Culiberg M (1995) Dezertifikacija in reforestacija slovenskega Krasa [Desertification and reforestation of the Karst in Slovenia]. Poročilo o raziskovanju paleolitika, neolitika in eneolitika v Sloveniji 22:201–217

Denton GH, Karlén W (1973) Holocene climate variations: their pattern and possible cause. Quat Res 3(2):155–174. https://doi.org/10.1016/0033-5894(73)90040-9

Ellenberg H (1988) Vegetation ecology of central Europe. Cambridge University Press, Cambridge

Feurdean A, Klotz S, Brewer S, Mosbrugger V, Tămas T, Wohlfarth B (2008) Lateglacial climate development in NW Romania—comparative results from three quantitative pollen-based methods. Palaeogeogr Palaeoclimatol Palaeoecol 265:121–133. https://doi.org/10.1016/j.palaeo.2008.04.024

Giesecke T, Brewer S, Finsinger W, Leydet M, Bradshaw RHW (2017) Patterns and dynamics of European vegetation change over the last 15,000 years. J Biogeogr 44(7):1441–1456. https://doi.org/10.1111/jbi.12974

Haas JN, Richoz I, Tinner W, Wick L (1998) Synchronous Holocene climatic oscillations recorded on the Swiss Plateau and at timberline in the Alps. Holocene 8(3):301–309. https://doi.org/10.1191/095968398675491173

Hänsel B, Mihovilić K, Teržan B (eds) (2015) Monkodonja: Istraživanje protourbanog naselja brončanog doba Istre, Knjiga 1: Iskopavanje i nalazi građevina = Forschungen zu einer protourbanen Siedlung der Bronzezeit Istriens, Teil 1: Die Grabung und der Baubefund. Arheološki muzej Istre, Pula

Jalut G, Esteban AA, Bonnet L, Thierry G, Fontugne M (2000) Holocene climatic changes in the Western Mediterranean, from south-east France to south-east Spain. Palaeogeogr Palaeoclimatol Palaeoecol 160:255–290. https://doi.org/10.1016/s0031-0182(00)00075-4

Kaligarič M, Culiberg M, Kramberger B (2006) Recent vegetation history of the North Adriatic grasslands: expansion and decay of an anthropogenic habitat. Folia Geobot 41:241–258. https://doi.org/10.1007/BF02904940

Kalis AJ, Merkt J, Wunderlich J (2003) Environmental changes during the Holocene climatic optimum in central Europe—human impact and natural causes. Quat Sci Rev 22(1):33–79. https://doi.org/10.1016/S0277-3791(02)00181-6

Kaniewski D, Marriner N, Morhange C, Faivre S, Otto T, Campo EV (2016) Solar pacing of storm surges, coastal flooding and agricultural losses in the Central Mediterranean. Sci Rep 6:197. https://doi.org/10.1038/srep25197

Kaniewski D, Marriner N, Morhange C, Rius D, Carre MB, Faivre S, Campo EV (2018) Croatia's mid-late Holocene (5200–3200 BP) coastal vegetation shaped by human societies. Quat Sci Rev 200:334–350. https://doi.org/10.1016/j.quatscirev.2018.10.004

Karger DN, Nobis MP, Normand S, Graham CH, Zimermann NE (2020) CHELSA-TraCE21k: downscaled transient temperature and precipitation data since the last glacial maximum. EnviDat. https://doi.org/10.16904/envidat.211

Karger DN, Nobis MP, Normand S, Graham CH, Zimmermann NE (2023) CHELSA-TraCE21k: high-resolution (1 km) downscaled transient temperature and precipitation data since the Last Glacial Maximum. Clim Past 19(2):439–456. https://doi.org/10.5194/cp-19-439-2023

Kroll H (2015) Biljni nalazi = Die Pflanzenfunde. In: Hänsel B, Mihovilić K, Teržan B (eds) Monkodonja: Istraživanje protourbanog naselja brončanog doba Istre, Knjiga 1: Iskopavanje i nalazi građevina = Forschungen zu einer protourbanen Siedlung der Bronzezeit Istriens, Teil 1: Die Grabung und der Baubefund. Arheološki muzej Istre, Pula, pp 105–108

Kutzbach JE, Guetter PJ (1986) The influence of changing orbital parameters and surface boundary conditions on climate simulations for the past 18,000 years. J Atmosph Sci 43(16):1726–1759

Kuzmanović N, Mihovilić K (2021) Novi podaci o gradinama na području Lanišća [New information about hillforts in the area of Lanišće]. In: Friš D, Matjašič Friš M (eds) Kaštelir: Prazgodovinska gradišča in etnobotanika za trajnostni turizem. Univerza v Mariboru, Maribor, pp 127–135. https://doi.org/10.18690/978-961-286-492-7

Lotter AF, Birks HJB, Eicher U, Hoffmann W, Schwander J, Wick L (2000) Younger Dryas and Allerød summer temperatures at Gerzensee (Switzerland) inferred from fossil pollen and cladoceran assemblages. Palaeogeogr Palaeoclimatol Palaeoecol 159(3–4):349–361. https://doi.org/10.1016/S0031-0182(00)00093-6

Magny M (2004) Holocene climate variability as reflected by mid-European lake level fluctuations and its probable impact on prehistoric human settlements. Quat Int 113:65–79. https://doi.org/10.1016/S1040-6182(03)00080-6

Magny M, Haas JN (2004) A major widespread climatic change around 5300 cal. yr BP at the time of the Alpine Iceman. J Quat Sci 19(5):423–430. https://doi.org/10.1002/jqs.850

Magny M, Combourieu-Nebout N, de Beaulieu J-L, Bout-Roumazeilles V, Colombaroli D, Desprat S, Francke A, Joannin S, Ortu E, Peyron O, Revel M, Sadori L, Siani G et al (2013) North-south palaeohydrological contrasts in the central Mediterranean during the Holocene: tentative synthesis and working hypotheses. Clim Past 9(5):2043–2071. https://doi.org/10.5194/cp-9-2043-2013

Mason P (1994) Neolitska in eneolitska naselja v Beli krajini: naselje v Gradcu in izraba prostora v času od 5. do 3. tisočletja BC [Neolithic and Eneolithic settlement in Bela krajina: the Gradac settlement and the use of space from 5th to 3th millenium BC]. Poročilo o raziskovanju paleolitika, neolitika in eneolitika v Sloveniji 22:183–199

Mason P (2001) Griblje in problem nižinskih arheoloških kompleksov v Sloveniji [Griblje and the problem of lowland archaeological complexes in Slovenia]. Varstvo spomenikov 39:7–27

Mason P, Andrič M (2009) Neolithic/Eneolithic settlement patterns and Holocene environmental changes in Bela krajina (South-Eastern Slovenia). Documenta Praehistorica 36:327–335. https://doi.org/10.4312/dp.36.21

Mayewski PA, Rohling EE, Stager JC, Karlén W, Maasch KA, Meeker LD, Meyerson EA, Gasse F, van Kreveld S, Holmgren K, Lee-Thorp J, Rosqvist G, Rack F, Staubwasser M, Schneider RR, Steig EJ (2004) Holocene climate variability. Quat Res 62(3):243–255. https://doi.org/10.1016/j.yqres.2004.07.001

McDermott F, Frisia S, Huang Y, Longinelli A, Spiro B, Heaton THE, Hawkesworth CJ, Borsato A, Keppens E, Fairchild IJ, van der Borg K, Verheyden S, Selmo E (1999) Holocene climate variability in Europe: evidence from $\delta^{18}O$, textural andextension-rate variations in three speleothems. Quat Sci Rev 18:1021–1038. https://doi.org/10.1016/S0277-3791(98)00107-3

Miracle PT, Forenbaher S (2005) Neolithic and Bronze Age herders of Pupićina Cave. Croatia. J Field Archaeol 30(3):255–281. https://doi.org/10.1179/009346905791072251

Miracle PT, Forenbaher S (eds) (2006) Prehistoric herders of northern Istria: the archaeology of Pupićina Cave. Arheološki muzej Istre, Pula

O'Brien SR, Mayewski PA, Meeker LD, Meese DA, Twickler MS, Whitlow SI (1995) Complexity of Holocene climate as reconstructed from a Greenland ice core. Science 270(5244):1962–1964. https://doi.org/10.1126/science.270.5244.196

Ogrin D (1996) Podnebni tipi v Sloveniji [The climate types in Slovenia]. Geografski vestnik 68:39–56

Peyron O, Bégeot C, Brewer S, Heiri O, Magny M, Millet L, Ruffaldi P, Van Campo E, Yu G (2005) Late-glacial climatic changes in Eastern France (Lake Lautrey) from pollen, lake-levels, and chironomids. Quat Res 64(2):197–211. https://doi.org/10.1016/j.yqres.2005.01.006

Rapuc W, Sabatier P, Andrič M, Crouzet C, Arnaud F, Chapron E, Šmuc A, Develle A-L, Wilhelm B, Demory F, Reyss J-L, Regnier E, Daut G, von Grafenstein U (2018) 6600 years of earthquake record in the Julian Alps (Lake Bohinj, Slovenia). Sedimentology 65(5):1777–1799. https://doi.org/10.1111/sed.12446

Reille M, Pons A (1992) The ecological significance of sclerophyllous oak forests in the west part of Mediterranean basin: a note on pollen analytical data. Vegetation 99(100):13–17

Roberts N, Brayshaw D, Kuzucuoğlu C, Perez R, Sadori L (2011) The mid-Holocene climatic transition in the Mediterranean: causes and consequences. Holocene 21(1):3–13. https://doi.org/10.1177/0959683610388058

Roberts N, Stevenson TC, Davis B, Cheddadi R, Brewer S, Rosen A (2004) Holocene climate, environment and cultural change in the circum-Mediterranean region. In: Battarbee RW, Gasse F, Stickley CE (eds) Past climate variability through Europe and Africa. Springer, Dordrecht, pp 343–362. https://doi.org/10.1007/978-1-4020-2121-3_17

Sabatier P, Wilhelm B, Ficetola GF, Moiroux F, Poulenard J, Develle A-L, Bichet A, Chen W, Pignol C, Reyss JL, Gielly L, Bajard L, Perrette Y, Malet E, Taberlet P, Arnaud F (2017) 6-kyr record of flood frequency and intensity in the western Mediterranean Alps: Interplay of solar and temperature forcing. Quat Sci Rev 170:121–135. https://doi.org/10.1016/j.quascirev.2017.06.019

Sadori L, Jahns S, Peyron O (2011) Mid-Holocene vegetation history of the central Mediterranean. Holocene 21(1):117–129. https://doi.org/10.1177/0959683610377530

Schmidt R, Müller J, Drescher-Schneider R, Robert K, Szeroczynska K, Barić A (2000) Changes in lake level and trophy at Lake Vrana, a large karstic lake on the Island of Cres (Croatia), with respect to palaeoclimate and anthropogenic impacts during the last approx. 16,000 years. J Limnol 59(2):113–130. https://doi.org/10.4081/jlimnol.2000.113

Schmidt R, Kamenik C, Tessadri R, Koinig KA (2006) Climatic changes from 12,000 to 4000 years ago in the Austrian Central Alps tracked by sedimentological and biological proxies of a lake sediment core. J Palaeolimnol 35:491–505. https://doi.org/10.1007/s10933-005-2351-2

Seppä H, John H, Birks B (2001) July mean temperature and annual precipitation trends during the Holocene in the Fennoscandian tree-line area: pollen-based climate reconstructions. Holocene 11(5):527–539. https://doi.org/10.1191/095968301680223486

Šilc U, Andrič M (2012) Dolgoročen vpliv človeka na biotsko raznovrstnost: primerjava fitocenoloških in palinoloških rezultatov (Bela Krajina) [Long-term impact of man on biodiversity: a comparison of phytocoenelogical and palynological results (Bela krajina)]. In: Andrič M (ed) Dolgoročne spremembe okolja 1, Opera Instituti Archaeologici Sloveniae 25. Založba ZRC, Ljubljana, pp 55–61. https://doi.org/10.3986/9789612545925

Tinner W, Lotter AF (2006) Holocene expansions of *Fagus sylvatica* and *Abies alba* in Central Europe: where are we after eight decades of debate? Quat Sci Rev 25(5–6):526–549. https://doi.org/10.1016/j.quascirev.2005.03.017

Tinner W, Hubschmid P, Wehrli M, Ammann B, Conedera M (1999) Long-term forest fire ecology and dynamics in southern Switzerland. J Ecol 87(2):273–289. https://doi.org/10.1046/j.1365-2745.1999.00346.x

Tolar T, Mason P, Kramberger B (2024) Prehistoric to mediaeval woodland management and agriculture in the westernmost Carpathian basin and the surrounding Northern Dinaric Karst: a case study of archaeobotanical analyses from the multiperiod site at the Obrežje International Border crossing. In: Fuerst-Bjeliš B, Mrgić J, Petrić H, Zorn M, Zwitter Ž (eds) Environmental Histories of the Dinaric Karst. Springer, Cham, pp 51–80

Toškan B, Achino KF, Becke C (2020) Životinjski ostaci iz brončanodobnog nalazišta Monkodonja: sirovi podaci = Animal remains from the Bronze Age settlement of Monkodonja: raw data. In: Hänsel B, Mihovilić K, Teržan B (eds) Monkodonja: Istraživanje protourbanog naselja brončanog doba Istre, Knjiga 3: Nalazi od metala, gline, kosti i kamena te ljudskih i životinjskih kostiju = Forschungen zu einer protourbanen Siedlung der Bronzezeit Istriens, Teil 3: Die Funde aus Metall, Ton, Knochen und Stein sowie die menschlichen und tierischen Knochen. Arheološki muzej Istre, Pula, pp 457–475

Vanniere B, Magny M, Joannin S, Simonneau A, Wirth SB, Hamann Y, Chapron E, Gilli A, Desmet M, Anselmetti FS (2013) Orbital changes, variation in solar activity and increased anthropogenic activities: controls on the Holocene flood frequency in the Lake Ledro area. Northern Italy. Clim Past 9(3):1193–1209. https://doi.org/10.5194/cp-9-1193-2013

von Grafenstein U, Eicher U, Erlenkeuser H, Ruch P, Schwander J, Ammann B (2000) Isotope signature of the Younger Dryas and two minor oscillations at Gerzensee (Switzerland): palaeoclimatic and palaeolimnologic interpretation based on bulk and biogenic carbonates. Palaeogeogr Palaeoclimatol Palaeoecol 159(3–4):215–229. https://doi.org/10.1016/S0031-0182(00)00086-9

Wanner H (2021) Late-Holocene: cooler or warmer? Holocene 31(9):1501–1506. https://doi.org/10.1177/09596836211019106

Wanner H, Solomina O, Grosjean M, Ritz SP, Jetel M (2011) Structure and origin of Holocene cold events. Quat Sci Rev 30(21–22):3109–3123. https://doi.org/10.1016/j.quascirev.2011.07.010

Wirth SB, Glur L, Gilli A, Anselmetti FS (2013) Holocene flood frequency across the Central Alps: solar forcing and evidence for variations in North Atlantic atmospheric circulation. Quat Sci Rev 80(2013):112–128. https://doi.org/10.1016/j.quascirev.2013.09.002

Wunsam S, Schmidt R, Muller J (1999) Holocene lake development of two Dalmatian lagoons (Malo and Veliko Jezero, Isle of Mljet) in respect to changes in Adriatic sea level and climate. Palaeogeogr Palaeoclimatol Palaeoecol 146:251–281. https://doi.org/10.1016/S0031-0182(98)00147-3

Maja Andrič Palynologist, PhD. Studies fossil pollen grains from lakes, wetlands and waterlogged archaeological sites to reconstruct vegetation history. Her research focuses on changes of Late glacial and Holocene vegetation, and multidisciplinary research to better understand which changes of the environment were triggered by climatic fluctuations and/or human impact on the environment in various archaeological time periods. Most of her research is being carried out at Ljubljansko barje, Julian Alps and Bela Krajina regions of Slovenia.

Dirk Nikolaus Karger Researcher, PhD at the interface of macroecology, biodiversity research, and climatology. His research focuses on how climate and biodiversity are linked in time and space. A focus of his recent work has been the development of algorithms to downscale climate data to the biologically relevant scales needed for ecological studies. He is currently studying how extreme weather and extreme events, such as megadroughts, affect vegetation and biodiversity.

Chapter 3
Prehistoric to Medieval Woodland Management and Agriculture in the Westernmost Carpathian Basin and the Surrounding Northern Dinaric Karst: A Case Study of Archaeobotanical Analyses from the Multiperiod Site at the Obrežje International Border Crossing

Tjaša Tolar, Philip Mason, and Bine Kramberger

Abstract Fruits/seeds and wood (charcoal) remains accumulated in archaeological deposits provide a valuable tool for reconstruction of past vegetation and its use. They can offer evidence complementary to pollen analysis or be the main source on past vegetation in areas where no pollen preservation is available. The study assembles seeds/fruits and charcoal evidence from the multiperiod site at the Obrežje, international border crossing between Slovenia and Croatia, in the immediate vicinity of the Gorjanci hills of the northern Dinaric Karst. The evidence is presented in the archaeological and archaeobotanical contexts of the region. The data is interpreted in terms of land use related to woodland management and exploitation of woodland resources during three chronological phases: the Late Bronze Age (cemetery and associated settlement foci), the Early Roman period (Middle Augustan vexillation fortress, small 1st-century rural settlement (farmstead) and associated cemetery) and Late Medieval period (settlement). The main vegetation type targeted by prehistoric population was thermophilous, mixed deciduous oak communities, which contained diverse undergrowth of fruit/nut bearing trees, shrubs and herbs. Those plant communities were the major source of fuel wood and gathered fruits/nuts. Agriculture is

T. Tolar (✉)
ZRC SAZU Institute of Archaeology, Ljubljana, Slovenia
e-mail: tjasa.tolar@zrc-sazu.si

P. Mason · B. Kramberger
Institute for the Protection of Cultural Heritage of Slovenia (ZVKDS), Centre for Preventive Archaeology (CPA), Ljubljana, Slovenia
e-mail: phil.mason@zvkds.si

B. Kramberger
e-mail: bine.kramberger@zvkds.si

B. Fuerst-Bjeliš et al. (eds.), *Environmental Histories of the Dinaric Karst*, Environmental History 17, https://doi.org/10.1007/978-3-031-56089-7_3

evidenced as well, with at least 11 cultural and some weed taxa together with some imported fruits from the Mediterranean (peach and possibly grape).

Keywords South-eastern Slovenia · Obrežje · Gorjanci hills (a part of Dinaric Karst) · Late Bronze Age · Early Roman period · Late Medieval period · Agriculture · Woodland exploitation

3.1 Introduction

The excavations on the Obrežje international border crossing between Slovenia and Croatia revealed the existence of a multiperiod site complex at an important strategic location in the Sava valley, one of the major communications corridors between Central and South-Eastern Europe, important for access to the Alpine ore sources, for exchange and dissemination of new technologies.[1] The site is located 5.5 km to the south-east of Dobova, in the area of the so-called Brežice gates, and it is positioned on the early Holocene terrace of the river Bregana, a right bank tributary of the river Sava. The fertile lowlands of the Lower Sava Valley in south-eastern Slovenia are known also as the Krško-Brežice plain. The uplands of Gorjanci hills (i.e. a part of Dinaric Karst; Croatian: Žumberačko gorje) define the southern boundary of the Brežice gates of the Sava and form a natural barrier in the Sava valley.

Sand and gravel present about half of the surface sediment strata in the Krško-Brežice basin, about half of which is composed of clay and silt. This is one of the westernmost regions of the Carpathian basin and as much as a third of the plain is a floodplain, where all three major rivers, Krka, Sava and Sotla flood. The Krško-Brežice basin was once dominated by extensive forests, however, over time they have been almost completely cleared for agriculture, so that forest only covers about a quarter of the landscape. Among the tree genera, willow, alder and poplar predominate, while oak and white hornbeam are present in the forests on clay deposits.[2] On the contrary, the Gorjanci hills, which rise west of Obrežje (as a part of Dinaric Karst), are a karst landscape, consisting mainly of thick deposits of limestone and dolomite, which is reflected in the rare surface watercourses and in the different tree composition: domination of beech (*Fagus sylvatica*) and hornbeam (*Carpinus betulus*) forests.[3]

The main axis of the site at Obrežje is the marshy valley of the former Struga stream. Here one of the largest Late Bronze Age cemeteries in Slovenia was excavated. The biritual cemetery, comprising four inhumation burials and 377 cremation burials, is dated to the Late Bronze Age (Br C2/D–Ha B2/3), or perhaps even earlier (Br C).[4] In addition, an isolated pit of the Early Copper Age Lasinja Culture, Late

[1] Mason (2003), pp. 202–203; Mason (2006), pp. 131–132; Mason and Kramberger (2022b).

[2] Perko (1998b), pp. 664–675.

[3] Perko (1998a), pp. 498–507.

[4] Mason and Kramberger (2022a), pp. 58–113.

Middle and Late Bronze Age settlement,[5] a Middle Augustan vexillation fortress, a small Early Roman farmstead and cemetery and a Late Medieval rural settlement were excavated to the north, east and south-east of the Late Bronze Age cemetery at Obrežje between 2001 and 2003.

The remains of a Middle to Late Bronze Age Virovitica group (Br B2/C1–Br D/Ha A1) settlement were excavated on the edge of the Pleistocene terrace of the river Sava to the north and north-east of the Struga stream. A Ha A and Ha B period Dobova-Velika Gorica group settlement was discovered to the south and south-east of this settlement, on the Early Holocene terrace of the river Bregana. However, this settlement was damaged by the middle Augustan vexillation fortress and Late Medieval settlement activity. Furthermore, systematic extensive surface collection has revealed as yet undated prehistoric settlement northwest of Obrežje, on the terrace between the Obrežje international border crossing and the village of Jesenice na Dolenjskem.[6]

Numerous archaeological sites show the repeated occupation of the fertile terraces and hills on the edges of the Lower Sava Valley with the appearance of the first farmers from the Late Neolithic and Copper Age onwards. However, despite the significant number of archaeological sites, it is evident that archaeobotanical research from the Lower Sava Valley is rather rare and scarce. The summarised archaeobotanical data (Table 3.1) shows that results have been recorded in different ways, making quantitative comparisons impossible between the sites. Therefore, only the presence and absence (ubiquity) of individual taxa is given in Table 3.1.

Only five sites from the region concerned have currently available archaeobotanical data (Fig. 3.1: sites: 1–5; Table 3.1). The earliest among them are from Čatež—Sredno polje, a multiperiod site, which yielded settlement remains dating to the Late Neolithic, Early Copper Age and Bronze Age, as well as the remains of Early Roman vexillation fort.[7] The archaeobotanical remains analysed were mostly found in the Late Neolithic pits dated to approximately between 4800 and 4600 BC, fewer are from a single pit that was dated to the Middle Bronze Age/transition to the Late Bronze Age, and the remaining archaeobotanical data originate from the furnaces of the Early Roman vexillation fortress.[8]

The Ajdovska jama cave, which was used as a cemetery approximately between 4350 and 4150 BC and as a cave shelter in the 4th millennium BC, has also yielded some plant remains.[9] The analysed carbonised grains were mostly scattered, or less often in larger concentrations near human skeletal remains, which could indicate that they served as mortuary deposits. On the other hand, wood (charcoal fragments) was scattered, in the hearths, or was preserved as carbonised logs or planks.[10]

[5] Mason and Kramberger (2022b).

[6] Djurić (2003), p. 204.

[7] Tomaž (2022).

[8] Culiberg (2022), pp. 133–135.

[9] Šercelj and Culiberg (1984), pp. 33–37; Culiberg and Šercelj (1986), 111–118; Horvat (1988), pp. 40–43; Culiberg et al. (1992), pp. 111–125; Bonsall et al. (2007), pp. 727–740; Horvat (2009), pp. 25–35; Sraka (2012); Sraka (2020), pp. 103–104.

[10] Culiberg and Šercelj (1986), pp. 111–112.

Table 3.1 Archaeobotanical data from Lower Sava Valley region sites (excluding the Obrežje international border crossing; see Fig. 3.1)

	Site and source of the data	Čatež–Sredno polje			Ajdovska jama	Col 1	
		Culiberg (2022)			Šercelj and Culiberg (1984); Culiberg and Šercelj (1986); Culiberg et al. (1992)	Culiberg (2020b)	
	No. of samples	26	2	3	no data	17	
	Period/Plant taxa	Late Neolithic	MBA or LBA	Roman period	Copper Age	Copper Age	LBA
Charcoal	oak (*Quercus* spp.)	65	10	5	xx	2	x
	service tree (*Sorbus* spp.)	2			xx	9	x
	black hornbeam (*Ostrya carpinifolia*)				1		
	common hornbeam (*Carpinus betulus*)	1	1	5			
	ash (*Fraxinus* spp.)	13			x		x
	beech (*Fagus sylvatica*)	69		2	2		x
	maple (*Acer* spp.)	19	15		x		x
	lime (*Tilia* spp.)	1			xx		
	alder (*Alnus* spp.)						x
	birch (*Betula* spp.)	2					
	elder (*Sambucus nigra*)				1		x
	walnut (*Juglans regia*)				6		
	elm (*Ulmus* spp.)	15					
	hazel (*Corylus avellana*)				x		
	apple/pear (*Malus/Pyrus* spp.)						
	cherry (*Prunus avium*)						
	willow (*Salix* spp.)				x		
	grapevine (*Vitis sp.*)	17 (one sample)					
	poplar (*Populus* spp.)	2		3	xx		

(continued)

Table 3.1 (continued)

	Site and source of the data	Čatež–Sredno polje			Ajdovska jama	Col 1	
		Culiberg (2022)			Šercelj and Culiberg (1984); Culiberg and Šercelj (1986); Culiberg et al. (1992)	Culiberg (2020b)	
	No. of samples	26	2	3	no data	17	
	Period/Plant taxa	Late Neolithic	MBA or LBA	Roman period	Copper Age	Copper Age	LBA
Seeds/Fruits	barley (*Hordeum vulgare*)				815; 270; 53; 106		
	wheat (*Triticum* spp.)				65; 17		
	einkorn (*Triticum monococcum*)				7; 226; 256; 88; 114; 306; 906		
	einkorn/emmer wheat (*T. mono/ dicoccum*)				120		
	emmer (*Triticum dicoccum*)				790; 1050		
	common wheat (*Triticum aestivum*)				2;		
	barley/wheat (*Hordeum/ Triticum*)						
	oat (*Avena*)				1; 1		
	broomcorn millet (*Panicum miliaceum*)						
	foxtail millet (*Setaria italica*)						
	bird vetch (*Vicia cracca*)				5; 2		
	peas/vetch (*Pisum sativum/ Vicia* spp.)				2		
	broad bean (*Vicia faba*)				32		
	lens (*Lens culinaris*)						
	hazel (*Corylus avellana*)						
	date (*Phoenix dactylifera*)						
	peach (*Prunus persica*)						

(continued)

Table 3.1 (continued)

	Site and source of the data	Cerklje ob Krki, OŠ			Cerklje ob Krki	Drnovo	
		Culiberg (2016)			Tolar (2016)	Culiberg (2007); Tolar (2017)	
	No. of samples	3	7	8	1	50	1
	Period/Plant taxa	LBA	Roman period	Early Medieval period	Roman grave	Roman graves	Hoard
Charcoal	oak (*Quercus* spp.)	6	21	19		x	
	service tree (*Sorbus* spp.)		11	6		x	
	black hornbeam (*Ostrya carpinifolia*)						
	common hornbeam (*Carpinus betulus*)						
	ash (*Fraxinus* spp.)		2			x	
	beech (*Fagus sylvatica*)	5				2	
	maple (*Acer* spp.)	4	2	1		4	
	lime (*Tilia* spp.)						
	alder (*Alnus* spp.)		1			x	
	birch (*Betula* spp.)			2			
	elder (*Sambucus nigra*)						
	walnut (*Juglans regia*)						
	elm (*Ulmus* spp.)		24	4			
	hazel (*Corylus avellana*)					x	
	apple/pear (*Malus/Pyrus* spp.)					3	
	cherry (*Prunus avium*)			6			
	willow (*Salix* spp.)						
	grapevine (*Vitis* sp.)						
	poplar (*Populus* spp.)		1	1			

(continued)

Table 3.1 (continued)

	Site and source of the data	Cerklje ob Krki, OŠ			Cerklje ob Krki	Drnovo	
		Culiberg (2016)			Tolar (2016)	Culiberg (2007); Tolar (2017)	
	No. of samples	3	7	8	1	50	1
	Period/Plant taxa	LBA	Roman period	Early Medieval period	Roman grave	Roman graves	Hoard
Seeds/Fruits	barley (*Hordeum vulgare*)						
	wheat (*Triticum* spp.)						
	einkorn (*Triticum monococcum*)						
	einkorn/emmer wheat (*T. mono/dicoccum*)						
	emmer (*Triticum dicoccum*)						
	common wheat (*Triticum aestivum*)						
	barley/wheat (*Hordeum/Triticum*)						x
	oat (*Avena*)						
	broomcorn millet (*Panicum miliaceum*)						x
	foxtail millet (*Setaria italica*)						x
	bird vetch (*Vicia cracca*)						
	peas/vetch (*Pisum sativum/Vicia* spp.)						
	broad bean (*Vicia faba*)						
	lens (*Lens culinaris*)					2	
	hazel (*Corylus avellana*)				1 charred frg	1	
	date (*Phoenix dactylifera*)				3 charred		
	peach (*Prunus persica*)				2 half-charred		

Source Authors' own analysis

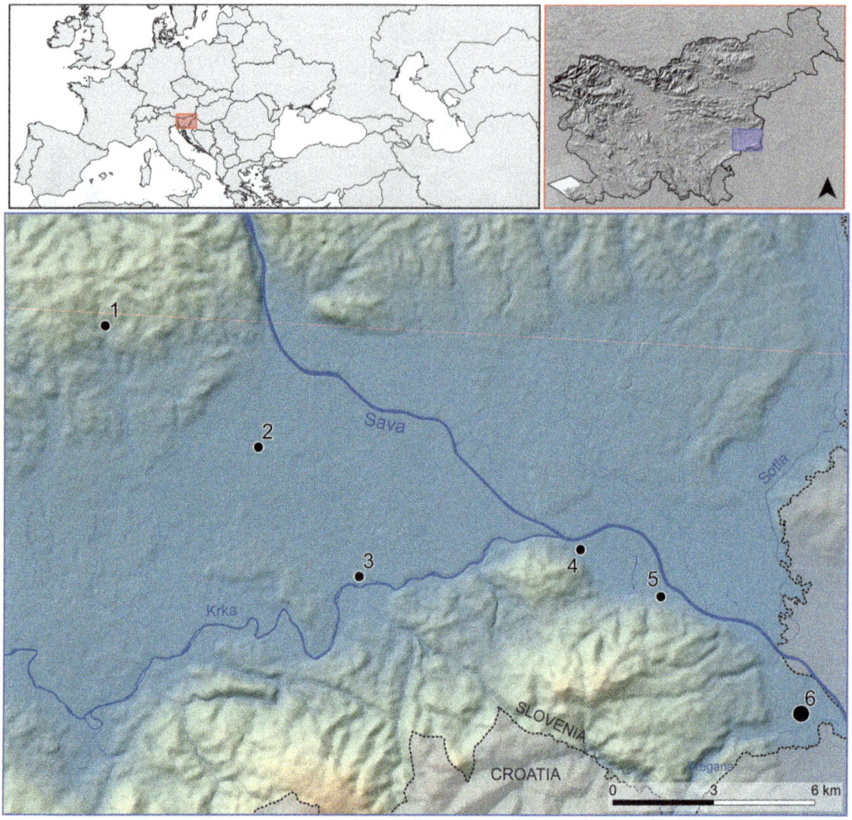

Fig. 3.1 Map of the study area showing some archaeological sites with currently available archaeobotanical data. 1 Ajdovska jama pri Nemški vasi; 2 Drnovo; 3 Cerklje ob Krki (the two different microlocations are presented separately in Table 3.1); 4 Čatež—Sredno polje; 5 Col 1 pri Podgračenem; 6 Obrežje international border crossing. *Source* Own elaboration. Figure prepared by Matjaž Mori. Base map data source: ©EuroGeographics for the European administrative boundaries; European Digital Elevation Model (EU-DEM), version 1.1; GURS, Hidrografija (2020); GURS, Državna meja (2018)

Archaeobotanical data are also available from the site Col 1 pri Podgračenem.[11] Charcoal remains were found here in the settlement layer that is assumed contemporary with the cemetery in Ajdovska jama, as well as in the settlement layer dated to the Late Bronze Age.[12]

Finally, the archaeobotanical analyses from the multiperiod site at Cerklje ob Krki, where the plant macro remains are related to the layers and features dating to the Late Bronze Age, Roman period and Early Medieval period,[13] and from the graves and a coin hoard in a pottery vessel at Drnovo, the site of the Roman municipium of

[11] Culiberg (2020b), pp. 112–113.

[12] Horvat (2020).

[13] Culiberg (2016), p. 131.

Neviodunum[14] can be mentioned among available archaeobotanical data so far from the Lower Sava Valley (Fig. 3.1; Table 3.1).

Analyses of charcoal samples from the archaeological sites in the Lower Sava Valley and surrounding hills (Table 3.1; Fig. 3.1) show that in some places (especially near the settlements) the forest vegetation has undergone anthropogenic alteration since the 5th millennium BC, as the elements of the climax beech-fir forests are being gradually replaced by light-demanding, pioneer and shrub species like oak, hornbeam, poplar, willow and hazel. These formed mixed, open oak-hornbeam forests. Wetland flood-plain forests were probably predominant beside the Sava River. These prehistoric forests were primarily composed of willow, alder and poplar, like those of today.

Although scarce, the summarised results in the Table 3.1 indicate a potentially greater presence and with it harvesting of beech (*Fagus sylvatica*) trees in the Late Neolithic period (i.e. at Čatež—Sredno polje site; 69 identified *Fagus* charcoal remains), what could indicate the existence of extensive beech forests in this earlier period. Maple (*Acer* sp.) often accompanies beech. It was present and used also in the Late Neolithic period, but the percentage of charcoal remains of maple at Čatež—Sredno polje are in comparison with beech significantly lower (i.e. 19 identified *Acer* charcoal remains). The amount of oak is higher (*Quercus* sp.; i.e. 65 identified *Quercus* charcoal remains), which together with significant amounts of ash (*Fraxinus* spp.; i.e. 13 identified charcoal remains) and elm (*Ulmus* spp.; i.e. 15 identified charcoal remains), and to a lesser extent charcoal remains of poplar (*Populus* spp.), birch (*Betula* spp.), common hornbeam (*Carpinus betulus*), lime (*Tilia* spp.) and service tree (*Sorbus* spp.) indicate a diverse wood exploitation at the Late Neolithic settlement. Even today, the forests in this area are diverse. Willow (*Salix* spp.), alder (*Alnus* spp.) and poplar (*Populus* spp.) predominate in the flood zone along the rivers in the Lower Sava Valley, while oak (*Quercus* spp.) and hornbeam (*Carpinus betulus*) are mainly found above the flood zone.[15] On the other hand, the higher parts of the Gorjanci hills, south-west of Čatež—Sredno polje, are almost entirely covered by extensive beech forests, while oak grows mainly in the western part of Gorjanci today, up to a height of 400–600 m. The periphery of the eastern part of Gorjanci is dominated by hornbeam (*Carpinus/Ostrya*) forests.[16]

The fields were most probably located on the cleared and burnt forest floor. Humans had already influenced the forest through their activities (foraging, agriculture and animal husbandry). The composition of the charred and waterlogged seeds clearly indicate that man has already regularly produced cereals and also legumes at least from ca. 4350 BC (the case of Ajdovska jama (cave); Table 3.1).[17] Barley (*Hordeum vulgare*), in addition to einkorn (*Triticum monococcum*) and emmer wheats (*Triticum dicoccum*) predominate at the Ajdovska jama site. All three taxa are

[14] Culiberg (2007), pp. 80–82; Tolar (2017).

[15] Perko (1998b), p. 668.

[16] Perko (1998a), p. 503.

[17] Culiberg et al. (1992), pp. 111–125.

considered to be the earliest cultivars in the area of modern Slovenia. At the Ljubljansko barje pile-dwellings from the 4th millennium BC, besides two types of wheat (*Triticum mono-* and *dicoccum*) and barley (*Hordeum vulgare*), also peas (*Pisum sativum*), poppy (*Papaver somniferum*) and flax (*Linum usitatissimum*) have been identified.[18] Vetch (*Vicia* spp.) most likely represents a weed taxon in the earliest periods.

3.2 Materials and Methods

Excavations at the Obrežje international border crossing took place from September 2001 to May 2003. The topsoil/ploughsoil was machine stripped. Mortuary contexts dating to different periods, were subject to total sampling—wet sieving and flotation. Contexts in smaller features (postholes) were also subject to total sampling (wet sieving/flotation), contexts in larger features (large pits, ditches) were subject to 25% sampling.

Additionally, one of the palaeochannels discovered at Obrežje was sampled for pollen analysis. This palaeochannel is of special importance, because it was filled with a 1.4 m deep stratigraphic sequence of fluvial deposits and an accumulated material. This included a layer of dark soil (Context no. 003), which yielded the largest assemblage of Middle to Late Bronze Age (Br B2/C1–Br D/Ha A1) pottery fragments on the site. It was interpreted as a midden deposit, containing domestic debris from the contemporary Virovitica group settlement in its immediate vicinity.

Archaeobotanical remains (seeds, fruits and charcoal) were obtained by the systematic sampling of settlement and mortuary contexts as well as deposits in paleochannels, from the Obrežje Slovenian/Croatian international border crossing (hereinafter referred to also as Obrežje, a multiperiod archaeological site). The collected sediment samples were processed using wet sieving/half-flotation techniques. The plant macro remains, caught on sieves, were sorted and analysed in the study, presented below (Fig. 3.2).

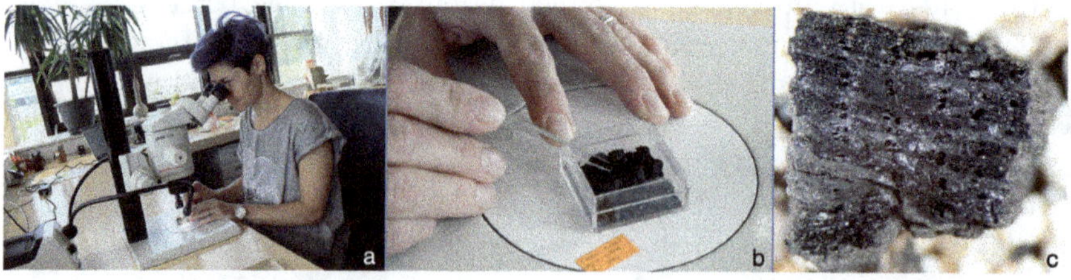

Fig. 3.2 **a** Archaeobotanical work with stereo microscope and **b**, **c** Charcoal samples. *Photos* Dragutin Valoh, ZRC SAZU (Scientific Research Centre of the Slovenian Academy of Sciences and Arts). Reproduced with permission

[18] Tolar et al. (2011), pp. 207–222.

Table 3.2 Numbers of archaeobotanically investigated, temporally and spatially defined samples from the archaeological site at the Obrežje international border crossing

Late Bronze Age		Roman period		Medival period	
Seeds/fruits samples	Charcoals	Seeds/fruits samples	Charcoals	Seeds/fruits samples	Charcoals
114	10	85	282	18	9
124		367		27	

Source Data by the authors

After wet sieving of such sampled sediment samples, the organic remains, trapped on sieves with a pore diameter up to 1 mm, were air-dried and plant macro remains were sorted out. Examinations were undertaken, using a stereo microscope Leica MZ75 with up to 50 magnification and a microscope Nikon Eclipse ME 600 with up to 600 magnification and plant taxa determined. We used our own reference collections of seeds, fruits, wood and charcoal and various identification keys.[19] The botanical nomenclature follows *Mala flora Slovenije.*[20]

All hand-sorted remains of seeds and fruits were counted and tried to identify to the species level. Charcoal fragments were, due to the abundance in individual sample, randomly subsampled by selecting 2–3 specimens from each sample for wood identification.

The archaeobotanical investigation of the excavations on the Obrežje international border crossing is one of the most extensive archaeobotanical research projects in Slovenia, with more than 500 archaeobotanically examined, temporally and spatially defined samples of charcoal and seeds/fruits remains. However, the analysis of the wet sieved material is still in progress and many samples from this site are still not chronologically defined and have been excluded from this study.

3.3 Results

The archaeological site at the Obrežje international border crossing is a multiperiod site, therefore the archaeobotanical results are divided into three units: Late Bronze Age, Roman period and Medieval period (Table 3.2).

In the case of the Late Bronze Age, only samples from graves were included in Table 3.3, because samples from contemporary settlement features have so far yielded only three fragments of charcoal: *Cornus* spp., *Pinus* spp. and a poorly preserved fragment of deciduous wood charcoal.

The analysis of the wet sieved material is still in progress, so far 3,193 plant macro remains (fragments of charcoal and seeds/fruits) were analysed from more than 500 samples (Table 3.2). Three samples (with only three plant macro remains) are derived

[19] Schweingruber (1990); Richter and Dallwitz (2002); Cappers et al. (2006).

[20] Martinčič et al. (1999).

Table 3.3 List of identified plant macro remains: charcoal and seeds/fruits remains at the Obrežje international border crossing in all three periods

	Period, Context/Plant Taxa		Late Bronze Age	Roman period		Medival period
			Graves	Settlement	Graves	Settlement
Charcoal	Diffuse porous wood			x	x	
	Ring porous wood			x		
	Oak	*Quercus* spp.	xx	xx		x
	Chestnut	*Castanea sativa*			x	x
	Ash	*Fraxinus* spp.		x	x	
	Ash/chestnut	*Fraxinus/ Castanea*				x
	Oak/chestnut	*Quercus/ Castanea*	xx	x	x	
	Rose family	Rosaceae	x	x		x
	Beech	*Fagus sylvatica*		x	x	
	Birch	*Betula* spp.		x	x	
	Service tree	*Sorbus* spp.		x	x	
	Apple/pear	*Malus/Pyrus* ssp.		x	x	
	Dogwood/ cornel	*Cornus* spp.		x		
	Maple	*Acer* spp.		x		x
	Hornbeam	*Carpinus betulus*		xx	x	
	Hop-hornbeam	*Ostrya carpinifolia*		x		x
	Elm	*Ulmus* spp.		x	x	
	Alder/hazel	*Alnus/Corylus*		x	x	
	Poplar/willow	*Populus/Salix*		x	x	x
	Walnut	*Juglans regia*		x		
	Lime	*Tilia* spp.		x		
Seeds/ Fruits	Barley	*Hordeum vulgare*	x	x	1	1
	Einkorn	*Triticum monococcum*				1
	Naked wheat	*Triticum aestivum*				1
	Cereals	Cerealia	x	x	1	
	Millet	*Panicum miliaceum*	x	x	1	xx

(continued)

Table 3.3 (continued)

Period, Context/Plant Taxa		Late Bronze Age	Roman period		Medival period
		Graves	Settlement	Graves	Settlement
Foxtail millet	*Setaria* spp.				1
Oat	*Avena* spp.		x		1
Rye	*Secale cereale*				1
Peas	*Pisum sativum*		x		
Faba bean	*Vicia faba*				xx
Lentil	*Lens culinaris*				1
Legumes	Fabaceae				1
Grapevine	*Vitis vinifera* ssp.	x	x		
Blackberry	*Rubus fruticosus* agg	1			
Peach	*Prunus persica*		2		
Hazel	*Corylus avellana*		1		1
Elderberry	*Sambucus* spp.				1

x—Present; xx—Present in greater numbers; 1, 2-Individual specimens

from Late Bronze Age settlement contexts, and 121 samples (with 1,281 plant macro remains) are derived from contemporary mortuary contexts. Four samples (with 30 plant macro remains) were obtained from prehistoric features that could not be dated more precisely. 367 samples (with 721 plant macro remains) are derived from the Early Roman vexillation fortress, among them 33 (58 plant macro remains) with contemporary graves. 27 samples (with 871 plant macro remains) are derived from the Late Medieval village, and the remaining samples are derived from contexts that are not yet chronologically defined.

3.3.1 The Late Bronze Age

Charcoal samples (7) from the Late Bronze Age (mortuary contexts), dated to between Ha A1 and Ha B2/3 phases (approximately 1200–800 BC), show slightly less species diversity than later samples (i.e. from the Roman and Medieval contexts) on the same site. In the LBA graves charcoal of chestnut (*Castanea;* 5) and oak (*Quercus;* 2) predominate, only one item is determined to Rosaceae family. The predominance of oak and chestnut could testify to the deliberate choice of these species in cremation of the deceased.[21]

Three more charcoal samples were analysed from the LBA settlement contexts (interpreted as waste material). They were of cornel (*Cornus*) and pine (*Pinus*).

Some remains of cultivated and gathered plants were also identified in the LBA grave samples: cereal grains, three grains of barley (*Hordeum vulgare*), one grain of millet (*Panicum milliaceum*) or foxtail millet (*Setaria italica*) and one blackberry (*Rubus fruticosus*) seed. Cereals grains were found in eight graves, which could be dated to different chronological phases based on the typo-chronological analysis, and the results of ^{14}C analysis of samples of cremated human bones. The earliest graves can be dated to Ha A1 phase and the latest to Ha B2/3 phase. Barley grains were identified in three graves, dated to phases Ha A2, Ha A2/B1 and Ha B1 (approximately 1100–800 BC), and a millet or foxtail millet seed in a grave that could be dated to Ha A2/B1 phase (1100–950 BC).[22]

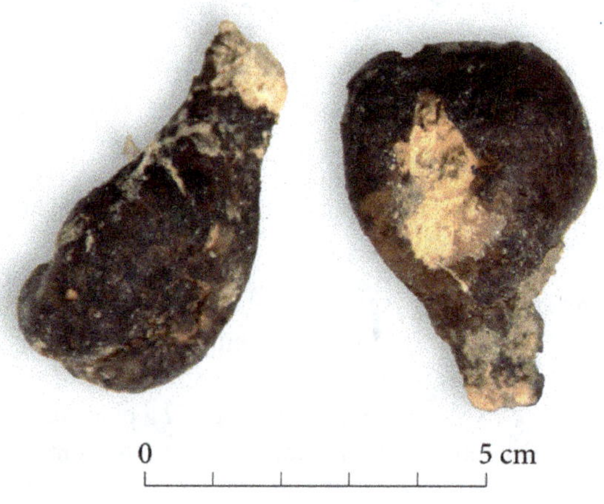

Fig. 3.3 Pips of probably cultivated grapevine (*Vitis vinifera* subsp. *vinifera*) from the LBA cremated grave (gr. no. 190). *Photo* Dragutin Valoh, ZRC SAZU. Reproduced with permission

Of interest are the grapevine seeds (*Vitis vinifera* subsp. *vinifera*) from five graves (gr. 22, gr. 63, gr. 137, gr. 150, gr. 190), preserved in a partially charred state. These were, presumably from fruit, deposited in both male and female graves, based on the results of anthropological analysis and gender-related objects (Fig. 3.3). These LBA graves could be dated to phases Ha A2 (gr. 22), Ha A2/B1 (gr. 63, gr. 150), Ha B2/3 (gr. 137), and one could not be dated precisely (gr. 190). These could be the earliest evidence for the cultivated vine subspecies that has yet been discovered in Slovenia.

[21] Tolar (2022), pp. 174–175.

[22] Mason and Kramberger (2022a), pp. 58–113; cf. Tolar (2022), pp. 174–190.

3.3.2 The Early Roman Period

The majority of Roman period finds are derived from the Augustan vexillation fort, which was occupied (reused) at least three times—this relates to a specific military population, probably with strategies of forest uses different from those of local population. Excavation has produced evidence of Early Roman period civilian settlement, dated to the 1st Century CE in the area of Obrežje Rondo outside the fort, but this represents the edge of a large, unexcavated settlement extending between the modern settlement of Jesenice na Dolenjskem and the Obrežje international border crossing. There was probably a single farmstead with an associated small cemetery within the former vexillation fort (1st Century CE).[23]

The early 1st-century CE contexts have produced by far the most samples with preserved and identified charcoal remains (Table 3.2), taken from different contexts. Species diversity was extremely high everywhere, except in contexts (fills) from ditches and in pits, containing smithing waste and semi-products. Such diversity indicates the exploitation of the forest as a habitat for various tree and shrub species.

The 25 samples taken from Early Roman period graves indicate great species diversity in contrast to LBA graves. At least 12 wood species have been identified: oak/chestnut (*Quercus* spp./*Castanea sativa;* 2), chestnut (1), ash (*Fraxinus* sp.; 4), beech (*Fagus sylvatica;* 3), birch (*Betula* spp.; 1), service tree (*Sorbus* ssp.; 2), apple/pear (Maloideae; 1), hornbeam (*Carpinus betulus;* 2), elm (*Ulmus* spp.; 1), alder/hazel (*Alnus* spp./*Corylus avellana;* 2), poplar/willow (*Populus* spp./*Salix* spp.; 2), not identified (NI) diffuse porous deciduous wood (DPW; 6).

The 10 samples taken from the fireplaces also indicate a variety of wood used for heating: oak (1), ash (1), apple/pear (1), hornbeam (1), alder/hazel (2), NI DPW (3) and NI ring porous wood (RPW; 1).

The majority, 215 samples, were derived from fills of refuse pits. Altogether, 150 identified fragments of charcoal belong to at least 17 different wood species. Most (50 items) belong to hornbeam, followed by: oak with 31 items, alder/hazel (with 16), dogwood/cornel (15), elm (13), ash (12), service tree (12), NI DPW (12), NI RPW (10), oak/chestnut (8), birch (6), poplar/willow (5), maple (*Acer* spp.; 5) and apple/pear (5). The others: walnut (*Juglans regia*), beech, lime (*Tilia* spp.), Rose fam. (Rosaceae), etc., were identified only with 4 to 1 specimens.

The six samples from the fills in the metallurgical waste pits contain mostly hornbeam charcoal (4). Individual specimens of dogwood/cornel (1) and beech (1) were additionally identified.

In the 13 samples from ditch fills, hazel charcoal (9) prevails, followed by individual (1–2) specimens of oak, ash, apple/pear, service tree, dogwood/cornel, elm, hornbeam and NI DPW and RPW. The identified charcoal is probably a deposit from stoves and fireplaces, and represents types of wood from the immediate vicinity of the settlement.

The Early Roman period seeds/fruits remains from Obrežje are dominated by the charred remains of cereal grains (wheat and barley), while the remains of oats (*Avena*

[23] Mason (2008), pp. 187–198.

sativa) and millet (*Panicum miliaceum*) (Fig. 3.4) are much less common. Among legumes, peas and vetch were recognised. Uncarbonised and half-carbonised grape pips were found as well. Peach (*Prunus persica*) and hazelnut (*Corylus avellana*) were present in two samples. Settlement samples from the fills of ditches, pits and postholes dominate among the samples with preserved seeds/fruits from this period. Only eight samples (out of 85; Table 3.2) are derived from mortuary contexts, in which by far the least archaeobotanical remains are preserved. Individual remains of wheat, barley and millet were only found in two graves.

Fig. 3.4 Remains of cultivated plants from the Early Roman period at the Obrežje international border crossing: **a** Sample no. 11741 (context no. 10776): barley (*Hordeum vulgare*), **b** Sample no. 4653 (context no. 3664): millet (*Panicum miliaceum*), barley (*Hordeum vulgare*), wheat (*Triticum* spp.) and **c** Sample no. 409 (context no. 3835): naked wheat (*Triticum durum/turgidum/aestivum*). *Photo* Dragutin Valoh, ZRC SAZU. Reproduced with permission

3.3.3 The Late Medieval Period

Five samples of charcoal gained from the fillings of Late Medieval refuse pits belong to five different taxa: oak (1), chestnut (1), ash/chestnut (2), black hornbeam (*Ostrya carpinifolia*; 1) and poplar/willow (1). We have also found a variety of charcoal in four samples from posthole fills: oak, ash/chestnut, maple and wood of the rose family (Rosaceae).

The remains of cultivated plants from the Late Medieval settlement at Obrežje have been preserved in the fills of pits, postholes and rubbish pits. Two large deposits of millet (623 grains in sample no. 2133) and faba bean (*Vicia faba*) (80 grains in sample no. 2141) were discovered (Fig. 3.5).

In addition to the two taxa mentioned, individual remains of a variety of cereals (barley, einkorn and naked wheat (*T. aestivum/durum/turgidum*), rye (*Secale cereale*) and/or oat; Fig. 3.6), foxtail millet, lentil (*Lens culinaris*) and legumes were also found in Late Medieval contexts. Wild fruits/nuts are represented by the remains of hazelnuts and elderberries (*Sambucus* spp.) in only two samples.

Fig. 3.5 Archaeobotanical remains from medieval pits (context nos. 8738 and 5731), sample nos. 2133 and 2141 at the Obrežje international border crossing: **a** Millet (*Panicum miliaceum*) and **b** Faba bean (*Vicia faba*). *Photo* Dragutin Valoh, ZRC SAZU. Reproduced with permission

Table 3.3 shows large species diversity of charcoal and seeds/fruits in the Early Roman period, especially settlement contexts. In contrast to the Late Bronze Age graves, where oak predominates. The species diversity of charcoal is also considerable in the Early Roman period graves (12 taxa identified). There are more nutritional plant remains preserved in the settlement layers than in the graves in all three periods. The diversity of plant-based diets is expected to increase over time, from wheat, barley and millet in prehistory to other cereals and legumes, as well as diverse fruits in the Roman and Late Medieval periods.

Fig. 3.6 Archaeobotanical remains from Late Medieval pit fill (context no. 8699), sample no. 7261 at the Obrežje international border crossing: **a** Various remains of cereal grains and **b** Naked wheat (*Triticum aestivum/durum/turgidum*). *Photo* Dragutin Valoh, ZRC SAZU. Reproduced with permission

3.4 Discussion

3.4.1 Charcoal Fragments and Forest Vegetation

The analyses and interpretation of charcoal and fruits/seeds remains from archaeological sites have been used as means for reconstructing past vegetation and its exploitation by the past communities in the Lower Sava Valley region in southeastern Slovenia. Plant macro remains from the multiperiod site at Obrežje international border crossing cover three periods: the Late Bronze Age, Roman period and Late Medieval period.

The current charcoal analyses indicate that oak wood for cremating the dead was preferred by the prehistoric (LBA) population at Obrežje, but this could not be the final and reliable interpretation as only seven samples were analysed. This practice is discussed, e.g. also by O'Donnell,[24] who interprets oak charcoal in graves as status graves of important individuals. Because of small numbers of charcoal samples analysed at the prehistoric (LBA) Obrežje site, the result is not statistically reliable and could not confirm the O'Donnell's assumption, but may open and continue the discussion in future research. There are also studies which deny the specific choice of wood for cremation in this period.[25] The oak tree was and still is a part of the natural vegetation of the Gorjanci region (a part of Dinaric Karst), mainly in its western and central part, up to a height of 400–600 m.[26] Pedunculate oak species (*Quercus robur*) thrives near rivers and lakes, and in the flood zones where the soil is more moist. In the past, lowland forests with this oak species, together with alder, willow and poplar, were more common, but they were cut down for agriculture.

The earliest among the compared prehistoric sites with archaeobotanical data in the Lower Sava Valley is the Late Neolithic settlement at Čatež—Sredno polje (206 identified charcoal remains; Table 3.1). The site is attributed to the Sava Group of the Lengyel Culture that represents so far the oldest known horizon of permanent settlements with agriculture and animal husbandry in the continental part of Slovenia (approximately 4800–4600 BC).[27] Beech (69) was the main vegetation type targeted by the then inhabitants, following by oak (65). To a somewhat lesser extent, also charcoal of maple (19), grapevine (17), elm (15) and ash (13) are present. Significantly, Late Neolithic settlement at Čatež—Sredno polje differs from the others compared sites in the Table 3.1. It is the only settlement at which beech was together with oak the main tree species identified among charcoal. It could be interpreted that beech forest was not far away from the settlement and therefore easily accessible or the inhabitants preferred beech wood for some reason. In general the compared prehistoric Lower Sava Valley region sites, i.e. Col 1 (7 taxa), Ajdovska jama (12

[24] O'Donnell (2016), pp. 161–171.

[25] Moskal-del Hoyo (2012), pp. 3386–3395.

[26] Perko (1998a), p. 503.

[27] E.g. Guštin (2005), pp. 7–22; Turk and Svetličič (2005), pp. 65–79.

taxa) and Čatež—Sredno polje (11 taxa) (Table 3.1) all show quite diverse charcoal species assemblages, which indicate the nearby forest vegetation type.

Roman layers from the Obrežje site with almost 300 analysed charcoal samples (Table 3.2 and 3.3) gave more representative results for the reconstruction of woodland vegetation and its exploitation. Firewood management seems to depend on the wood that thrived in the nearby mixed deciduous forest and lowland floodplain. Apparently, the Roman period population at this site did not choose a specific type of wood for fuel, while a large variety of wood species were present in hearths and ovens, and in graves. On the other hand, possible use of oak and hornbeam wood could be ascertained in the metallurgical waste pits. The today distribution of hornbeam is the outskirts of the eastern part of Gorjanci.[28] It was probably also more easily accessible, unlike for example beech, which probably grew on more remote and higher lying areas in the central part of the Gorjanci hills. The wood of hornbeam is together with beech and oak wood one of the most suitable firewood and their charcoal is first classed.[29] Nevertheless, a considerable charcoal species diversity in the Roman-era layers at Obrežje (17 arboreal taxa) indicates the main vegetation type targeted by the then inhabitants. It was thermophilous, mixed decidous oak communities, which contained diverse undergrowth of fruit/nut bearing trees, shurbs and herbs. Those plant communities were the major source of fuel wood and gathered fruits/nuts. The same can be observed at two other Lower Sava Valley region Roman-era archaeological sites (Cerklje ob Krki (7 taxa) and Drnovo (8 taxa); see Table 3.1).

It is often the core of debate among scientists, whether the observed frequencies of individual tree and shrub taxa in the charcoal assemblages reflect accurately their actual proportions in the past vegetation, and whether they can be used to infer environmental conditions in the past. Any such correlation is likely to be distorted by the influence of differential wood combustion alongside ecological and cultural gradient (i.e. wood selection; fuel collection)—they all affect species availability and representation.[30] Today's vegetation is the result of natural processes as well as human activities in the past. Larger quantities of wood were needed for the construction of settlements and fuel, while the population also deliberately cleared the forests for pasture and arable land. Wood was needed in daily life and nonetheless also for the pyres, on which the dead were cremated. Pollen analysis of lacustrine or marsh sediments detects vegetation changes and provides insight into the development of forest vegetation, but could not determine the composition of the forest community and vegetation in the immediate vicinity and in the time of the settlement living, as it gives a picture of a wider area in longer time period.[31] Moreover, the yield of pollen material through water channels, along with the sediment erosion, makes

---

[28] Perko (1998a), p. 503.

[29] Društvo oglarjev Slovenije (https://www.drustvo-oglarjev.si/oglje.html); Zavod za gozdove Slovenije, Energijski ekvivalenti med različnimi gorivi (http://www.zgs.si/delovna_podrocja/lesna_biomasa/izracunajte_si_sami/energijski_ekvivalenti_med_razlicnimi_gorivi/index.html).

[30] Asouti and Austin (2005), pp. 1–18.

[31] See also Chap. 2, in this volume: Andrič and Karger (2024).

reconstruction of local palaeovegetation more difficult. In contrast, charcoal and other plant macro remains from archaeological sites, which are also well chronologically defined, reflect the vegetation of the immediate vicinity of the settlement. Selection of certain species over a long period of time (as long as the settlement lived), such as, e.g. oak for building houses or the selective choice of wood for heating could result in certain tree species being depleted and forest vegetation changed. Pollen analysis is of great help in detecting this phenomenon.

There are two to three main forest types represented at Obrežje today. These are *Castaneo sativae-Fagetum* (i.e. beech (*Fagus sylvatica*) and chestnut (*Castanea sativa*) in community with *Quercus petraea* at 100–700 m a. s. l.) and *Ostryo-Fagetum* (i.e. hop-hornbeam (*Ostrya carpinifolia*) and beech (*Fagus sylvatica*) in community with *Fraxinus ornus*, *Acer pseudoplatanus*, *Sorbus aria*, *Acer campestre* and *Quercus petraea* at 200–1000 m a. s. l.) with *Vaccinio myrtilli-Carpinetum betuli* (i.e. hornbeam (*Carpinus betulus*) in community with *Quercus petraea*, *Picea abies*, *Castanea sativa*, *Prunus avium* at 200–450 m a. s. l.),[32] which is partly consistent with the charcoal assemblages found at Obrežje site, especially in the later, i.e. Roman and Medieval periods (see Table 3.3; *Castanea sativa*, *Quercus* spp., *Fagus sylvatica*, *Ostrya carpinifolia*, *Fraxinus* spp., *Acer* spp., *Sorbus* spp., *Carpinus betulus*). The results from Obrežje may indicate that main plant associations characteristic for today's forest vegetation in the region were formed at least as early as in the Roman period, though the question of vegetational (dis)continuities in Late Antiquity and Early Medieval period and, to a certain degree, later remains open. It is, however, clear that not all forest communities from Roman or Medieval times are still present in the landscape, due to late modern drainage works, for instance. Today's conditions around Obrežje and in the Lower Sava Valley in general are the result of extensive regulation and melioration work in the last centuries.[33] Originally, the Sava and Krka in the Lower Sava Valley, and the Sotla from Bizeljsko to Obrežje had pronounced meandering courses, which were in the nineteenth and twentieth centuries anthropogenically modified by strengthening the river banks, and more recently by building new artificial sections of the river channels. As a result, the rivers were almost completely straightened, the floodplains were drained, and the forest was over time almost completely cleared for agriculture. Of the former extensive "flood forests" between Krka, Sava and Sotla, only one primeval forest has been preserved, located within the Krakovski gozd about 22 km west of Obrežje, where the oldest oaks are more than 300 years old and over 40 m high.[34] On the other hand, beech forests and hornbeam forests now dominate the karst landscape of Gorjanci hills, west of Obrežje with rare surface watercourses.[35]

The nearest palynological data have been obtained from a core at the site of Mlaka in Bela krajina, located in the Dinaric Karst region, 54 km south-west of

[32] Marinček and Čarni (2002).

[33] Šebek (2009).

[34] Perko (1998b), pp. 664–675.

[35] Perko (1998a), pp. 498–507. See also Chap. 2, in this volume: Andrič and Karger (2024), pp. 29–50.

the Obrežje site.[36] The geographical and environmental conditions at Mlaka are not comparable with the current conditions around Obrežje. The low hills surrounding the karst landscape of Mlaka are covered today by open woodlands of *Quercus petraea* and *Carpinus betulus* or *Pinus sylvestris*, *Betula pendula*, *Juniperus communis* and *Pteridium aquilinum*, with *Fagus sylvatica* forest predominant at higher altitudes. However, the palynological data from Mlaka—where the surroundings are from the geological point of view comparable with the Gorjanci near Obrežje—show the most important tree taxon beech (*Fagus sylvatica*), which due to constant small-scale forest felling and burning greatly reduced its presence in the landscape many times.[37] The appearance of taxa characteristic of meadows, fields and pastures at Mlaka site is associated with increased forest biodiversity, microcharcoal peaks and known archaeological sites in the area. Vegetation development in Bela krajina was very dynamic as a result of frequent forest clearance and burning events (from app. 7500 cal BP at Mlaka), which were associated with increased biodiversity and followed rapid forest recovery through phases *Pteridium/Betula*, *Corylus*, *Alnus—Pinus—Quercus—Carpinus betulus* to *Fagus* again.[38] The present-day open landscape formed only ca. 1000 years ago; the forest vegetation in Bela krajina today consists of: *Abio albe-Carpinetum betuli*, *Pteridio-Betuletum pendulae* and varied beech forests (*Fagetum*).[39]

High taxa diversity obtained by charcoal analyses at three Roman settlements (Obrežje (17 taxa), Cerklje ob Krki (7) and Drnovo (8); Table 3.1 and 3.3; the somewhat lower diversity in the latter two is the result of a smaller number of examined samples) in the Lower Sava Valley region shows the non-specific choice of wood for domestic fuel and diverse forest vegetation. Oak wood (at Cerklje ob Krki and Obrežje), elm (at Cerklje ob Krki) and hornbeam (at Obrežje) do stand out with an increased numbers, what could indicate the popularity (i.e. specific choice) or easier accessible (i.e. more common) wood in the vicinity of the settlements. In general "mixed" charcoal from the Roman contexts (graves and settlement) at the Obrežje international border crossing, without such significant predominance of oak (such as in LBA graves at Obrežje) neither of beech (like at Soča Valley region sites)[40] might indicate: (1) net reduction of beech and oak/chestnut in forest vegetation by the Roman period, and/or (2) different wood supply areas, including, for instance, a changed role of Dinaric Karst and wetland forests, and/or (3) that the Roman population in this region did not pay special attention to the type of wood used for cremating the dead.

The relatively high diversity of charcoal is also observed for the compared earlier prehistoric settlements (Čatež—Sredno polje (11 taxa) and Col 1 (7); Table 3.1) and for the cave site Ajdovska jama (12 taxa) which was used as an inhumation cemetery between approximately 4350 and 4150 BC and as a cave shelter in the

---

[36] Andrič and Willis (2024), p. 809.

[37] See also Chap. 2, in this volume: Andrič and Karger (2024), pp. 29–50.

[38] Andrič (2007), pp. 763–776.

[39] Marinček and Čarni (2002),18–49.

[40] Culiberg (2020a), pp. 243–260.

4th millennium BC. Here, the predominance of beech and oak at the Late Neolithic settlement Čatež—Sredno polje stands out in numbers of identified charcoal samples. In contrast, the LBA Obrežje-graves samples do show a specific choice of wood for cremation, i.e. oak/chestnut (Table 3.3).

The results from the Lower Sava Valley (SE Slovenia) differs from, e.g. Middle and Upper Soča Valley region (NW Slovenia, e.g. at Most na Soči—Pucarjev rob and Repelc, 160 m a. s. l.; Jerovca on Šentviška planota, 790 m a. s. l.; Čadrg—Laze, 700 m a. s. l.; Srpenica, 364 m a. s. l.) and the site Jelenšek nad Godovičem (W Slovenia; 817 m a.s.l.) where Culiberg[41] reports of the predominance of beech (*Fagus sylvatica*) wood in the Iron Age cremation graves. The only exception is the Iron Age cemetery in Tolmin (lowland Middle Soča Valley in NW Slovenia; 201 m a.s.l.), where not a single specimen of beech charcoal was found at the burial site, but lots of others, like hornbeam and ash (*Fraxinus* sp.), and to a lesser extent also maple, poplar, willow, service tree (*Sorbus* sp.) and chestnut. Oak occurs only in individual cases.[42] The above-mentioned cemeteries from NW Slovenia are located in the Julian Alps, more precisely above the Soča valley, which is covered today by the alpine beech forest with spruce (*Picea abies*) and fir (*Abies alba*). On the surrounding hills and plateaus beech forest actually dominates,[43] what could also be a potential natural forest in prehistory, and is supported by the current archaeobotanical data.[44] Beech wood still is and probably was most likely easier accessible in the region of Middle and Upper Soča Valley, the same as at the Jelenšek cemetery (Rovte Hills; W Slovenia, 817 m a. s. l.). By reforesting the forest, man has caused alpine beech forests to be characterised by two layers of tree crowns: beech in the lower part and larch (*Larix decidua*) and spruce in the upper part. With increasing altitude, the conditions for the growth of spruce and larch deteriorate. They are replaced by a belt of bush growth with creeping pine (*Pinus mugo*). The forest line ranges between 1200 and 1800 m a. s. l. On some smaller areas, other tree-communities may appear, such as for example on the sandbanks next to the alpine rivers, extensive communities with European hop-hornbeam (*Ostrya carpinifolia*) is common.[45]

3.4.2 Seeds/Fruits Remains and Nutritional Habits

All identified cultivar species on the prehistoric site at the Obrežje international border crossing are considered among the earliest cultivated plants in the territory of modern Slovenia, as proven by available data from other so-far excavated sites. This is especially so in the case of barley, einkorn and emmer, which are proven

[41] Culiberg (2020a), pp. 243–260.

[42] Culiberg (2020a), p. 256.

[43] Kunaver (1998), pp. 61–63; Gabrovec (1998), p. 78.

[44] Culiberg (2020a), pp. 258–259.

[45] Kunaver (1998), pp. 62–63.

to occur more massively from circa 4000 BC onwards.[46] Late Neolithic sites, which have been so far excavated at the Krško-Brežice plain like Čatež—Sredno polje site (see Table 3.1) from the first half of the 5th mill. BC yielded so far no archaeobotanical results to prove cultivated plants.[47]

The earliest traces of millet in Slovenia are much younger, i.e. from circa 1000 BC,[48] what confirms also the charred seed identified as *Panicum/Setaria* found in one grave (Gr. No. 150, SU 11136) at the Late Bronze Age cemetery at Obrežje, which was dated to Ha A2/B1, i.e. ca. 1100–950 BC.[49] However, the appearance of millet cultivation is reported from neighbouring countries (Hungary, Austria, Italy and Croatia) several centuries earlier, i.e. from ca. 1600–1300 BC,[50] which means that we have probably not discovered the oldest traces of millet in Slovenia yet. Beside blackberry seed, which was found in the urn of one grave (Gr. No. 231, SU 13133), the cultivated grapevine seeds (see Fig. 3.3) are probably of greater interest. They come from the Late Bronze Age graves (Gr. No. 22, SU 2442; Gr. No. 63, SU 12643; Gr. No. 137, SU 3401; Gr. No. 150, SU 11137; and Gr. No. 190, SU 252) at Obrežje site (i.e. from the Ha A2, A2/B1 and Ha B2/3 phases, i.e. ca. 1100–800 BC). This is an interesting find, as the existing archaeobotanical knowledge anticipates the appearance and spread of the cultivated form of grapevine (*Vitis vinifera* subsp. *vinifera*) in the territory of Slovenia only with the Romans. Although finds of wild grapevine seeds (*Vitis vinifera* subsp. *sylvestris*) and, therefore, the popularity of this fruit among prehistoric people, have been reported several times already from the Copper Age (i.e. 4th mill. BC) Ljubljansko barje pile-dwellings.[51] It is important to mention here the larger amount of *Vitis* charcoal at the Late Neolithic sample from the settlement at Čatež—Sredno polje (Table 3.1), what is the evidence that *Vitis* tree was grown nearby, not only fruits were eaten. A total of 17 pieces of charcoal of *Vitis* were found in the Late Neolithic feature PO 105 at Čatež—Sredno polje. This pit contained also a significant amount of Late Neolithic ceramic fragments, stone tools and stone implements, as well as several other pieces of charcoal, which were identified as beech (4 identified *Fagus* remains), ash (2 identified *Fraxinus* remains), maple (1 identified *Acer* remain), and perhaps birch (1 identified cf. *Betula* remain).[52] One of the samples of charcoal from this pit was also radiocarbon dated: KIA17858, 5782 ± 30 BP, 4712–4547 cal BC (95.4 probability).[53]

Similar archaeobotanical finds from prehistoric layers are evidenced also further east, in Croatia, e.g. from the Late Bronze Age site at Kalnik-Igrišče, which is located circa 70 km north-east of Obrežje. Millet, barley, faba bean, common wheat,

---

[46] Tolar et al. (2011), pp. 207–222; Tolar (2018), pp. 461–498.

[47] E.g. Culiberg (2022), pp. 133–135.

[48] Tolar and Pavlin (2022), pp. 99–120.

[49] Tolar (2022),184.

[50] Filipović et al. (2020), pp. 1–17.

[51] Culiberg (1984), pp. 91–100; Jeraj (2004), pp. 56–64; Tolar et al. (2008), pp. 93–102; Tolar et al. (2011), pp. 207–222.

[52] Culiberg (2022), pp. 134–135.

[53] Tomaž (2022), pp. 139, 474–493.

emmer and spelt (*Triticum spelta*), lentil, peas, and to a lesser extent also oat and rye (for the last it is not clear yet whether it represents weed flora during this period) were also found at Kalnik-Igrišče. Among the wild fruits/nuts, wild apple, dogwood/cornel and acorn, but no grape pips, were found at Kalnik-Igrišče site.[54] Current archaeobotanical evidence on early farming in continental Croatia, obtained on the basis of 11 Early/Middle Copper Age settlements, also shows that cereal production in continental Croatia was based on growing barley, emmer and einkorn,[55] what is the same as in Slovenia (e.g. Eneolithic Ljubljansko barje pile-dwellings).[56] However, sporadic finds of other cereals, such as naked wheat and millet have also been obtained in Croatia, although in smaller quantities, which is perhaps indicative of their lesser importance. In addition to cereals, pulses (i.e. legumes; Fabaceae) were also likely to have been regularly grown as crops, but their evidences are limited within the Croatian assemblage.[57]

All identified taxa of presumably nutritional plants from the Roman period contexts at Obrežje site (i.e. cereals, barley, oat, millet, pea, grapevine, peach and hazelnut) are also usually present at other Roman sites in Slovenia (e.g. Vrhnika (*Nauportus*), Cerknica, Školarice, Mustar, Hrušica, Ig, Ljubljana (*Emona*)—Kongresni trg, Slovenska cesta, Tribuna, Erjavčeva).[58] In addition to the nutritional plant taxa from the Obrežje site (Table 3.3), lentil and date (*Phoenix dactylifera*) were additionally found at two other sites in the Lower Sava Valley region (i.e. Cerklje ob Krki and Drnovo sites; see Table 3.1).[59] However, this is far from the full spectrum of the plant diet at that time period, as many other remains of cultivated plants are known from other Roman sites in Slovenia mentioned in this paragraph, including several types of wheat, rye, beans, and other fruit trees (cherries and other *Prunus* species, olives, figs and pomegranates),[60] which were introduced to the local environment from the Mediterranean, either as imported fruits or as planted trees.

Two major deposits of millet (more than 600 grains) and faba bean (more than 80 beans) are worth mentioning from the Medieval features at Obrežje. They were found in pits (context nos. 8738 and 5731), which were located in close proximity to a house. These pits have initially been interpreted as rubbish pits.

The nearest more comprehensively archaeobotanically investigated sites in Slovenia from time period as Obrežje are as follows: Velike njive near Krško (circa 19 km NW of Obrežje) and Metlika-Mestni trg (circa 36 km to the SW of Obrežje); both are unpublished at the moment and have not yet been exactly defined chronologically, i.e. from the prehistoric period to the Late Medieval period. At Velike njive (Early Bronze Age, Middle Bronze Age, Late Bronze Age and Early Medieval),

[54] Essert (2019), pp. 160–166.

[55] Reed, (2018), pp. 239–255.

[56] Tolar et al. (2011), pp. 207–222.

[57] Reed (2018), pp. 239–255.

[58] Tolar (2016, 2020); Toškan et al. (2022), pp. 281–312.

[59] Tolar (2016, 2017).

[60] Šercelj (1990), pp. 309–312; Culiberg (1999), pp. 323–331; Toškan et al. (2022), pp. 281–312; Tolar (2017).

precisely the same remains of cultivated plants were identified as those at Obrežje: wheat, barley, a significant amount of millet, foxtail millet, peas, vetch, grapevine and peach. This is also reflected in the charcoal spectra. High wood diversity with at least 14 taxa was identified; oak and chestnut predominate.[61] At Metlika-Mestni trg (also LBA settlement contexts) the cereal grains were even better preserved, beside barley and some millet grains also common wheat and spelt were identified. Among legumes, lentil, faba bean, vetch and pea vines (*Lathyrus* spp.) were identified. Oak charcoal is also very common at that site.[62]

3.5 Conclusion

Very diverse numbers of samples and preserved plant macro remains from different occupations and mortuary deposits of the multiperiod site at the Obrežje international border crossing give representative results with regard to vegetation composition and exploitation of woodland resources over time in the Lower Sava valley, in the immediate vicinity of the Gorjanci hills of the Dinaric Karst.

Late Bronze Age charcoal samples are rare, only seven samples from LBA cremation graves were analysed in contrast to 282 samples from the early Roman period contexts at Obrežje site. Among them oak and chestnut significantly predominate, therefore it seems that particularly selected wood (i.e. oak) was used by the LBA community for cremation. We did not find such a species specification in the Roman period contexts, neither in mortuary nor in settlement deposits. At least 17 different wood taxa were identified in the Early Roman period contexts. Some more hornbeam and oak charcoals were identified in the fills of refuse pits. Great species diversity of charcoal in Roman period contexts suggests the existence of mixed, open oak-hornbeam forests with diverse shrub undergrowth. Plant communities with oak are common for the modern vegetation of the Gorjanci region (a part of Dinaric Karst), especially for its western and central part up to a height of 400–600 m, and on the other hand, the pedunculate oak species (*Quercus robur*) is, together with alder, willow, poplar, common for the lowland floodplain, which is typical for the Lower Sava Valley. So, not surprisingly, oak was one of the more readily available woods. Hornbeam (*Carpinus betulus*) seems to have been particularly exploited by the Romans, as were other tree taxa (i.e. beech, maple, elm, lime, etc.; see Table 3.3). Albeit on a smaller scale, they indicate a great variety of tree/bush taxa in forests of that time, similar to today's.

Analysis of charcoal from Obrežje perhaps indicates a more rigid selection of wood by the prehistoric population than by the Roman military garrison and the inhabitants of the Roman farmstead. However, it is not possible to say whether this is due to a habit, a lack of "good" wood or a different wood supply area. The latter option potentially suggests a stronger reliance on wood from the surrounding

[61] Tolar (2021b).

[62] Tolar (2021a).

Dinaric Karst in prehistory and potentially more strictly local wood supplies by the Roman vexillation fort. Might this be the reason why hornbeam became one of more popular species then? Another option could be that the Roman army was less selective in the exploitation of local resources during military operations, than previous prehistoric populations with a tradition of living there. Furthermore, in the case of graves, differences in the preferred wood for cremating the dead between the Romans and the LBA settlers could reflect the differences in the funeral rituals and beliefs. But these are only the first results, more such comparisons and especially more comparable amounts of samples should be analysed in future.

By far the most samples for seed/fruit analyses derive from the Late Bronze Age graves (Table 3.2), but not the most nutritional plant taxa were found in them (Table 3.3; 5 taxa in LBA vs. 8 taxa in Roman and 12 taxa in Medieval contexts). Similarly, to wood, the seeds/fruits spectrum was more heterogeneous in the Roman and Medieval settlement layers (Table 3.3), as is expected. Besides barley and millet, blackberries and grape pips were found in the LBA graves. In the Augustan vexillation fort and the Early Roman civilian graves the remains of wheat, oat, pea, vetch, peach and hazelnut join to the seeds/fruits spectra, and the medieval samples at Obrežje site additionally supplemented the set of food plants with rye, faba bean and lentil.

Acknowledgements The authors would like to thank the Institute for the Protection of Cultural Heritage of Slovenia, Centre for Preventive Archaeology (IPCHS, CPA), which allowed the partial publication of the results of the analyses of archaeobotanical material from the excavations at the Obrežje site in the presented paper. We would also like to thank the staff of the Institute of Archaeology ZRC SAZU and IPCHS, CPA for assistance with material processing, GIS data and photography. Our thanks also goes to the Slovenian Research and Innovation Agency (ARIS) for funding this study (P6-0064).

References

Sources

Online Resources

Društvo oglarjev Slovenije [Slovenian Charcoal Association]. https://www.drustvo-oglarjev.si/oglje.html

© EuroGeographics. https://ec.europa.eu/eurostat/web/gisco/geodata/reference-data/administrative-units-statistical-units/countries

European Digital Elevation Model (EU-DEM), version 1.1. https://land.copernicus.eu/imagery-in-situ/eu-dem/eu-dem-v1.1

GURS [Surveying and Mapping Authority, Slovenia]. Državna meja 2018 [State border 2018]. https://ipi.eprostor.gov.si/jgp/data

GURS. Hidrografija 2020 [Hydrography 2020]. https://ipi.eprostor.gov.si/jgp/data

Zavod za gozdove Slovenije, Energijski ekvivalenti med različnimi gorivi [Slovenian Forest Service, Fuel energy calculator]. http://www.zgs.si/delovna_podrocja/lesna_biomasa/izracunajte_si_sami/energijski_ekvivalenti_med_razlicnimi_gorivi/index.html

Literature

Andrič M (2007) Holocene vegetation development in Bela krajina (Slovenia) and the impact of first farmers on the landscape. Holocene 17(6):763–776

Andrič M, Willis KJ (2003) The phytogeographical regions of Slovenia: a consequence of natural environmental variation or prehistoric human activity? J Ecol 91:807–821

Andrič M, Karger DN (2024) The Holocene vegetation and land-use history in the northern Dinaric Karst. In: Fuerst-Bjeliš B, Mrgić J, Petrić H, Zorn M, Zwitter Ž (eds) Environmental histories of the Dinaric karst. Springer, Cham, pp. 29–50

Asouti E, Austin P (2005) Reconstructing woodland vegetation and its exploitation by past societies, based on the analysis and interpretation of archaeological wood charcoal macro-remains. Environ Archeol 10:1–18

Bonsall C, Milena H, Kathleen M, Muriel M, Higham Thomas FG, Catriona P, Cook Gordon T (2007) Chronological and dietary aspects of the human burials from Ajdovska cave, Slovenia. Radiocarbon 49(2):727–740

Cappers R, Bekker R, Jans JEA (2006) Digitale Zadenatlas van Nederland. Barkhuis Publishing & Groningen University Library, Groningen

Culiberg M (1984) Karpološke in ksilotomske raziskave kolišča na Partih. Izkopavanja 1981 [Karpologische und xylotomische Forschungen im Pfahlbau auf Parti – Ausgrabungen 1981] PRPNES XII:91–100

Culiberg M (1999) Palaeobotany in Slovene archaeology. Arheološki vestnik 50:323–331

Culiberg M (2007) Analize rastlinskih ostankov [Analyses of plant remains] In: Miškec A (ed) Našli smo zaklad: o novčni zakladni najdbi z Drnovega. Narodni muzej Slovenije, Ljubljana, pp 80–82

Culiberg M (2016) Poročilo o identifikaciji lesnih vrst [Report on identification of wood species]. In: Vojaković P (ed) Končno poročilo o arheoloških raziskavah na lokaciji OŠ Cerklje ob Krki, par. št. 3124/11, 3126/1 in 3217/11 k. o. Cerklje ob Krki. Unpublished excavation report, Ljubljana, Archive of ZVKDS CPA [Institute for the Protection of Cultural Heritage of Slovenia, Centre for Preventive Archaeology]

Culiberg M (2020a) Arheobotanične raziskave na izbranih najdiščnih pososke železnodobne skupnosti [Archaeobotanical evidence from selected sites of the Posočje Iron Age community] In: Tecco Hvala, S (ed) Most na Soči. Arheološke raziskave v letih 2000–2016 na levem bregu Idrijce [The 2000–2016 archaeological investigations on the left bank of the Idrijca]. OIAS, vol 43. Založba ZRC, Ljubljana, pp 243–260

Culiberg M (2020b) Paleovegetacijske raziskave [Palaeovegetation research]. In: Horvat M (ed) Col 1 pri Podgračenem. Arheologija na avtocestah Slovenije, vol 85. ZVKDS, Ljubljana, pp 112–113

Culiberg M (2022) Paleobotanične ostaline [Paleobotanical remains]. In: Tomaž A (ed) Čatež – Sredno polje. Arheologija na avtocestah Slovenije, vol 98. ZVKDS, Ljubljana, pp 133–135

Culiberg M, Šercelj A (1986) Karpološke in antrakotomske analize iz Ajdovske jame pri Nemški vasi [Karpologische und Anthrakotomische analysen aus der Zentralhalle in der Höhle Ajdovska jama bei Nemška vas]. PRPNES XIV:111–118

Culiberg M, Horvat M, Šercelj A (1992) Karpološke in antrakotomske analize rastlinskih ostankov iz neolitske jamske nekropole Ajdovska jama [Karpologische und Anthrakotomische Analysen Pflanzlicher Überreste aus der Neolithischen Höhlennekropole Ajdovska jama] PRPNES XX:111–125

Djurić B (2003) Obrežje – Draga-Goričko [Obrežje – Draga-Goričko]. In: Prešeren D (ed) Zemlja pod vašimi nogami – Arheologija na avtocestah Slovenije – Vodnik po najdiščih. ZVKDS, Ljubljana, p 204

Essert S (2019) Rezultati biljnih ostataka [The results of the plant remains analysis]. In: Karavanić S, Kudelić A (eds) Kalnik-Igrišče, Late Bronze Age settlement. Monographiae Instituti Archaeologici, vol 14. Institut za arheologiju, Zagreb, pp 160–166

Filipović D, Meadows J, Corso MD, Kirleis W, Alsleben A, Akeret Ö, Bittmann F, Bosi G, Ciuta B, Dreslerova D, Effenberger H, Gyulai F, Heiss A, Hellmund M, Jahns S, Jakobitsch T, Kapcia M, Klooss S, Kohler-Schneider M, Kroll H, Makarowicz P, Marinova E, Märkle T, Medović A, Mercuri AM, Mueller-Bieniek A, Nisbet R, Pashkevich G, Perego R, Pokorny P, Popieszny Ł, Przybyla M, Reed K, Rennwanz J, Stika H-P, Strobbe A, Tolar T, Wasylikowa K, Wiethold J, Žerl T (2020) New AMS 14C dates track the arrival and spread of broomcorn millet cultivation and agricultural change in prehistoric Europe. Sci Rep 10:1−17

Gabrovec M (1998) Cerkljansko, Škofjeloško, Polhograjsko in Rovtarsko hribovje [Cerkljansko, Škofjeloško, Polhograjsko and Rovtarsko hills]. In: Perko D, Orožen Adamič M (eds) Slovenija. Pokrajine in ljudje. Založba Mladinska knjiga, Ljubljana, pp 72−82

Guštin M (2005) Savska skupina lengyelske kulture [The *Sava* group of the Lengyel culture]. In: Guštin M (ed) Prvi poljedelci, Savska skupina lengyelske culture. Annales Mediterranea, Koper, pp 7–22

Horvat M (1988) Ajdovska jama pri Nemški vasi : nekropola v jami [Ajdovska cave at Nemška vas : cemetery in a cave]. Arheološki pregled 29:40–43

Horvat M (2009) Prvi prebivalci v Posavju [The first inhabitants of Posavje]. In: Peternel J (ed) Ukročena lepotica – Sava in njene zgodbe. Mohorjeva družba (Celovec), Sevnica, pp 25–35

Horvat M (ed) (2020) Col 1 pri Podgračenem [Col 1 at Podgračeno]. Arheologija na avtocestah Slovenije, vol 85. ZVKDS, Ljubljana, pp 20–46

Jeraj M (2004) Paleobotanične raziskave na kolišču Hočevarica [Paleobotanical analyses of the Hočevarica pile dwelling]. In: Velušček A (ed) Hočevarica. Eneolitsko kolišče na Ljubljanskem barju. OIAS, vol 8. Založba ZRC, Ljubljana, pp 56−64

Kunaver J (1998) Julijske Alpe [Julijan Alps]. In: Perko D, Orožen Adamič M (eds) Slovenija. Pokrajine in ljudje. Založba Mladinska knjiga, Ljubljana, pp 54−70

Marinček L, Čarni A (2002) Komentar k vegetacijski karti gozdnih združb Slovenije. Merilo 1:400,000 [Commentary on the vegetation map of the forest communities of Slovenia. Scale 1:400,000]. Založba ZRC, Ljubljana, pp 18–49

Martinčič A, Wraber T, Jogan N, Ravnik V, Podobnik A, Turk B, Vreš B (1999) Mala flora Slovenije [Small flora of Slovenia]. Tehniška založba Slovenije, Ljubljana

Mason P (2003) Obrežje MMP [Obrežje MMP (International Border Crossing)]. In: Prešeren D (ed) Zemlja pod vašimi nogami – Arheologija na avtocestah Slovenije – Vodnik po najdiščih. ZVKDS, Ljubljana, pp 202–203

Mason P (2006) Obrežje-arheološko najdišče Obrežje [Obrežje-archaeological site Obrežje]. Varstvo spomenikov, poročila 39–41:131–132

Mason P (2008) The Roman Fort at Obrežje and Augustan Military Activity in the Sava Valley in Slovenia. In: Kühlborn J-S (ed) Rom auf dem Weg nach Germanien: Geostrategie, Vormarschtrassen und Logistik. Internationales Kolloquium in Dellbrück-Anreppen vom 4. bis 6. November 2004. Bodenaltertümer Westfalens, vol 45. Verlag Philip von Zabern, Mainz, pp 187–198

Mason P, Kramberger B (2022a) Bronastodobno grobišče [Bronze Age cemetery.]. In: Mason P, Kramberger B (eds) Obrežje. Prazgodovina. Arheologija na avtocestah Slovenije, vol 105/1. ZVKDS, Ljubljana, pp 58−113

Mason P, Kramberger B (eds) (2022b) Obrežje. Prazgodovina [Obrežje. Prehistory] Arheologija na avtocestah Slovenije, vol 105/1. ZVKDS, Ljubljana

Moskal-del Hoyo M (2012) The use of wood in funerary pyres: random gathering or special selection of species? Case study of three necropolises from Poland. J Archaeol Sci 39(11):3386−3395

O'Donnell L (2016) The power of the pyre—A holistic study of cremation focusing on charcoal remains. J Archaeol Sci 65:161−171

Perko D (1998a). Gorjanci [Gorjanci]. In: Perko D, Orožen Adamič M (eds) Slovenija. Pokrajine in ljudje. Založba Mladinska knjiga, Ljubljana, pp 498−507

Perko D (1998b) Krška ravan [Krško plain]. In: Perko D, Orožen Adamič M (eds) Slovenija. Pokrajine in ljudje. Založba Mladinska knjiga, Ljubljana, pp 664−675

Reed K (2018) The development of farming in the Eneolithic: the archaeobotanical remains. In: Balen J, Miloglav I, Rajković D (eds) Back to the past, Copper Age in northern Croatia. Archaeological Museum in Zagreb, University of Zagreb and Archaeological museum Osijek, Zagreb, pp 239–255

Richter HG, Dallwitz MJ (2002) Commercial timbers: descriptions, illustrations, identification, and information retrieval. Computer Program. https://www.delta-intkey.com/wood/index.htm

Sraka M (2012) 14C calendar chronologies and cultural sequences in 5th millennium cal BC in Slovenia and neighbouring regions. Documenta Praehistorica 39:349–376

Sraka M (2020) Calendar chronologies and cultural dynamics in the 5th millennium BC in Slovenia. Case studies in Bayesian chronological modeling of the settlements at Gradec pri Mirni and Bukovnica and cave necropolis at Ajdovska jama. In: Gutjahr C, Tiefengraber G (eds) Beiträge zur Kupferzeit am Rande der Südostalpen, Akten des 4. Wildoner Fachgesprächs am 16. und 17. Juni 2016 in Wildon/Steiermark (Österreich), Materialhefte zur Archäologie des Südostalpenraumes 1, Hengist-Studien 5 and ISBE-Forschungen, vol 1. Verlag Marie Leidorf GmbH, Rahden/Westf, pp 91–132

Schweingruber FH (1990) Mikroskopische Holzanatomie. Eidgenössische Forschungsanstalt für Wald, Schnee und Landschaft, Birmensdorf

Šebek Ž (2009) Krško – življenje z reko Savo [Krško – Life with the Sava river]. Zavod Neviodunum, Krško

Šercelj A (1990) Zanimive karpološke najdbe v dveh rimskih grobovih v Rabelčji vasi pri Ptuju [Interessante Karpologische Funde in zwei Römischen Gräbern in Ptuj (Poetovio)]. Arheološki vestnik 41:309–312

Šercelj A, Culiberg M (1984) Rastlinski ostanki iz Ajdovske jame pri Nemški vasi [Pflanzenreste aus der Ajdovska jama bei Nemška vas]. PRPNES XII:33–37

Tolar T (2016) Cerklje ob Krki - arheobotanična analiza, neobjavljeno poročilo [Cerklje ob Krki - archaeobotanical analysis, unpublished report]. Archive IOA [Institute of Archaeology] ZRC SAZU

Tolar T (2017) Rastlinski makroostanki z najdišča Drnovo 2003, neobjavljeno poročilo [Plant macroremains from Drnovo 2003 site, unpublished report]. Archive IOA ZRC SAZU

Tolar T (2018) Primerjava različnih metod vzorčenja in priprave arheobotaničnih vzorcev z eneolitskih kolišč Strojanova voda in Maharski prekop na Ljubljanskem barju [Comparison of different sampling and treatment methods in order to reconstruct plant economies at the Eneolithic pile-dwellings of Strojanova voda and Maharski prekop at Ljubljansko barje]. Arheološki vestnik 69:461–498

Tolar T (2020) Arheobotanične analize [Archaeobotanical analyses] In: Žerjal T, Novšak M (2020) Školarice pri Spodnjih Škofijah. Arheologija na avtocestah Slovenije, vol 86. ZVKDS, Ljubljana, pp 243–248

Tolar T (2021a) Arheobotanične analize na najdišču Metlika-Mestni trg, neobjavljeno poročilo [Archaeobotanical analyses at Metlika-Mestni trg site, unpublished report]. Archive IOA ZRC SAZU

Tolar T (2021b) Rastlinski makroostanki z najdišča Velike njive 2003, neobjavljeno poročilo [Plant macroremains from Velike Njive 2003 site, unpublished report]. Archive IOA ZRC SAZU

Tolar T (2022) Arheobotanične analize [Archaeobotanical analyses]. In: Mason P, Kramberger B (eds) Obrežje. Prazgodovina. Arheologija na avtocestah Slovenije, vol 105/1. ZVKDS, Ljubljana, pp 174–190

Tolar T, Jakše J, Korošec-Koruza Z (2008) The oldest macroremains of *Vitis* from Slovenia. Vegetation History Archaeobotany 17(1):93–102

Tolar T, Jacomet S, Velušček A, Čufar K (2011) Plant economy at a late Neolithic lake dwelling site in Slovenia at the time of the Alpine Iceman. Vegetation History Archaeobotany 20(3):207–222

Tolar T, Pavlin P (2022) The earliest finds of millet and possible associated changes in material culture in Slovenia. In: Kirleis W, Corso MD, Filipović D (eds) Millet and what else? The wider context of the adoption of millet cultivation in Europe. Scales of transformation, vol 14. Sidestone Press Academics, Leiden, pp 99–120

Tomaž A (ed) (2022) Čatež – Sredno polje [Čatež – Sredno polje]. Arheologija na avtocestah Slovenije, vol 98. ZVKDS, Ljubljana

Toškan B, Kovač M, Tolar T (2022) Bioarheološke raziskave rimskodobnega grobišča Marof na Igu: antropologija, arheozoologija, arheobotanika [Bioarchaeological investigations of the Roman cemetery in Marof at Ig (Slovenia): anthropological, archaeozoological and archaeobotanical evidence]. Arheološki vestnik 73:281–312

Turk P, Svetličič V (2005) Neolitska naselbina v Dragomlju [The Neolithic settlement in Dragomelj]. In: Guštin M (ed) Prvi poljedelci, Savska skupina lengyelske kulture. Annales Mediterranea, Koper, pp 65–79

Tjaša Tolar Archaeobotanist. Ph.D., research associate and assistant professor. Her research focuses on the identification of food plants in the past, from the Palaeolithic to the Middle Ages. In addition to the remains of seeds/fruits of cultivated and gathered plants important for nutrition, she is also interested in the environmental vegetation of the past, the remains of which (i.e., wood/charcoal) are also preserved in archaeological layers.

Philip Mason Archaeologist and conservation councillor, D.Phil. Oxon. The topic of his doctoral thesis was social organisation in the Late Bronze Age and Early Iron Age in the southeastern Alpine region. His field research included excavations in historic urban centres in southeastern Slovenia, as well as field surveys, evaluations, and excavations in advance of housing and infrastructure construction. This included excavation of the multi-period site in advance of the construction of the Obrežje international border crossing as part of the SAAS project. He has also been involved in the presentation of excavated sites in situ in urban contexts.

Bine Kramberger Archaeologist and conservation councillor, Ph.D. He is mainly concerned with the study of human activities through the recovery and analysis of material culture and with the publication of archaeological sites excavated as part of the national highway construction programme in the Republic of Slovenia. He is also interested in the lifestyle and settlement dynamics of prehistoric societies, burial practices and rituals. His author- and co-authorships on several monographs and interdisciplinary studies makes him one of more recognized archaeologists in Slovenia.

Chapter 4
The Formation of the Kras Landscape from Prehistory to the Early Modern Period

Dimitrij Mlekuž Vrhovnik and Tomaž Fabec

Abstract The chapter explores the *longue durée* perspective of transformations of the Kras landscapes from prehistory to the present, viewing the Kras landscape as a result and medium of a longstanding and complex interplay between the history, social and economic developments and ecological dynamics. The landscape is treated as a matter of history; the chapter focuses on the series of its transformations and modifications. We introduce new datasets in the discussion of the long-term history of the Kras landscape by confronting two complementary lines of evidence. First, we examine the elements of past landscapes that are still preserved in the modern landscape seen on airborne laser scanning imagery and face it with the data from doline infills, which provide a diachronic view of the development of prehistoric landscapes. The history of landscape transformation from prehistory to the present is not a smooth, unbroken development; instead, it is marked by qualitative remakings and reorderings. Our concern here is with two significant episodes of modification. The second and third millennium BC is the period of large-scale clearance and formation of the cultural landscape of the Kras plateau, structured around the rearing of domestic animals. The subsequent major landscape reorganization occurred since Middle Ages, but especially after 1500 AD. These accumulations can be associated with the historically documented process of formation of the modern Kras cultural landscape, with the time of intensive deforestation, overgrazing and farming, which turned some parts of the Kras into a stone desert.

Keywords Cultural landscape · Archaeology · Prehistory · Doline · Airborne laser scanning · Kras

D. Mlekuž Vrhovnik (✉)
Department of Archaeology, University of Ljubljana's Faculty of Arts, Ljubljana, Slovenia
e-mail: dimitrij.mlekuz@ff.uni-lj.si

D. Mlekuž Vrhovnik · T. Fabec
Institute for the Protection of the Cultural Heritage of Slovenia, Ljubljana, Slovenia
e-mail: tomaz.fabec@zvkds.si

B. Fuerst-Bjeliš et al. (eds.), *Environmental Histories of the Dinaric Karst*, Environmental History 17, https://doi.org/10.1007/978-3-031-56089-7_4

4.1 Introduction

Any cultural landscape can be seen as an artefact, a cumulative result of long-term interaction between the environment and human cultural practices. Any landscape should be considered an archaeological object in itself, and this especially rings true with the Kras. The landscape played a central role in the subsistence of pre-industrial societies, as growing crops and raising animals were the focus of daily life of most people and communities in the past, especially in prehistory, but also in other study periods. Life was lived in a landscape. If we want to understand the prehistoric and historic cultural landscapes, the patterns of land use, their formation, development and their importance for prehistoric societies, we must approach the landscapes mainly as traces of how societies related to the natural world. Traces of prehistoric and historical land use are not just numb records of the ways prehistoric people utilized the land but fragments of media through which social life was organized and practised.[1]

Kras is a cultural landscape; it is a heavily modified and transformed landscape. There is certain inertia of the karstic landscape, a notion that legacies of the past are preserved in a landscape and influence life, which Meinig describes as "one aspect which is so pervasive as to be easily overlooked: the powerful fact that life must be lived amidst that which was made before."[2] The modern Kras landscape thus has a time depth, its own history of formations and transformations. Landscape transformations involve a series of reorderings, reuses and representations of the landscape. There is a strong sense of the multi-layered nature of the landscape.

Landscape surface is often compared to a palimpsest. Palimpsest is parchment, where older text was erased to make place for the new. However, traces of the old text are still visible and can be deciphered. In the same way, we can recognize traces of past activities in a landscape, despite being covered by modern land use. We can recognize several episodes in the development of the Kras landscape palimpsest, which have effectively reshaped the organization of the landscape, episodes that can be described as the "rasing" of palimpsest. Thus, the Kras landscape can be understood as a palimpsest of different reorganizations of the landscape. However, while palimpsests were seldom erased and reused more than twice, the landscapes are always in the process of transformation and change. Traces from different periods exist simultaneously and are part of the landscape where people live and act.[3]

Our main goal is to explore the *longue durée* perspective of transformations of the Kras landscapes from prehistory to the present, viewing the Kras landscape as a result and medium of a longstanding and complex interplay between the history, social and economic developments and ecological dynamics. We treat the landscape as a matter of history, focussing on the series of transformations and modifications of the landscape. Everything in the landscape has historical roots, and even the most

[1] Barrett (1988), p. 9.

[2] Meinig (1979), p. 44.

[3] Mlekuž (2013), pp. 122–126.

recent examples of landscape features are part of a long chain of processes and events that stretches back centuries or millennia.

However, the history of landscape transformation from prehistory to the present is not a smooth, unbroken development; instead, it is marked by qualitative remakings and reorderings.

Traditionally, the key evidence for the Kras prehistory was either Mesolithic, Neolithic or Copper Age cave sites or Bronze and Iron Age hillfort settlements.[4] However, those sites were isolated points in an empty space, discussed without reference to the wider landscape.

This chapter focuses on the space between the sites, to the Kras landscape itself. The landscape becomes a source of its history instead of deducing its history from isolated places. Our concern here is with two major episodes of modification. The second and third millennium BC is the period of large-scale clearance and formation of the first landscape pattern. This restructuring is contrasted with the Early Modern reorganization of the landscape, which formed the traditional cultural landscape of the Kras, as perceived today.

We introduce new datasets in the discussion of the long-term history of the Kras landscape; we confront two complementary lines of evidence. Together, they provide a record of the formation and long-term development of the Kras landscape.

First, we examine the elements of past landscapes that are still preserved in the modern landscape using airborne laser scanning. We focus on the traces of past life in the landscape, preserved in a landscape matrix. Life in a landscape, with the daily activities and tasks that constantly modify and rearrange the landscape, has its own inherent temporality and development. Unfortunately, this synoptic view from a distance sees long-term development activities collapsed in one layer, the surface of the Kras.

Therefore, we combine this line of evidence with the data from doline infills, which provide a diachronic view of the development of prehistoric landscapes. The variable accumulation of doline infills records the changes in their environment, giving us a continuous view of the changes in a landscape.

4.2 Landscape Pattern

On the Kras, natural processes act slowly; there are no fast episodes of erosion and accumulation that would destroy or bury older remains, periodically resetting the landscape. Generally, erosion is slow and removes mostly fine sediments, accumulating in depressions, dolines and caves. Therefore, human interventions leave almost permanent scars on the surface as topographic anomalies, either in the form of positive features, such as banks, cairns and walls, or negative features, such as cuts, pits and ditches.

[4] Vončina (2018); Montagnari Kokelj et al. (2020); Peresani and Tozzi (2018); Ferrari et al. (2018).

We mapped these features using airborne laser scanning. Laser scanning is a technology which accurately and repeatedly measures distance using a laser pulse, by precise measurement of time needed for the laser pulse to travel from the object and back, and transforms these measurements into a series of points, or a point cloud, from which information on the morphology of the object being scanned may be derived. Airborne laser scanning (ALS; also known as LiDAR, Light Detection And Ranging) is thus an active remote sensing technique which records the morphology surface of the Earth using laser scanning.[5]

ALS allows us to systematically cover large areas and enable us to detect and record the ground beneath the vegetation canopy. ALS records landscape in an indiscriminate way; every place, every feature, every trace, and every square metre is, in principle, treated with the same attention and resolution. This means that all traces preserved on the surface, created by humans or by natural processes, are recorded in the same way.[6]

To be detectable using ALS, the features must be preserved as topographic anomalies in the landscape surface. In the Kras, these traces are extremely well preserved, making the Kras surface a record of its own past.

The latest, a more recent layer of landscape reorganization, appears very prominently in airborne laser scanning imagery. We can discern the field division, materialized in a network of dry stone walls that cover the landscape that corresponds to the field division from cadastral maps from the early-nineteenth century.[7] There is a pattern of cultivated fields around villages, especially in the rare patches of thicker soils. Most of the dolines in the vicinity of villages were transformed into fields or gardens, with dry stone walls surrounding the cultivated soil in the doline bottom, creating the so-called cultivated doline. The outfields were used mostly as pastures and meadows, partly cleared of stone, accumulated in cairns and divided and bounded by stone walls.

However, underlying this layer seems to be the older layer of the Kras palimpsest, the landscape that predates the Early Modern reorganization of the landscape (Fig. 4.1). This layer can be discerned as low, faint features that consist of several typical forms.

One of the most common features that can be recognized on ALS high-resolution topographic data are numerous circular mounds, between 5 and 25 m in diameter and up to 1.5 m high. They are bell-shaped in sections, with smooth transitions to the ground. They can be interpreted as clearance cairns. A clearance cairn is an irregular and unstructured accumulation of fieldstones removed from arable land or pasture and collected into a low mound or cairn (Fig. 4.2).[8]

These cairns are obviously older than modern landscapes, as they are more common in forests and pastures, where modern land use has not destroyed them;

---

[5] Opitz (2012), p. 13.

[6] Mlekuž (2012), pp. 102–104.

[7] Panjek (2015); Moritsch (1969).

[8] Mlekuž (2015), pp. 679–680.

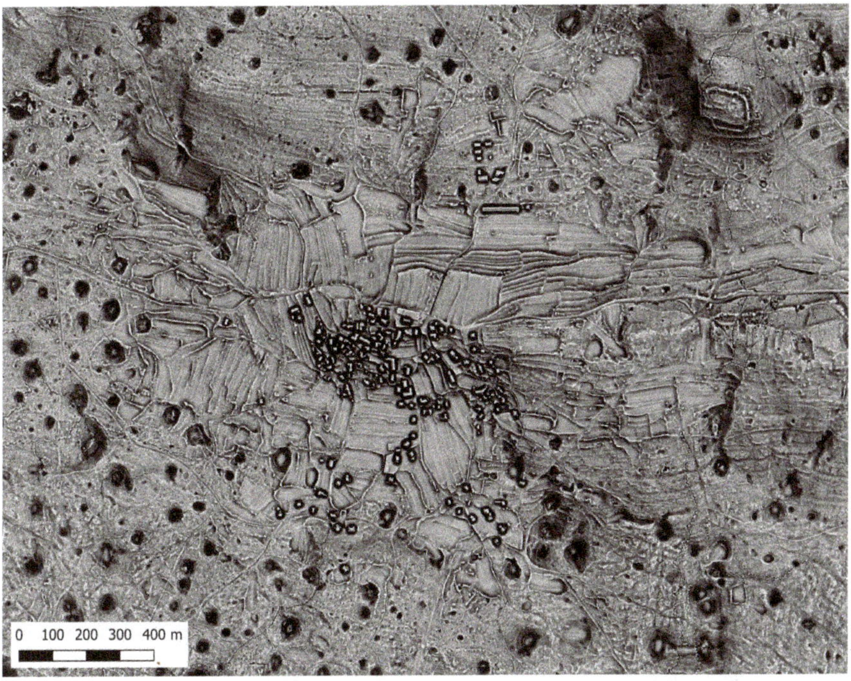

Fig. 4.1 The Kras landscape palimpsest around the village Kostanjevica na Krasu, as seen on a high-resolution digital elevation model produced from ALS data. *Source* authors

Fig. 4.2 A clearance cairn on Drenje Hill near the village of Kazlje. *Source* authors

they are much rarer on meadows and fields, where they were either destroyed by ploughing, cleared or entirely covered by turf and grass.

A test excavation of cairns proved that they consist of around 50–60% of limestone, from up to 30–40 cm large stones to small rubble. Stones are usually sorted, with smaller stones on the top of the cairn. Cairns are usually erected directly on bedrock, as if areas of protruding bedrock were selected as starting places for the accumulation of stones. Cairns contain very little cultural material; most of the excavated cairns were completely empty of pottery or bone. Voids between stones are filled with soil. Relatively large quantities of soil in the cairn can be attributed to the deposition of wind-transported soil and litter.

Cairns usually appear in groups and cover extensive areas, creating a regular pattern of closely spaced cairns that covers areas up to a few hectares (Fig. 4.3). Cairnfields are the result of land clearance. Cairnfields point to past cleared areas for pastures or meadows and can be therefore interpreted as traces of land use connected with animal husbandry, cattle rearing or preparation of winter fodder required for winter feeding of animals kept in byres.[9] From their volume—which ranges between 30 and 80 m^3 per cairn, and their density, which is around thirty cairns per hectare, we can calculate that, on average, around 200–250 kg of stones and rubble was collected from each square metre around cairns. This is close to Early Modern cairns, where densities of between 150 and 380 kg of collected rubble per square metre were estimated.[10]

Some cairns are exceptional, either due to their size, shape or isolated positions. These cairns are usually found on hilltops or ridges in visually exposed positions. We can understand them as barrows, either as funeral monuments, boundary markers, navigational markers or even remnants of defensive towers.[11]

Cairnfields are often bounded by low linear positive features, which could be interpreted either as collapsed dry stone walls or linear cairns. This may indicate piles of material dumped on the edge of the cleared area that delineated the field or grassland from the forest. Internal dumped walls that divide cairnfields can be a result of the gradual expansion of cleared areas, documenting the dynamic process of clearance.

The process of clearance thus also divided and bounded the cleared surface (Fig. 4.4). One of the most common features that can be observed are enclosed areas, enclosures or fields bounded by walls.[12] Their shape varies. We can find irregular forms that adapt to the topography and enclose areas between dolines. More common are oval, circular, semi-rectangular and even rectangular forms of enclosures. There is also huge variability in size, from a few hundred square metres to a few hectares. Enclosures often contain clearance cairns.

It is difficult to distinguish between fields and enclosures. Where a "field" suggests a bounded place where cultivation and controlled grazing took place, "enclosures"

[9] Johnson (2001), pp. 104–107.

[10] Gams (1992), p. 28.

[11] Teržan and Turk (2005).

[12] Mlekuž (2015), pp. 680–682.

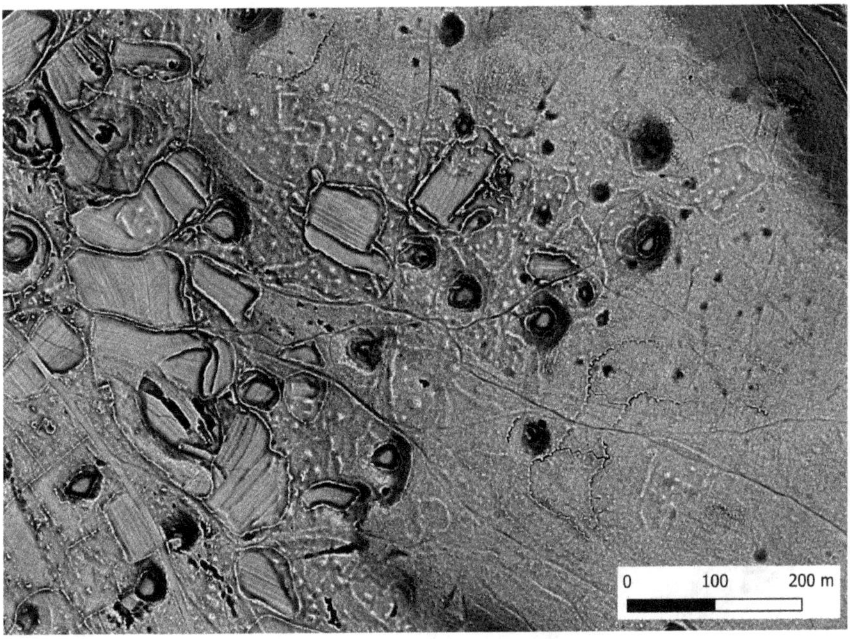

Fig. 4.3 The cairnfield on Drenje Hill near the village of Kazlje on ALS high-resolution topographic data. *Source* authors

were the areas designated for animal keeping.[13] Most of the enclosures could be therefore understood as structures for the confinement and shelter of animals. Based on shapes and sizes, we might divide enclosures into two classes, either simple enclosures, pens, or larger, structured compounds. Pens are usually circular or oval in shape and up to around 50 m in diameter. We can understand them as folds for keeping the animals.

Rounded designs are more common than rectangular ones. Rounded formations are not necessarily less sophisticated forms of construction. Not only as they require less boundary construction and maintenance works to enclose a given area, but curvilinear forms, given the behavioural characteristics of animals, have been demonstrated to allow ease of livestock handling, reducing the distance walked and the number of people needed to manage animals.[14]

Compounds are larger enclosures; they reach a size up to a hectare or more. They are usually structured and divided into smaller compartments or pens and folds. Their entrances can be complex and are often associated with a walled track leading to the enclosure. Complex entrances can serve the separation of animals.[15] This suggests that these were used for larger herds, maybe herds of whole communities, for

[13] Johnson and Rose (1994), p. 59.

[14] Weller (1982), p. 67.

[15] Pryor (1996), pp. 317–322.

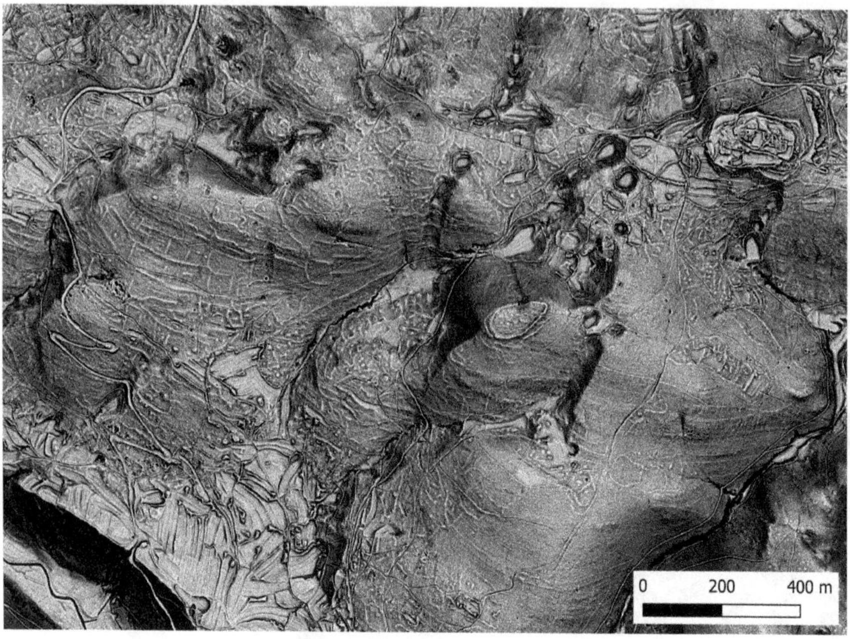

Fig. 4.4 A cultural landscape around the hillfort Tabor pri Vrabčah, with clearance cairns, field boundaries, walled tracks and enclosures. *Source* authors

wintering, as a shelter from predators and riding parties, as areas for collecting dung and as places for animal management, such as castration, shearing and milking.[16] Larger enclosures might be used as pastures for optimizing grazing resources.

The large compounds are more common on exposed hilltops. Rounded enclosures located on hilltops have often been interpreted as hillforts, fortified prehistoric settlements located on exposed hilltops. Enclosures usually have much less massive boundary walls and contain no evidence of inhabitation, such as house platforms or settlement terraces. Compounds are usually also smaller than hillforts. An excellent example of such an enclosure is Stare Ograde, close to the Tabor pri Vrabčah hillfort (Fig. 4.4). First hillforts were established in the early Bronze Age. On the Kras, around 120 hillforts are recorded.[17] However, there are considerable differences in their size, position, amount of work invested in ramparts, and internal structure.[18] Not all stone enclosures are hillforts; there is enormous variability in shapes, forms, and sizes. Thus, some sites might have served as large central settlements, while others were smaller satellite hillforts or just animal pens and enclosures.

[16] Pryor (1996).

[17] Marchesetti (1903).

[18] Slapšak (1995), pp. 24–27.

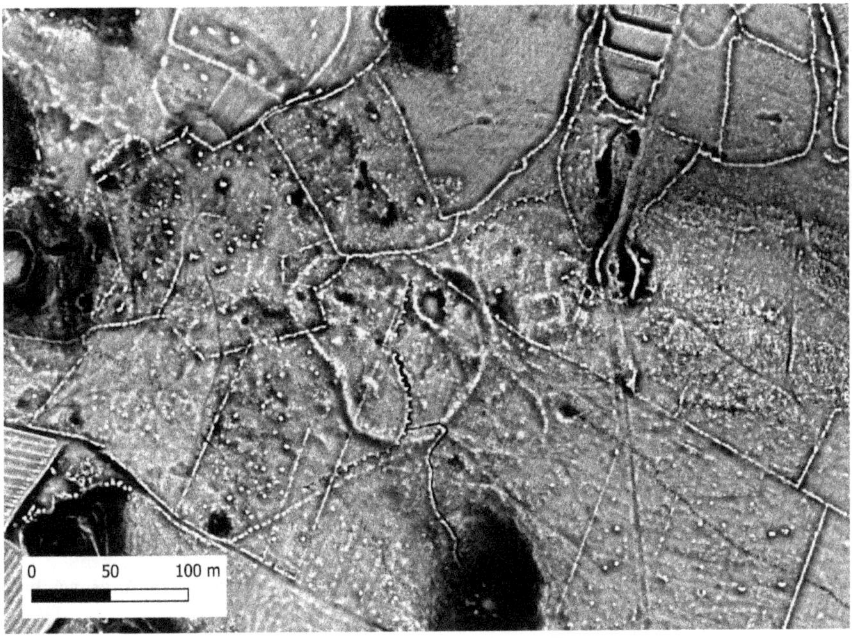

Fig. 4.5 A large enclosure on the Vrhki hill, near the village of Kazlje. *Source* authors

Sometimes, the division between enclosures and hillforts is not so straightforward. This might point to the pastoral origins of hillforts; some of them might have developed from enclosures, or their use may relate to the development of focal places in the landscape.[19] However, arguing from lack of evidence is difficult, and how these enclosures fit within their contemporary landscape and land use needs further clarification.

The function of compounds away from hillforts is to optimize the use of grazing resources, protect animals and provide a sheltered place for the daily care of animals. An example of such a large compound is located under Vrhki Hill, near Kazlje (Fig. 4.5). It covers an area of ~ 8000 m^2 and is divided into several folds. Within the compound, several cairns can be recognized. The entrance to the compound is complex, obviously made to separate animals. A few hundred metres north of the compound lies a large cairnfield with a funnel that empties into a walled track.

Other linear features can be interpreted as field boundaries.[20] Stone walls and fences not only made the time-consuming livestock herding unnecessary but also enabled better husbandry of the available land.[21] In addition, they played a role in controlling damage by wild species. Drystone walls can be made stock-proof where there is sufficient stone and a labour force. Boundaries have to be of a certain scale

[19] Fowler (1983), pp. 193–195.

[20] Mlekuž (2015), pp. 682–684.

[21] Seymour (2003), p. 138.

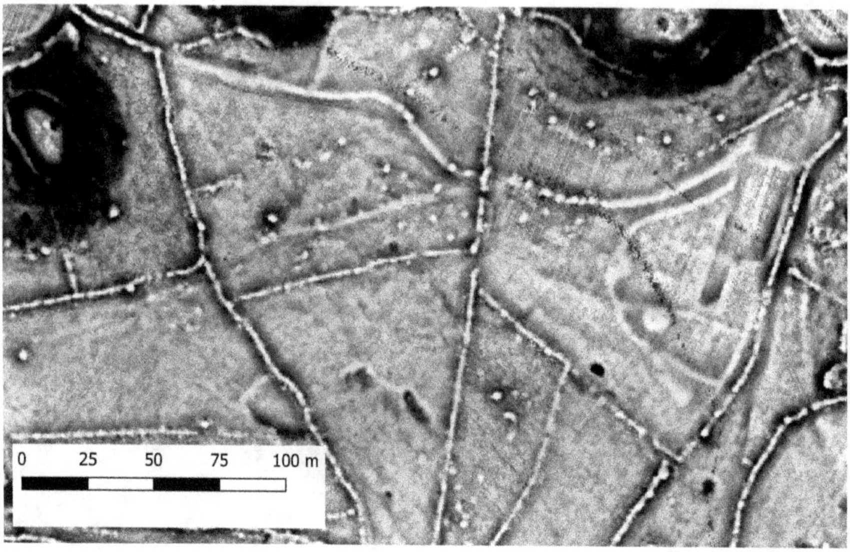

Fig. 4.6 A "funnel" leading into the walled track. Zlobčev golobnjak near Kazlje. *Source* authors

and robustness to withstand stock and provide adequate control over them. Some of the walls are substantial, larger than they need to be if necessary to delineate plots and fields or keep animals from straying.

Some of the structures were not built high enough. Without additional fences and hedges, they could not be used as stock-proof barriers, indicating their roles as conceptual boundaries indicating land tenure.[22]

Pairs of parallel walls a few metres apart that run across the landscape and connect the pastures with the hillforts can be interpreted as bounded paths or tracks (Fig. 4.6). Walls were erected as barriers for livestock, preventing them from entering the fields or meadows while moving from pens to pastures.[23] Bounded tracks demonstrate the regular movement of animals and the need to direct and control the flow of animals across the landscape. Track *termini* are often funnel-shaped in the form of entrances that condense flows and animals and lead them to the walled paths. Funnels allow the management of herds and their travel from pastures to a hillfort.[24]

Most of these traces in a landscape suggest a landscape pattern structured around animal rearing. This layer of palimpsest was probably a result of a long process of development, adjustment, alteration and accretion.[25] Fields and structures were only in rare cases laid out in one event in the form that we now perceive them; they have their own biographies, constantly developing and changing structure and meaning.

[22] Fowler (1983), p. 110; Johnston (2005), p. 219.

[23] Yates (2007), pp. 16–17; Chadwick (2007), pp. 137–146; Pryor (1996), pp. 314–316.

[24] Pryor (1996), pp. 317–322.

[25] Chadwick (2008a), p. 219; Chadwick (2008b), p. 7.

The main issue in explaining those traces is the question of their age. When were they made and used in the landscape? In our case, these features are difficult to date, as there is only sporadic material in a secure context. However, it is quite obvious that those traces represent the older layer in a landscape palimpsest. They are preserved only in marginal areas, where they were not touched and modified by very labour-intensive farming and land use characteristics for the Early Modern Period, the period of most intensive human impact on the Kras landscape. These traces have quite straightforward stratigraphic relations to the recent traces. Thus, prehistoric traces are covered by the Early Modern meadows and pastures and overlain by dry stone walls that mark the parcels recognized on cadastral maps from the nineteenth century.

Only in some isolated cases are the older traces incorporated into the Mediaeval to Early Modern field division and land use. This is not due to the continuity of land use or landscape organization but due to the sheer materiality and mass of these traces. It was simpler and cheaper to incorporate them than to completely destroy them, although they did not optimally fit in with the desired new arrangement of the landscape. Especially massive hillfort walls were left in place, and modern land use adapted to them, as in the case of several massive hillforts, like Debela Griža near Volčji grad.

In rare cases, we have encountered cultural material associated with these features. In the enclosure near Kazlje (Fig. 4.3), we have directly dated a cattle bone associated with Late Bronze pottery to the thirteenth century BC. Traces discussed in the paper are, therefore, older than the modern cultural landscape that emerged in the Middle Ages and the Early Modern Era. Therefore, some of them were already there in the Late Bronze Age or earlier. But when did the first organization of cultural landscape that we see on ALS data begin?

4.3 Phases of Most Significant Landscape Change

These traces of landscape clearance and rearrangement can be confronted with the dynamics of sedimentary infill of dolines.

Dolines are one of the most typical karstic landforms. Dolines are subcircular bowl or funnel-like depressions in limestone caused by karst processes, either the chemical dissolution of carbonate rocks, collapse or suffusion processes. The Kras landscape is packed with dolines; there are often tens or hundreds of dolines per square kilometre.

Dolines are one of the rare surface karst landforms where the net accumulation of sediments is greater than erosion and can be considered sedimentation basins due to the continuous accumulation of sediments. The depositions in a doline result from the interaction of natural and anthropogenic factors at the local (doline) and/or broader (landscape) level. The presence and effect of anthropogenic factors in doline infills are reflected not only in archaeological remains such as finds, structures, layers, etc., but also in anthropogenic anomalies perceptible in soil profiles, such as composition, magnetic susceptibility curves, geochemical data, etc.

Our study investigated 20 dolines in the Divača Kras (Fig. 4.7), where we mechanically excavated 30 test trenches along the entire longitudinal axis of the doline bottom. In larger dolines, test trenches were dug in several places. Test trenches reached the bottom of the fine-grained sedimentary infills, reaching compact limestone or thick-grained sediments. Fine-grained sedimentary infills rarely exceeded a depth of 3 m.[26]

Cross-sections of every individual doline were recorded and analysed. In the sections, the soil's magnetic susceptibility was measured in three columns at 5-cm vertical intervals, where soil samples were also collected. In total, about 3000 measurements were taken, and about 1000 soil samples were collected for further laboratory analyses. Where we detected archaeological traces and structures, we also documented them in the ground plan.

The first observation is that the stratigraphic sequence in all sampled dolines is similar and follows the same pattern; layers with the same properties occur in almost all dolines in similar stratigraphic positions, thus enabling us to generalize a stratigraphic sequence and processes behind it, displaying a striking pattern that enables us to generalize the sedimentary dynamics of the doline catchments (Fig. 4.8).

This pattern is further supported by the radiocarbon dating of individual layers from different dolines, as it showed that these were horizons that actually formed in the same periods. We can conclude that at least some of the factors that conditioned the sedimentary infills in the dolines are the results of processes that operated at the regional level. Integration with archaeological data suggests that analysed soil horizons are the result of anthropogenic impact and modifications within the doline catchments during the Holocene.

Based on radiocarbon dating and archaeological material in dolines, we can identify two periods of discernible human impact in the doline sedimentary sequences: increased human impact in the Bronze Age and another in the Early Modern Period. Compared to these two periods, the human impact in the remaining periods is almost imperceptible.

In the Early Modern Period, dolines were increasingly cultivated; this resulted in the formation of a so-called cultivated doline, the doline whose shape and soil conditions were modified and improved for cultivation.[27] This is an integral part of forming the agrokarst landscape,[28] a well-researched and well-known process.[29] This impact was clearly visible in the investigated dolines, both in the field and in the analysed soil profiles. Modern layers usually differed from the others in the presence of sharp-edged limestone gravel, pottery fragments, bricks, metal objects and, less

---

[26] Fabec (2018).

[27] Gams (1974); Gams (1992); Gams and Gabrovec (1999); Radinja (1987); Panjek (2015).

[28] Nicod (1987).

[29] Gams (1992); Panjek (2006); Panjek (2015).

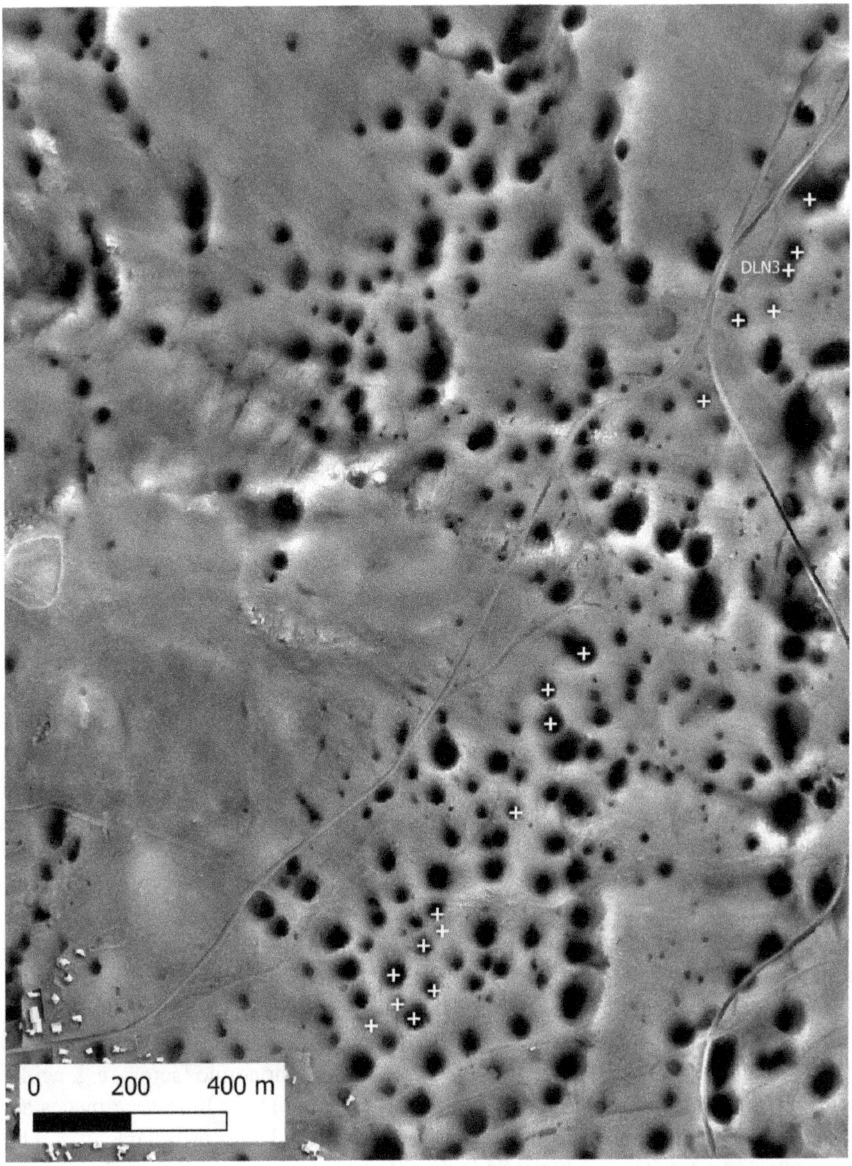

Fig. 4.7 Doline in the Divača Kras, analysed in the study, marked by crosses. *Source* authors

frequently, animal bones. Gravel and manuring garbage in the sediment are the result of agricultural practices to improve arable land.[30]

[30] Gams (1992), p. 35.

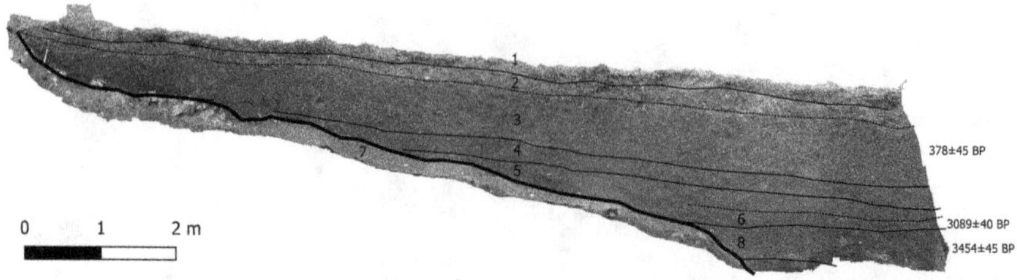

Fig. 4.8 A cross-section from the DLN3 doline (see its location in Fig. 4.7) illustrates a typical stratigraphic infill sequence. *Source* authors

Those anthropogenic impacts were clearly detected in all granulometric curves; Early Modern layers are the only ones with a high fraction of coarse-grained sediments.

Geochemical analyses showed that Early Modern layers are highly carbonate and organic, which is a result of the practices of liming and fertilization or bioaccumulation, while pre-Early Modern layers are generally highly silicate. Magnetic susceptibility measurements showed relatively high values, increased charcoal content and increased sedimentation intensity. The Early Modern human reorganization is clearly visible in the doline sedimentary infills.

In the same way, traces of human impact dated to the Bronze Age were found in most of the excavated dolines and were identified on the basis of small finds as well as traces of anthropogenic impact in sedimentation. In several dolines, we have discovered prehistoric pottery fragments, stone artefacts and animal bones. Bronze Age pottery is relatively common in the Kras. On average, up to 10 potsherds were found in an individual doline; there is also one doline with 30 potsherds. The pottery mostly belonged to larger vessels, probably pots and bowls, with wide rims. There were far fewer smaller and finer vessels. Rare typologically identifiable ceramics can be attributed to the Bronze Age. This is in line with the results of excavations for new vineyards across the Kras, which found pottery fragments in 47 out of 54 excavated dolines; all pottery fragments can be attributed to the Bronze Age.[31]

Pottery finds are not evidence of settlement midden. Animal remains a typical component of settlement waste are rare. However, sparse finds of bone certainly indicate that their absence is not the result of purely taphonomic factors, e.g. acidic soil reaction. Larger limestone pieces are also present. These limestone pieces act as an outlier in layers with Bronze Age pottery, as these layers are characterized by very low stoniness. Thus, pottery finds do not indicate that people settled and lived in dolines. Instead, they are evidence of off-site activities.

Another important component of Bronze Age layers is the presence of bright orange or yellow silt lumps, occurring together with charcoal fragments. These elements are the most common and visible indicators of anthropogenic impact. Clay lumps are most likely a sedimentary remnant of burnt soil. They lay scattered in layers, together with pottery shards and charcoal. Charcoal was the most abundant

[31] Fabec (2018), pp. 18–27.

Table 4.1 Generalized reference stratigraphic sequence of doline infills derived from the analysis of dolines presented in Fig. 4.7

Colour	Thickness	Finds	Dates
Dark brown, Yellowish brown	0.6 m	Early modern finds, stones	1700 AD–1800 AD
Dark brown, Yellowish brown		Early modern finds, stones	
Dark brown, Yellowish brown		Early modern finds	
Dark Grey, brown, dark brown	0.3 m		600 AD–1700 AD
Dark brown	0.2 m		500 BC–600 AD
Reddish brown	0.4 m	Prehistoric finds	1000 BC–500 BC
Reddish brown		Prehistoric finds, burnt clay lumps	2500 BC–1000 BC
Brown, dark yellowish brown	1 m		10,000 BC–2500 BC
Dark reddish brown			
Reddish brown, dark yellowish brown			
Scree			Pleistocene

material in layers with burnt clay lumps, much more than in all other layers combined. The charcoal fragments are larger; in one case, even more than 10 cm large pieces of charred branches were preserved. Most charcoal is produced in rapid fires, so it is possible that the relatively high proportion of charcoal fragments in these layers reflects an above-average fire frequency as the result of human-induced fires.

Radiocarbon dating of charcoal samples from a layer with silty lumps of burnt soil shows a fairly uniform picture (Table 4.1). Just under half of the radiocarbon dates fall into the first half of the 2nd millennium BC; 10 of 12 available dates fall from the end of the third to the middle of the second half of the second millennium BC. Radiocarbon dates thus show the beginning of human impact in the Early Bronze Age with increased human impact in the Middle Bronze Age.

Layers with a large number of silty lumps dated to the Bronze Age correlate with the highest magnetic susceptibility values. Considering the complexity of interpreting magnetic susceptibility measurements, these observations reinforce the explanation that layers with silt lumps are the result of burning events or that the orange–yellow silty lumps are, in fact, a trace of burnt soil. Prolonged exposure of the soil to fire results in the conversion of weakly magnetic iron oxides and hydroxides (e.g. hematite) into more magnetic forms (magnetite and maghemite).[32] Other studies in the Kras have shown that high values of soil magnetic susceptibility are mainly the

[32] Mušič (1997), pp. 37–38.

result of human use of fire, both in dolines and in intensively populated areas.[33] Differences between Bronze Age sediments and sediments of other prehistoric strata were not recognized in geochemical analyses, which is probably due to the fact that the effects of fires on soil geochemical properties are very short-lived.[34]

The increased human impact on the landscape in the Bronze Age and the Early Modern Period was also visible in the fluctuations in sedimentation intensity. In the last millennium, from the Mediaeval Period to the present day, up to 0.9 m of soil has been deposited on average in the surveyed dolines, of which 0.6 m of soil was deposited in the three centuries of the Early Modern Period, indicating that this was the period with the fastest sedimentation rate. Bronze Age layers are half as thin, as their average thickness is about 0.45 m. Given that this is a longer period, around 3000 years, the average sedimentation rate was approximately 17% of the one determined for the last millennium (Table 4.1).

Surprisingly, however, the sedimentation rate slowed down significantly after both Bronze Age and Early Modern peaks in human impact. In 6 dolines, prehistoric strata lie directly below the Mediaeval layers, while in the other 6 cases, they are separated by a deposit up to slightly < 0.4 m thick. Generally, between both periods of increased human impact, therefore, during the Iron Age and Roman period, up to 20 cm of predominately aeolian sediment was deposited in the dolines; half of the amount that was deposited during the Bronze Age and less than a quarter of the amount from 1000 AD to the present day.

Estimating the sedimentation rate before Bronze Age is problematic because, in most cases, we did not reach the doline bedrock. However, the thickness of the deepest part of the sedimentary infills of the studied dolines indicates that sedimentation was significantly slower. Thus, in the 8000 years of the Holocene before the Bronze Age, just as much sediment is estimated to have been deposited as in the last 500 years from the beginning of the Early Modern Period to the present day.

Increased sedimentation intensity in the Bronze Age can also be observed on a supraregional scale. In the core from Polje Čepić in Istria, the sedimentation rate more than tripled in the second millennium BC.[35] In the second millennium BC, there was also an increased accumulation in Lake Vrana on Cres.[36] The Central Adriatic Marine core RF93-30 shows increased soil erosion in the Po Valley at about the same time.[37] In all cases, the increased sedimentation is thought to be due to anthropogenic vegetation disturbances.

The vegetation during the Late Glacial shows the predominance of pine (*Pinus*), which suggests that the Kras in the late glacial was overgrown by very open, mostly pine-birch forest. Pine was soon joined by oak (*Quercus*) in the early Holocene.[38]

---

[33] Mušič (1997); Mušič and Dimic (1994).

[34] Chandler et al. (1983); Ollesch and Vacca (1999).

[35] Balbo et al. (2006).

[36] Schmidt et al. (2000); Miko et al. (2005).

[37] Oldfield et al. (2003).

[38] Legnani (1967); Gallizia Vuerich et al. (1994); Nisbet (2000).

Oak, and other temperate deciduous trees begin to show more visibly at the transition from the Late Glacial to the Holocene[39] or even somewhat later.[40]

In a long period of about 9000–4000 cal BP, the oak-hornbeam light forest was fully established in the Kras.[41] A major change in vegetation dates back to the beginning of the 7th millennium BC. In the southwest of the territory of present-day Slovenia, the abundance of linden (*Tilia*) increased highly significantly.[42]

In the Mid-Holocene, but perhaps even earlier,[43] the composition of the forest was influenced by human impact reflected in the growth of heliophyte shrubs, i.e. partridge (*Sorbus*), dogwood (*Cornus*), blackthorn (*Prunus spinosa*) or hazel (*Corylus*).[44] In the doline infills from this period, oak charcoal predominates, which, together with the otherwise relatively lower representation of hornbeam, probably indicates the existence of an oak-hornbeam light forest. The relatively high proportion of beech (*Fagus*) is surprising. Beech charcoal was discovered in one of the dolines, and radiocarbon dated to the 4th millennium BC. It suggests that beech expanded on the Kras plateau in the Early Holocene[45] and, in some parts of the Kras, it remained until the 4th millennium BC. The abundance of beech is a phenomenon with a clear parallel in present-day central Slovenia, where beech became one of the most frequent tree species since the beginning of the 7th millennium BC.[46] The presence of ash and heliophyte dogwood indicates the open forest at that time. Then between 2000 and 1500 BC, there was a period of intensive deforestation of the Kras.

During the Late Roman Period and Early Middle Ages, due to demographic decline and consequently lower agricultural pressure on the forest, oak is expected to expand strongly and rapidly and then shrink somewhat more gradually to its previous, relatively modest extent.

The process of forest cutting to establish new fields and meadows has led to almost complete deforestation of the region during the Modern Period.[47] In the anthracitic record from doline sediment infills of this time, oak no longer occupies the predominant place that is occupied by hornbeam. Especially in this phase, the lowest diversity of the tree community is shown, as only the two taxa mentioned are represented in the record. This could speak even more clearly of deforestation in favour of agricultural land and the huge reorganization of the landscape.

---

[39] Culiberg (1991); Šercelj (1996).

[40] Legnani (1967), pp. 83–85; Nisbet (2000), p. 166.

[41] Turk et al. (1993), pp. 70–71; Culiberg (1991); Nisbet (2000), p. 166.

[42] Andrič (2004), p. 510.

[43] Mlekuž (2005b), pp. 167–176.

[44] Turk et al. (1993), p. 70; Andrič (2004), pp. 511–514.

[45] Legnani (1967), pp. 85–94.

[46] Andrič (2004), p. 510.

[47] Panjek (2015), pp. 77–105.

4.4 Formation of the Kras Landscape

The first human reorganization of the landscape can be situated within a long-term process that started with the adoption of domestic animals by local Mesolithic hunter-gatherers around 6000 BC and the fast process of formation of Neolithic pastoral systems.[48]

The evidence from cave sites clearly shows that caves were used as seasonal herding camps for highly mobile and small nomadic pastoralist communities, who moved around the landscape following the specialized herds composed mainly of small stock.[49] There is evidence of short, seasonal visits to the caves. The seasonal patterns suggest that cave sites in the Neolithic were not merely outstations of a more extensive pastoral system, with home sites elsewhere, but they comprised a complete yearly cycle of seasonal mobility. Neolithic pastoralists were probably nomads moving around the landscape and using caves as shelters and pastoral camps.[50]

Deposits in caves form distinctive "layer cake" sediments of alternating white and black layers.[51] Those layers are thin, and their clusters form stacks that can be up to several metres thick and cover large areas of the caves. The layer cake sediments result from the periodic burning of the dried and trampled dung on the cave floor, indicating that domestic herds were penned in the cave. Probably one of the main reasons was to reduce the volume of manure deposits, as dung loses about 97% of its volume during burning. Other possible reasons include the disinfection of caves and the protection of animals from parasites in the dung.

In the Neolithic, herds of sheep and goats were exploited mainly for meat.[52] Although animals were milked, as evidence of milk lipids preserved in pots testify,[53] animal management strategies were not optimized and aimed towards the satisfaction of immediate needs.[54]

The whole Dinaric Alps were settled by small, autarchic and mobile groups. Although pure pastoralism is rare in the ethnographic record, Neolithic pastoralists of the Kras and eastern Adriatic hinterland were specialized pastoralists who moved around with their flocks of sheep and goats.

In the Late Neolithic and in the Copper Age, in the 4th and 3rd millennium BC, the composition of the domestic herds changed. Bone assemblages demonstrate trends towards the optimization of sheep and goats meat production, which is a result of mixed herding strategies, where a small number of a variety of animals is kept for a

---

[48] Mlekuž (2005a), pp. 16–44.

[49] Ibid., pp. 42–44; Mlekuž (2007), pp. 78–80.

[50] Mlekuž (2005a), pp. 37–38.

[51] Brochier (1983); Brochier (1991); Boschian and Montagnari-Kokelj (2000).

[52] Mlekuž (2005a), pp. 34–36; Mlekuž (2006).

[53] Budja et al. (2013).

[54] Mlekuž (2005a), pp. 27–39.

mixture of products, mainly meat and milk, principally for domestic use. Cattle have become more common, obviously for the use of milk and milk products.[55]

This resulted first in reduced mobility[56] demonstrated in a changed seasonal pattern, suggesting less seasonal use of caves. This could mean more frequent visits to the caves that could indicate more intensive use of the landscape around the cave and increased territoriality of the herding groups associated with specific places. All these communities left very few traces in the landscape. This changed in the Bronze Age, as demonstrated in Sects. 4.2 and 4.3. Both lines of evidence presented in Sects. 4.2 and 4.3 show two episodes of large reorganization of the landscape, two main layers of the Kras landscape palimpsest.

The first thorough prehistoric reorganization of the landscape can be dated to the Bronze Age, a period after 2500 BC. Anthropogenic horizons of burnt ground and charcoal that appear in doline infills suggest large-scale vegetation burning and soil disturbance, a result of the forest clearance and removal of tree stumps. In some cases, these interventions are accompanied by sparse finds of prehistoric pottery. This process can reflect the formation of a cultural landscape with meadows, pastures, fields, animal pens and tracks that can be recognized as the oldest layer of the Kras landscape palimpsest. This period of a massive reorganization of the Kras landscape lasted for a few centuries and was followed by a stable, open cultural landscape.

Moreover, in the Early Bronze Age, there was a marked change in depositional practices in the caves; the burning of dung deposits stopped, and deposition was much slower, indicating that caves were no longer used as sheep pens. In the Mala Triglavca cave, located a few hundred metres from the doline, described in Sect. 4.3, deposition of layer cake deposits stopped around 2100 BC.[57]

The faunal record, mainly from hillforts, shows a general trend towards more specialized herds (Fig. 4.9). The proportion of sheep and goats decreases considerably, while the proportions of cattle (and pigs) increase sharply. The more significant share of cattle and pigs could be understood as a consequence of greater diversification of animal management strategies, where each species has its own role in the pastoral system and a greater focus on secondary animal products, especially milk.[58]

The impact on the landscape and the resulting landscape pattern testify that the landscape was modified largely for pastoralism. Most of the landscape was organized around the basic needs of livestock. This demonstrates the importance of animals and animal husbandry in the prehistory of Kras. Animals played a crucial role in the subsistence economy of communities and households; and the nexus of daily activities and practices. As such, they were tightly woven into the identity of the community. These traces demonstrate that people and animals moved on predictable daily and seasonal rhythms, moving from settlements to communal pastures where animals were kept in large compounds, pastured on cleared pastures and then funnelled into the walled paths that moved herds past the fields back to the settlements.

---

[55] Ibid., pp. 29–32.

[56] Ibid., pp. 37–38; Miracle and Pugsley (2006), pp. 331–336.

[57] Mlekuž and Pergar (2008).

[58] Fabec (2018), pp. 107–115.

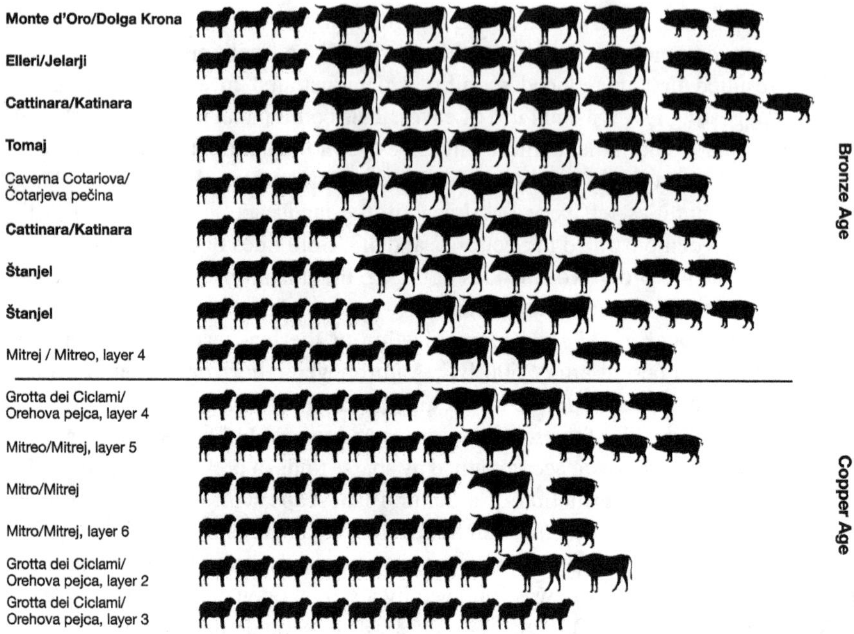

Fig. 4.9 Trends in the Kras faunal record during the Copper and Bronze Ages. Hillfort sites are typeset in bold. *Source* authors

Extensively husbanded sheep and goats have been widely regarded as one of the most effective uses of marginal land in the Modern Period. Sheep and goats can graze on very stony, marginal land, while cattle need higher quality grass. A cow needs much more fodder than a single sheep and requires a daily supply of large quantities of water. Rearing cattle, therefore, requires more hay and more labour investment in a landscape. Surface stones and stumps can impede hay collection on meadows. To improve pastures and meadows, the land must be cleared of surface gravel and stumps, as the traces of prehistoric cairnfields demonstrate. In the nineteenth century, pastures with protruding bedrock and surface rubble in Kras were worth only a third of cleared, improved pastures.[59]

Bronze Age traces of vegetation burning and soil erosion in dolines can therefore be understood as a result of land clearing, linked to the expansion and increasing quality improvement of pasture areas for growing cattle herds. These are collective works which demonstrate group cohesion and large investment into the landscape. This cultural landscape and way of life continued into the Iron Age. The landscape was developed, altered, and adjusted; the features, such as cairns, boundary walls and pens, accreted; however, it was the development of the landscape that was formed in the Early Bronze Age.

[59] Gams (1992), p. 20.

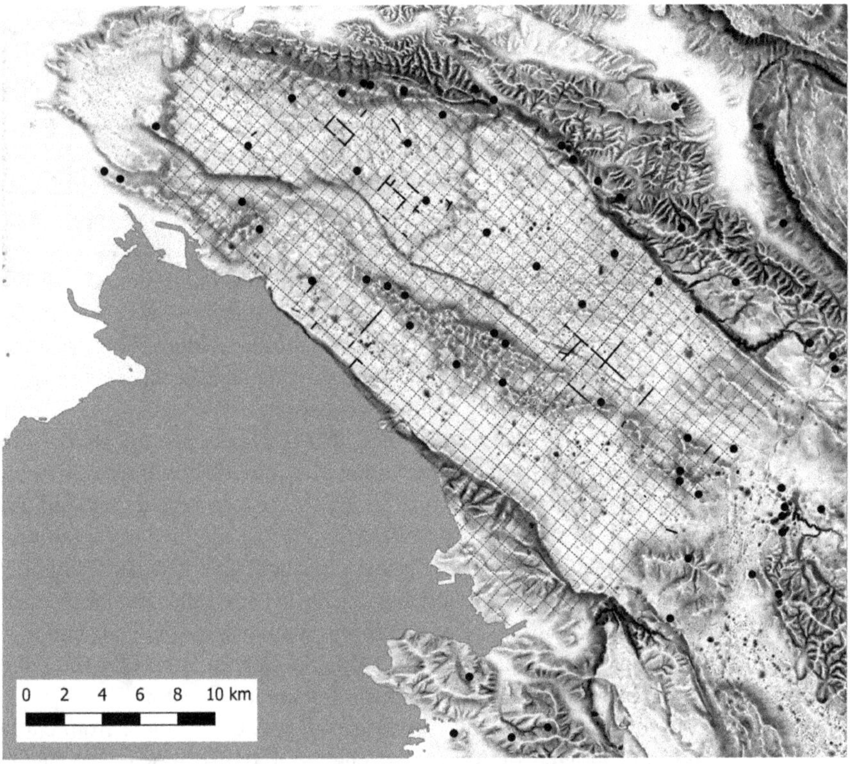

Fig. 4.10 Map of hillforts (dots) on the Kras with superimposed traces of Roman centuriation (black lines). *Source* authors

This development was cut by the Roman conquest in the first century BC.[60] The incorporation of the Kras into the Roman state primarily meant the establishment of an urban centre, the colony of Tergeste (present-day Trieste, Italy). The Kras became integrated into the territory of Tergeste (Fig. 4.10). It established control and economic infrastructure (military outposts, public roads and road stations, customs stations) and economic stability through various forms of agriculture, either within the framework of traditional indigenous communities but also in close connection with urbanization, with the appropriation of the territory and the allocation of state land to colonists and privileged elites, and with the (re) organization of the use of other resources.[61]

We can recognize this intervention in the area of the Kras plateau through the traces of a regular Roman land division called *centuriatio* or *limitatio*.[62] It is the division of land into square fields or *centuria*, delimited by orthogonal and equidistant *limites*

[60] Slapšak (2003), pp. 245–249.

[61] Ibid., p. 248.

[62] Mlekuž (2018).

or boundaries. The size of a centuria is 20 × 20 actus (around 710 × 710 m), and the area of a centuria is 200 *iugera* (about 50.4 ha; Fig. 4.10).

Centuriation arranged the territory for the new settlements, demarcated the fields, established a legal framework for the distribution of land to the colonists and prepared space for infrastructure in the form of roads, bridges and canals.

This radical and sudden rearrangement of the landscape and the establishment of new spatial patterns and relationships deeply cut into the existing indigenous landscape established throughout prehistory. It meant a dissolution of an old way of life and traditional communities and probably a much lower agricultural density, so the agricultural density in the region was probably significantly lower than in the Bronze Age. However, the slow rate of sedimentation recorded in the dolines between the Bronze Age and the Modern Period does not indicate significant interventions into the landscape associated with the Roman occupation.

It was probably during the Mediaeval period that the landscape got its present form, its main contours, at least, with the formation of a network of villages and their cultural landscape.[63] However, it was during the Early Modern Period that land use became increasingly labour-intensive, significantly altering the mediaeval cultural landscape. This was influenced by broader processes characteristic of the integrated peasant economy[64] in times of early capitalism.[65] In this case, the development of the port of Trieste was particularly relevant. It was acquired by the Habsburgs in 1382, developed into their principal Adriatic port, and then grew rapidly since the early eighteenth century when Emperor Charles VI made it a free port.[66]

Especially from the nineteenth century onwards, it was the pressure from cattle herds that strongly encouraged the landscape change. Extensive clearance of the rocky surface into pastures and, where possible, into meadows for mowing the grass to feed the cattle in winter. The adoption of cattle rearing even influenced the vernacular architecture, which underwent significant structural changes.[67]

4.5 Conclusion

The formation of the Kras landscape is a long historical process which started with the formation of Neolithic pastoral systems around 6000 BC and the increased adoption of cattle in the Copper Age for the use of milk and milk products. This corresponded with social changes in the Early Bronze Age, which can be recognized in more significant social cohesion and territoriality of communities. This process started an extensive modification of human impact on the landscape after 2500 BC, especially

[63] Slapšak (1995), pp. 31–33, 43–47.

[64] Panjek (2017).

[65] Gestrin (1991).

[66] Bajt and Vidic (2011); Kalc (2008); Moritsch (1969).

[67] Panjek (2015), pp. 54–56; Gams (1992), pp. 60–68. See also Chap. 8, in this volume: Remec (2024), pp. 203–237.

between 2000 and 1500 BC. It can be recognized in doline infills with anthropogenic horizons of burnt ground and charcoal, which suggest the burning of vegetation and removal of tree stumps in the frame of forest clearance. The period of intensive modification of the landscape is followed by the slower accumulation of predominately aeolian sediments, which suggests a much more open landscape.

This process coincides with the formation of the first open cultural landscape, including permanent stone structures, such as monumental hillforts, barrows, cairns and stone walls, creating new landscape patterns and artificially co-created landscapes structured mainly around the needs of domestic animals.

The incorporation into the Roman Empire disrupted this general pattern but did not create a new one.

The next major reorganization in dolines occurred during the Middle Ages, especially after 1500 AD. This process can be seen in dolines, where almost up to a metre of deposit was accumulated in a few hundred years. These accumulations can be associated with the historically documented process of formation of the modern Kras cultural landscape, with the time of intensive deforestation, overgrazing and farming, which turned limited parts of the Kras into a stone desert.

The Kras landscape has historical roots, and a modern landscape is a result of a long chain of events that stretches back across millennia. There is no purely natural landscape on the Kras. The Kras landscape is an artefact; it has been, to some extent, culturally shaped and transformed.

Karst landscape testifies that pre-industrial societies did not have only local and transitory environmental impacts and that human impact on landscape did not begin decades ago but has been profoundly influencing animals, stones, soil, and billions of other entities for thousands of years.

References

Andrič M (2004) Paleookolje v Sloveniji in severnemu delu hrvaške Istre v pozni prazgodovini [The vegetation of Slovenia and northern Istria in the late prehistory]. Arheološki vestnik 55:509–525

Bajt D, Vidic M (eds) (2011) Slovenski zgodovinski atlas [Slovenian historical atlas]. Nova revija, Ljubljana

Balbo LA, Andrič M, Rubinić J, Moscariello A, Miracle PT (2006) Palaeoenvironmental and archaeological implications of a sediment core from Polje Čepić, Istria, Croatia. Geol Croat 59(2):109–124. https://doi.org/10.4154/GC.2006.08

Barrett CJ (1988) Fields of discourse: reconstituting a social archaeology. Critique Anthropol 7(3):5–16. https://doi.org/10.1177/0308275X8800700301

Boschian G, Montagnari-Kokelj E (2000) Prehistoric shepherds and caves in the Trieste Karst (Northeastern Italy). Geoarchaeology 15(4):331–371

Brochier EJ (1983) Bergeries et feux de bois néolithiques dans le Midi de la France. Quartär 33–34:181–194. https://doi.org/10.7485/QU33_08

Brochier EJ (1991) Géoarchéologie du monde agropastoral. In: Guilaine J (ed) Pour une archéologie agrarie. A. Colin, Paris, pp 303–322

Budja M, Ogrinc N, Gašparič AZ, Potočnik D, Žigon D, Mlekuž D (2013) Transition to farming: transition to milk culture: a case study from Mala Triglavca, Slovenia. Document Praehist 40:97–117. https://doi.org/10.4312/dp.40.9
Chadwick MA (2007) Trackways, hooves and memory days: human and animal memories and movements around Iron Age and Romano-British rural landscapes. In: Cummings V, Johnston R (eds) Prehistoric journeys. Oxbow, Oxford, pp 131–152
Chadwick MA (2008a) Fields for discourse? Toward more self-critical, theoretical and interpretive approaches to the archaeology of field systems and land allotment. In: Chadwick AM (ed) Recent approaches to the archaeology of land allotment. Archaeopress, Oxford, pp 205–237
Chadwick MA (2008b) Introduction. In: Chadwick AM (ed) Recent approaches to the archaeology of land allotment. Archaeopress, Oxford, pp 1–24
Chandler CC, Cheney P, Thomas P, Trabaud L, Williams D (1983) Fire in forestry, vol. 1: forest fire behavior and effects. Wiley, New York
Culiberg M (1991) Late glacial vegetation in Slovenia. SAZU, Ljubljana
Fabec T (2018) Arheologija vrtač na Krasu [Archaeology of dolines on Kras Plateau]. ZVKDS, Ljubljana
Ferrari A, Forenbaher S, Micheli R, Montagnari Kokelj E, Pessina A, Velušček A, Visentini P (2018) Neolithic and Eneolithic of Caput Adriae. In: Borgna E, Guida Càssola P, Corazza S (eds) Preistoria e protostoria del Caput Adriae. Istituto italiano di preistoria e protostoria, Firenze, pp 61–74
Fowler JP (1983) The farming of prehistoric Britain. Cambridge University Press, Cambridge
Gallizia Vuerich L, Laetizia S, Princivalle F (1994) Studio mineralogico e palinologico su alcuni sedimenti della Grotta dell'Edera (Carso triestino): un tentativo di ricostruzione paleoclimatica. Il Quaternario 7(2):569–576
Gams I (1974) Kras. Slovenska matica, Ljubljana
Gams I (1992) Sistemi prilagoditve primorskega dinarskega krasa na kmetijsko rabo tal [Systems of adapting the littoral Dinaric Karst to agrarian use]. Geografski zbornik 31:5–106
Gams I, Gabrovec M (1999) Land use and human impact in the Dinaric karst. Int J Speleol 28(1):55–70. https://doi.org/10.5038/1827-806X.28.1.4
Gestrin F (1991) Slovenske dežele in zgodnji kapitalizem [Slovenian Lands and the Early Capitalism]. Slovenska matica, Ljubljana
Johnson R (2001) Breaking new ground: land tenure and fieldstone clearance during the Bronze Age. In: Brück J (ed) Bronze Age landscapes: tradition and transformations. Oxbow, Oxford, pp 99–109
Johnson N, Rose P (1994) Bodmin moor: an archaeological survey, volume 1: the human landscape to c.1800. English Heritage, Royal Commission on the Historical Monuments of England and Cornwall Archaeological Unit, London
Johnston R (2005) Social archaeology of garden plots in the Bronze Age of northern and western Britain. World Archaeol 37(2):211–223
Kalc A (2008) Tržaško prebivalstvo v 18. stoletju: priseljevanje kot gibalo demografske rasti in družbenih sprememb [Population of Trieste in the 18th Century: Immigration as a Factor of Demographic Growth and Social Change]. Založba Annales, Koper
Legnani F (1967) La caverna dei Ciclami nel Carso Triestino. Studio paleoclimatologico del riempimento. Atti e Memorie della Commissione Grotte "Eugenio Boegan" 7:75–94
Marchesetti C (1903) I castellieri preistorici di Trieste e della Regione Giulia. Museo civico di Storia naturale, Trieste
Meinig WD (1979) The beholding eye: ten versions of the same scene. In: Meining DW (ed) The interpretation of ordinary landscapes: geographical essays. Oxford University Press, New York, pp 33–47
Miko S, Mesić S, Forenbaher S, Belak M (2005) Archeometric analysis of Bronze Age potsherds from Vrana Lake (Cres Island): a study of provenance and dating of slope processes. Abstracts book of 3rd Croatian geological congress. Croatian Geological Society, Opatija, pp 99–100

Miracle TP, Pugsley L (2006) Vertebrate faunal remains from Pupićina cave. In: Miracle PT, Forenbaher S (eds) Prehistoric herders of northern Istria: the archaeology of Pupićina cave, vol 1. Arheološki muzej Istre, Pula, pp 259–399

Mlekuž D (2005a) The ethnography of the Cyclops: Neolithic pastoralists in the eastern Adriatic. Document Praehist 32:16–51. https://doi.org/10.4312/dp.32.2

Mlekuž D (2005b) Trajektorije sprememb mezolitskih in neolitskih krajin dinarske Slovenije. [Trajectories of Mesolithic and Neolithic Landscapes of North Dinarides]. PhD Thesis. University of Ljubljana, Ljubljana

Mlekuž D (2006) Milk or meat? Neolithic economies of Caput Adriae. In: Pessina A, Visentini P (eds) Preistoria dell'Italia Settentrionale. Comune di Udine and Museo Friulano di storia naturale, Udine, pp 453–458

Mlekuž D (2007) Who were the cyclopes? Odyssey and Neolithic, Eneolithic and Bronze Age pastoralists from the east Adriatic coast and Dinarides. In: Blečić M (ed) Scripta praehistorica in honorem Biba Teržan. Narodni muzej Slovenije, Ljubljana, pp 69–89

Mlekuž D (2012) Messy landscapes: LiDAR and practices of landscaping. In: Opitz RS, Cowley DC (eds) Interpreting archaeological topography: 3D data, visualisation and observation. Oxbow, Oxford, pp 102–116. https://doi.org/10.2307/j.ctvh1dqdz.12

Mlekuž D (2013) Skin deep: LiDAR and good practice of landscape archaeology. In: Corsi C, Slapšak B, Vermeulen F (eds) Good practice in archaeological diagnostics. Springer, Cham, pp 113–129. https://doi.org/10.1007/978-3-319-01784-6_6

Mlekuž D (2015) Oblike prazgodovinske poljske razdelitve na Krasu [Forms of prehistoric fields division in the Kras]. Kronika 63(3):675–690

Mlekuž D (2018) Sledovi rimske zemljiške razdelitve na Krasu [Traces of the Roman field division in the Kras]. In memoriam Iva Mikl Curk: Nova odkritja med Alpami in Črnim morjem, rezultati raziskav rimskodobnih najdiščv obdobju med leti 2005 in 2015. ZVKDS, Ljubljana, pp 65–76

Mlekuž D, Pergar M (2008) Mala Triglavca near Divača. In: Teržan B, Črešnar M (eds) Absolute dating of the Bronze and Iron Ages in Slovenia. Narodni muzej Slovenije, Ljubljana, pp 581–586

Montagnari Kokelj E, Bernardini F, Leghissa E (2020) Eneolitico e Carso triestino: dati e problemi aperti. In: Muscio G, Visentini P (eds) Antichi abitatori delle Grotte in Friuli. Civici Musei, Museo Archeologico e Museo Friulano di Storia Naturale, Udine, pp 211–224

Moritsch A (1969) Das nahe triester Hinterland: zur wirtschaftlichen und sozialen Entwicklung vom Beginn des 19. Jahrhunderts bis zur Gegenwart. H. Böhlaus Nachf., Graz

Mušič B (1997) Magnetic susceptibility measurements in dolinas. Annales 7(10):37–42

Mušič B, Dimic F (1994) Magnetna susceptibilnost kot kvantitativni kriterij za razvrščanje arheoloških materialov [Magnetic susceptibility as a quantitative criterion for the classification of archaeological materials]. Arheo 16:37–44

Nicod J (1987) Amenagements agraires dans de petites depressions karstiques en Provence et dans les Causes, et dans quelques regions de comparaison en Italie et Yougoslavie. In: Kunaver J (ed) Karst and man, proceedings of the international symposium on human influence in Karst. Filozofska fakulteta, Ljubljana, pp 97–110

Nisbet R (2000) Nota preliminare sull'antracologia dei depositi olocenici della Grotta dell'Edera, Carso Triestino (scavi 1990–1999). In: Biagi P (ed) Studi sul Paleolitico, Mesolitico e Neolitico del bacino dell'Adriatico in ricordo di Antonio M. Radmilli. Società per la Preistoria e Protostoria della Regione Friuli-Venezia Giulia, Trieste, pp 161–170

Oldfield F, Ascoli A, Accorsi CA, Mercuri AM, Juggins S, Langone L, Rolph T, Trincardi F, Wolff GA, Gibbs Z, Vigliotti L, Frignani M, van der Post K, Branch N (2003) High resolution late Holocene palaeo environmental record from the central Adriatic Sea. Quat Sci Rev 22(2–4):319–342. https://doi.org/10.1016/S0277-3791(02)00088-4

Ollesch G, Vacca A (1999) Turnover dei fosfati e dei processi di degradazione in un ecosistema a pascolo in Sardegna. Bollettino Della Società Italiana Della Scienza Del Suolo 48:329–348

Opitz R (2012) An overview of airborne and terrestrial laser scanning in archaeology. In: Opitz RS, Cowley DC (eds) Interpreting archaeological topography: 3D data, visualisation and observation. Oxbow, Oxford, pp 13–31. https://doi.org/10.2307/j.ctvh1dqdz.7

Panjek A (2006) Človek, zemlja, kamen in burja: zgodovina kulturne krajine krasa [Man, earth, stone and bora: history of the cultural landscape of Kras]. Založba Univerze na Primorskem, Koper

Panjek A (2015) Kulturna krajina in okolje Krasa: o rabi naravnih virov v novem veku [The cultural landscape and environment of kras: on the uses of natural resources in the modern period]. Založba Univerze na Primorskem, Koper

Panjek A (2017) The integrated peasant economy as a concept in progress. In: Panjek A, Larsson J, Mocarelli L (eds) Integrated peasant economy in a comparative erspective: Alps, Scandinavia and beyond. University of Primorska Press, Koper, pp 11–49

Peresani M, Tozzi C (2018) Il Paleolitico e il Mesolitico del Friuli Venezia Giulia: quarant'anni di studi e ricerche. In: Borgna E, Guida Càssola P, Corazza S (eds) Preistoria e protostoria del Caput Adriae. Istituto italiano di preistoria e protostoria, Firenze, pp 45–59

Pryor F (1996) Sheep, stockyards and field systems: Bronze Age livestock populations in the Fenlands of Eastern England. Antiquity 70(268):313–324. https://doi.org/10.1017/S0003598X00083290

Radinja D (1987) Modern agricultural land improvement in Slovene Dinaric Karst. In: Kunaver J (ed) Karst and man, proceedings of the international symposium on human influence in Karst. Filozofska fakulteta, Ljubljana, pp 123–135

Remec M (2024) Afforestation with non-native *Pinus nigra* in the karstic areas of southwestern Slovenia since the mid-nineteenth century: environmental implications. In: Fuerst-Bjeliš B, Mrgić J, Petrić H, Zorn M, Zwitter Ž (eds) Environmental histories of the Dinaric Karst. Springer, Cham, pp 203–237

Schmidt R, Muller J, Dresher-Schneider R, Krisai R, Szeroczynska K, Barić A (2000) Changes in lake level and trophy at Lake Vrana, a large karstic lake on the Island of Cres (Croatia), with respect to palaeoclimate and anthropogenic impacts during the last approx. 16,000 years. J Limnol 59(2):113–130. https://doi.org/10.4081/jlimnol.2000.113

Šercelj A (1996) Začetki in razvoj gozdov v Sloveniji [The Beginning and Development of Forests in Slovenia). SAZU, Ljubljana

Seymour J (2003) The new complete book of self-sufficiency: the classic guide for realists and dreamers. Dorling Kindersley, London

Slapšak B (1995) Možnosti študija poselitve v arheologiji [Possibilities of the settlement studies in archaeology]. Arheo 17:1–90

Slapšak B (2003) The end of prehistoric communities in the Karst region. Arheološki vestnik 54:243–257

Teržan B, Turk P (2005) The Iron Age tower upon Ostri vrh. In: Bandelli G, Montagnari Kokelj E (eds) Carlo Marchesetti e i castellieri 1903–2003: Atti del Convegno Internazionale di Studi. Edizioni Quasar, Trieste, pp 339–353

Turk I, Zvezdana M, Prus T, Culiberg M, Šercelj A, Perko V, Dirjec J, Pavlin P (1993) Podmol pri Kastelcu: novo večplastno arheološko najdišče na Krasu, Slovenija [Podmol near Kastelec: a new multi-layered archaeological site on the Karst in Slovenia]. Arheološki vestnik 44:45–96

Vončina D (ed) (2018) Gradišča v zahodni in osrednji Sloveniji: zbornik s posvetovanja o gradiščih [Hillforts in western and central Slovenia]. Zavod Krasen Kras, Gorjansko

Weller J (1982) History of the farmstead. Faber and Faber, London

Yates TD (2007) Land power and prestige: Bronze Age field systems in Southern England. Oxbow, Oxford

Dimitrij Mlekuž Vrhovnik Associate professor of archaeology at the University of Ljubljana and a heritage consultant at the Institute for the Protection of the cultural heritage of Slovenia. His research interests are archaeological theory, landscape archaeology and the archaeology of the Neolithic.

Tomaž Fabec Archaeologist, conducting fieldwork in western Slovenia. His main research interests include the prehistory and protohistory of Kras, geoarchaeology and the archaeology of the First World War. Currently, he is working on a comprehensive study of cave sites from Kras.

Chapter 5
Landscape Change and Fire Risk in the Croatian Dinaric Karst: Looking Back and Moving Forward

Ivan Tekić, Borna Fuerst-Bjeliš, Anamarija Durbešić, Petra Radeljak Kaufmann, and Marin Cvitanović

Abstract In recent decades, there has been a strong trend of abandonment of traditional land uses in the Mediterranean region, leading to changes in vegetation cover and an increase in the risk of wildfires. One of the areas where anthropogenic changes in land cover have been especially prominent is the Croatian Dinaric Karst. Focusing on the last two centuries, this research contributes to the understanding of historical links between socio-economic, demographic and landscape changes that occurred in three main Dinaric socio-environmental zones (the hinterland, the littoral and the islands) and explores how they have shaped the development of fire-prone landscapes in the study region. Through a mixed-method approach, different data sources are utilised, including the analysis of existing literature, archival sources, recent comparative forestry, satellite and fieldwork data. Finally, the paper also focuses on possible future changes in the context of three development scenarios, aiming to explore how to manage and protect these landscapes at a time when both population and climate are quickly changing.

I. Tekić (✉)
Oikon Ltd. - Institute of Applied Ecology, Zagreb, Croatia
e-mail: itekic@oikon.hr

B. Fuerst-Bjeliš · P. Radeljak Kaufmann
Department of Geography, University of Zagreb Faculty of Science, Zagreb, Croatia
e-mail: bornafb@geog.pmf.unizg.hr

P. Radeljak Kaufmann
e-mail: radeljak@geog.pmf.hr

A. Durbešić
Department of Forestry, Sector for Planning, Analysis, Forestry and Informatics, Hrvatske šume d.o.o., Zagreb, Croatia
e-mail: anamarija.durbesic@hrsume.hr

M. Cvitanović
Department of Life and Environmental Sciences, Faculty of Science and Technology, Bournemouth University, Poole, UK
e-mail: mcvitanovic@bournemouth.ac.uk

B. Fuerst-Bjeliš et al. (eds.), *Environmental Histories of the Dinaric Karst*, Environmental History 17, https://doi.org/10.1007/978-3-031-56089-7_5

Keywords Fire risk · Landscape change · Dinaric Karst · Croatia · Mediterranean · Development scenarios · Environmental history

5.1 Introduction

In the last few years, the Mediterranean experienced extreme wildfires, particularly affecting South and Southeast Europe. The hardest hit areas include southern Italy, Greece, Turkey and the Croatian Dinaric Karst region in the Eastern Adriatic. The air temperature of 48.8 °C measured in Sicily on 11 August 2021 is considered the highest temperature ever measured in Europe. According to the latest IPCC report,[1] the Mediterranean is a hot spot of increasing temperatures, wildfires and other related hazards. Several factors have contributed to this increase in wildfires and associated hazards, including climate change, and major land use and land cover changes over recent decades.

One of the Mediterranean areas that has experienced significant anthropogenic changes in land cover is the Croatian Dinaric Karst. The region's sensitive environment, characterised by permeability of rock and soil and regular summer droughts, has been altered by various processes, including rapid littoralisation, agricultural abandonment, economic transition and armed conflict.

In many aspects, the Croatian Dinaric Karst, is a part of the wider Euro-Mediterranean region, both environmentally and culturally. Its position within the global atmospheric system, combined with its local topography and landforms, has contributed to the development of climate patterns that are found throughout the Mediterranean, whilst still retaining distinct local characteristics which have shaped a specific landscape structure and co-shaped a way of life. For example, regular summer droughts are a unifying feature of the Mediterranean climate, and throughout millennia, they have conditioned the development of specific agricultural practices that favour summer crops which can withstand the dry conditions, as well as arboriculture, which involves crops like the olive tree that have deep roots. Pastoralism has also played a significant role in the region, particularly in its distinctive form of transhumance, as the most convenient subsistence economy till the recent time.[2]

Throughout history, fire has played a distinctive role in maintaining this subsistence economy as well as the landscape structure of the region. However, it is important to note the difference between these controlled fires, which were purposefully used to maintain the economy and landscape structure, and uncontrolled fires that occur as a consequence of natural causes or human negligence.[3] In addition to its destructive power, there is evidence that fire also plays an important role in the ecological functioning of the region.[4]

---

[1] Tollefson (2021), pp. 171–172.

[2] Braudel (1992), pp. 53–55; McNeill (1992), p. 277; Simmons (2008), p. 73.

[3] McNeill (1992), p. 275; Pyne (2010), pp. 81–116.

[4] Touchan et al. (2012), pp. 45–50.

Recently, there has been a strong trend of abandonment of traditional land uses in the region, which, in line with the change of the vegetation structure and the documented climate changes, has resulted in an increased fire risk.[5] This is particularly expressed in the areas of increased fuel load in the wildland–urban interface (WUI) as a consequence of de-agrarianisation and reforestation, which are further threatened by recent rising temperatures and reduced precipitation. Research shows that wildfires in WUI have recently taken on alarming dimensions in the Mediterranean countries.[6]

Focusing on the last two centuries, this research contributes to the understanding of the historical links between socio-economic, demographic and landscape changes that occurred in three main Dinaric socio-environmental zones: the hinterland, the littoral and the islands. The study explores how these changes have contributed to the continued development of fire-prone landscapes in the region. A mixed-method approach is utilised, incorporating different data sources such as existing literature, historical maps and narrative sources, remotely sensed images and interviews with local firefighters and farmers. Since the present state is the consequence of our past actions in and towards the environment, and the future will depend on our current actions, the paper also focuses on possible future changes through three development scenarios, aiming to explore how to manage and protect these landscapes at the time when both population and climate are quickly changing.

5.2 Croatian Dinaric Karst: Environmental History Setting

The Dinaric Karst makes up almost half of the Croatian territory (46.6%). Environmentally, it mostly falls within the Mediterranean and sub-Mediterranean regions, characterised by corresponding climate types and patterns, vegetation and landscape structure (Fig. 5.1). Carbonate bedrock, mostly limestones, and karst landscapes dominate the region and condition the drainage. Coastal mountain ranges receive the highest precipitation, but most of the water disappears underground due to fractured carbonate complexes. The Mediterranean climate–ecology region encompasses the area of Mediterranean climate (Csa) in the narrow coastal belt of Dalmatia and Dalmatian islands, as well as transitional sub-Mediterranean climate (Cfa) in the hinterland and on higher mountain altitudes. The karst hydrology in Mediterranean Dinaric Croatia causes the majority of surface (precipitation) water to sink underground, leading to a significant water scarcity augmented with strong evaporation. This is particularly evident in hot summer months, when drought conditions are frequent, and during the winter, when the dry and cold wind *bora* contributes to the overall

[5] Pyne (2010), pp. 81–116.

[6] Lovreglio et al. (2014), p. 156.

dryness. Vegetation in the region is therefore mainly composed of xerophytes, which are acclimatised to summer droughts, high temperatures and high levels of solar radiation. It consists primarily of evergreen holm oak forest (*Quercus ilex*) together with coniferous plants such as different pines (*Pinus*) and cypress (*Cupressus)* in the hottest and driest belts of littoral and islands, and sub-Mediterranean deciduous oak forest (*Quercus pubescens*), hornbeam (*Carpinus orientalis*) and beech forest (*Fagus sylvatica*) mainly in the hinterland and at higher altitudes. However, human activities such as (over)grazing, farming and logging have transformed the original natural forest into the Mediterranean and sub-Mediterranean scrubland varieties over thousands of years. Maquis is a Mediterranean scrubland biome, typically consisting of a dense growth of evergreen shrubs in the hottest and driest Mediterranean belt; on islands and in the narrow littoral zone. Sub-Mediterranean scrubland consists mostly of deciduous continental plant species, and it is mainly in the hinterland and at higher altitudes.[7]

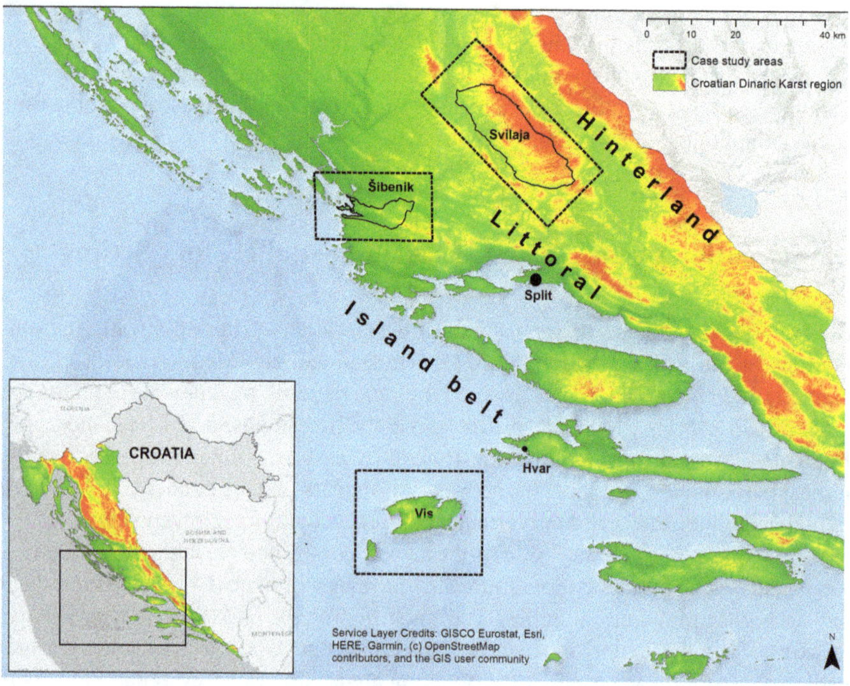

Fig. 5.1 Schematic map of the socio-environmental zones of the Croatian Dinaric Karst and the location of the case study areas. *Source* Authors' own elaboration, data source: The area and approximate northern border of the Dinaric Karst in Croatia according to Roglić[8] and Gams[9]

[7] Fuerst-Bjeliš and Glamuzina (2021), pp. 13–18.

[8] Roglić (1965), pp. 79–87; Roglić (1974), pp. 1–69.

[9] Gams (1974), p. 359; Gams (2004), p. 27.

For centuries, the population of the Dinaric Karst region has relied heavily on pastoralism as a convenient subsistence economy and a way of life,[10] adapting to the region's climate constraints and karst environment, which are characterised by scarce water, soil, and arable land.[11] Of the other economic activities, agriculture, logging and fuelwood gathering, together with pastoralism, had the greatest impact on the land.[12] The basic structure of the Mediterranean rural landscape has been defined by French geographers,[13] using a Roman trilogy of general land use—the *ager*, which consisted of both seeds (arable farming) and trees (arboriculture); outside these arable areas, which are highly localised in the Dinaric Karst, is the *saltus*, i.e. the scrubland used for grazing; and the forest—*silva*, which was fragile as it was constantly threatened by shepherds or loggers.

There is ample evidence of forest clearing and burning throughout history. Although anthropogenic pressure in the Dinaric Karst region can be traced back to millennia, most authors believe that the process of deforestation, primarily associated with (over)grazing but also with other land uses, began intensively in the fourteenth century.[14] Although there have been examples of sustainable management of resources,[15] many cultivation techniques were untenable due to political insecurity and frequent wars which were particularly evident in the Croatian Dinaric Karst, a combat zone and frontier for centuries.[16] This led to ignorance of laws and agreements, ultimately resulting in rapid depletion of forest or overgrazing of pastures. The pastoral economy in the Mediterranean has always tended towards overgrazing, which haunted the landscapes for centuries.[17]

Fire played an important role in maintaining *saltus* and pastures by preventing them from reverting to forests and promoting grassland regeneration. Post-fire plants have a much higher mineral and protein content (as much as 300–700%), and these nutritional properties, together with the ash salts, were important for the quality of feed for grazing animals.[18] This was a common practice of the cattle breeders, which was also known in other European regions,[19] but the costs of pastoralism in the fragile Mediterranean mountain ecosystem were far higher than similar practices in northern and southern Europe.[20]

[10] Fuerst-Bjeliš (2020), p. 78.

[11] Fuerst-Bjeliš (2020), pp. 82–86.

[12] McNeill (1992), p. 272.

[13] Claval (2007), pp. 23–24.

[14] McNeill (1992), pp. 272–282; Simmons (1998), pp. 335–361; Braudel (1992), pp. 3–32; 53–64; Zorn et al. (2015), pp. 563–564; Touchan et al. (2012), pp. 45–50.

[15] Panjek (2018), p. 68.

[16] Fuerst-Bjeliš (2020), pp. 82–86; Braudel (1992), p. 30; McNeill (1992), p. 279.

[17] McNeill (1992), pp. 274, 280–281.

[18] Simmons (2008), p. 36.

[19] Simmons (1998), pp. 335–361.

[20] McNeill (1992), p. 281.

According to cadastral[21] and narrative sources[22] from the eighteenth and nineteenth centuries, it appears that overgrazing in this fragile, sensitive Mediterranean karst environment had negative impact, ultimately leading to deforestation and unsustainable grassland use. As a result, the landscape was perceived as barren, dry and rocky, with little vegetation. In contrast, more humid hinterland mountains were perceived as green islands in the midst of a rocky desert.[23]

The special distinctiveness of Croatian Dinaric Karst lies in its longitudinal zonation (NW–SE direction). This zonation is reflected not only in the environmental factors, including geomorphology, climate and vegetation, but also in the cultural and ecological contexts. This interplay between the environmental zonation and human influences co-affects the way of life and subsistence economy. The Croatian Dinaric Karst features three distinct zones, ranging from the hottest and driest Euro-Mediterranean belt to the sub-Mediterranean. These zones include the island belt, the littoral and the hinterland (Fig. 5.1).

The Eastern Adriatic littoral has maintained a degree of continuity since Antiquity, with urban and economic networks developing with Greek poleis in the central Adriatic, and later with Roman and mediaeval Croatian cities along the entire coast. This long-term characteristic of the Littoral has persisted despite significant changes over time. Although the level of connection between the littoral and the hinterland has varied over history, it has remained relatively strong for the most part. However, the configuration and longitudinal zonation of Croatian Dinaric Karst—with its high mountain barrier to the hinterland—have contributed to the hinterland remaining peripheral in relation to the main littoral development fluxes. As a result, the hinterland has primarily developed as remote, peripheral rural areas.[24]

In the second half of the twentieth century, the process of intense littoralisation brought significant economic and demographic growth to the littoral zone, mostly due to the tourism and related industries. The demographic progress in the littoral zone is directly dependent on population outflow from the hinterland and islands—two other rural marginal zones—which served as the workforce reservoir for the expanding economic activities in the littoral. Such a massive population outflow resulted in the dissolution of a traditional lifestyle that was centred around pastoralism and localised tillage zones in the karst. As a consequence, there was widespread land abandonment and considerable environmental change, primarily expressed through vegetation succession. These landscape changes directly influence the fire risk. Our research, as well as some previous findings for other Mediterranean areas, shows that the wildfire risk is closely related to land cover,[25] whilst it is also linked to some

[21] Fuerst-Bjeliš (2003), pp. 47–62.

[22] De Concina (1809), p. 99; Matutinović (2009), p. 292; Petter (1834), p. 250; Tommaseo (1847), p. 152; de Visiani (1847), p. 268; Yriarte (1883), p. 575; Modrich (1892), p. 507.

[23] Fuerst-Bjeliš et al. (2011), pp. 117–130; Fuerst-Bjeliš and Kale (2018), pp. 553–556.

[24] Fuerst-Bjeliš (2020), p. 76.

[25] Bajocco and Ricotta (2008), p. 241; Moreira et al. (2009), p. 563; Fuerst-Bjeliš et al. (2016), p. 161.

societal factors, such as the loss of traditional knowledge of fire management due to depopulation.[26]

5.3 Research Overview

The characteristics of fire regimes in the Mediterranean have been the topic of considerable research in recent decades.[27] Many of these studies point out the complexity of the issue of wildfires, with much of the debate centring on the relative contribution of climate and fuel (i.e. natural vegetation cover available for combustion). This issue has been studied through a range of case studies across the Mediterranean, attracting strong arguments from both sides.[28] Hill[29] argued that the 'syndrome' of rural exodus and land abandonment has been a major contributing factor to the increased fire hazards in all Mediterranean countries, with the exception of coastal regions where tourism and urbanisation dominate. Pausas and Keeley[30] have also concluded that the reduction of agriculture, grazing and controlled fires due to the depopulation of rural areas has led to fuel build-up. This, coupled with increased ignitions at the wildland–urban interface, has caused an increase in large high-intensity fires. Other studies also emphasise that land use and land management changes are the Mediterranean fire regime.[31]

On the other hand, it has been established that some of the annual variability in burned area can be explained by climatic parameters,[32] whilst others stress the interplay of both climatic and non-climatic factors. Turco et al.[33] conducted a quantitative assessment of climate, environmental and socio-economic changes since the 1970s and concluded that positive trends in fire frequency can be explained by both recent socio-economic transformations and climate warming in recent decades. In a comprehensive study that included data on the 130-year fire history in the Spanish province of Valencia,[34] fire occurrence data were correlated with rural population censuses and climate data. The results have indicated that a major shift in the fire regime happened in Spain in the 1970s due to the accumulation of fuel resulting from decades of land abandonment. However, data showed that after the massive fires in the 1970s, there was a stronger correlation between climate parameters and

---

[26] Pyne (2010), pp. 169–200; Tedim et al. (2015), pp. 79–80.

[27] Slocum et al. (2010), pp. 828–840.

[28] Falk et al. (2007), pp. 809–823; Fernandes et al. (2012), pp. 380–384; Koutsias et al. (2012), pp. 41–53; McWethy et al. (2013), pp. 900–912; Turco et al. (2016), p. e0150663.

[29] Hill et al. (2008), pp. 146–157.

[30] Pausas and Keeley (2014), pp. 1109–1120.

[31] Moreira and Russo (2007), pp. 1461–1476; Moreira et al. (2011), pp. 2389–2402; Dubinin et al. (2011), pp. 547–562; Curt et al. (2013), pp. 150–161.

[32] Pausas (2004), pp. 337–350; Pereira et al. (2005), pp. 215–226.

[33] Turco et al. (2016), p. e0150663.

[34] Pausas and Fernández-Muñoz (2012), pp. 215–226.

fire occurrence than with socio-economic changes. In another study,[35] the fire regime data for the agriculturally developed Rif region in north Morocco were compared with the Valencian one prior to the 1970s shift, when the traditional rural economy collapsed, and the fire regime became fuel-limited. The findings indicated that the current fire regime in the Rif matches the pre-1970s fire regime observed in Valencia.

In addition to the study of fire regimes, different studies have also analysed the susceptibility of different land cover types to fires. For instance, research[36] concluded that in Mediterranean Europe, scrublands and conifer stands are more prone to large fires, whilst agricultural areas are less susceptible. This has been confirmed by Barros and Pereira[37] who showed that fire selectivity is higher for scrublands, pine stands and eucalyptus plantations compared to agricultural crops and agroforestry lands. Similarly, Moreira et al.[38] found that scrublands were the only land cover type that burned much more than expected. Finally, the preference of fire for a specific land cover type changes with the size of the fire, with the effect diminishing as the size of the fire increases.[39]

Whilst recent research on fire regimes focuses on the period since the 1970s, when Earth observation satellites became available for quick and efficient tracking of global and regional coverage of fire occurrences,[40] a longer historical perspective is often needed to fully understand changes in these regimes. In Croatia, research on this topic is still relatively underdeveloped, and much of it has been based on analysing the characteristics and spatial distribution of contemporary fires. For instance, Pavlek et al.[41] analysed environmental and socio-economic factors that influence fire occurrence in Mediterranean Croatia in 2013 and concluded that large fires correlated with scrubland and other types of degraded vegetation which highlights the importance of socio-economic and demographic processes such as depopulation and de-ruralisation for the fire regime. A similar conclusion was reached by Jajtić et al.[42] when comparing the fire seasons of 2011 and 2013. Mamut[43] confirmed the importance of vegetation cover for fire risk in the period from 1998 to 2008, but also stressed the importance of the Mediterranean climate and particularly dominant winds.

The long-term historical perspective of the relationship between environmental changes and fire risk in the Croatian Dinaric Karst is mainly studied by Cvitanović, Durbešić, Fuerst-Bjeliš and Tekić[44] for the hinterland and littoral regions of the Eastern Adriatic. Their findings indicate that changes in subsistence farming over the last two centuries have contributed to an increase in fire risk. Specifically, land

[35] Chergui et al. (2018), pp. 619–628.

[36] Nunes et al. (2005), pp. 661–673; Silva et al. (2009), pp. 597–608.

[37] Barros and Pereira (2014), p. e84760.

[38] Moreira et al. (2009), p. 563.

[39] Barros and Pereira (2014), p. e84760.

[40] Chuvieco et al. (2019), pp. 45–64.

[41] Pavlek et al. (2017), pp. 22–35.

[42] Jajtić et al. (2019), pp. 323–341.

[43] Mamut (2011), pp. 37–49.

[44] Fuerst-Bjeliš et al. (2016), pp. 161–172; Tekić et al. (2015), pp. 207–221.

abandonment has led to vegetation succession, resulting in an increase in the most fire-prone vegetation cover types such as sub-Mediterranean scrub and evergreen forest.

5.4 Scope and Methodology

To investigate the interdependence between fire risk in the Croatian Dinaric Karst area and the landscape changes resulting from socio-economic processes and climate change, three distinct case studies have been analysed through two comparative periods—the nineteenth century and the present. The case studies were chosen to represent three zones of Croatian Dinaric Karst, each marked by different environmental conditions and socio-economic development through history. The littoral zone, situated along the Eastern Adriatic littoral development axis, encompasses an area in close proximity to the city of Šibenik. This zone serves as a prime example of a semi-urbanised coastal landscape that has transitioned from an economy based on wine and olive oil production to one focused on tourism. It also maintains strong connections with the nearby urban centre. The hinterland zone, centred around Svilaja mountain, is characterised by a rural, agro-pastoral landscape. Lastly, the island socio-environmental zone is represented by Vis island, one of the more remote and sparsely populated islands in the region. Due to its isolation, the island has been able to retain its traditional islander lifestyle for a longer period (Fig. 5.1). Each case study was assessed through analysis and comparison of historical records from the nineteenth century and contemporary data on land use, land cover and climate, which was followed by the calculation of the fire risk for both periods.

The selection of parameters for fire risk assessment is defined by the Ministry of the Internal Affairs of Croatia and is provided by the official Regulation on the Forest Fire Protection (2014)[45] and Fire Risk assessment measures (2003).[46] Each parameter is evaluated and assigned a score based on its potential fire risk. Factors such as land cover and land use data, climate parameters and anthropogenic factors are considered variable and can change over time. Other parameters such as lithological characteristics, soil, relief and exposure are assumed to be constant in this analysis. Based on these assessment documents, the riskiest parameters include coniferous forests and degraded forests due to their high resin content and dryness; forests near arable land where weeds are frequently burned; and recreational areas such as wildland–urban interface (WUI). Additionally, areas with high mean annual temperature, low mean annual precipitation and low humidity also pose a higher fire risk, as do rock layers of sands, limestone or dolomite; southern exposure; low elevation; and low forest structure.

This research made use of detailed nineteenth-century cadastral plans and accompanying documentation gathered during the Franciscan cadastral survey of Dalmatia

[45] "Pravilnik o zaštiti šuma" [Regulation on the Forest Fire Protection], NN 33/2014.

[46] "Mjerila za procjenu opasnosti" [Fire Risk assessment measures], NN26/2003.

between 1823 and 1838, which provided comprehensive land use data for the area. These data were organised according to administrative sections called communes, which corresponded to the territory of individual settlements. To conduct a thorough analysis of the land use categories, the cadastral plans were digitised. This was accompanied by an analysis of narrative historical sources that provided contextual information to complement the data. For contemporary land use and land cover, a combination of sources was used, including CLCCro (Corine Land Cover) Datasets, management programmes for management units provided by Croatian Forests Ltd[47] and interpretation of landscape through satellite imagery and fieldwork assessment.

To enable a valid and clear comparison of land use and land cover between the two periods, adjustment and unification (merger) of the landscape categories were necessary due to the different sources used for evaluation. The sources varied in their type, methodology and origin (cadastre, CLCCro, forestry documentation, fieldwork). To achieve this, the study defined fewer main landscape categories (instead of the more detailed categories used in the nineteenth-century cadastre) that are relevant for fire risk assessment. These categories include forest (evergreen, deciduous), scrubland (maquis and sub-Mediterranean), pastures, arable land (including olive groves) and vineyards. In the nineteenth-century cadastral records, scrubland/maquis was not present as a land use category but was included in the forest category '*bosco di stanghe*' or forest with scrubland, whilst the categories '*young forest*', '*forest with mid-sized trees*' and '*forests with tall trunks*' were included in the forest category. In contrast, pastures were those land parcels where the main activity was animal grazing and were marked with prevalence of grassy vegetation, although woody vegetation in the form of shrubs may have been present. The category of 'pasture' for modern period taken from CLCCro Dataset also implies existence of shrubs and woody vegetation, but it had to be covering < 30% of the ground. All other categories, such as settlements, roads and waters, that are not relevant for fire risk and/or cover only small sections of study areas are included in the category 'other'.

Climate parameters were obtained from the Croatian Meteorological and Hydrological Service. Given the limited data available for the nineteenth century, we relied on the meteorological station of Hvar, which recorded temperature, precipitation and humidity from 1858 to 1888. Due to the territorial proximity of all case studies to this station (25 km from Vis island, and approximately 65 km from the city of Šibenik and Svilaja mountain areas: Fig. 5.1), it represents a valid source for average temperature, precipitation and humidity values, key climate parameters for assessing the fire risk. Due to the higher elevation of Svilaja mountain, the temperature gradient was taken into consideration and data corrected accordingly.

In the course of a previous study, three scenarios of spatial development were constructed for the region of Croatian Dinaric Karst (Dalmatia), covering the period until 2031. Thereby, results of multivariate statistical analysis (factor and cluster analysis) were used to isolate underlying factors of development characteristics and develop a regional typology. These results were then combined with two rounds of the Delphi method, where an interdisciplinary panel of experts examined possible paths

[47] Croatian Forests Ltd, "Programi gospodarenja" [Management programs for management units].

for the future development of Dalmatia and its subregions. The resulting scenarios showed processes leading from the contemporary situation towards a future state, meaning that they had explorative character.[48] In this study, the research team used these scenarios as a framework to explore how socio-economic, demographic and landscape changes could influence the wildfire incidence in the three main Dinaric socio-environmental zones: the hinterland, the littoral and the islands.

5.5 Results and Discussion

5.5.1 Landscape Change Since the Nineteenth Century

According to travel accounts from the eighteenth century, the Croatian Dinaric Karst landscape was predominantly treeless and dotted with cultivated valleys that were nestled between the barren hills. These descriptions have been corroborated by detailed statistical data on land use in the late eighteenth century.[49] The available historical data on the economy of the region confirm that agriculture and especially pastoralism were the primary means of livelihood. As agricultural density increased over time, this led to greater pressure on the landscape. Woodlands in particular were affected, with frequent cutting of firewood and poorly regulated[50] or even sustainable[51] agroforestry practices, but also by flocks of sheep and goats for grazing contributing to their decline.[52] Given the poor quality of karst pastures, woodlands played a crucial role in sustaining the region's sheep and goat populations. During harsh periods, the leaves, buds and small branches of trees offered a vital source of food. As a result, the history of pastoralism in the Croatian Dinaric Karst is closely linked to the loss of woodlands, as the need for grazing land and fodder drove the clearing and conversion of forested areas. The cadastral survey of the Austrian Empire, carried out in Dalmatia between 1823 and 1838, provides the most detailed and precise information on land use in the nineteenth century. Woodland areas were abundant on the islands of southern Dalmatia, where the shipbuilding industry mandated strict regulations for their preservation.[53] On the mainland, they covered the sparsely populated hilly and semi-mountainous areas in the hinterland. In contrast, the coastal zone was characterised by a pronounced lack of forests, reflecting the increased pressure on the landscape due to high population density and economic development. However, it should be noted that in Dalmatia, there was a lack of proper forests with older and developed trees. As an Austrian official from the

[48] Radeljak Kaufmann (2016), p. 56.

[49] Foretić (1963), p. 300.

[50] McNeill (1992), p. 275.

[51] Panjek (2018), p. 68.

[52] Wessely (1877), pp. 57–109.

[53] Jedlowski (1975), pp. 34–70.

late eighteenth century described it, most 'forest' was 'scrubland rather than forest, bushes rather than trees'.[54]

Cadastral records for three case studies clearly emphasise the prevalent landscape categories (Table 5.1; Fig. 5.2). Pastures were the most dominant land use category in the littoral and hinterland areas, covering 83% (Šibenik—littoral) and 67% (Svilaja—hinterland) of the area, respectively. This reflects the centuries-old lifestyle of the communities living in the area, which has developed as an adaptation to the climate and karst characteristics of the region, such as lack of water, lack of arable land and regular summer drought. The category of pastures includes several subcategories, such as pastures with bushes, pastures with coppiced trees and pastures with vineyards/olive/fruit trees. According to cadastral records, these areas were used for livestock grazing and unsupervised firewood collection, indicating the prevalence of an open-type landscape with some elements of maquis/sub-Mediterranean scrubland, but not in continuous cover. On Vis island, pastures covered 29% of the area, whilst maquis was the most widespread category with a share of almost 50%. This reflects the fact that Vis was the furthest populated island from the coast, and its relatively small population placed more emphasis on shipbuilding, fishing and viticulture. Forests as a class were significantly present only in the hinterland socio-environmental zone (case study Svilaja mountain) where they covered 17% of the area.

Table 5.1 Land use categories in the nineteenth century

Land use categories	Šibenik—littoral area		Vis island		Svilaja—hinterland	
	(ha)	(%)	(ha)	(%)	(ha)	(%)
Deciduous forest	16.1	0.3	N/A	N/A	4659.0	17.3
Sub-mediterranean scrubland	343.6	6.0	N/A	N/A	N/A	N/A
Maquis	N/A	N/A	4916.2	49.3	N/A	N/A
Pastures	4715.0	82.6	2861.5	28.7	18,034.5	67.0
Arable land	462.2	8.1	52.4	0.5	3894.5	14.5
Vineyards	165.4	2.9	1974.9	19.8	87.5	0.3
Other	6.7	0.1	176.1	1.8	224.3	0.9
Total	5709.0	100	9981.1	100	26,899.8	100

Source Authors' own analysis[55]

[54] Novak (1960), p. 486.

[55] HR DAST-152, Baljke (1820), (1846); Boraja (1825), (1844); Bračević (1830); (1831); Crivac (1831a), (1831b); Čavoglave (1829), (1830); Gradac (1836), (1837); Kadina Glavica (1829), (1830); Kljake (1829), (1830); Krapanj (1825), (1844); Milešine (1830), (1831); Miočić (1828), (1846); Mirlović Polje (1829), (1830); Donji Muć (1831), (1832); Gornji Muć (1831), (1832); Donje Ogorje (1831), (1841); Otavice (1829), (1830); Parčić (1829), (1830); Donje Postinje (1830), (1831); Gornje Postinje (1831a), (1831b); Pribude (1830), (1831); Sutina (1831), (1832); Umiljanović (1829), (1830); Vis (1876), (1835).

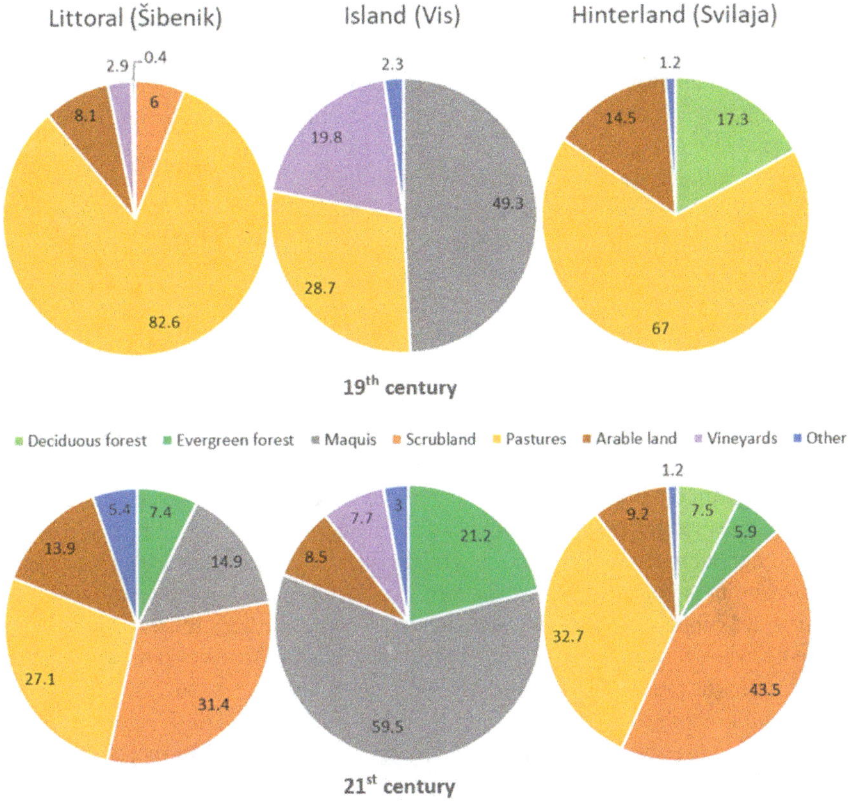

Fig. 5.2 Land use categories in the nineteenth and twenty-first centuries. *Source* Authors' own elaboration, data source: Tables 5.1 and 5.2

Agricultural activities have been declining in Croatia throughout the whole twentieth century, but the process intensified especially after the Second World War. Rural emigration was emphasised by the rapid industrialisation of the littoral and the development of tourism there. The decline of agricultural activities in Croatia began in the twentieth century and was further intensified after the Second World War. The planned development of industry in the coastal region, coupled with the growth of tourism, led to a shift in focus from the islands and hinterland to the coast.[56] This process of littoralisation resulted in rural emigration, a collapse of pastoralism and a decrease in the share of the agricultural population in Dalmatia from 72.1% in 1948 to just 3.4% in 1991.[57] The availability of employment outside agriculture was a contributing factor to this decline. Although some exploitation of woodlands by the local people continued until the 1990s, the scale of it was insignificant compared to the earlier periods. The Croatian War of Independence further contributed to rural

[56] Klempić-Bogadi, and Podgorelec (2009), p. 221.

[57] Glamuzina et al. (1991), p. 125; Nejašmić and Toskić (2013), p. 89.

abandonment, particularly in the hinterland. Extensive land abandonment led to rapid vegetation succession, with open areas quickly being overgrown by pine trees that spread from plantations through wind-dispersed seeds.

The impacts of economic and demographic changes on land use in Croatia are clearly visible in the land use categories of 2021 (Table 5.2; Fig. 5.2). In the hinterland, pastures have decreased considerably to only 32% of the total area, whilst sub-Mediterranean scrubland has increased to 43%. This reflects the process of rural abandonment and vegetation succession, which has primarily affected areas historically used for grazing, as the share of arable land has not decreased significantly. The emergence of evergreen forest with almost 6% of land cover in the hinterland is due to reforestation efforts, where black pine (*Pinus nigra*) was planted in order to prevent slope erosion and floods. Littoral areas have experienced dramatic decreases in the coverage of pastures, which have decreased from 82% to just 27% of the land area. This process of succession has resulted in an increase in maquis to almost 15% of the area and sub-Mediterranean scrubland to over 31%, whilst arable land increased by almost 6%. Similar to the hinterland area, the emergence of evergreen forest is attributed to extensive reforestation efforts that established pine monocultures throughout the twentieth century with the difference being that *Pinus halepensis* was the dominant pine species planted here. *Pinus halepensis* now covers almost 40% of the wider Šibenik area with dense stands, whilst a further 40% is covered with sparse formations. This is a yearly increase of 600% from 170 years ago when only a single stand was present on the island of Krapanj.[58] The decline of vineyards on the islands and in the littoral is a direct result of depopulation and de-agrarianisation;[59] however, the lower resolution of the twenty-first century land use maps, compared to detailed cadastral survey from the nineteenth century, has led to an exclusion of very small parcels with vineyards which makes viticulture underrepresented in the twenty-first-century statistics.

5.5.2 *Fire Risk Assessment and Change in the Course of Two Centuries: From the Nineteenth to Twenty-First Century*

Fire risk for all three case studies was evaluated according to the Croatian methodology established by the official *Fire risk assessment measures* (2003). The parameters of lithology, pedology, orography and exposure were considered constant and did not change between comparative periods, whilst climate, anthropogenic factors and especially vegetation cover represent changing variables that directly influence the changing fire risk vulnerability. According to the official *Fire risk assessment measures*, the relevant climate parameters for fire risk assessment are mean temperature, precipitation and humidity. To compare the values between different periods, we

---

[58] Tekić et al. (2015), pp. 214–219.

[59] See also Chap. 6, in this volume: Blaće et al. (2024).

Table 5.2 Land use categories in the twenty-first century

Land use categories	Šibenik—littoral			Vis island			Svilaja—hinterland		
	(ha)	(%)	(%) change	(ha)	(%)	(%) change	(ha)	(%)	(%) change
Deciduous forest	84.1	1.5	↑1.2	N/A	N/A	N/A	2005.7	7.5	↓9.8
Evergreen forest	422.9	7.4	↑7.4	1908.1	21.2	↑21.2	1589.5	5.9	↑5.9
Maquis	848.5	14.9	↑14.9	5361.7	59.5	↑10.2	N/A	N/A	N/A
Sub-mediterranean scrubland	1793.1	31.4	↑25.4	N/A	N/A	N/A	11,695.8	43.5	↑43.5
Pastures	1546.8	27.1	↓55.5	30.3	0.3	↓28.4	8798.9	32.7	↓34.3
Arable land	788.9	13.9	↑5.8	768.7	8.5	↑8.0	2475.4	9.2	↓5.3
Vineyards	N/A	N/A	↓2.9	698.5	7.7	↓12.1	N/A	N/A	↓0.3
Other	224.9	3.9	N/A	249.8	2.7	N/A	334.5	1.2	N/A
Total	5709.0	100		9017.1	100		26,899.8	100	

Source Authors' own analysis[60]

measured the change in climate parameters between 1858–1888 and 1991–2018[61] at the Hvar station. This station was selected as the only representative station for the study areas due to the availability of data for the nineteenth century (as described in the Scope and Methodology section). Between the two comparison periods, mean temperature increased by 0.9 °C, mean precipitation decreased by 17.5 mm, whilst mean air humidity remained constant at 66%. The observed increase in temperature and decrease in precipitation are generally consistent with climate change in the Mediterranean region, and both contribute to the increased fire risk.

To estimate the values of the nineteenth century for all case studies and respective socio-environmental zones, we applied the determined values of change to other meteorological stations in the study area. Specifically, we used the Knin meteorological station for the hinterland, the Šibenik station for the littoral area of Šibenik and the Komiža station on the island of Vis. Despite the observed increase in temperature and decrease in precipitation, which generally leads to higher fire risk, our analysis suggests that these changes in climate parameters are not the most critical factors for increased fire risk. This is because all corresponding values of climate parameters were already in the highest category of fire risk assessment (mean T above 12 °C and mean precipitation below 800 mm) in the nineteenth century, with the exception of slightly higher mean precipitation in the hinterland (station Knin and Svilaja mountain; mean precipitation above 1000 mm), which may reduce overall fire risk.

[60] CLCro (2021); Croatian Forests Ltd, "Programi gospodarenja" [Management programs for management units]; Google Earth; fieldwork assessment.

[61] Available data set provided by Croatian Meteorological and Hydrological Service.

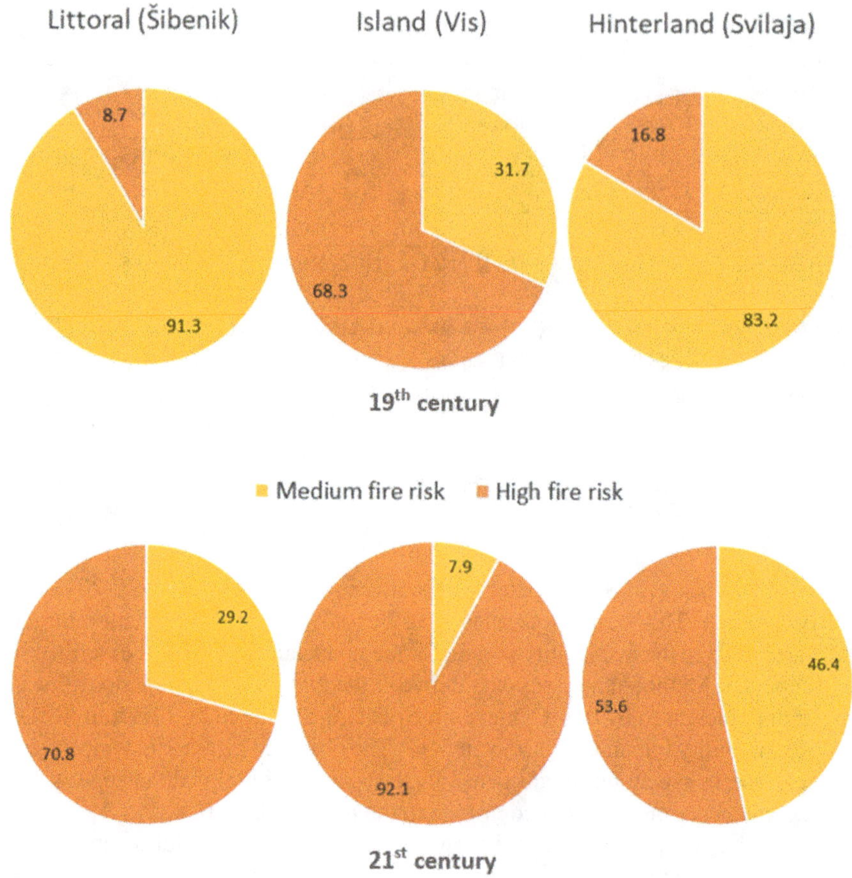

Fig. 5.3 Fire risk assessment in the nineteenth and twenty-first centuries. *Source* Authors' own elaboration, data source: Tables 5.3 and 5.4

In the nineteenth century, pastures dominated the mainland landscape, with a low share of either forest or degraded forms of forest such as scrubland or maquis. This open type of landscape was mostly characterised by medium fire risk in 91% of the area in the littoral case study and 83% of the area in the hinterland. However, the island of Vis had almost 70% of the area under high fire risk, mostly due to the larger share of maquis, which is proven to be the most fire-prone vegetation type,[62] and a lower share of open pastures in comparison with the mainland (Table 5.3; Fig. 5.3).

The increase of fire risk observed in the twentieth and twenty-first century in all case studies is closely linked to landscape changes that occurred in the twentieth century, specifically, the succession of vegetation types of maquis and garrigue on former pastures. The share of the category of high fire risk in the twenty-first century has increased to 53% in the hinterland, 70% in the coastal area and to 92% on the

[62] Fuerst-Bjeliš et al. (2016), p. 161.

Vis island. The data for the island of Vis, although indicative, should be viewed with caution as there are no official government data (Croatian Forests Ltd) on land use in private ownership for the twenty-first century. The share of land use in private ownership amounting to 63.83% of the total area was assessed through analysis and calculation from the Corine Land Cover Dataset (CLCCro 2021).The ratio of moderate to high risk in the hinterland is more favourable compared to the other case studies due to the beneficial effect of orography (higher elevation and related higher mean precipitation), as well as the lack of human impact due to heavy depopulation and the absence of agricultural burning in the second half of the twentieth century (Table 5.4; Fig. 5.3).

Table 5.3 Fire risk assessment in the nineteenth century

Fire risk	Šibenik—littoral area		Vis island		Svilaja—hinterland	
	(ha)	(%)	(ha)	(%)	(ha)	(%)
Medium	3446.4	91.3	3107.2	31.7	16,353.8	83.2
High	329.7	8.7	6697.9	68.3	3306.2	16.8
Extremely high	N/A	N/A	N/A	N/A	N/A	N/A
Total	3776.1	100	9981.1	100	19,660.1	100

Source Authors' own analysis[63]

Table 5.4 Fire risk assessment in the twenty-first century

Fire risk	Šibenik—littoral area			Vis island			Svilaja—hinterland		
	(ha)	(%)	(%) change	(ha)	(%)	(%) change	(ha)	(%)	(%) change
Medium	1102.2	29.2	↓62.1	217.4	7.9	↓23.8	9582.5	46.4	↓36.8
High	2673.9	70.8	↑62.1	2522.2	92.1	↑23.8	11,052.9	53.6	↑38.5
Extremely high	N/A	N/A	N/A	N/A	N/A	N/A	N/A	N/A	N/A
Total	3776.1	100		2739.6	92.1		20,635.3	100	

Source Authors' own analysis[64]

63 "Pravilnik o zaštiti šuma" [Regulation on the Forest Fire Protection], NN 33/2014; "Mjerila za procjenu opasnosti" [Fire Risk assessment measures], NN26/2003; Croatian Forests Ltd, "Programi gospodarenja" [Management programs for management units]; CMHC, "Meteorological station Hvar"; CMHC, "Meteorological station Knin"; CMHC, "Meteorological station Šibenik"; CMHC, "Meteorological station Komiža"; HR DAST-152, *idem.*

64 "Pravilnik o zaštiti šuma" [Regulation on the Forest Fire Protection], NN 33/2014; "Mjerila za procjenu opasnosti" [Fire Risk assessment measures], NN26/2003; Croatian Forests Ltd, "Programi gospodarenja" [Management programs for management units], CLCro (2021); Croatian Forests Ltd.

According to recent research,[65] wildfires in the Croatian Dinaric Karst show similar patterns to other Mediterranean areas, primarily in the north and west, where there has been a rapid increase in burned area in the second half of the twentieth century. However, the southern and eastern Mediterranean has experienced relatively lower levels of fire activity, possibly due to different patterns of land use (overgrazing and high pressure on forest in terms of subsistence clear-cutting).[66]

From 1958 to 1967, an average of 2407 ha of woodland area burned each year in Croatia, and this increased to 3685 ha in 1968 and to almost 5000 ha in the 1980s.[67] Most of these fires took place in the Dinaric Karst during the summer months. As depopulation and land abandonment continued to intensify in the 1990s, because of general economic and social changes that were further emphasised by the Croatian War of Independence (1991–1995), forest fires became even more problematic.

More detailed research focusing on types of vegetation and contributing factors of fire occurrence in littoral counties in Croatia shows that scrubland and grasslands in the Croatian Dinaric Karst (Dalmatian region) are characterised by both a large number of fires and a large total area burned, although the fires are smaller on average on a statistically significant level,[68] which is consistent with other Mediterranean areas.[69] On the other hand, forests have a lower number of fires overall, but the burned area tends to be 3–4 times larger than the average.

The connection between wildfire occurrence on one side and grassland and scrubland vegetation cover on the other is usually explained through agricultural abandonment, which resulted in an accumulation of surface fuel.[70] Additionally, grassland and scrubland can also be seen as highly intermingled vegetation types at different stages of the process of agricultural abandonment and secondary succession. Interviews with firefighters and farmers in the area have also confirmed the links between the occurrence of wildfires and the process of agricultural abandonment.[71]

Whilst the link between population changes and agricultural abandonment is evident, establishing a direct link between population change and fire occurrence has proven more challenging. Research has highlighted a connection between fire occurrence and a general decline in population, but only in rural areas with a significant share of the agricultural population.[72] The research on forest fires causes and motivations in the South and Southeast Europe, including Croatia, has shown that agricultural burnings are the most prevalent cause of forest fires; most of them stem

[65] Pavlek et al. (2017), p. 22; Jurjević et al. (2009), pp. 63–72.

[66] Viedma et al. (2017), p. 1.

[67] Vajda (1970), p. 101; Bertović (1987), p. 34.

[68] Fuerst-Bjeliš et al. (2016), p. 161; Pavlek et al. (2017), pp. 22–35; Jajtić et al. (2019), p. 323.

[69] Nunes et al. (2005), p. 661; Silva et al. (2009), p. 597; Barros and Pereira (2014), p. e84760; Moreira et al. (2009), p. 563.

[70] Turco et al. (2016), p. e0150663; Elia et al. (2013), p. 149.

[71] Jajtić et al. (2019), p. 323.

[72] Jajtić et al. (2019), p. 340.

from negligence.[73] It seems that traditional agricultural practices of burning agricultural waste and old grass are not a useful wildfire prevention method anymore, likely due to an overall increase in vegetation cover which represents easily flammable fuel.[74] The advanced age of farmers practising this management method together with the loss of traditional knowledge could play a role as well. Some research, although inconclusive and spatially variable, suggests that this might be also a consequence of climate change which changed the patterns of days suitable for adapted agricultural burning, either moving them to earlier spring and later autumn months, or reducing them overall.[75]

5.5.3 *Wildland–Urban Interface: Potential New Fire Risk Areas?*

Urbanisation, peri-urbanisation and littoralisation, coupled with tourism and recreation, have intensified human activities in environments that are increasingly prone to wildfires due to the abandonment of traditional farming practices. All this, together with the widespread afforestation with pine trees along the coast, contributes significantly to the increase in fire risk.

The littoral zone of the Croatian Dinaric Karst is a development axis, attracting economic activities and population mainly from the hinterland and the island zone. This has led to the creation of wildland–urban interface (WUI) zones that represent new fire risk areas due to the combination of land abandonment and the proliferation of scrubland and maquis vegetation. These conditions result in a concentration of fuel, along with increasing peri-urban and recreational activities, which increase the risk of fire. The expansion of WUI zones has been a global trend in recent decades and has a direct relationship with the risk of forest fires, as human activities increase the likelihood of ignition and the spread of wildfires.[76] The map of WUI areas in Europe suggests that the trend of urban growth in the Croatian Dinaric Karst is still only moderate in comparison with other Mediterranean regions. In addition, the correlation of large fires with the extent of WUI shows that large fires occur slightly less frequently near WUI areas here than away from WUI areas, but the difference is not statistically significant.[77] However, recent research in the Croatian Dinaric Karst has identified that areas in close vicinity to the largest littoral cities of Zadar, Šibenik, Split and Dubrovnik are indeed at a higher risk of fire occurrence,[78] pointing to the increasing importance of WUI.

---

[73] Tedim et al. (2022), p. 562.

[74] Romero-Calcerrada and Perry (2004), pp. 217–232.

[75] Clarke et al. (2019), pp. 148–157.

[76] Chas-Amil et al. (2013), p. 127; Tedim et al. (2015), pp. 81–83.

[77] Modugno et al. (2016), pp. 112–126.

[78] Pavlek et al. (2017), pp. 22–35; Jajtić et al. (2019), p. 323.

5.5.4 *Moving Forward: Wildfire Risk Within the Scope of Development Scenarios*

Without knowing and understanding the relationships between humans and the environment in the past, we are unable to understand the current state of the environment and its processes, including the risk of wildfires. On the other hand, if we know and understand the past and present of our actions and human–environment relationships, we can use this knowledge to predict and possibly (re)manage undesirable and unsustainable developments in the future.

Three development scenarios for Dalmatia, as a region of Croatian Dinaric Karst, and its subregions examine possible future developments from different perspectives: 'Monoculture of tourism' (MoT), 'Sustainable tourism' (ST) and 'Integral development' (ID).[79] In the MoT scenario, tourism exclusively dominates the economic and employment structure of Dalmatia, with high seasonal activity. To a limited degree, tourism development is tied with local agriculture, and to a higher degree with complementary manufacturing, services and the construction sector. It is also followed by infrastructural development, but levels of environmental protection, as well as innovation and technology, will be low. Tourism will not help in reversing depopulation and population ageing. The population will continue concentrating in the coastal zone, whereby hinterland and islands will be used as secondary, 'weekend' places of residence.

In the ST scenario, tourism supply is highly diversified, with strong effects on the development of agriculture, manufacturing, small-scale industries, trade and various services. Compared to the monoculture of tourism, the institutional framework will be stronger and levels of knowledge and innovation higher. A higher degree of employment and economic activities will still be connected with bigger cities, spreading in the cities' surroundings. Infrastructural development will, along with the increasing role of smaller- and middle-sized urban centres, influence more balanced and spatially dispersed development. Smaller towns on the coastline, as well as islands and the hinterland, will experience an increasingly seasonal way of living.

The ID scenario examines development based on reindustrialisation. Knowledge, innovation, information and communication technologies, and technological development in general, will bring about the increased significance of production sectors, similar to the level of importance held prior to the 1990s, before the political–economic change and the 1991–1995 Croatian War of Independence took place. In some parts of Dalmatia, tourism will remain the most important economic activity. Complementary development of agriculture and tourism will contribute to the overall development of islands and the hinterland. Although demographic ageing will remain a major issue throughout the region, new year-round job opportunities will influence the migration of highly educated workers. In parts of the hinterland and the more easily accessible islands, depopulation will decrease.[80]

---

[79] Radeljak Kaufmann (2016), p. 56; Radeljak Kaufmann (2017), p. 581.

[80] Radeljak Kaufmann (2016), p. 56; Radeljak Kaufmann (2017), p. 581.

As explained previously in the chapter, several main factors were recognised as possible influences in the wildfire incidence. Within the framework of each scenario, the level of risk could increase/decrease in the period until 2031 or remain unchanged. It is also expected that the risk level will differ amongst the three socio-environmental zones (Table 5.5).

In the period until 2031, the unfavourable demographic processes are not expected to change significantly, apart from the ID scenario. Given that all scenarios include infrastructural development, which should counteract the consequences of population concentration in the coastal zone, depopulation in other zones and the aged population structure overall, the wildfire risk remains the same from the demographic point of view. The exception is the ID scenario, where the more favourable population dynamics and structure, especially in better accessible areas, is expected to decrease the wildfire incidence in all three zones.

With regard to economic activities and land use, the wildfire incidence remains unchanged in the littoral zone for all scenarios. In the MoT scenario, it is expected to increase in the hinterland and on the islands due to the overall lack of advancements in sectors other than tourism and underutilisation of spatial resources. The ST and ID scenarios bring a decrease in the level of wildfire risk in the island and hinterland zones due to a more balanced economic development and more sustainable use of resources.

In the case of climate change, the wildfire incidence is expected to depend on the general development rationale and institutional framework. In the littoral, there is no significant change in the wildfire incidence across all scenarios. In the ST scenario, with a stronger institutional framework that counteracts the level of risk generally increased by climate change, the wildfire incidence remains the same for the other

Table 5.5 Change in the level of risk of wildfire within the framework of three development scenarios until 2031

Possible influences in the wildfire incidence	Monoculture of tourism scenario	Sustainable tourism scenario	Integral development scenario
Population dynamics and structure	I (o) L (o) H (o)	I (o) L (o) H (o)	I (−) L (−) H (−)
Economic activities	I (+) L (o) H (+)	I (−) L (o) H (−)	I (−) L (o) H (−)
Land use	I (+) L (o) H (+)	I (−) L (o) H (−)	I (−) L (o) H (−)
Climate change	I (+) L (o) H (+)	I (o) L (o) H (o)	I (−) L (o) H (−)

Note +/−/o refer to increased/decreased/unchanged risk of wildfire. Letters I, L, H refer to the three main zones: I (islands), L (littoral) and H (hinterland)
Source Authors' own analysis

two zones as well. In the MoT scenario, wildfire incidence could increase for islands and the hinterland because of the weaker institutional response. It is the opposite in the ID scenario, where a strong institutional framework supports the overall integral development, which should attribute to a lower wildfire incidence.

5.6 Conclusion

The differences in fire risk resulting from the prevalence of fire-prone vegetation between the case study areas are already evident in the nineteenth century. The island of Vis has the highest proportion of land exposed to high fire risk compared to the other two study areas in the coastal and inland areas, where fire risk is medium.

Throughout the twentieth century, a decrease in the total (agricultural) population and the decline in traditional agricultural activity have been observed not only in Croatia but also in other parts of the Mediterranean. These processes, coupled with littoralisation and abandonment of pastoralism, have been identified as contributing factors to the increase in forest fire risk and changes in fire regime and behaviour, with an increase in larger and more intense fires.[81] Our statistical analysis of land cover changes confirmed that the increased fire risk is mainly due to the homogenisation of land cover in terms of the prevalence of maquis and scrubland, the most fire-prone vegetation types, at the expense of abandoned pastures, which has led to an increase in highly flammable fuels.

As a result, the fire risk has increased dramatically in all three socio-environmental zones. High risk is absolutely prevalent in all zones, exceeding 92% on Vis island. However, the increase in high risk since the nineteenth century has been highest in the littoral zone, where it increased from 8.7% to over 70% of the study area.

The analysis has also shown that the increase in temperature and decrease in precipitation—which typically contribute to an elevated risk of fire—were not significant factors in the observed increase in fire risk in the Croatian Dinaric Karst. This is because the climate parameters were already at the highest category of fire risk assessment in the nineteenth century. However, it is important to note that the continuation of current socio-economic processes, combined with the already observed and predicted effects of climate change in the Mediterranean region (such as higher temperatures and longer droughts in summer), could contribute to an increase in the frequency of forest fires in the future. The wildland–urban interface has been identified as a potential new fire risk area due to the accumulation of fuel and the increase in peri-urban and recreational activities. However, in the development scenarios, the littoral zone is predicted to have a consistent (unchanged) wildfire incidence, mainly due to better infrastructure development and institutional support. Out of the three possible development scenarios, tourism monoculture proves to be the most likely to increase the risk of forest fires, whilst the integral development scenario and (slightly

[81] Azevedo et al. (2011), pp. 329–351; Pyne (2010), pp. 169–200.

less) the sustainable tourism scenario would lead to a reduction in fire risk in both the hinterland and littoral zones of the Dinaric Karst.

Acknowledgements This research was supported by the University of Zagreb Grant for research on the Influence of palaeo and recent climate changes on processes in marine and continental environments in 2020–2023 (No. 20285705; 20286205; 20286474; 20286585). We would like to express our gratitude to Prof. D. Perica and Prof. N. Jasprica as well as the anonymous reviewers for their valuable advice.

References

Published Sources

CLCCro (Corine Land Cover Croatia) database (2021) Ministry of Environment and Energy. © European Union, Copernicus Land Monitoring Service 2021 European Environment Agency (EEA). http://corine.haop.hr/map-page

Croatian Forests Ltd., Programi gospodarenja za gospodarske jedinice Borovača (2014–2023), Crivac (2020–2029), Jelinjak (2014–2023), Konjička Draga (2018–2027), Kozjak, Svilaja I (2015–2024), Drvenjak-Kamensko (2017–2026), Zelovo (2021–2030), Vis (2019–2028)

Croatian Meteorological and Hydrological Service database (CMHC). Meteorological station Hvar. Precipitation, Temperature and Humidity, time series 1858–1888 and 1991–2018. Croatian Meteorological and Hydrological Service

Croatian Meteorological and Hydrological Service database (CMHC). Meteorological station Knin. Precipitation, Temperature and Humidity, time series 1991–2018. Croatian Meteorological and Hydrological Service

Croatian Meteorological and Hydrological Service database (CMHC). Meteorological station Šibenik. Precipitation, Temperature and Humidity, time series 1991–2018. Croatian Meteorological and Hydrological Service

Croatian Meteorological and Hydrological Service database (CMHC). Meteorological station Komiža Precipitation, Temperature and Humidity, time series 1991–2018. Croatian Meteorological and Hydrological Service

Google Earth 7.3.4., *Vis, Grebaštica, Boraja, Svilaja*. Historical imagery year 2021. http://www.google.com/earth/index.html. Accessed Feb 2022

Mjerila za procjenu opasnosti od šumskog požara. Pravilnik o zaštiti šuma od požara: Prilog 1. (2003). Ministry of the Internal Affairs of Croatia. NN 26/2003. https://narodne-novine.nn.hr/clanci/sluzbeni/2003_02_26_382.html

Pravilnik o zaštiti šuma od požara (2014) Ministry of the Internal Affairs of Croatia. NN3/2014. https://narodne-novine.nn.hr/clanci/sluzbeni/2014_03_33_599.html

Archival Sources: State Archives Split (HR-DAST)

HR DAST-152, Baljke (1820), i.e. HR DAST-152 (Državni Arhiv u Splitu - Arhiv mapa za Istru i Dalmaciju), Litografato nell' i.r Justituto litografatico del Catastro, Balke, K.O. 8, Baljke, folder 8, 1820

HR DAST-152, Boraja (1825), i.e. HR DAST-152 (Državni Arhiv u Splitu - Arhiv mapa za Istru i Dalmaciju), Litografato nell' i.r Justituto litografatico del Catastro, Boraia, K.O. 52, Boraja, folder 52, 1825

HR DAST-152, Krapanj (1825), i.e. HR DAST-152 (Državni Arhiv u Splitu - Arhiv mapa za Istru i Dalmaciju), Litografato nell' i.r Justituto litografatico del Catastro, Crappano, K.O. 279, Krapanj, folder 279, 1825

HR DAST-152, Miočić (1828), i.e. HR DAST-152 (Državni Arhiv u Splitu - Arhiv mapa za Istru i Dalmaciju), Litografato nell' i.r Justituto litografatico del Catastro, Miocich, K.O. 355, Miočić, folder 355, 1828

HR DAST-152, Čavoglave (1829), i.e. HR DAST-152 (Državni Arhiv u Splitu - Arhiv mapa za Istru i Dalmaciju), Litografato nell' i.r Justituto litografatico del Catastro, Zavoglave, K.O. 81, Čavoglave, folder 81, 1829

HR DAST-152, Kadina Glavica (1829), i.e. HR DAST-152 (Državni Arhiv u Splitu - Arhiv mapa za Istru i Dalmaciju), Litografato nell' i.r Justituto litografatico del Catastro, Cadinaglavizza, K.O. 218, Kadina Glavica, folder 218, 1829

HR DAST-152, Kljake (1829), i.e. HR DAST-152 (Državni Arhiv u Splitu - Arhiv mapa za Istru i Dalmaciju), Litografato nell' i.r Justituto litografatico del Catastro, Kliake, K.O. 242, Kljake, folder 242, 1829

HR DAST-152, Mirlović Polje (1829), i.e. HR DAST-152 (Državni Arhiv u Splitu - Arhiv mapa za Istru i Dalmaciju), Litografato nell' i.r Justituto litografatico del Catastro, *Mirilovich in Campagana*, K.O. 353, Mirlović Polje, folder 353, 1829

HR DAST-152, Otavice (1829), i.e. HR DAST-152 (Državni Arhiv u Splitu - Arhiv mapa za Istru i Dalmaciju), Litografato nell' i.r Justituto litografatico del Catastro, Ottavizez, K.O. 421, Otavice, folder 421, 1829

HR DAST-152, Parčić (1829), i.e. HR DAST-152 (Državni Arhiv u Splitu - Arhiv mapa za Istru i Dalmaciju), Litografato nell' i.r Justituto litografatico del Catastro, Parcich, K.O. 431, Parčić, folder 431, 1829

HR DAST-152, Umiljanović (1829), i.e. HR DAST-152 (Državni Arhiv u Splitu - Arhiv mapa za Istru i Dalmaciju), Litografato nell' i.r Justituto litografatico del Catastro, Umglianovich, K.O. 675, Umiljanović, folder 675, 1829

HR DAST-152, Bračević (1830), i.e. HR DAST-152 (Državni Arhiv u Splitu - Arhiv mapa za Istru i Dalmaciju), Litografato nell' i.r Justituto litografatico del Catastro, Bracevich, K.O. 52, Bračević, folder 52, 1830

HR DAST-152, Čavoglave (1830), i.e. HR DAST-152 (Državni Arhiv u Splitu - Arhiv mapa za Istru i Dalmaciju), Protocollo delle particelle dei terreni Zavoglave, K.O. 81, Čavoglave, folder 81, 1830

HR DAST-152, Donje Postinje (1830), i.e. HR DAST-152 (Državni Arhiv u Splitu - Arhiv mapa za Istru i Dalmaciju), Litografato nell' i.r Justituto litografatico del Catastro, Postigne Inferior, K.O. 479, Donje Postinje, folder 479, 1830

HR DAST-152, Kadina Glavica (1830), i.e. HR DAST-152 (Državni Arhiv u Splitu - Arhiv mapa za Istru i Dalmaciju), Protocollo delle particelle dei terreni Cadinaglavizza, K.O. 218, Kadina Glavica, folder 218, 1830

HR DAST-152, Kljake (1830), i.e. HR DAST-152 (Državni Arhiv u Splitu - Arhiv mapa za Istru i Dalmaciju), Protocollo delle particelle dei terreni Kliake, K.O. 242, Kljake, folder 242, 1830

HR DAST-152, Milešine (1830), i.e. HR DAST-152 (Državni Arhiv u Splitu - Arhiv mapa za Istru i Dalmaciju), Litografato nell' i.r Justituto litografatico del Catastro, Milessine, K.O. 348, Milešine, folder 348, 1830

HR DAST-152, Mirlović Polje (1830), i.e. HR DAST-152 (Državni Arhiv u Splitu - Arhiv mapa za Istru i Dalmaciju), Protocollo delle particelle dei terreni Mirilovich in Campagana, K.O. 353, Mirlović Polje, folder 353, 1830

HR DAST-152, Otavice (1830), i.e. HR DAST-152 (Državni Arhiv u Splitu - Arhiv mapa za Istru i Dalmaciju), Protocollo delle particelle dei terreni Ottavizez, K.O. 421, Otavice, folder 421, 1830

HR DAST-152, Parčić (1830), i.e. HR DAST-152 (Državni Arhiv u Splitu - Arhiv mapa za Istru i Dalmaciju), Protocollo delle particelle dei terreni Parcich, K.O. 431, Parčić, folder 431, 1830
HR DAST-152, Pribude (1830), i.e. HR DAST-152 (Državni Arhiv u Splitu - Arhiv mapa za Istru i Dalmaciju), Litografato nell' i.r Justituto litografatico del Catastro, Pribude, K.O. 494, Pribude, folder 494, 1830
HR DAST-152, Umiljanovići (1830), i.e. HR DAST-152 (Državni Arhiv u Splitu - Arhiv mapa za Istru i Dalmaciju), Protocollo delle particelle dei terreni Umglianovich, K.O. 675, Umiljanovići, folder 675, 1830
HR DAST-152, Bračević (1831), i.e. HR DAST-152 (Državni Arhiv u Splitu - Arhiv mapa za Istru i Dalmaciju), Protocollo delle particelle dei terreni Bracevich, K.O. 52, Bračević, folder 52, 1831
HR DAST-152, Crivac (1831a), i.e. HR DAST-152 (Državni Arhiv u Splitu - Arhiv mapa za Istru i Dalmaciju), Litografato nell' i.r Justituto litografatico del Catastro, Zrivaz, K.O. 96, Crivac, folder 96, 1831
HR DAST-152, Crivac (1831b), i.e. HR DAST-152 (Državni Arhiv u Splitu - Arhiv mapa za Istru i Dalmaciju), Protocollo delle particelle dei terreni Crivaz, K.O. 96, Crivac, folder 96, 1831
HR DAST-152, Donji Muć (1831), i.e. HR DAST-152 (Državni Arhiv u Splitu - Arhiv mapa za Istru i Dalmaciju), Litografato nell' i.r Justituto litografatico del Catastro, Much Inferiore, K.O. 373, Donji Muć, folder 373, 1831
HR DAST-152, Donje Ogorje (1831), i.e. HR DAST-152 (Državni Arhiv u Splitu - Arhiv mapa za Istru i Dalmaciju), Litografato nell' i.r Justituto litografatico del Catastro, Ogorje Inferiore, K.O. 401, Donje Ogorje, folder 401, 1831
HR DAST-152, Donje Postinje (1831), i.e. HR DAST-152 (Državni Arhiv u Splitu - Arhiv mapa za Istru i Dalmaciju), Protocollo delle particelle dei terreni Postigne Inferior, K.O. 479, Donje Postinje, folder 479, 1831
HR DAST-152, Gornji Muć (1831), i.e. HR DAST-152 (Državni Arhiv u Splitu - Arhiv mapa za Istru i Dalmaciju), Litografato nell' i.r Justituto litografatico del Catastro, Much Superiore, K.O. 374, Gornji Muć, folder 374, 1831
HR DAST-152, Gornje Ogorje (1831), i.e. HR DAST-152 (Državni Arhiv u Splitu - Arhiv mapa za Istru i Dalmaciju), Litografato ncll' i.r Justituto litografatico del Catastro, Ogorje Superiore, K.O. 402, Gornje Ogorje, folder 402, 1831
HR DAST-152, Gornje Postinje (1831a), i.e. HR DAST-152 (Državni Arhiv u Splitu - Arhiv mapa za Istru i Dalmaciju), Litografato nell' i.r Justituto litografatico del Catastro, Postigne *Superiore*, K.O. 480, Gornje Postinje, folder 480, 1831
HR DAST-152, Gornje Postinje (1831b), i.e. HR DAST-152 (Državni Arhiv u Splitu - Arhiv mapa za Istru i Dalmaciju), Protocollo delle particelle dei terreni Postigne Superiore, K.O. 480, Gornje Postinje, folder 480, 1831
HR DAST-152, Milešine (1831), i.e. HR DAST-152 (Državni Arhiv u Splitu - Arhiv mapa za Istru i Dalmaciju), Protocollo delle particelle dei terreni Milessine, K.O. 348, Milešine, folder 348, 1831
HR DAST-152, Pribude (1831), i.e. HR DAST-152 (Državni Arhiv u Splitu - Arhiv mapa za Istru i Dalmaciju), Protocollo delle particelle dei terreni Pribude, K.O. 494, Pribude, folder 494, 1831
HR DAST-152, Sutina (1831), i.e. HR DAST-152 (Državni Arhiv u Splitu - Arhiv mapa za Istru i Dalmaciju), Litografato nell' i.r Justituto litografatico del Catastro, Suttina, K.O. 633, Sutina, folder 633, 1831
HR DAST-152, Donji Muć (1832), i.e. HR DAST-152 (Državni Arhiv u Splitu - Arhiv mapa za Istru i Dalmaciju), Protocollo delle particelle dei terreni Much Inferiore, K.O. 373, Donji Muć, folder 373, 1832
HR DAST-152, Gornji Muć (1832), i.e. HR DAST-152 (Državni Arhiv u Splitu - Arhiv mapa za Istru i Dalmaciju), Protocollo delle particelle dei terreni Much Superiore, K.O. 374, Gornji Muć, folder 374, 1832
HR DAST-152, Gornje Ogorje (1832), i.e. HR DAST-152 (Državni Arhiv u Splitu - Arhiv mapa za Istru i Dalmaciju), Protocollo delle particelle dei terreni Ogorje Superiore, K.O. 402, Gornje Ogorje, folder 402, 1832

HR DAST-152, Sutina (1832), i.e. HR DAST-152 (Državni Arhiv u Splitu - Arhiv mapa za Istru i Dalmaciju), Protocollo delle particelle dei terreni Suttina, K.O. 633, Sutina, folder 633, 1832

HR DAST-152, Vis (1835), i.e. HR DAST-152 (Državni Arhiv u Splitu - Arhiv mapa za Istru i Dalmaciju), Protocollo delle particelle dei terreni Lissa, K.O. 698, Vis, 1835

HR DAST-152, Gradac (1836), i.e. HR DAST-152 (Državni Arhiv u Splitu - Arhiv mapa za Istru i Dalmaciju), Litografato nell' i.r Justituto litografatico del Catastro, Gradaz, K.O. 185, Gradac, folder 185, 1836

HR DAST-152, Gradac (1837), i.e. HR DAST-152 (Državni Arhiv u Splitu - Arhiv mapa za Istru i Dalmaciju), Protocollo delle particelle dei terreni Gradaz, K.O. 185, Gradac, folder 185, 1837

HR DAST-152, Donje Ogorje (1841), i.e. HR DAST-152 (Državni Arhiv u Splitu - Arhiv mapa za Istru i Dalmaciju), Protocollo delle particelle dei terreni Ogorje Inferiore K.O. 401, Donje Ogorje, folder 401, 1841

HR DAST-152, Boraja (1844), i.e. HR DAST-152 (Državni Arhiv u Splitu - Arhiv mapa za Istru i Dalmaciju), Protocollo delle particelle dei terreni Boraia, K.O. 52, Boraja, folder 52, 1844

HR DAST-152, Krapanj (1844), i.e. HR DAST-152 (Državni Arhiv u Splitu - Arhiv mapa za Istru i Dalmaciju), Protocollo delle particelle dei terreni Crappano K.O. 279, Krapanj, folder 279, 1844

HR DAST-152, Baljke (1846), i.e. HR DAST-152 (Državni Arhiv u Splitu - Arhiv mapa za Istru i Dalmaciju), Protocollo delle particelle dei terreni Balke, K.O. 8, Baljke, folder 8, 1846

HR DAST-152, Miočić (1846), i.e. HR DAST-152 (Državni Arhiv u Splitu - Arhiv mapa za Istru i Dalmaciju), Protocollo delle particelle dei terreni Miocich, K.O. 355, Miočić, folder 355, 1846

HR DAST-152, Vis (1876), i.e. HR DAST-152 (Državni Arhiv u Splitu - Arhiv mapa za Istru i Dalmaciju), Litografato nell' i.r Justituto litografatico del Catastro, Lissa, K.O. Vis, 1876

Literature

Azevedo JC, Moreira C, Castro JP, Loureiro C (2011) Agriculture abandonment, land-use change and fire hazard in mountain landscapes in north-eastern Portugal. In: Li C, Raffaele L, Chen J (eds) Landscape ecology in forest management and conservation: challenges and solutions for global change. Springer, Heidelberg, pp 329–351. https://doi.org/10.1007/978-3-642-12754-0_14

Bajocco S, Ricotta C (2008) Evidence of selective burning in Sardinia (Italy): which land-cover classes do wildfires prefer? Landsc Ecol 23:241–248

Barros A, Pereira J (2014) Wildfire selectivity for land cover type: does size matter? PLoS ONE 9(1):e84760. https://doi.org/10.1371/journal.pone.0084760

Bertović S (1987) Osnove zaštite šuma od požara [Basics of forest fire protection]. CIP, Zagreb

Blaće A, Cvitanović M, Čuka A, Faričić J (2024) Land use/land cover changes on Croatian islands since the beginning of the twentieth century: drivers and consequences. In: Fuerst-Bjeliš B, Mrgić J, Petrić H, Zorn M, Zwitter Ž (eds) Environmental histories of the Dinaric karst. Springer, Cham, pp 141–165

Braudel F (1992) The Mediterranean and the Mediterranean world in the age of Philip II. Translated by Siân Reynolds; Abridged by Richard Ollard. HarperCollins Publishers, London

Chas-Amil M-L, Touza J, García-Martínez E (2013) Forest fires in the wildland–urban interface: a spatial analysis of forest fragmentation and human impacts. Appl Geogr 43:127–137. https://doi.org/10.1016/j.apgeog.2013.06.010

Chergui B, Fahd S, Santos X, Pausas J (2018) Socioeconomic factors drive fire-regime variability in the Mediterranean Basin. Ecosystems 21(4):619–628. https://doi.org/10.1007/s10021-017-0172-6

Chuvieco E, Mouillot F, van der Werf G, Miguel JS, Tanasse M, Koutsias N, García M, Marta D et al (2019) Historical background and current developments for mapping burned area from satellite Earth observation. Remote Sens Environ 225:45–64. https://doi.org/10.1016/j.rse.2019.02.013

Clarke H, Tran B, Boer MM, Price O, Kenny B, Bradstock R (2019) Climate change effects on the frequency, seasonality and interannual variability of suitable prescribed burning weather conditions in south-eastern Australia. Agricultural and Forest Meteorology 271:148–157.

Claval P (2007) About rural landscapes: the invention of the Mediterranean and the French school of geography. In: Roca Z, Speck T, Terkenli T, Plieningerand T, Hoechtl F (eds) European Landscapes and Lifestyles: The Mediterranean and Beyond. Edições Universitarias Lusofonas, Lisbon, pp 15–31

Curt T, Borgniet L, Bouillon C (2013) Wildfire frequency varies with the size and shape of fuel types in south-eastern France: implications for environmental management. J Environ Manag 117:150–161. https://doi.org/10.1016/j.jenvman.2012.12.006

de Concina G (1809) Viaggio nella Dalmazia Litorale. Liberale Vendrame, Udine

de Visiani R (1847) Flora Dalmatica. Hofmeister, Leipzig

Dubinin M, Luschekina A, Radeloff V (2011) Climate, livestock, and vegetation: what drives fire increase in the arid ecosystems of Southern Russia? Ecosystems 14:547–562. https://doi.org/10.1007/s10021-011-9427-9

Elia M, Lovreglio R, Ranieri NA, Sanesi G, Lafortezza R (2013) Cost-effectiveness of fuel removals in Mediterranean wildland-urban interfaces threatened by wildfires. Forests 7:149. https://doi.org/10.3390/f7070149

Falk D, Miller C, McKenzie D, Black A (2007) Cross-scale analysis of fire regimes. Ecosystems 10:809–823. https://doi.org/10.1007/s10021-007-9070-7

Fernandes P, Loureiro C, Magalhães M, Ferreira P, Fernandes M (2012) Fuel age, weather and burn probability in Portugal. Int J Wildland Fire 21(4):380–384. https://doi.org/10.1071/WF10063

Foretić D (1963) Tabella enciclopedica del regno di Dalmazia. Radovi Instituta Jugoslavenske Akademije Znanosti i Umjetnosti u Zadru 10:291–307

Fuerst-Bjeliš B (2003) Reading the Venetian cadastral record: an evidence for the environment, population and cultural landscape of the 18th century Dalmatia. Hrvatski Geografski Glasnik 65(1):47–62

Fuerst-Bjeliš B (2020) Environmental response to marginality: between borderlands and Littoralization in the Eastern Adriatic. In: Fuerst-Bjeliš B, Leimgruber W (eds) Globalization, Marginalization and Conflict. Political, economic and social processes. Springer, Cham, pp 75–87

Fuerst-Bjeliš B, Cvitanović M, Durbešić A (2016) Fire risk incidence over the last 200 years: case study in the Mediterranean Croatia. In: Pina H, Remoaldo P, Ramos C (eds) The overarching issues of the European space/rethinking socioeconomic and environmental problems, repositioning territorial development policies. FLUP, Porto, pp 161–172

Fuerst-Bjeliš B, Glamuzina N (2021) The historical geography of Croatia. Territorial change and cultural landscapes. Springer, Cham. https://doi.org/10.1007/978-3-030-68433-4

Fuerst-Bjeliš B, Kale J (2018) Povijesna dinamika kulturnih krajolika šibenskog područja [Historical dynamics of the cultural landscapes of the Šibenik area]. In: Kurelac I (ed) Šibenik od prvog spomena. Šibenik City Museum, Šibenik-Zagreb, pp 545–561

Fuerst-Bjeliš B, Lozić S, Cvitanović M, Durbešić A (2011) Promjene okoliša središnjeg dijela Dalmatinske zagore od 18. stoljeća [Environmental changes in the central part of the Dalmatian Zagora since the eighteenth century]. In: Matas M, Faričić J (eds) Zagora između stočarsko-ratarske tradicije te procesa litoralizacije i globalizacije. University of Zadar and Matrix Croatica, Zadar, pp 117–130

Gams I (1974) Kras—zgodovinski, naravoslovni in geografski oris [Kras—historical, natural history and geographical outline]. Slovenska matica, Ljubljana

Gams I (2004) Kras v Sloveniji v prostoru in času [Karst in Slovenia]. Založba ZRC, ZRC SAZU, Ljubljana

Glamuzina M, Glamuzina N (1996) Promjene u biološkoj i ekonomskoj strukturi stanovništva Južne Hrvatske (Dalmacije) od 1948. do 1991. godine [Changes in the biological and economic structure of the population in southern Croatia (Dalmatia) from 1948 to 1991]. Geoadria 1(1):17-34

Hill J, Stellmes M, Udelhoven T, Röder A, Sommer S (2008) Mediterranean desertification and land degradation: mapping related land use change syndromes based on satellite observations. Global Planet Change 64(3–4):146–157. https://doi.org/10.1016/j.gloplacha.2008.10.005

Jajtić K, Galijan V, Žafran I, Cvitanović M (2019) Analysing wildfire occurrence through a mixed-method approach: a case study from the Croatian Mediterranean. Erdkunde 73(4):323–341. https://doi.org/10.3112/erdkunde.2019.04.05

Jedlowski D (1975) Venecija i šumarstvo Dalmacije od 15. do 18. vijeka [Venice and forestry in Dalmatia from the fifteenth to the eighteenth century]. (PhD dissertation). Institut za jadranske kulture i melioraciju krša, Split

Jurjević P, Vuletić D, Gračan J, Seletković G (2009) Šumski požari u Republici Hrvatskoj (1992–2007) [Forest fires in the Republic of Croatia (1992–2007)]. Šumarski List 133:63–72

Klempić-Bogadi S, Podgorelec S (2009) Sociodemografske značajke i procesi u hrvatskim obalnim gradovima [Socio-demographic characteristics and processes in Croatian coastal cities]. Geoadria 14(2):221–247

Koutsias N, Arianoutsou M, Kallimanis A, Mallinis G, Halley J, Dimnopolous P (2012) Where did the fires burn in Peloponnisos, Greece, the summer of 2007? Evidence for a synergy of fuel and weather. Agricult for Meteorol 156:41–53. https://doi.org/10.1016/j.agrformet.2011.12.006

Lovreglio R, Danna E, Mou G (2014) Mitigazione del pericolo incendi nelle aree di interfaccia urbano/foresta: an esempio di selvicoltura preventiva in Sardegna. Forest 11:156–167. https://doi.org/10.3832/efor1187-011

Mamut M (2011) Veza prirodnogeografske i sociogeografske osnove Dalmacije s ugroženošću otvorenog prostora požarom [Ties between the geographical and social geographical features of Dalmatia with the endangerment of forest fires]. Šumarski List 135(1–2):37–49

Mantero G, Morresi D, Marzano R, Motta R, Mladenoff D, Garbarino M (2020) The influence of land abandonment on forest disturbance regimes: a global review. Landsc Ecol 35:2723–2744. https://doi.org/10.1007/s10980-020-01147-w

Matutinović L (2009) Ogled o Ilirskim provincijama i Crnoj Gori [An essay on the Illyrian provinces and Montenegro]. Translated by Jagoda Milinković. Školska knjiga, Zagreb

McNeill JR (1992) The mountains of the Mediterranean world: an environmental history. Cambridge University Press, Cambridge

McWethy D, Higuera P, Whitlock C, Veblen T, Bowman D, Cary G, Haberle S, Keane R, Maxwell B, McGone M et al (2013) A conceptual framework for predicting temperate ecosystem sensitivity to human impacts on fire regimes. Glob Ecol Biogeogr 22:900–912. https://doi.org/10.1111/geb.12038

Modrich G (1892) La Dalmazia romana-veneta-moderna: note e ricordi di viaggio. L. Roux e C., Torino/Roma

Modugno S, Baltzer H, Cole B, Borelli P (2016) Mapping regional patterns of large forest fires in wildland–urban interface areas in Europe. J Environ Manag 172:112–126. https://doi.org/10.1016/j.jenvman.2016.02.013

Moreira F, Russo D (2007) Modelling the impact of agricultural abandonment and wildfires on vertebrate diversity in Mediterranean Europe. Landsc Ecol 22:1461–1476. https://doi.org/10.1007/s10980-007-9125-3

Moreira F, Vaz P, Catry F, Silva JS (2009) Regional variations in wildfire susceptibility of land-cover types in Portugal: implications for landscape management to minimize fire hazard. Int J Wildland Fire 18(5):563–574. https://doi.org/10.1071/WF07098

Moreira F, Viedma O, Arianoutsou M, Curt T, Koutsias N, Rigolot E, Barbati A, Corona P, Vaz P, Xanthopoulos G, Mouillot F, Bilgili E (2011) Landscape: wildfire interactions in southern Europe: implications for landscape management. J Environ Manag 92(10):2389–2402. https://doi.org/10.1016/j.jenvman.2011.06.028

Nejašmić I, Toskić A (2013) Starenje stanovništva u Hrvatskoj–sadašnje stanje i perspektive [Ageing of the population in Croatia–the current situation and perspectives]. Hrvatski Geografski Glasnik 75(1):89–110. https://doi.org/10.21861/HGG.2013.75.01.05

Novak G (1960) Poljoprivreda Dalmacije u drugoj polovini XVIII stoljeća [Agriculture in Dalmatia in the second half of the eighteenth century]. Starine (JAZU) 50:461–518

Nunes M, Vasconcelos M, Pereira J, Dasgupta N, Alldredge R, Rego F (2005) Land cover type and fire in Portugal: do fires burn land cover selectively? Landsc Ecol 20:661–673. https://doi.org/10.1007/s10980-005-0070-8

Panjek A (2018) Against the desert in the karst. A paradigm-shift from ruined landscape to cultural savannah. Ekonomska i Ekohistorija 21(14):52–71

Pausas J (2004) Changes in fire and climate in the eastern Iberian peninsula (Mediterranean Basin). Climat Change 63:337–350. https://doi.org/10.1023/B:CLIM.0000018508.94901.9c

Pausas J, Fernández-Muñoz S (2012) Fire regime changes in the Western Mediterranean Basin: From fuel-limited to drought-driven fire regime. Climat Change 110(1–2):215–226. https://doi.org/10.1007/s10584-011-0060-6

Pausas J, Keeley J (2014) Abrupt climate-independent fire regime changes. Ecosystems 17(6):1109–1120. https://doi.org/10.1007/s10021-014-9773-5

Pavlek K, Bišćević F, Furčić P, Grđan A, Gugić V, Malešić N, Moharić P, Vragović V, Fuerst-Bjeliš B, Cvitanović M (2017) Spatial patterns and drivers of fire occurrence in a Mediterranean environment: a case study of southern Croatia. Geografisk Tidsskrift-Danish J Geogr 117(1):22–35. https://doi.org/10.1080/00167223.2016.1266272

Pereira M, Trigo R, da Camara C, Pereira J, Leite S (2005) Synoptic patterns associated with large summer forest fires in Portugal. Agricult Forest Meteorol 129(1–2):11–25. https://doi.org/10.1016/j.agrformet.2004.12.007

Petter F (1834) Compendio geografico della Dalmazia, con un' appendice sul Mentenero. Fratelli Battara, Zara

Pyne SJ (2010) Vatra: sažeta povijest [Fire: a brief history]. Translated by Olga Šarić. Prosvjeta, Zagreb

Radeljak Kaufmann P (2016) Integrating factor analysis and the Delphi method in scenario development: a case study of Dalmatia, Croatia. Appl Geogr 71:56–68. https://doi.org/10.1016/j.apgeog.2016.04.007

Radeljak Kaufmann P (2017) Izrada i analiza scenarija u prostornom planiranju: scenariji prostornog razvoja Dalmacije do 2031. godine [Scenario making and analysis in spatial planning: scenarios of spatial development for Dalmatia until 2031]. Annales-Anali za Istrske in Mediteranske Studije Ser Hist Sociol 27(3):581–598. https://doi.org/10.19233/ASHS.2017.41

Roglić J (1965) The delimitations and morphological types of the Dinaric Karst. Naše Jame 7(1–2):79–87

Roglić J (1974) Prilog hrvatskoj krškoj terminologiji [Contribution to Croatian karst terminology]. Krš Jugoslavije (carsus Iugoslaviae) 9(1):1–72

Romero-Calcerrada R, Perry GLW (2004) The role of land abandonment in landscape dynamics in the SPA 'Encinares del rio Alberche y Cofio, Central Spain, 1984–1999. Landsc Urban Plann 66:217–232

Silva J, Moreira F, Vaz P, Catry F, Godinho-Ferreira P (2009) Assessing the relative fire proneness of different forest types in Portugal. Plant Biosyst 143(3):597–608. https://doi.org/10.1080/11263500903233250

Simmons IG (1998) Towards an environmental history of Europe. In: Butlin RA, Dodgshon RA (eds) An historical geography of Europe. Clarendon Press, Oxford, pp 335–361

Simmons IG (2008) Global environmental history: 10,000 BC to 2000 AD. Edinburgh University Press, Edinburgh

Slocum M, Beckage B, Platt W, Orzell S et al (2010) Effect of climate on wildfire size: a cross-scale analysis. Ecosystems 13(6):828–840. https://doi.org/10.1007/s10021-010-9357-y

Tedim F, Leone V, Lovreglio R, Xanthopoulos G, Chas-Amil M-L, Ganteaume A, Efe R, Royé D, Fuerst-Bjeliš B, Nikolov N et al (2022) Forest fire causes and motivations in the southern and south-eastern Europe through experts. Percept Appl Curr Polic for 13(4):562. https://doi.org/10.3390/f13040562

Tedim F, Xanthopoulos G, Leone V (2015) Forest fires in Europe: facts and challenges. In: Shroder JF, Paton D (eds) Wildfire hazards, risks and disasters. Elsevier, Amsterdam, pp 77–99. https://doi.org/10.1016/B978-0-12-410434-1.00005-1

Tekić I, Fuerst-Bjeliš B, Durbešić A (2015) The impact of Aleppo pine afforestation on the structure and dynamics of landscape in Mediterranean Croatia. In: Pina H, Martins F (eds) The overarching issues of the European space: spatial planning and multiple paths to sustainable and inclusive development. FLUP, Porto, pp 207–221

Tollefson J (2021) IPCC climate report: earth is warmer than it's been in 125,000 years. Nature 596:171–172

Tommaseo N (1847) Intorno a cose dalmatiche e triestine. Trieste, I Papsch & C. del Lloyd Austr

Touchan R, Baisan C, Mitsopoulos I, Dimitrakopoulos AP (2012) Fire history in European black pine (*Pinus Nigra* Arn.) forests of the Valia Kalda, Pindus Mountains, Greece. Tree-Ring Res 68(1):45–50

Turco M, Bedia J, Di Liberto F, Fiorucci P, von Hardenberg J, Koutsias N, Llasat M-C, Xystrakis F, Provenzale A (2016) Decreasing fires in Mediterranean Europe. PLoS ONE 11(3):e0150663. https://doi.org/10.1371/journal.pone.0150663

Vajda Z (1970) Problem zaštite šuma od požara u SR Hrvatskoj [The problem of protecting forests from fires in SR Croatia]. Šumarski List 94(3–4):92–105

Viedma O, Moreno JM, Güngöroğlu C, Cosgun U, Kavgaci A (2017) Recent land-use and land-cover changes and its driving factors in a fire-prone area of southwestern Turkey. J Environ Manag 197:1–13. https://doi.org/10.1016/j.jenvman.2017.02.074

Wessely J (1877) Kras hrvatske krajine i kako da se spasi, za tiem kraško pitanje uploške [The karst of the Croatian landscape and how to save it]. Šumarski List 1(2):57–109

Yriarte C (1883) La rive dell' Adriatico e il Montenegro: Venezia, Chioggia, Trieste, l'Istria, il Quarnero e sue isole, la Dalmazia, il Montenegro, Ravenna, Ancona, Loreto, Foggia, Brindisi, Lecce, Otranto. Fratelli Treves, Milano

Zorn M, Kumer P, Ferk M (2015) Od gozda do gozda ali kje je goli, kamniti Kras? [From forest to forest or where is the bare rocky karst?]. Kronika 63(3):561–574

Ivan Tekić Associate expert, Ph.D. His work and research is focused on remote sensing, environmental history and land cover and land use changes, specially in the Mediterranean context. His Ph.D. with the title "Forestry and traditional woodland management in North Dalmatia c. 1790 to 1990: an environmental history" was received from the University of Nottingham, UK.

Borna Fuerst-Bjeliš Senior researcher and full professor. Her research interests include environmental history and landscape change, focusing spatially mainly on the Mediterranean karst region, including the development of research methods (historical cartography and GIS technology), borderlands, spatial perception and spatial (regional) identities. She is (co-)author/(co-) editor of books on topics of Historical Geography of Croatia, Mediterranean Identities, and Globalization, Marginalization and Conflict. She served as Vice-President of the European Society for Environmental History (2017–2019) and she is currently Chair of the Commission on Marginalization, Globalization and Regional and Local Responses of the International Geographical Union (2020-present).

Anamarija Durbešić GIS associate expert. She received her Ph.D. in 2012 from the University of Zagreb Faculty of Science, Department of Geography, with the dissertation titled "Landscape change of the southern slopes of the Svilaja Mountains—GIS approach". Her main research interest is the application of the GIS method in the analysis of cultural landscapes. Since 2010 she has been a lecturer at the Department of Geography of the University of Zagreb Faculty of Science, where she teaches the course *Historical GIS*.

Petra Radeljak Kaufmann Associate professor. She received her Ph.D. from the same department in 2014 with the dissertation titled "Scenario Method in Spatial Planning: A Case Study of the Southern Croatian Littoral". Her main research interests relate to specific methods and techniques in regional and spatial planning (especially scenario method, Delphi method, and spatial analysis methods) and development challenges of rural and karst areas. She teaches several courses at the undergraduate, graduate, integrated, and doctoral levels.

Marin Cvitanović Principal academic in human geography. His main areas of interest are socio-environmental systems, with an emphasis on human drivers of land use/cover changes, and cultural ecosystem services. His work is rooted in geography, but his interests extend to environmental history, biology, cultural studies, social psychology and political science. His previous research focused on post-socialist landscape transformation in Croatia, human-induced mangrove changes in East Africa, endangered species' habitat fragmentation in the UK, and the evolving significance and functions of urban allotments in Zagreb, Croatia.

Chapter 6
Land Use/Land Cover Changes on Croatian Islands Since the Beginning of the Twentieth Century—Drivers and Consequences

Ante Blaće, Marin Cvitanović, Anica Čuka, and Josip Faričić

Abstract Croatian islands form an archipelago of 1244 islands, islets, rocks, and rocks awash along the eastern Adriatic coastline, only 50 of which are permanently populated. Karst environment and climate conditions predetermined them as areas of traditional Mediterranean agriculture. Even though some of the islands have been inhabited since prehistoric times, existing historical documents enable us to reconstruct specific land use types from the nineteenth century onwards. The aim of this paper was to determine the main factors that influenced land use changes and landscape development during the twentieth century. The methodology was based on a comparison of land use categories from different periods and sources like archival cadastral data and CORINE land cover. Results indicate the peak of agriculture activities and land use at the turn of the twentieth century as a result of a population growth and a conjuncture in wine production. Islands' population started to decline during the 1920s, which resulted in the land abandonment and secondary succession. Nowadays, only a small share of land is being cultivated and mainly by cash crops such as olive trees and grapes. As a result, Mediterranean maquis and forests cover the majority of islands' surface which increases the risk for summer fires and threatens traditional agricultural landscape. This research contributes to the understanding of complex social processes on the Croatian islands during the twentieth century and their influence on the formation of different land use patterns.

A. Blaće (✉) · A. Čuka · J. Faričić
Department of Geography, University of Zadar, Zadar, Croatia
e-mail: anblace@unizd.hr

A. Čuka
e-mail: acuka@unizd.hr

J. Faričić
e-mail: jfaricic@unizd.hr

M. Cvitanović
Department of Life and Environmental Sciences, Faculty of Science and Technology, Bournemouth University, Poole, UK
e-mail: mcvitanovic@bournemouth.ac.uk

B. Fuerst-Bjeliš et al. (eds.), *Environmental Histories of the Dinaric Karst*, Environmental History 17, https://doi.org/10.1007/978-3-031-56089-7_6

Keywords Agriculture · Croatian islands · Depopulation · Land use/land cover changes · Mediterranean landscape · Tourism · Environmental history

6.1 Introduction

Humans have been changing the Earth's environment for thousands of years. Archaeological research in certain parts of Europe has shown that significant changes of the environment under human influence can be dated back to 8500 years BP, with fire being the first powerful tool used to change the physical characteristics of Earth.[1] Together with the increase of the human population and technological and scientific advances, the intensity of human-induced environmental change has increased dramatically. Today, even the most remote parts of the planet are affected by human activity through global atmospheric circulation, sea currents, or global warming.[2]

People live on land and a significant part of their activities focuses especially on terrestrial ecosystems. Human use of land can include food production, provision of shelter, recreation, extraction and procession of materials, and so on. The land sub-system is therefore central to the study of interaction between people and the environment.[3] FAO (1995) defines[4] "land" as a set of attributes of the biosphere immediately above (and below) the surface, containing evidence of (among other things) past and present human activity. The main characteristic of this land sub-system is its changeability. Changes in the uses of land are a result of environmental and human dynamics and of their interactions mediated by land.[5]

The study of the changes in land cover and human use of land (land use/cover change, LULC) has received much attention during the past decade, because of the pivotal role it plays in many urgent societal and environmental issues such as climate change, food security, soil degradation, and biodiversity.[6] Because of its complexity, there is no general theory of LULC, but different theories help us to analyse different aspects of change in specific situation. The main difference between the concept of land use and land cover is the aspect of human use of land. Land cover is characterised by physical attributes of land surface (soil, vegetation cover, water, man-made structures) and is usually analysed via remote sensing, while land use is a more complex term, including social and economic purposes for which land is managed.[7] However, these concepts are often used interchangeably.

There is extensive research on LULC in general (see e.g., Meyer and Turner 1994; Lambin and Geist 2006; for an extensive overview on the subject). However,

---

[1] Simmons (2008), p. 31.

[2] McKibben (1989).

[3] Kuemmerle (2020) and Briassoulis (2000).

[4] Food and Agriculture Organization (FAO) (1995).

[5] Briassoulis (2000), p. 5.

[6] Turner et al. (1995) and Geist and Lambin (2002).

[7] Lambin et al. (2006), p. 4.

we still cannot say that we fully understand how and why the environment changes. Therefore global, national, and regional caste studies are still the most usual and most useful way in exploring the factors and circumstances that influence LULC.

One of the main goals of LULC research is to identify the main factors which play a role in these changes. They can be broadly divided into two groups—proximate and underlying causes. While proximate causes include physical changes of the land surface, underlying causes are formed by a variety of variables and require additional research.

Land degradation like overgrazing and soil erosion which are the result of intensified land use over the history, can be noticed in many Mediterranean areas.[8] However, after a long period of overusing, many traditional agrarian societies transformed their economies by mostly shifting to tertiary activities and abandoning fertile land. Some regions became places of intensive emigration and depopulation which also led to secondary succession. That is why Mediterranean parts of Europe have been detected as hotspots of land abandonment.[9] The same processes occurred on Croatian islands which recorded intensive emigration at the turn of the twentieth century.[10] Many islands, mostly small and remote, are still facing emigration and depopulation, but even on the islands with more favourable socio-economic situation, land abandonment is a long-lasting process. The main objective of this paper was to investigate the dynamics of land use/land cover changes on Croatian islands and to determine the drivers influencing landscape changes. Such an all-encompassing approach for all Croatian islands has not been done before.

6.2 Research Area

The Croatian islands are a scattered and fragmented archipelago off the Croatian part of the Adriatic coast (Fig. 6.1). They consist of 1244 island units: islands, islets, and reefs with a total area of 3259 km^2.[11] Island groups in Croatia are grouped according to regional affiliation, not according to inter-island connections. The island groups are as follows: Istrian, Kvarner, and Dalmatian. The Istrian island group is the smallest and has no permanently inhabited islands. The Dalmatian island group is the largest and is further divided into the North Dalmatian, Central Dalmatian, and South Dalmatian subgroups.

The Croatian islands are parts of geological megastructures formed at the contact of the Adriatic microplate as part of the African lithosphere plate and the Eurasian lithosphere plate.[12] They were separated from the mainland at the end of the Upper Pleistocene and the beginning of the Holocene when the sea level of the Adriatic

[8] Sluiter and de Jong (2007), p. 560.

[9] Sluiter and de Jong (2007), p. 575.

[10] Nejašmić (1991), p. 81; Nakićen and Čuka (2016), p. 326; Čuka et al. (2020), p. 54.

[11] Duplančić Leder et al. (2004), p. 11.

[12] Vlahović et al. (2005), p. 335.

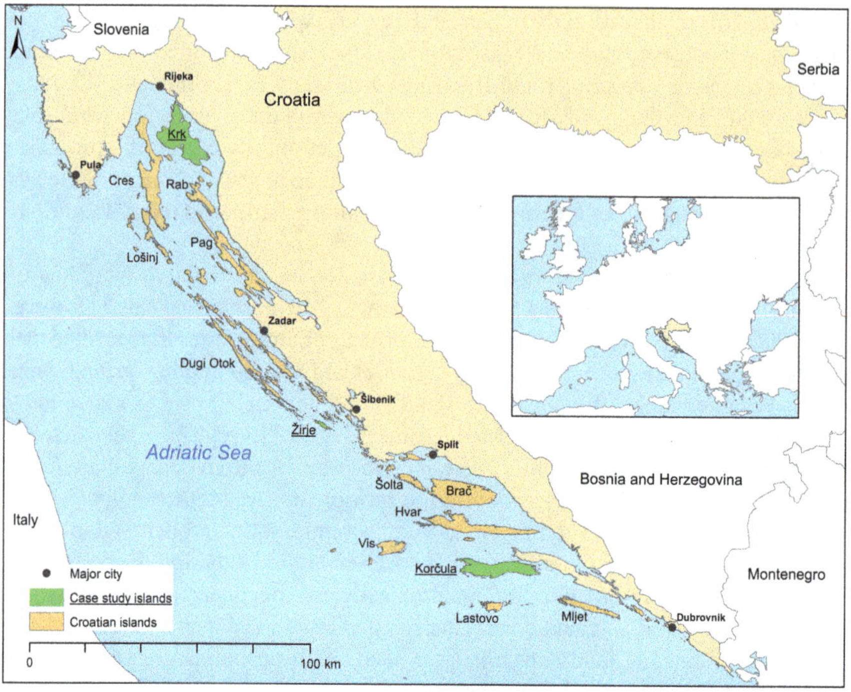

Fig. 6.1 Research area—Croatian islands. *Source* Based on: *Central Registry of Spatial Units in the Republic of Croatia (GIS shapefiles)*, Croatian Geodetic Administration, 2016, *Europe's Seas*, Geodatabase European Environmental Agency, accessed 5 May 2021, http://cdr.eionet.europa.eu/hr/eea/clc

Sea rose by 121 ± 5 m.[13] They consist mainly of carbonate rocks (limestones and dolomites) of Cretaceous and Eocene age, and there is less occurrence of flysch and unbound clasts (Aeolian sands in Susak and Unije and pebbles caused by accumulation of alluvium and subsequent abrasion in most island bays). Very rare and insignificant are occurrences of magmatic rocks (Komiža on the island of Vis, the islets of Jabuka and Brusnik).[14] Limestone was mined on almost all the islands for construction and renovation of houses and port infrastructure, for drystone fencing, and for the production of lime, which was used in construction on the islands and on the neighbouring mainland with sand extracted from the seabed. The exploitation of stone was most intensive on the island of Brač, where even today that activity is an important factor in shaping the island landscape.

Relief forms and hydrographic features form a complex karst phenomenon. Although precipitation is relatively high, especially in the late autumn and winter, the carbonate lithological base is mostly permeable, so there is intensive sinking and seepage and underground runoff. Surface occurrences of fresh water are very rare.

[13] Fairbanks (1989), p. 637.

[14] Velić and Vlahović (2009).

The exception is the large natural accumulation of fresh water—Vrana Lake on the island of Cres. Lack of surface fresh water has for centuries been an aggravating factor for the development of agriculture and other economic activities, especially during summer droughts when water needs are greatest due to high air temperatures. Households mostly met their water needs by collecting rainwater. The integration of islands in the immediate vicinity of the mainland into regional water supply systems during the second half of the twentieth century enabled greater resilience of the island economy to seasonality and variability of precipitation. On most islands, the water supply is still not sufficiently developed, so it depends on the delivery of water by water carriers from the mainland, and less often on desalination, which occurs only on three islands (Lastovo, Mljet and Dugi Otok).

Most of the Dalmatian islands are located in the Mediterranean climate zone (Csa), while further north are the Istrian and Kvarner islands in the area of moderately warm humid climate with hot summers (Cfa).[15] Winters are mild and humid, and summers are hot and dry, while the transitional seasons are marked by frequent changes of anticyclones and cyclones with variable amounts of precipitation. The average annual air temperature in the climate period 1971–2000 ranged from 15.3 °C in Mali Lošinj on the island of Lošinj (Kvarner island group) to 16.4 °C in Hvar on the island of the same name (Dalmatian island group), with the coldest month being January with an average temperature of 7.9 °C in Mali Lošinj and 9.1 °C in Hvar, and the warmest month is July with an average temperature of 24.1 °C in Mali Lošinj and 25.0 °C in Hvar. The average annual rainfall ranges from 713.7 mm in Hvar to 930.0 mm in Mali Lošinj, with the highest rainfall falling in November (94 mm in Hvar and 125.1 mm in Mali Lošinj), and the least in July (26.4 mm in Hvar and 30.9 mm in Mali Lošinj).[16]

On the geological substrate, mostly shallow horizons of brown and red soils developed, formed by chemical dissolution of limestone and mechanical wear of dolomite and flysch deposits. The area of arable land is greatest on the flattened parts of larger islands (Krk, Rab, Brač, Hvar, Korčula, Mljet) and on gentle slopes (Ugljan, Pašman, Dugi Otok, Lošinj and Šolta), while pedological resources on most other islands are modest. In order to preserve the soil as a valuable natural resource, the islanders terraced the slopes with dry stone walls, which often form complex lace as one of the most recognisable landscape patterns.

Climatic and pedological conditions have enabled the development of olive growing and viticulture combined with vegetable growing. Karst rocks and lower Mediterranean vegetation with essential and aromatic plants enabled the development of livestock breeding, while holm oak forests and associated floral elements (myrtle, strawberry tree, mastic tree, prickly juniper, etc.) were used in housing and shipbuilding, for firewood, for production of agricultural and fishing tools and for the production of lime in lime kilns.[17] These activities contributed to the degradation of indigenous vegetation and the introduction of pedologically favourable

[15] Filipčić (2001), p. 14.

[16] Zaninović (2008), p. 189.

[17] Faričić and Juran (2021), p. 303.

zones for the cultivation of Mediterranean plants and the stripping of island slopes used in pasture livestock breeding.[18] However, with the depopulation and land abandonment that intensified in the second half of the twentieth century, a reversible process occurred: the indigenous vegetation is restored, such as myrtle and holm oak forest community (*Myrto-Quercetum ilicis*) and Aleppo pine (*Pinus halepensis*) and occupies increasing shares in the island areas.

Various economic activities have facilitated the development of a network of settlements, the shaping of landscapes and the development of various traditional knowledge and skills. This island culture consists of many elements of tangible and intangible heritage: from urban cores in small island towns with many historical layers of architecture (Cres, Krk, Osor, Mali Lošinj, Rab, Pag, Hvar, Korčula, Vis, etc.) to various customs and cultural expressions in rural communities, which are often associated with agricultural and maritime motives, all under the great influence of Christianity to which, at least declaratively, the vast majority of the island's population belongs. This cultural heritage together with a rich natural heritage is a fundamental shaping element of the islands' identity. Furthermore, it is the attractive basis of modern evaluation and economic use on islands, especially in tourism. Accordingly, many traditional forms of land use are being transformed from existential economic activities into cultural patterns in the collective memory of islanders. Therefore, for many islanders, olives and olive oil, grapevines, and wine became a part of intangible heritage, not topics of economic interest.

Administrative-territorial fragmentation of Croatian islands does not contribute to socio-economic development. The islands belong to coastal regions that are organised as seven counties with headquarters in urban centres that are located on the mainland: Pazin (Pula), Rijeka, Gospić, Zadar, Šibenik, Split, and Dubrovnik. Local self-government units (cities and municipalities) are organised on larger islands (for example, the island of Brač with an area of 395.4 km^2 has eight municipalities, the island of Krk with an area of 405.2 km^2 has seven municipalities), while smaller islands belong to local self-government units whose headquarters are situated on larger islands (for example islands of Unije, Srakane, Susak, and Ilovik belong to the City of Mali Lošinj on Lošinj Island) or in cities on the mainland (for example islands of Premuda, Silba, Olib, Ist, Molat, Iž, and Rava belong to the City of Zadar). This multiple fragmentation hinders the synergy of natural and human resources and the complementarity of island economic microsystems. Given the growing decline in population, the islands mostly function as a periphery of coastal regions which is dependent on coastal cities in various spheres of social and economic life. The peripherality increases with distance from the mainland, with a reduced frequency of passenger ship connections and with the length of the voyage. It is the strongest on small and the most remote islands. In contrast, the islands in the immediate vicinity of the mainland are in a much more favourable socio-economic position, some of which are connected to the mainland by a bridge (Krk, Pag, Vir, Murter, and Čiovo) and have largely lost their islandness in functional terms.

[18] Faričić (2012), p. 364.

6.3 Materials and Methods

We used several different sources for analysing the data on land use. The oldest were publications containing data on revision of the cadastre from 1896 that was published for Istria and Dalmatia (administrative units within Austria-Hungary) as two separate booklets.[19] They contain information on land classes and surface area, livestock numbers, and the results of the 1900 census.

The 1950s data were collected from the book[20] on general geographic features of Croatian Islands. Land cover changes in the most recent period (1980–2018) have been analysed by using the CORINE land cover inventory from Copernicus Land monitoring Service of the European Environment agency.[21] The land cover data have a minimum mapping unit of 25 ha and anniversary years are 1980, 1990, 2000, 2006, 2012, and 2018. The changes have been identified for the entire island area of Croatia at the second and third level of CORINE land cover classification nomenclature.

As a case study we decided to analyse land use changes on three Croatian islands—Korčula island, Krk island and Žirje island. According to Croatian classifications[22] Korčula (279 km^2) and Krk (406 km^2) are classified as large islands while Žirje (15 km^2) is a small island. Krk is situated near the mainland and has been connected to it via bridge. Krk and Korčula recorded mostly positive economic and demographic trends in the last couple of decades while Žirje is inhabited by a small, mostly elderly population. These features also reflect different trends in land use.

6.4 Drivers Influencing LULC Changes

Natural factors are important drivers affecting land use pattern in a certain area, but socio-economic factors are mostly responsible for land use changes.[23] Islands are highly sensitive to those changes because they are remote, isolated, often vary in size and shape or in geomorphological, ecological, and demographic setting.[24]

This research covers LULC changes on Croatian islands since the beginning of the twentieth century and drivers which led to those changes. They are predominantly socio-economic and connected to main long-lasting processes such as depopulation, deruralization, ageing, abandonment of agriculture, intensification of tourism, intensive construction of second homes, and apartments for rent and economic transformation from primary-to-tertiary sector of activity. Various studies give examples of grouping the drivers in certain categories which helps in their systematisation and makes easier the explanations on their influence to LULC change. Van Vliet

[19] *Gemeindelexikon* (1906), *Općinski rječnik* (1908).

[20] Rubić (2017)

[21] *Corine Land Cover*.

[22] Zimmerman (1997).

[23] Plieninger et al. (2016), p. 209.

[24] Walsh et al. (2020), pp. 63–88.

et al. (2015)[25] did research on manifestation and underlying drivers of agricultural land use change in Europe where they presented six groups of underlying drivers within a well-developed conceptual framework. While studying the driving forces of landscape changes in Europe, Plieninger et al. (2016)[26] grouped the drivers in two main categories: underlying and proximate. Following the similar concept, the main drivers of change on Croatian islands have been identified (Table 6.1) Following text brings more thorough explanation of key drivers.

6.4.1 Crisis in Winegrowing

Starting from the early twentieth century, when agriculture and fishery were the most important economic activities, the first underlying driver of change was the crisis in winegrowing and spread of phylloxera all over Croatian coast and islands. Other than permanently ruining almost all vineyards in two decades, it was one of the main reasons for strong emigrational wave. Emigration had a dual effect on the islands' landscape. There was a lack of young, strong workforce to cultivate the land but the long-term effect was the decrease in biological reproduction since the birth rate fell, causing natural decrease and affecting population ageing. All these processes were followed by abandonment of arable land and caused considerable landscape changes.[27] Vineyards were firstly replaced with pastures (and also olive groves and sometimes orchards) but the second half of the twentieth century brought important socio-economic changes. Transformation from primary to secondary and tertiary activities intensified after the 1960s. Coastal cities developed industry which lacked the workforce, while housing and quality of life also attracted people from the surrounding areas. Most islands remained on spatial and economic periphery, with no significant investments so many islanders moved to work on the coast.[28]

6.4.2 Land Ownership Structure and Subdivision

Land ownership and subdivision play an important role in land use opportunities, therefore it may be considered as one of the key drivers of landscape changes. The law-based inheritance in Croatia is such that if the deceased has not left a will, he is legally inherited by all of his/her heirs in succession. So, the land was often inherited by all heirs who did not always divide property in situ, but only became legal co-owners. After some time, even islanders who remained on the land did not know which part of the plots belonged to them and which belonged to legal owners, their

[25] van Vliet et al. (2015), p. 27.

[26] Plieninger et al. (2016), p. 209.

[27] Čuka et al. (2017), p. 114.

[28] Klempić Bogadi and Podgorelec (2020), pp. 89–110.

Table 6.1 The main drivers influencing LULC change on Croatian islands since early twentieth century

Underlying drivers	Proximate drivers	Explanation	Manifestation of change
Political	Political system changes	From the beginning of the twentieth century islands were part of different political systems: from Austro-Hungarian Empire (up to 1918), Kingdom of Serbs, Croats and Slovenians/Kingdom of Yugoslavia (1918–1945), Socialist Federal Republic of Yugoslavia (1945–1991) to Republic of Croatia (from 1991). Some islands were also part of Italy in the first half of the twentieth century	***First half of the twentieth century*** Intensification: increase of pastures Extensification: land abandonment, more complex landscape structure, decrease of vineyards ***Second half of the twentieth century*** Intensification: increase of built-up areas, increase of wooded areas Extensification: decrease of pastures, decrease of total agricultural area
	Political unrest	Many islanders (mostly male) were killed during the First and Second World War. Croatian War of Independence (1991–1995) did not directly affect islands and their infrastructure but led to economic changes because of the decline in tourism	
	Political decisions	Agrarian Reform (1919–1941; abolition of remaining feudal relations) and Agrarian Reform and Colonisation Law in 1945 affected land use changes	
	Land inheritance	The historical practice was often patrimonial but lately all heirs were included in land division. The land was also inherited by those who did not live on islands and it was often neglected and abandoned (especially fertile land and pastures)	
	Accession to the EU in 2013	New investments in infrastructure, intensive building of new second homes and houses for rent due to the growth of tourism	
	Natural conservation policy	There are three national parks (Brijuni, Kornati, Mljet), two nature parks (Telašćica, Lastovo) and many other protected natural (NATURA 2000) sites on islands. In those areas human activities are limited	

(continued)

Table 6.1 (continued)

Underlying drivers	Proximate drivers	Explanation	Manifestation of change
Economic	Economic transition	Transition from primary to secondary and finally tertiary activities in the second half of the twentieth century. Today, islands are places with predominantly developed tertiary activities	***At the turn of the twenty-first century*** Intensification: increase of built-up areas, increase of woodlands, increase of large vineyards plantations, and olive groves Extensification: decrease of pastures, decrease of total agricultural area
	Changes of economic systems	Transition from peasantry/free market to planned economy and to free market again	
	Agricultural extensification	Abandonment of agriculture (specially cattle breeding) due to tourism development and emigration	
	Subsidies in agriculture	Increase of subsidies after accession of Croatia to the EU in 2013	
	Strong growth of tourism	At the turn of the twenty-first century tourism became the most important activity on islands	
Demographic	Depopulation	Process of depopulation on many islands started at the end of the nineteenth century (migration, demographic transition). It continued up to today, especially on small and remote islands	
	Migration	The process of emigration is present for over a century, and it has affected all aspects of social-economic development and land use change	
	Ageing	Ageing is present in many islands' settlements, and it directly affects land use	
Spatial and natural	Small plots	Plots on islands are rather small which is a result of inheriting tradition and the fact that fertile land is very limited so agricultural land was divided in a way that all families had a small share of it	
	Unresolved ownership issues	Landownership changes were not regularly updated in ownership books	
	Environmental	Reforestation and fires	

Source Authors' own elaboration

relatives or siblings, who had left the island. It led to a chaos in ownership books and resulted in dividing the land into progressively smaller tracts which made the land use even more complicated.[29]

6.4.3 *Development of Services (Tourism)*

In the second half of the twentieth century an important driver that generated strong changes in land use was the transformation of economic structure and the development of services, especially tourism. It started developing on islands already in nineteenth century but its more dynamic development can be followed only after 1980s.[30] It caused the depopulation of central parts of the islands and the abandonment of primary activities: agriculture and fisheries. It also stimulated the construction of second homes and apartments used only during the summer season which resulted in a paradox: multiple increase of dwellings in the period of intensive depopulation. Larger islands, and especially the bridged ones, experienced significant changes of land use around former settlements when old vineyards, pastures, and gardens were replaced by apartment buildings and second homes which led to a formation of a typical second home landscape.[31]

Areas with fertile soil located in the vicinity of the coast, which are very rare in the predominant karstic island area, are repurposed for various tourist activities, especially on those islands that are closer to the mainland or are better connected with it. This contributes to the creation of bipolar landscape patterns on the Croatian islands: intensively used coastal zone near island settlements and most of the remaining island areas which are left to the succession of vegetation. In both cases, traditional landscape patterns of terraced island slopes with olive groves and vineyards are gradually disappearing.[32]

6.4.4 *Population*

According to 2011 Census data, total population of Croatian islands was near to 125,000,[33] while the preliminary results from the newest 2021 Census shows that the number of islanders decreased to around 121,000.[34] The main demographic processes present on the islands during the twentieth century were emigration, depopulation,

---

[29] "Zakon o nasljeđivanju".

[30] Mikačić (1994), p. 519.

[31] Opačić (2008), p. 59.

[32] Faričić et al. (2010), p. 183.

[33] Klempić Bogadi and Podgorelec (2020), pp. 89–110.

[34] "Popisane osobe, kućanstva i stambene jedinice".

and population ageing.[35] The demographic peak was reached in 1921 (cca 167,000) and after that year, the number of islanders has been rapidly decreasing, generating important landscape changes. Demographic decline started at the turn of the twentieth century because of the crisis in grape-growing and wine production which were the most important economic activities on the islands. The fertile land had been used to the maximum before phylloxera arrived in the late nineteenth century and within a few decades destroyed almost all the vineyards.[36] So, the main push factor for emigration at the beginning of the twentieth century was the poor economy and the majority of emigrants were young males aged 15–26, mostly engaged in agriculture before leaving the country.[37] Due to gender-selective emigration and the demographic transition, the islands' population decreased during the following decades. Besides emigration to foreign countries which continued through the twentieth century, rapid industrialization which began in littoral Croatia after the Second World War also strongly affected the islands' population. From 1953 to 1981, the islands lost one quarter of their inhabitants and many of them moved to coastal cities where they got employed in industry. The result of this long-term emigration was population ageing which caused additional population problems on many Croatian islands after the 1960s.[38] Nowadays, elderly people who are engaged in agriculture mostly do it for personal needs. They cultivate only those land parcels that are closer to settlements, those that are more easily accessible and those with the highest quality of pedological features.

Economic transition to services and strong development of tourism in the last few decades led to socio-economic development of larger islands, urban island settlements and near, especially bridged islands. Tourism was the main driver of rapid building of second homes. Regardless of the falling population number on islands (2.3% decline in period 1971–2011), the number of dwellings is continuously increasing (3.4 times in the same period).[39] The small islands and small rural settlements are being transformed from a working–residential area to occasionally/seasonally used area which reflects in land use/cover changes.[40] Along with economic transition and abandonment of traditional sectors of activity, utilised agricultural area is decreasing. However, the built-up area in coastal islands' zones is expanding. Islander's number in the last few censuses is in stagnation or slightly increasing but demographic growth is ostensible.

---

[35] Mišetić (1961), p. 304.

[36] Faričić (2012); Čuka et al. (2017); Faričić and Čuka (2020).

[37] Čuka et al. (2020), p. 49.

[38] Lajić and Mišetić (2013), p. 169.

[39] Faričić and Čuka (2020), pp. 55–88.

[40] Faričić et al. (2010), p. 176.

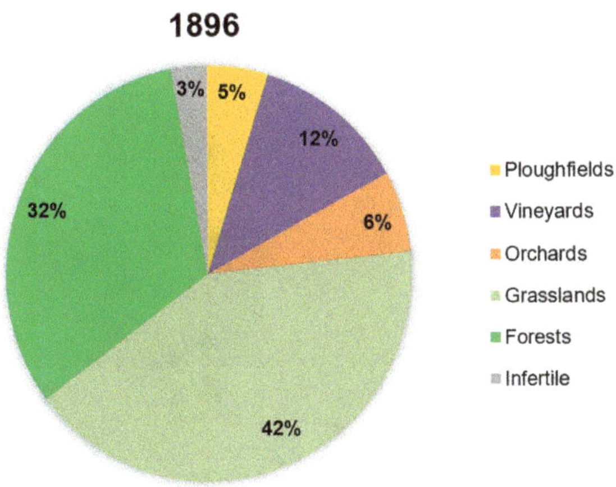

Fig. 6.2 Land use on Croatian islands in 1896. *Source* Based on *Gemeindelexikon* (1906); *Općinski rječnik* (1908)

6.5 Land Use/Land Cover Changes on Croatian Islands

6.5.1 Changes Until 1950

Late nineteenth century data indicates that the majority of the agricultural land on all islands was used for grasslands and pastures, one third of the islands surface was covered by forests and 12% by vineyards (Fig. 6.2). This large share of vineyards is the result of conjuncture in wine growing in Croatia after phylloxera destroyed the French, the Spanish, and the Italian vineyards. Croatian farmers, especially in littoral areas and the islands used the opportunity and converted other land plots (previously covered in olive trees or in macchia) to vineyards. On some islands like Brač, vineyards covered up to 30% of the total island area and on the small Prvić Island almost 37%. Olive trees still covered 6% and ploughfields just 5%. Due to shallow soils, there were not many suitable locations to grow grain, it was, therefore, usually imported from the mainland.

During the next half a century, arable land experienced the largest changes (Fig. 6.3). After the conjuncture in the 1880s and 1890s, in the early twentieth century phylloxera devastated vineyards on Croatian islands. They never recovered to their previous size. Other changes were the increase of the infertile land due to the increase of the built-up areas because of the demographic growth and the increase in pastures.

6.5.2 Changes in 1980–1990 Period

In total, 145 km^2 of land cover changes have been recorded (4.4% of total area surface), making this the period with the most dynamic land cover changes. The

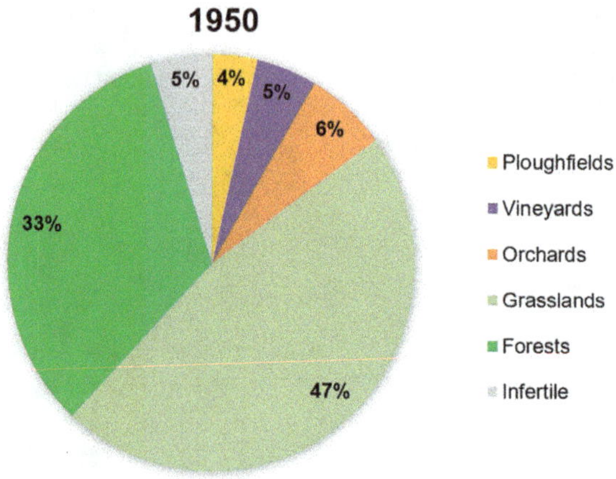

Fig. 6.3 Land use on Croatian islands in 1950. *Source* Made according to Rubić (2017)

most common changes were different stages of secondary succession (CORINE class 3.2 Shrub and/or herbaceous vegetation associations into 3.1 Forests) which account for 37.8% of all changes. Additional 15.9% of changes is linked to changes from agricultural areas (class 2) to forests and semi-natural areas (class 3), most often class 2 (Pastures) into 3.2 (Shrub and/or herbaceous vegetation associations). This is possibly linked to the processes of agricultural abandonment.

It is followed by changes from 2.4 Heterogeneous agricultural areas to 3.3 Open spaces with little or no vegetation, which account for 37.5% changes, mostly possibly due to wildfires. The increase in artificial surfaces (class 1) is relatively minor and accounts for 0.14 km^2 per year, or less than 1% of all recorded changes in the study period. The changes in vineyards and olive groves, a category often linked with the Adriatic islands, were very small. Some processes of abandonment and overgrowing were recorded, but the overall change during the decade was a reduction of only 0.87 km^2.

Looking at specific islands, there is an obvious difference between north and south, where southern (Dalmatian) islands recorded stronger changes compared to islands in the north (bay of Kvarner) area. Brač island recorded changes on 43.4 km^2 (11% of its surface), followed by Hvar (25.9 km^2 8.7%) and Korčula (24.2 km^2, 8.9%). On the other hand, changes were recorded on only 1.9 km^2 or 0.4% of the surface of Krk Island, and 6.1 km^2 or 1.5% of Cres Island. Most changes follow the dominant trends established in the area, but on Brač 56% of all changes are different types of forest degradation, probably linked to fires, while on the island of Vir 14.5% changes are linked to expansion of artificial surfaces.

The situation in 1990: the most common land cover on Croatian islands is Shrub and/or herbaceous vegetation associations (3.2) covering 940.2 km^2 or 28.7% of total area surface. It is followed by forests (3.1) which cover 867.7 km^2 or 26.5%, and heterogeneous agricultural areas (2.4) covering 560.2 km^2 or 17.1%. Smaller categories include pastures (2.3) on 336.2 km^2 or 10.3% and permanent crops 2.2 on 189 km^2 or 5.7%. Artificial surfaces cover 2.9% of the entire area.

6.5.3 Changes in 1990–2000 Period

The changes in the 1990–2000 period were much smaller in scale—only 25.2 km^2 or 0.8% of the total land cover has changed, with the major trajectories of change remaining similar to the previous period. It includes different types of forest degradation (possibly linked to fire)—the change from 3.1. Forests to 2.4 heterogeneous agricultural areas or to 3.2 Shrub and/or herbaceous vegetation associations accounted for 15.3 km^2 (60.7%) of all changes. Other major change is linked to secondary succession, where 3.2 Shrub and/or herbaceous vegetation associations changed to 3.1. Forests on 3.2 km^2 or 12.7% of total area surface. Even though the change in artificial surfaces are small in absolute numbers (1.2 km^2, 0.12 km^2 per year), there is a noted relative increase (4.8%) in the total share of all recorded changes.

Looking at specific islands, again most changes follow the dominant trend established in the period (changes in natural vegetation and agriculture, linked mostly to wildfires and secondary succession). In absolute numbers, the highest rate of changes were recorded on Korčula (10 km^2), Brač (3.9 km^2) and Krk 3.4 km^2, Lošinj 3.2 km^2 and Vir 1.9 km^2. The only thing that stands out in terms of the processes recorded is Brač Island where 23% of all changes are linked to increase in artificial surfaces.

6.5.4 Changes in 2000–2006 Period

Similar processes from the previous periods are observed in the 2000–2006 period where 21.9 km^2 of land cover changed, with 16.7 km^2 or 76.3% linked to different types of degradation of natural vegetation, probably linked to fires. Increase in artificial surfaces accounts for 5.3% of all recorded changes, and with an annual increase of 0.2 km^2, this is the period with the strongest changes in this category. Expansion of vineyards and olive groves has been recorded on 0.74 km^2 or 3.4% of all recorded changes, which is a noted increase from the previous periods.

Looking at the islands specifically, the highest absolute changes were recorded on Brač (6.7 km^2 or 1.6% of island's total area surface), Hvar 6.2 km^2 or 2.1%, and Pag 1.6 km^2 or 0.6% of island's area surface. Most changes are similar to the main established trends, except on Pag where 13% of all changes are increase in artificial areas, with the percentage going up to 65% of all recorded changes on Ugljan Island.

6.5.5 Changes in 2006–2012 Period

In the 2006–2012 period, the changes were recorded on 22.7 km^2, with 17.6 km^2 or 77.7% of all changes are different types of forest and shrub degradation, probably linked to wildfires. Interestingly, 12.63 km^2 or 54% of all changes are recorded on only one island—Brač, largely linked to the 2011 Milna fire which was one of the largest recorded fires in Croatian history. Artificial surfaces increased by only 0.5% in the study period, but olives and vineyards increased by 1.6 km^2 (7% of all recorded changes). Apart from Brač, stronger changes were recorded on the Pag island (1.7 km^2) where 91% of changes were different stages of vegetation succession, and Hvar (1.5 km^2), where 51% of changes were expansion of olive groves and vineyards.

6.5.6 *Changes in 2012–2018 Period*

In the 2012–2018 period, the changes were recorded on 12 km^2, again following the trends from previous periods. The strongest change was linked to forest degradation (Forests 3.1 mostly to categories 3.2 and 3.3) which were observed on 3.3 km^2 or 27.8% of all changes. It is followed by changes from 3.2 Shrub and/or herbaceous vegetation associations to 3.3 Open spaces with little or no vegetation which were recorded on 2.6 km^2 or 21.6%. Both these processes can be linked to the occurrence of wildfires. Increase in olive groves and vineyards was recorded on 1.8 km^2 or 15%, making this the period with the strongest increase in permanent agricultural areas. The increase in artificial areas encompassed 1.3 km^2 or 10.8%.

Looking at islands specifically, the largest absolute changes were recorded on Brač (2.8 km^2), Čiovo (2.2 km^2), Pag (2.2 km^2) and Hvar 1.7 km^2. The changes are similar to dominant trends established in the period, with some exceptions—on Čiovo 65% of recorded changes is forest succession, and on Korčula 27% is expansion of olive groves.

6.5.7 *The Situation in 2018 and Comparison with 1990*

The most common land cover on Croatian islands in 2018 is still 3.2 Shrub and/or herbaceous vegetation associations, which covers 1041 km^2 or 31.8% of total area surface. It increased by 101 km^2 from 1990, mostly due to secondary succession and forest degradation, often linked to wildfires. Forests are still the second largest type of land cover, covering 839.1 km^2 or 25.7% of total area surface. Their size has reduced by 28.6 km^2 compared to 1990. The third most common land cover type is heterogeneous agricultural areas 2.4 which cover 490.3 km^2 or 15% of the surface of Croatian islands. It has been reduced by 69.9 km^2. There is a noted increase in 3.3 Open spaces with little or no vegetation which have increased by 42 km^2, in 2.2 permanent crops which have increased by 24.7 km^2 and 1. Artificial surfaces which increased by 36 km^2.

Most common processes of land cover changes are linked to natural vegetation. The earlier period (1980–1990) is dominated by secondary succession, and in later periods, forest degradation linked to wildfires is more prevalent. Furthermore, expansion of olive groves and vineyards and artificial surfaces is much more prominent

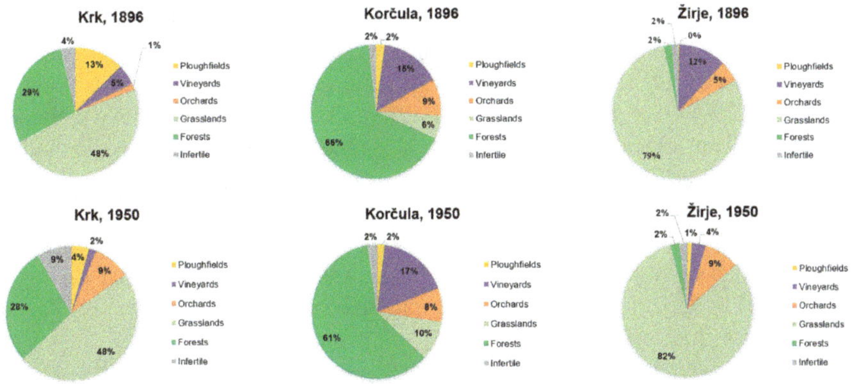

Fig. 6.4 Land use on Krk, Korčula and Žirje islands in 1896 and 1950. *Source* Based on *Gemeindelexikon* (1906); *Općinski rječnik* (1908); Rubić (2017)

after the year 2000. Generally, southern (Dalmatian) islands have recorded stronger rates of change, again mostly linked to wildfire occurrence.

6.5.8 Land Use Changes in Krk, Korčula, and Žirje Islands

During the same half century period (1896–1950) (Fig. 6.4), Krk displayed the largest change in reduction of ploughfields and vineyards and increase of orchards—olive grows. On the other hand, Korčula, a traditionally strong winegrowing region kept the large share of vineyards throughout the first half of the twentieth century. Phylloxera appeared on Korčula Island approximately 20–25 years later than on the islands in the northern Adriatic and Korčula's viticulture managed to recover from the plague rather successfully. It is also noticeable that in both periods forests covered around 2/3 of the Korčula Island. From antiquity, Korčula wears an eponymy *Corcyra Nigra* (Black Korčula) referring to its wide-spread black pine forests. Throughout the centuries, forests were managed and preserved until the studied period. Namely, the medieval town of Korčula reached laws that protected the forests from mismanagement and cutting.[41]

Žirje Island seems to have experienced a somewhat similar situation to Krk Island, a decrease in vineyards and an increase in olive groves. However, due to limited areas of arable land, Žirje was by far primarily covered in karst pastures and grasslands.

Krk Island.

From 1990 to 2000 land cover changes have been recorded on 3.3 km^2 which is < 1% of the total island surface. Out of these 3.3 km^2, the largest change (66.6%

[41] Anić et al. (2012), p. 170.

of all recorded changes) is from transitional woodland shrub (324) to coniferous forests (312). This could be linked to advanced secondary succession processes due to cessation of pasturing or traditional agricultural practices in coastal Croatia.[42] Another 24.7% of recorded changes were very similar, where transitional woodland/shrub (324) went into broad-leaved forests (311). The remaining 8.7% was the change from inland water bodies (512) to heterogeneous agricultural areas (243) linked to changes in water levels in Ponikve Lake.

From 2000 to 2006, the changes have been recorded on 3.4 km^2, which is again < 1% of the total surface. The changes were from different types of semi-natural vegetation (323) or agriculture (242 and 243) into discontinuous urban fabric (112) and mineral extraction sites (131).

In the period from 2006 to 2012, the changes have been recorded on 2.2 km^2, consisting of a transition from open space with little or no vegetation (333) into sclerophyllous vegetation (323), which could indicate a possible regrowth of vegetation after a fire.

The largest changes have been recorded in the period from 2012 to 2018, where 10.7 km^2 or 2.6% of total area surface registered land cover changes. Out of these 10.7 km^2, the largest change (39%) was from different types of forests into burnt areas (334). It is closely followed by a change from heterogeneous agricultural areas (243) and sclerophyllous vegetation (323) into artificial surfaces, namely industrial, commercial, and transport units (12) and mine, dump, and construction sites (13) which encompassed 38% of all recorded changes. The remaining 13% was a change from transitional woodland shrub (324) into arable land (211).

Korčula Island.

From 1990 to 2000, land cover changes have been recorded on 10 km^2, which is 3.7% of the total island surface. Out of all recorded changes, 90.5% is from broad-leaved forests (311) to heterogeneous agricultural areas (243), which is probably linked to wildfires. The remaining changes are from transitional woodland shrub (324) to heterogeneous agricultural areas (243).

From 2000 to 2006, land cover changes have been recorded on 10.5 km^2, which is 3.9% of the island surface. All changes are from coniferous forests (312) into shrub and/or herbaceous vegetation associations (324), again suggesting wildfires as a possible cause.

From 2006 to 2012, land cover changes have been recorded on 4.4 km^2, which is 1.6% of the total island surface. Most changes (43%) were from coniferous forests (312) into olive groves (223). This is closely followed by the change from mixed forests (313) into transitional woodland shrub (324), encompassing 41% of all changes recorded. The remaining 16% was from transitional woodland shrub (324) into olive groves (233), meaning that 59% of all recorded changes was olive groves expansion.

From 2012 to 2018, the changes were recorded on 12.9km^2, which is 4.7% of the total island surface. The strongest change was from forests (312) into transitional

[42] Jajtić et al. (2019), p. 323.

woodland shrub (324), encompassing 55.8% of all the recorded changes. This again suggests wildfires as a potential cause of change. The changes from forests (31) or arable land (21) into olive groves (223) encompassed 31% of recorded changes, and the remaining 13.2% of changes was from transitional woodland shrub (324) into forests (31).

Summary: Comparatively, Krk island has recorded relatively minor land cover changes, where in the 1990 to 2018 period, only 4.8% of the island experienced changes. They are mostly linked to secondary succession and wildfires, but changes linked to economic activities of mineral extraction are also noted. Korčula has recorded more intensive changes with 13.9% of the island experiences changes in land cover. Most changes are linked to wildfires, but in the later periods, expansion of olive groves is also noted.

No land cover changes have been recorded on Žirje during the studied period. Although the CORINE dataset was not precise enough to record changes, we assume that grasslands, macchia, and forests continued to spread.

6.6 Discussion

Results we obtained were compared with other research performed on LULC changes in Mediterranen islands. It is noticeable that the research on Greek islands are prevailing, in comparison with islands in the western Mediterranean. This is not unexpected, Greek islands are numerous and negative socio-economic processes that are taking place, especially on smaller islands, have been in the focus of scientists for the last couple of decades.

Results of the research on landscape change on Sifnos Island (Greece)[43] show that the maquis vegetation of *Juniperus phoenicea* expanded over time on former cropland area with hotspots of expansion mainly located away from settlements. The main drivers of landscape change were agricultural decline and tourism. This is similar to processes that occurred on Croatian islands.

Two studies of Lesvos Island, one on landscapes in general[44] and another in wood-pasture landscape[45] show that landscape changes have been caused by developments in economy and social structure. Formal agrosilvopastoral land-use system in the interior has been replaced by a mere grazing system. The abandonment of crop cultivation and intensification of grazing led to an opposite transition of oak woodlands with area expansion and loss in different parts and surprisingly to a general increase in oak forests. As a consequence, a more heterogeneous landscape evolved.

[43] Tzanopoulos and Vogiatzakis (2011), p. 58.

[44] Kizos and Koulouri (2006), pp. 330–342.

[45] Schaich et al. (2015), p. 110.

On Lemnos Islands,[46] the extensification of agriculture has been present since the 1960s. Changes have not been so intensive as on some other islands resulting in a specific rural landscape and abandonment of the marginal areas.

Methodologically, more complex studies have been performed for the Ionian Island, using future land use models and Markov Chain in one research,[47] and global (GLM) and local (GWR) regression approaches in another.[48] The results showed that the landscape of the Ionian Islands is expected to develop into two major patterns, one being the increase of forests and scrublands and another urbanisation of lowland and coastal areas, primarily due to tourism. In this case, tourism was positively perceived because it reduced the abandonment of agricultural land, leading to a combination of agricultural production with tourism activities.

The studies in Italian Islands of Elba,[49] Giannutri[50] and Sicily,[51] and on French Corsica[52] despite the difference in size and population, revealed the same prevailing processes. Land abandonment and increase of maquis and forests, especially in the hinterland, and antropogenization of the coastal belt due to the strengthening of service activities.

Land abandonment since the nineteenth century occurred on Corsica which led to expansion of shrublands and forests, particularly in mountain areas.

A group of authors[53] performed a meta-analysis and reviewed more than 900 papers on drivers of LULC on islands on global scale. Many of the LULC drivers are tourism related[54] but, according to their findings, the primary cause of the changes on islands in the Atlantic Ocean is urbanisation together with deforestation and agriculture.[55]

In conclusion, the LULC changes and processes occurring on the Croatian island are similar to those on other Mediterranean islands, especially smaller ones with similar demographic and economic issues.

In our research, we used several types of data—cadastral revisions and CORINE Land Cover derived from satellite imagery. It is noticeable that there is a different level of reliability of data, cadastral revision from 1896 being most reliable, 1950 data that were estimated through sampling and CORINE (especially from 1980 and 1990) least reliable due to the generalisation and minimal mapping unit covering 25 ha. Difficulty in analysis is that Croatian islands are numerous but small in size and any type of remote sensing analysis will lack absolute correctness. Nonetheless, the acquired data showed the general trend of land abandonment and reforestation during the last

---

[46] Dimopoulos and Kizos (2020), p. 103894.

[47] Kefalas et al. (2020), p. 1.

[48] Kefalas et al. (2019), p. 9.

[49] Carta et al. (2018), p. 78.

[50] Foggi et al. (2015), p. 437.

[51] La Mela Veca et al. (2016), p. 3.

[52] San Roman Sanz et al. (2013), p. 1.

[53] Walsh et al. (2020), p. 63.

[54] Walsh et al. (2020), p. 64.

[55] Walsh et al. (2020), p. 80.

120 years[56] which is the prevailing land cover process on Croatian islands. On the other hand, we managed to detect land use specifies such agricultural intensification in planting new vineyards and olive groves on some islands like Korčula and Hvar.

6.7 Conclusion

The results of the LULC analyses on Croatian islands show that in the last century the area has been faced with significant changes caused by various drivers. The main process that is occurring from the second half of the twentieth century is the abandonment of agriculture followed by succession of natural vegetation on all islands. The drivers of LULC change are predominantly socio-economic and connected to the main long-lasting processes such as depopulation, deruralization, ageing, abandonment of agriculture, intensification of tourism, intensive construction of second homes and apartments for rent, and economic transformation from primary to tertiary sector of activity.

In the late nineteenth century, the most valuable agricultural land on islands (fields) was covered with vineyards and wheat while the slopes were often used as olive grows or grazing areas. The arrival of grape disease phylloxera on the islands at the turn of the twentieth century, caused a huge wave of emigration of young islanders, while people who remained on islands were predominantly involved in agricultural activities up to the second half of the twentieth century. The main land use change in that period was decrease of vineyards and increase of olive grows, pastures, and grassland.

The second half of the twentieth century was a period of huge socio-economic changes. Industrialization and urbanisation on the mainland attracted more young islanders, males, and females, who decided to emigrate from predominantly rural settlements on islands. However, in the last decades of the twentieth century tourism developed in many island settlements and became the main economic activity. It led to the over building of second homes and apartments, especially on bridged islands, which influenced the increase of the share of the infertile area.

Even though most islands are faced with continuous decrease of utilised agricultural land and increase of forests (mostly maquis, scrub, and Aleppo pine), some islands recorded an increase in agricultural area. Increase in olive groves and vineyards has been recorded even during the last decade, mostly on larger islands with higher share of the agricultural land. Agricultural intensification has partially been a result of the high subsidies introduced in agriculture after Croatian accession in the EU. However, after 2012, the strongest change in land cover was linked to forest degradation which is the result of the wildfires, the main natural driver of LULC change.

---

[56] See also Chap. 5, in this volume: Tekić et al. (2024).

References

Sources

Corine Land Cover, Copernicus, European Environment Agency. https://land.copernicus.eu/pan-european/corine-land-cover. Accessed 23 Sept 2021

Croatian Geodetic Administration (2016) *Central Registry of Spatial Units in the Republic of Croatia (GIS shapefiles)*

Europe's Seas, Geodatabase, European Environmental Agency. http://cdr.eionet.europa.eu/hr/eea/clc. Accessed 5 May 2021

Gemeindelexikon der im Reichsrate vertretenen Königreiche und Länder, VII. Österreichisch-illyrisches Küstenland (Triest, Görz und Gradiska, Istrien). K. K. Statistischen Zentralkommission, Wien (1906)

Općinski rječnik za kraljevine i zemlje zastupane u carevinskom vijeću, XIV. Dalmacija. C. KR. središnja statistička komisija, Beč (1908)

"Popisane osobe, kućanstva i stambene jedinice, prvi rezultati Popisa 2021. po naseljima" [Census of population, households and dwellings in 2021—First results]. Croatian Bureau of Statistics (2021). https://view.officeapps.live.com/op/view.aspx?src=https%3A%2F%2Fpopis2021.hr%2Fassets%2Fxls%2Fpopis_2021_prvi_rezultati.xlsx&wdOrigin=BROWSELINK. Accessed 2 Feb 2022

"Zakon o nasljeđivanju" [Inheritance Law]. https://www.zakon.hr/z/87/Zakon-o-naslje%C4%91ivanju. Accessed 15 Feb 2022

Literature

Anić I, Meštrović Š, Matić S (2012) Značajniji događaji iz povijesti šumarstva u Hrvatskoj [Important events in the history of forestry in Croatia]. Šumarski list 136(3–4):169–176

Briassoulis H (2000) Analysis of land use change: theoretical and modeling approaches. West Virginia University. https://researchrepository.wvu.edu/cgi/viewcontent.cgi?article=1000&context=rri-web-book

Carta A, Taboada T, Müller JV (2018) Diachronic analysis using aerial photographs across fifty years reveals significant land use and vegetation changes on a Mediterranean Island. Appl Geogr 98:78–86. https://doi.org/10.1016/j.apgeog.2018.07.010

Čuka A, Mirošević L, Faričić J, GraovacMatassi V (2017) Phylloxera revisited: the spread of grapevine disease in Dalmatia and its influence on socio-economic development and agricultural landscape. Annales Series historia et sociologia 27(1):101–118. https://doi.org/10.19233/ASHS.2017.09

Čuka A, Matassi VG, Blaće A (2020) Emigration from Dalmatia (Croatia) to the United States from 1892 to 1924—analysis of the Ellis Island Database. Društvena istraživanja 29(1):49–69. https://doi.org/10.5559/di.29.1.03

Dimopoulos T, Kizos T (2020) Mapping change in the agricultural landscape of Lemnos. Landsc Urban Plan 203:103894. https://doi.org/10.1016/j.landurbplan.2020.103894

Duplančić Leder T, Ujević T, Čala M (2004) Coastline lenghts and areas of islands in the Croatian part of the Adriatic Sea determined from the topographic maps at the scale of 1:25000. Geoadria 9(1):5–32. https://doi.org/10.15291/geoadria.127

Fairbanks RG (1989) A 17000-year glacio-eustatic sea level record: influence of glacial melting rates on the Younger Dryas event and deep-ocean circulation. Nature 342:637–642. https://doi.org/10.1038/342637a0

Faričić J (2012) Geografija sjevernodalmatinskih otoka [Geography of North Dalmatian Islands]. Školska knjiga, Zagreb

Faričić J, Graovac V, Čuka A (2010) Croatian small islands—residential and/or leisure area. Geoadria 15(1):145–185. https://doi.org/10.15291/geoadria.548

Faričić J, Čuka A (2020) The Croatian Islands: an introduction. In: Starc N (ed) The notion of near islands: the Croatian archipelago. Rowman and Littlefield, Lanham, Boulder, New York, London, pp 55–88

Faričić J, Juran K (2021) Human footprints in the Karst landscape: the influence of lime production on the landscape of the Northern Dalmatian Islands (Croatia). Geosciences 11(8):303. https://doi.org/10.3390/geosciences11080303

Filipčić A (2001) Razgraničenje Köppenovih klimatskih tipova Cf i Cs u Hrvatskoj [Boundary separation of Köppen's climatic types Cf and Cs in Croatia]. Acta Geogr Croat 35:7–18

Foggi B, Lastrucci L, Geri F, Rocchini D (2015) Recent landscape changes on a small Mediterranean Island. Landsc Res 40(4):437–450. https://doi.org/10.1080/01426397.2014.902920

Food and Agriculture Organization (FAO) (1995) Planning for sustainable use of land resources. FAO Land and Water Bulletin 2. Food and Agriculture Organization of the United Nations, Rome

Geist HJ, Lambin EF (2002) Proximate causes and underlying driving forces of tropical deforestation. Bioscience 52(2):143–150

Jajtić K, Galijan V, Zafran I, Cvitanović M (2019) Analysing wildfire occurrence through a mixed-method approach: a case study from the Croatian Mediterranean. Erdkunde 73(4):323–341

Kefalas G, Kalogirou S, Poirazidis K, Lorilla RS (2019) Landscape transition in Mediterranean Islands: the case of Ionian Islands, Greece 1985–2015. Landsc Urban Plan 191(August):103641. https://doi.org/10.1016/j.landurbplan.2019.103641

Kefalas G, Poirazidis K, Xofis P, Kalogirou S, Chalkias C (2020) Landscape dynamics on insular environments of south-east Mediterranean Europe. Geocarto Int. https://doi.org/10.1080/10106049.2020.1790677

Kizos T, Koulouri M (2006) Agricultural landscape dynamics in the Mediterranean: Lesvos (Greece) case study using evidence from the last three centuries. Environ Sci Policy 9(4):330–342. https://doi.org/10.1016/j.envsci.2006.02.002

Klempić Bogadi S, Podgorelec S (2020) The demography of the archipelago—migration as a way of life. In: Starc N (ed) The notion of near islands: the Croatian archipelago. Rowman and Littelfield, Lanham, Boulder, New York, London, pp 89–110

Kuemmerle T (2020) Post-socialist land use change in the Carpathians (PhD dissertation). Humboldt University, Berlin

Lajić I, Mišetić R (2013) Demografske promjene na hrvatskim otocima na početku 21stoljeća [Demographics changes on Croatian islands at the beginnings of the 21st Century]. Migracijske i etničke teme 29(2):169–199. https://doi.org/10.11567/met.29.2.3

La Mela Veca DS, Cullotta S, Sferlazza S, Maetzke FG (2016) Anthropogenic influences in land use/land cover changes in Mediterranean Forest landscapes in Sicily. Land 5(1):3. https://doi.org/10.3390/land5010003

Lambin EF, Geist HJ (2006) Land-use and land-cover change. Springer, Berlin

Lambin EF, Geist HJ, Rindfuss RR (2006) Local processes with global impacts. In: Lambin EF, Geist HJ (eds) Land use and land-cover change. Springer, Berlin, pp 1–8

McKibben B (1989) The end of nature. Random House Trade Paperbacks, New York

Meyer WB, Turner BL II (1994) Changes in land use and land cover: a global perspective. Cambridge University Press, New York

Mikačić V (1994) Otočni turizam Hrvatske [Island tourism in Croatia]. Društvena istraživanja 3(4–5):517–529

Mišetić R (2010) Srednja Dalmacija: prostor diferenciranoga demografskog razvitka (1961–2001) [Central Dalmatia: an area with differentiated demographic development (1961–2001)]. Migracijske i etničke teme 26(3):297–319

Nakićen J, Čuka A (2016) Demografski razvoj otoka Brača i sklonost otočana iseljavanju [Demographic development of the Island of Brač and islanders' tendency to emigrate]. Migracijske i etničke teme 32(3):319–351. https://doi.org/10.11567/met.32.3.2

Nejašmić I (1991) Depopulacija istočnojadranskih otoka i izumiranje kao moguća demografska perspektiva [The depopulation of the East Adriatic islands and the dying out as a possible demographic perspective]. Migracijske i etničke teme 7(1):77–99

Opačić VT (2008) Second homes of the island of Krk—spatial distribution and structural characteristics of their owners. Geoadria 13(1):41–80. https://doi.org/10.15291/geoadria.567

Plieninger T, Draux H, Fagerholm N, Bieling C, Bürgi M, Kizos T, Kuemmerle JP, Verburg PH (2016) The driving forces and landscape change in Europe: a systematic review of the evidence. Land Use Policy 57:204–214. https://doi.org/10.1016/j.landusepol.2016.04.040

Rubić I (2017) *Naši otoci na Jadranu* [Yugoslav islands in the Adriatic]. Matica hrvatska, Ogranak Zadar, Zadar

San Roman Sanz, Fernandez C, Mouillot F, Ferrat L, Istria D, Pasqualini V (2013) Long-term forest dynamics and land-use abandonment in the Mediterranean mountains, Corsica, France. Ecology and Society 18(2):38. https://doi.org/10.5751/ES-05556-180238

Schaich H, Kizos T, Schneider S, Plieninger T (2015) Land change in eastern Mediterranean wood-pasture landscapes: the case of deciduous oak woodlands in Lesvos (Greece). Environ Manag 56(1):110–126. https://doi.org/10.1007/s00267-015-0496-y

Simmons IG (2008) Global environmental history: 10,000 BC to 2000 AD. Edinburgh University Press, Edinburgh

Sluiter R, de Jong SM (2007) Spatial patterns of Mediterranean land abandonment and related land cover transitions. Landscape Ecology 22:559–576. https://doi.org/10.1007/s10980-006-9049-3

Tekić I, Fuerst-Bjeliš B, Durbešić A, Radeljak Kaufmann P, Cvitanović M (2024) Landscape change and fire risk in the Croatian Dinaric Karst—looking back and moving forward. In: Fuerst-Bjeliš B, Mrgić J, Petrić H, Zorn M, Zwitter Ž (eds) Environmental histories of the Dinaric Karst. Springer, Cham, pp 109–140

Turner BL II, Skole DL, Sanderson SE, Fischer G, Fresco L, Leemans R (1995) Global change report: land-use and land-cover change. Science/Research Plan. IGPB Report No. 35

Tzanopoulos J, Vogiatzakis IN (2011) Processes and patterns of landscape change on a small Aegean Island: the Case of Sifnos, Greece. Landsc Urban Plan 99(1):58–64. https://doi.org/10.1016/j.landurbplan.2010.08.014

van Vliet J, de Groot HLF, Rietveld P, Verburg PH (2015) Manifestations and underlying drivers of agricultural land use change in Europe. Landsc Urban Plan 133:24–36. https://doi.org/10.1016/j.landurbplan.2014.09.001

Velić I, Vlahović I (eds) (2009) Tumač geološke karte Republike Hrvatske 1:300.000 [Explanatory notes of the geological map of the Republic of Croatia in 1:300.000 scale]. Hrvatski geološki institut, Zagreb

Vlahović I, Tišljar J, Velić I, Matičec D (2005) Evolution of the Adriatic carbonate platform: palaeogeography, main events and depositional dynamics. Palaeogeogr Palaeoclimatol Palaeoecol 220(3–4):333–360. https://doi.org/10.1016/j.palaeo.2005.01.011

Walsh SJ, Brewington L, Laso F, Shao Y, Bilsborrow RE, Arce Nazario J, Mattei H, Page PH, Frizzelle BG, Pizzitutti F (2020) Social-ecological drivers of land cover/land use change on islands: a synthesis of the patterns and processes of change. Springer, Berlin, pp 63–88. https://doi.org/10.1007/978-3-030-43973-6_3

Zaninović K (ed) (2008) Climate atlas of Croatia 1961–1990, 1971–2000. Državni hidrometeorološki zavod, Zagreb

Zimmerman R (1997) Strategija razvitka otoka s prostorno-planskih gledišta [Island development strategy from a spatial planning point of view]. In: Starc N, Kaštelan-Macan M, Ćurlin S (eds) Nacionalni program razvitka otoka. Ministarstvo razvitka i obnove, Zagreb, pp 33–126

Ante Blaće Associate professor, PhD. His research focuses on land-use changes, landscape development and implementation of quantitative methods in geography. He has collaborated on several projects that dealt with different aspects of landscape and land-use changes. He previously served as a technical editor for Geoadria, a scientific journal published by the University of Zadar and

Croatian Geographical Society—Zadar, from 2015 to 2022. Since 2022, he has been serving as the executive editor of the same journal.

Marin Cvitanović Principal academic in human geography. His main areas of interest are socio-environmental systems, with an emphasis on human drivers of land use/cover changes, and cultural ecosystem services. His work is rooted in geography, but his interests extend to environmental history, biology, cultural studies, social psychology and political science. His previous research focused on post-socialist landscape transformation in Croatia, human-induced mangrove changes in East Africa, endangered species' habitat fragmentation in the UK, and the evolving significance and functions of urban allotments in Zagreb, Croatia.

Anica Čuka Researcher and professor, PhD. She teaches courses on Agricultural and Rural Geography, Regional Geography of Australia, Oceania and Antarctica, Geography of Croatian Islands, and Croatian Emigration. She obtained her doctoral degree with a thesis on the transformation of the Dugi otok Island Landscape changes. Her scientific research focuses on topics within the field of social geography. She frequently examines Croatian islands in her work, particularly their socio-economic development, and landscape and land use changes. She is currently collaborating on a project at the University of Zadar that deals with island temporality.

Josip Faričić Professor, PhD. His main areas of interest are cartography, historical geography and geography of Croatian islands. He studies old maps as sources of spatial data, which, together with archival and other sources, enable research into historical changes in spatial functions and thus the landscape. He is the principal researcher of the scientific project Early Modern Nautical Charts of the Adriatic Sea: Information Sources, Navigation Means and Communication Media, funded by the Croatian Science Foundation. From 2023, he is the Rector of the University of Zadar.

Part II
Woodland and Forests

Chapter 7
Deforestation and Reforestation Attempts of the Military Frontier's Coastal Dinaric Karst (1762–1881), Croatia

Filip Pavelić, Hrvoje Petrić, and Mislav Radošević

Abstract The Dinaric Karst is the dominant landscape of the eastern Adriatic coast that was under the administration of the Habsburg Military Frontier. In the second half of the eighteenth century, the forests located on the coast of the Military Frontier came under increased pressure from the local population and the Habsburg authorities. This, in cooperation with harsh environmental conditions and their impacts (wind, rainfall, soil erosion), contributed to the accelerated deforestation of the Dinaric Karst forests. The process is most visible in the narrow coastal strip squeezed between the Adriatic Sea and the Velebit and Kapela mountains. Even today, their slopes show a distinct difference in forest areas—the east-facing slope in the region of Lika is characterized by rich and dense forests, while the western slope, impoverished and barren, is primarily covered with low vegetation, and rare forest patches. This article deals both with the deforestation of the coastal Dinaric Karst in the eighteenth and the first half of the nineteenth century and with attempts at reforestation in the second half of the nineteenth century. The period of deforestation was marked by the unsustainable forest management by the local population, which was affected by poverty and poor infrastructure caused by the difficult Dinaric Karst environment, with simultaneous attempts by the state to impose its control over the forests, with the idea of economic exploitation of their resources. The second half of the nineteenth century was the time of the beginnings of organized reforestation in that area, but also in the whole of Croatia.

F. Pavelić
Ph.D. Student, Catholic University of Croatia, Zagreb, Croatia
e-mail: filippavelic1@gmail.com

H. Petrić (✉)
Department of History, University of Zagreb Faculty of Humanities and Social Sciences, Zagreb, Croatia
e-mail: hrvoje.petric@ffzg.unizg.hr

M. Radošević
Institute of Social Ecology (SEC), University of Natural Resources and Life Sciences Vienna (BOKU), Vienna, Austria
e-mail: radosevicmislav@gmail.com

B. Fuerst-Bjeliš et al. (eds.), *Environmental Histories of the Dinaric Karst*, Environmental History 17, https://doi.org/10.1007/978-3-031-56089-7_7

Keywords Habsburg military frontier · Deforestation · Reforestation · Sustainability · Karst expansion

> As far as Velebit's seabed enjoys southern-italian heat and vegetation, at mid-height Velebit is the temperature and vegetation of northern Germany or Prussia, and at the tips of the cold of Swedish or northern Russia. This is also due to the fact that the inhabitants of this area live as nomads, i.e. they are continuously moving.[1]

7.1 Introduction

The Dinaric Karst of the coastal part of the former Habsburg Military Frontier,[2] in the smaller, northern part, includes the coastal slopes of Mount Kapela (highest peak 1536 m), and further on the elongated coastal slope of Mount Velebit (1757 m) on the eastern Adriatic coast.[3] This area is intersected by numerous elongated valleys formed in less resistant rocks, the lowest parts of which are submerged, so today they can be recognized as sea bays.[4] Most of the coastline is difficult to access, and harbors and settlements have been formed in the more accessible bays.[5] The following map (Fig. 7.1) shows the researched coastal area that follows the mountain ranges of Velebit between the port town of Senj in the north, and the village of Tribanj Sv. Magdalena in the south, in today's Croatia (in between the black lines, the islands and continental part of Velebit interior are not included).

Parts of the coast of the Velebit and Kapela mountains (between Senj and Tribanj Sv. Magdalena in today's Lika-Senj County in Croatia), which belonged to the Military Frontier until 1881, even today leave an impression of harsh barrenness unmatched on the mainland part of the Croatian part of the Adriatic coast. The fundamental characteristic of the steep coastal slope climate is dryness despite relatively large amounts of annual precipitation, which is unevenly distributed throughout the year. The main interconnected factors of dryness in the growing season are low rainfall in the summer, permeable limestone substrate, sunny slopes and high evapotranspiration in the warm part of the year. Consequently, there are almost no surface water flows. In the colder part of the year, there is an abundance of precipitation, although mostly in the form of rainfall. On the western slopes of Velebit, snow is a rare occurrence. Probably the most important factor of dryness is also the *bora*, the wind that dominates the researched area with strong intensity.[6] The vegetation cover is also adapted to this.[7] These characteristics of the karst in the steep Velebit

---

[1] *Danica horvatska, slavonska i dalmatinska*, 5 February 1848, p. 22.

[2] Kaser (1997); Rothenberg (1960); Dávid and Fodor (2021); Ingrao (1994).

[3] Fuerst-Bjeliš and Lozić (2006).

[4] Rogić (1975), p. 126.

[5] Klemenčić (2017), p. 11.

[6] Kužić (2020), pp. 55–80.

[7] Rogić (1975), p. 127; Penzar et al. (2001), pp. 119, 124–133; Rogić (1957, p. 80); Klemenčić (2017), p. 12.

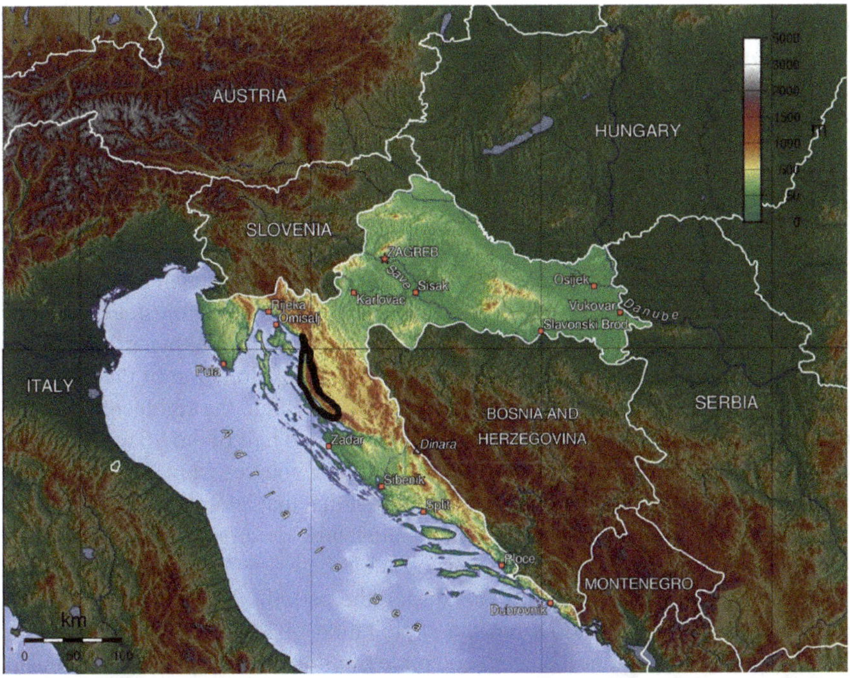

Fig. 7.1 Modern day Croatia. Black lines mark the area of coastal karst that used to be a part of the Military Frontier. Map obtained via www.croatiamap360.com

coastline were also analyzed in the work of the famous Serbian geographer and karst specialist, Jovan Cvijić.[8]

During the Habsburg administration of the Military Frontier, in addition to natural factors, the appearance of the natural vegetation cover was also influenced by communities with economy based on the traditional sheep and goat herding, as well as the preindustrial use of forests. From the foothills by the sea to the border of the forest in the higher areas, the landscape is dominated by rock because the proportion of exposed rock is greater than the area covered by vegetation. Part of the vegetation belongs to a degraded mixed forest, i.e., thickets of pubescent oak (honey oak, *Quercus pubescens Willd.*) and hop-hornbeam (association *Querco pubescenti-Ostryetum carpinifoliae*). In the last years of the Military Frontier administration, black pine (Pinus nigra Arnold) was used for afforestation of the coastal Dinaric Karst because of its resistance to wind (bora), dryness of the shallow soil cover and air.[9]

Coastal Dinaric Karst of Velebit western slope does not have equal width. It is widest around Jablanac or Stinica opposite the southern part of the island Rab and it

[8] Cvijić (1893), p. 98.

[9] Rogić (1965), p. 35; Rogić (1957), pp. 97–98; Rogić (1958), pp. 97–98; Vukelić et al. (2018), p. 102; Forenbacher (2002).

narrows toward Karlobag. In the northern zone between Ledenice and Sv. Juraj, the slope above the sea ends with a higher bend. The lowest parts of the soil are clearly separated from the interior by steep slopes, which are mostly bare or overgrown with scrub or artificially cultivated pine forest. The steep slopes are marked by a series of elevations that mark the second (higher) fold at 700–800 m above sea level, and it consists of a series of karst coves separated from each other by low partitions. Above the second (higher) fold rises the central Velebit ridge, whose height varies from 1200 to 1700 m above sea level.[10]

Special attention in research so far has been focused on Croatian Mediterranean forests,[11] especially those around the town of Senj,[12] i.e., their deforestation through different historical epochs and the attempt to find its causes. By looking at the previous research, it is possible to conclude the following: The exploitation of the forests of Senj is a process that can be monitored from the thirteenth to the eighteenth century; more intensive exploitation was present from the second half of the fifteenth century to the middle of the eighteenth century; cartographic sources from the second half of the sixteenth and seventeenth centuries suggest that the deforestation process and stripping of the coastal and parts of the mountain landscape around Senj were already present. The economic relations and wood trade between Senj and the Republic of Venice had short-term interruptions during the period of more intensive Uskok presence in Senj (1537–1617), but the exploitation of forests for the purpose of Uskok shipbuilding, war needs and wood trade by "domestic" merchants continued. Although Uskoks were displaced after the Uskok War (1615–1617), the immigration of a large number of Bunjevci-Vlach herder settlers to the Senj surroundings meant the process of deforestation continued, not only through the wood trade or the increased pressure put on forests due to the demographic growth of the population, but also due to the increased need for cattle grazing in an already vegetation-depleted environment, where young forests often perished because of goats and sheep, thus preventing the natural regeneration process of forests.[13]

In our research factors of deforestation, thus, degeneration of Dinaric Karst will be viewed as opposed by factors of reforestation and regeneration. We divided the factors of deforestation into environmental (relief, soil, climate, influence of the sea and winds, vegetation cover), economic (trade, extensive animal husbandry and overgrazing, transport infrastructure, presence of poverty and hunger as a result of communities socio-cultural status and agro-ecological policies) and political–social factors (absence/presence of central government, administrative challenges and problems, the relationship between the local community and the state government, legal regulations, traditional rights and administrative reforms). On the other hand, we

[10] Rogić (1975), pp.126–127.

[11] Fuerst-Bjeliš (2018), pp. 136–144.

[12] Kovačević (1981), pp. 175–196.

[13] Štefanec (2003), pp. 337–365; Bara (2012), pp. 11–31; Marković (1980), pp. 5–139; Bracewell (1992); Rukavina (1970), pp. 99–124; Grüenfelder (1999–2000), pp. 355–363; Sabljak (1990), pp. 267–272; Severinski (1965), pp. 232–245; Mlacović (2014), pp. 86–101.

presented the factors of reforestation primarily through administrative and educational measures and processes by the Military Frontier authorities, whose development can be traced back to the first Teresian Forest Code in 1765, until the establishment of the Karst Reforestation Inspectorate in 1878. Documents originated in this period use traditional units of measurement (such as acres and fathoms); therefore, they are here converted into today's metric measurement system.[14]

The paper used different types of sources as evidence in our environmental history research of the coastline of the Military Frontier and the forests located in that area. In the eighteenth century, these are maps of the Lika and Otočac regiments from ca. 1764–1766 and documents related to the adoption of the first Forest Code for the area of the Karlovac Generalate. Among them, the report of engineer Major Pierker stands out, who was tasked with surveying all the forests of the Lika and Otočac regiments in 1764 and 1765. Maps from the end of the eighteenth century (ca. 1774–1775) and a military map from ca. 1865–1869, which are available in digital format, were used as well.[15] In addition, a number of documents (reports, decisions, proposals) collected by Alexander Buczynski and created in preparation for the introduction of cantonal organization in 1787 in the Military Frontier were used.[16] For insight into microhistorical demographic processes, death registers from Roman Catholic parishes from Karlobag and Cesarica in the eighteenth and nineteenth centuries were also investigated.[17] Among the narrative sources for the eighteenth century, the description of the Karlovac Generalate from the pen of Ivan Dominik Vukasović (1728–1799), a priest who served in a number of Lika and Otočac regimental parishes, was used.[18] For the first half of the nineteenth century, the works of well-known officers and the first modern proto-statisticians of the Habsburg Monarchy, Johann Andreas Demian (1765–1845)[19] and Karl Bernhard Hictzinger (1786–1864),[20] were used as primary sources. For the later period of the second half of the nineteenth century, a study of the famous Austrian forester and one of the founders of the Imperial Forestry Association (*Reichsforstverein*) Joseph Wessely (1814–1898) was

---

[14] Hosch (1875), pp. 33–37. Available online at: http://anno.onb.ac.at/cgi-content/anno-plus?aid=fwk&datum=1877&page=163&size=45.

[15] HR-HDA-902, B.III.3, Karta Ličke pukovnije (Map of Lika Regiment), 1764–1765; HR-HDA-902, B.III.10, Karta Otočke pukovnije (Map of Otočac Regiment) ca. 1766; "Arcanum Maps", Maps of Europe, Europe in the XVIII. [!] century: Karlstädter Generalat (1774–1775)—First Military Survey. https://maps.arcanum.com/en/map/europe-18century-firstsurvey/?layers=163%2C165&bbox=1453482.967312411%2C5487111.036163977%2C2153319.4590046923%2C5754360.205598402ž; "Arcanum Maps", Maps of Europe, Europe in the XIX. [!] century: Croatia (1864–1865)—Second military survey of the Habsburg Empire. https://maps.arcanum.com/en/map/europe-19century-secondsurvey/?layers=158%2C164&bbox=1483604.285170887%2C5502028.354307513%2C2084093.5793792314%2C5731339.439163041.

[16] Buczynski (2011).

[17] HR-HDA-883, Parish Karlobag: Register of deaths 1774–1804;HR-HDA-883, Parish Cesarica: Register of deaths 1795–1839. Available online at: www.familysearch.org.

[18] Vukasović (2005).

[19] Demian (1806).

[20] Hietzinger (1817–1823).

used.[21] It should be emphasized that the statistical data provided by the sources are not fully verifiable. According to statistical methods of analysis, data for the eighteenth century cannot be considered completely accurate and can be viewed as indicators of certain trends, but not as definitive facts. The quality of statistical data in the nineteenth century was much better; since in 1807, the triangulation method was introduced in the measurement of land areas.

7.2 The Coastal Dinaric Karst of Military Frontier: Population, Forests and Administration in the Eighteenth and Nineteenth Centuries

7.2.1 *The Eighteenth Century*

7.2.1.1 Everyday Life of the Local Population

The life of the population of the coastal Dinaric Karst, a narrow piece of land squeezed between the Adriatic Sea and the steep western slopes of the Velebit and Kapela mountains, was difficult throughout the observed period, as well as deep into the twentieth century. In the period observed in this paper, that area was under the administration of the Habsburg Military Frontier, which is divided into generalates as larger administrative units and regiments as smaller ones. Its westernmost administrative unit is the Karlovac Generalate, and the coast of the Adriatic Sea was touched by Lika, Otočac and, to a small extent, the Ogulin Regiment.[22]

The weather conditions in the Velebit Podgorje (Velebit foothills, the name for the western/coastal slope of Velebit from the town of Senj to the river Zrmanja) are bad, often extreme—summers are hot and often dry, winters are cold with abundant snow at higher altitudes, and all of the above is accompanied by strong winds, especially the bora strengthened by the direction of the Velebit. Its influence, along with the process of erosion, is the most destructive for the vegetation of the Velebit Podgorje. Pierker (1766) writes that the wind frequently uproots trees and blows the fertile soil into the sea.[23] It is not only disastrous for the vegetation, but occasionally for the population as well, as shown by various extracts from the death register of the parish in Karlobag. In one such example, the pastor wrote in 1789 that four people drowned after their boat capsized due to an extremely strong gust of bora.[24]

This karst area is prone to soil depletion, as explained, and the carrying capacity of the land is very low. Hence, the fruits of the land do not meet the needs of the

[21] Wessely (1876).

[22] Moačanin (1991–1992), pp. 157–163; Holjevac (2014), pp. 553–571.

[23] Kosović (2015), p. 13.

[24] HR-HDA-883, Parish Karlobag: Register of deaths 1774–1804;HR-HDA-883, Parish Cesarica: Register of deaths 1795–1839.

population, and cycles of relative overpopulation are short. There is little fertile soil in that extremely karst area, and the population cannot use it to meet their nutritional needs. Pierker, Vukasović and Demian, narrative sources of the eighteenth and the very beginning of the nineteenth century, refer to folklore and state that in the "former times," the sea slope of Velebit was rich in vineyards, olive groves and forests. According to them, the coastal part consists of large stone masses that are under constant blows of rain and wind because there is no forest cover that protected them as in the past.[25] According to Rukavina, even in the twentieth century, there were folk tales that at the end of the nineteenth century, it was possible to descend from Velebit to the seaside with the help of the branches of trees, so dense were the woods.[26]

In the Velebit Podgorje, there are no areas where farming is possible, especially not to the extent required to feed the population.[27] Also, Velebit is a mountain poor in ores that could be exploited economically. In order to adapt to the environment in which it lives, the population was dominantly oriented toward extensive animal husbandry. Since there are no surface watercourses in that area, raising cows and horses was impossible, and the only domestic animals that were successfully raised were sheep and goats. Especially important was the goat, raised primarily for its milk and leather which was used for shoes and bags.[28] The Croatian Colored Goat, a historic local breed, is hardy, mobile and adaptable, accustomed to the karst landscape and able to take advantage of the nutritionally weak vegetation.[29] In its search for food, the goat is more than happy to graze all the vegetation it can get to, including tree saplings on which it gnaws buds and tops.[30] For this reason, goats were often targeted by laws that either restricted their grazing in the forest or required their extermination. One of the examples can be found in Friuli, in the village of Attimis, where it was forbidden to take goats into the forest as early as 1320.[31] In Pierker's report in 1766, which served as a template for the Forest Code, a complete ban on keeping goats in the Velebit Podgorje was proposed, explaining that it is impossible to create a valid forest as long as the population depends on them.[32] The basis of the orders for the Forest Code for the Karlovac Generalate states that within six years, it is necessary to remove all goats from that area, and during that period, they may no longer be allowed into the forest at any cost.[33] That ban lasted until 1803. A comparison of the number of goats in the Lika and Otočac regiments nevertheless indicates that the ban was unsuccessful. Although the number of goats decreases from 1770 to 1802, it is still high—in 1770, 21,613 lived in the Lika regiment on 50.33 square miles (2896.5 km^2), and in Otočac on 31.6 square miles (1818.6 km^2), 18,360 goats lived;

[25] Demian (1804–1807), pp. 38–39; Vukasović (2005), p. 124; Kosović (2015), p. 13.

[26] Rukavina (1990), p. 282.

[27] Kosović (2015), p. 14.

[28] Hietzinger (1817–1823), p. 187.

[29] Mioč, et al. (2008), p. 440.

[30] Vaniček (1875), p. 605.

[31] Degrassi (2014), p. 77.

[32] Kosović (2015), p. 14.

[33] Ibid., p. 77.

in 1802, there were 17,286 goats in the Lika regiment, and 12,513 in Otočac.[34] It should be emphasized that these are official statistics in which the number of heads is probably lower than the real situation. Some of the goats were probably "hidden" by the population in light of the ban and the desire to avoid additional taxation.

The forests of the coastal Dinaric Karst were not only suffering because of the population's unsustainable herding habits. In the area of Velebit, wood is a resource that is comparable to food and water in its importance for the life of the inhabitants. First of all, wood gives the inhabitants of mountain regions a roof over their heads and warmth.

Fire was one of the key elements in human lives, from prehistoric times until the twentieth century. Its importance was further emphasized in the mountainous regions of the Mediterranean, including Velebit, due to harsh climatic conditions. Modest dwellings where large families lived with many children required a constant flame on the hearth for most of the year. The fire was under no circumstances to be extinguished, as this often meant several hours of walking to the next house where live embers could be found.[35] Accordingly, the cutting of trees for firewood was also a matter of vital importance, so this resource was often overexploited. The consumption of wood was also affected by the infrastructural deficiencies of houses and mountain livestock farmsteads. Those mountain dwellings were the residence of herders who came with their cattle to higher altitudes in the summer, but in the Velebit Podgorje area, they were often converted into permanent residences, and there were no differences in the appearance of mountain and foothills houses. They were usually built in valleys that are somewhat protected from strong north or south winds, and where there is some land for cultivation. It was usually just one room that was used for sleeping, living and storing food, with a low ceiling and no floor. In the middle of the house, there was a hearth lined with stone slabs that retained the heat. The dwelling itself had no windows, but only one door. Some of the apartments were plastered with loam for thermal insulation, or moss was pressed into the space between the stones that make up the walls, with only rare exceptions being properly plastered.[36] The hearth in such dwellings used a lot of wood to generate the heat needed for heating all family members, cooking and baking bread, and in that time period, there were no energy-efficient stoves on the grounds of the Karlovac Generalate.[37]

7.2.1.2 Commercial Aspects of Forest Exploitation

An important element in the exploitation of timber from the coastal Dinaric Karst was trade. Several sides with different viewpoints were interested in selling wood from the forests of the Military Frontier coast. First of all, for the common people of the Velebit western slope, the sale of timber meant the possibility of making money,

[34] Roksandić (1988), pp. 87–92.

[35] Rukavina (1990), p. 283.

[36] Rukavina (1970), pp. 110–111.

[37] Buczynski (2011), p. 99.

either in terms of cash or goods such as grain and salt.[38] They also extracted trees from the forests on both slopes of Velebit, primarily focusing on more easily accessible trees, regardless of their age or possible planned forest renewal. However, bringing whole logs down from the mountain to the sea was an extremely difficult task in the eighteenth century due to the lack of quality roads. Sources state that the interior of Lika was often under a thick snow cover suitable for the transport of trees, but we do not have similar data for the coastal Dinaric Karst area and cannot estimate the precipitation in that area. Another option was to cut smaller amounts of wood and bring them down to the shore. At the same time, the beginnings of regulated forest management meant the state started to impose its influence in trade processes. The construction of state warehouses for wood enabled Frontier people to sell wood at a set minimum purchase price. Bringing order in the forest affairs, especially the directives on what can and cannot be cut down, was not to everyone's liking. This also led to the emergence of wood smuggling from the Velebit forests as a common economic activity.[39]

Military authorities frowned upon the local methods of forest management, or specifically the methods of felling trees. First sawmills appeared in the territory of the Karlovac Generalate only in 1763, and they were not yet widespread in the eighteenth century.[40] Until then, as well as for a long time after, felling was done with an axe. This was an exhausting job in which thinner trees often suffered—large and old trees, ideal for felling, represented a very demanding undertaking to cut down with an axe. Also, such a way of cutting wood produces much more "waste" than sawing—the felled wood had to be cut into smaller pieces, thus creating a larger number of splinters.[41] Sawmills, too few in number compared to the available timber, were built only in the continental part of Lika because there was a lack of streams that could move them in coastal Dinaric Karst. The Forest Code (*Waldordnung*) proposes the construction of sawmills for spruce and fir in Brušane and Pazarište (interior Lika) in order to stop their deterioration.[42]

Considering the lack of arable land in Lika, forest clearing was the most common way of acquiring new land, especially in the area of the Dinaric Karst under Velebit. The size of the estates of families squeezed between the Adriatic and Velebit was many times smaller than the estates of families in the continental parts of the Karlovac Generalate. Although this model of land expansion could bring some short-term help to poor families, in the long term, it had the effect of further washing away the soil and impoverishing the quality of the land. After a while storms and rains blew and washed away the agricultural land, mainly gardens, acquired at the cost of deforestation, now unprotected by trees and their roots. The rate of soil erosion and desertification was thus accelerated, which is attested in available statistical data: Between 1746 and 1764, in some coastal villages, the amount of arable land even decreased, despite the

[38] Grüenfelder (2002), p. 129.

[39] Rukavina (1990), p. 284.

[40] Kaser (2003), p. 393.

[41] Rukavina (1990), p. 284.

[42] Kosović (2015), p. 80.

process of forest clearing.[43] In its place came mostly bare karst of very limited use, where only goats could find food.

Despite the bans and regulations during the eighteenth century, forest clearing was not completely curbed due to the low levels of state government control in the area. The fact that from 1764 to 1787, the Karlovac Generalate recorded an increase in acres of usable land from 216,152 acres (1243.74 km^2) to 327,137 acres (1882.35 km^2) speaks of the long-term scope and trend of expanding arable land at the expense of forests. For the sake of comparison, between 1787 and 1895, this number will increase to 462,700 acres (2662.37 km^2), which means that the increase in the number of arable areas in the twenty years of the eighteenth century was comparable to the increase in the number during the entire nineteenth century.[44]

Roads across Velebit are an important part of the timber exploitation story. Velebit is a difficult mountain to traverse with a small number of usable passes. Historical maps give evidence that the Romans also tried to bypass Velebit as much as possible, and the only road that crossed the mountain in their time was the one that leads from Senj to Arupium near today's Otočac. Pavao Ritter Vitezović on his map from 1699 draws three roads across Velebit: Brinje–Žuta Lokva–Vratnik–Senj, Novi–Oštarije[45]–Karlobag and Lovinac–Sveti Rok–Mali Alan–Obrovac. In addition to these, the route Žuta Lokva–Otočac–Perušić–Novi–Gračac–Zrmanja–Knin[46] is also planned. The lack of quality roads is one of the reasons for the increased deforestation of the coastal forests of the Habsburg Military Frontier. Although there were undoubtedly larger and better-quality forests on the continental side, due to logistical difficulties, the inhabitants of the coastal Dinaric Karst area more often decided to cut down trees on their side of the mountain. The few groves that survived were located in hard-to-reach areas where it was almost impossible to extract the wood without damaging it completely.

The transport of logs from the slopes of Velebit to the coast was a big undertaking. The poor general condition of the roads leading from continental Lika to the seaports is evident from the fact that Pierker in his report put forward 10 proposals for the repair of the existing or construction of completely new roads on the territory of the Lika and Otočac regiments. He also states that not even a single log will be able to be carried out without considerable damage if those projects are not carried out.[47] His proposals were rejected because they were deemed unprofitable due to the price and logistical demands.[48] Because of the demanding nature of transport across the Velebit, in the second half of the eighteenth century, wood from the Military Frontier area was 20 percent more expensive than wood that came to port cities from other parts of the Austrian coast.[49] The road connection with the interior is the main

[43] Kaser (2003), pp. 386–387.

[44] Kaser (1997), pp. 486–488.

[45] Fuerst-Bjeliš et al. (2000), pp. 111–132.

[46] Feletar (2016), p. 48.

[47] Kosović (2015), pp. 66–67.

[48] Ibid., pp. 86–89.

[49] Gruenfelder (2002), p. 137.

reason why Senj was a town of greater importance to the authorities than Karlobag in this period. Although Karlobag has a better and more protected port than Senj, the delivery of goods to Karlobag was much more difficult, while Senj, located closer to central Croatia, profited from the construction of the *Jozefina* (Josephina) Road from Karlovac, which allowed Senj's development into the main Croatian seaport.

7.2.1.3 Preparations for the Adoption of Forest Code—An Insight into the Forest Conditions

The middle of the eighteenth century is characterized by the aspirations of the Habsburg Monarchy to increase its control over its peripheral parts. In 1762, Maria Therese, planning to use the rich timber of the Karlovac Generalate to develop the Monarchy's naval fleet, forbade the export of oak from the coastal part of the Military Territory and gave the Trieste-Commercial-Intendance (*Suprema Intendenza Commerciale per la Provincia mercantile del Litorale)* supreme control over the forests of the Karlovac Generalate. Oak was the most valuable tree because of its structure and toughness, as well as its various uses in shipbuilding. Beech wood could be used for oars and barrels, but it was more useful as firewood, which was very necessary in the cold conditions of Velebit anyway.[50]

In the same year, delegates from various court cabinets met to give their expert opinion on the possibility of improving the state of the forests in the Karlovac Generalate. The conclusions can be summarized as follows—it is necessary to create an accurate detailed report of the Karlovac Generalate forests, and an appropriate and consistently implemented forest order. The year 1762 marks the beginning of first professional forestry works, which consist of descriptions of forests, field surveys and mapping. At the beginning of 1765, forester Dragutin Franzoni was appointed governor of the first forest administration of the Karlovac Generalate, thus beginning the organized management of forests in the Croatian Military Frontier.[51] Unfortunately, no information about his education has been preserved, but given that the first forestry schools in Germany were just being founded at that time, it can be assumed he was self-taught.

Major Pierker carried out a precise survey of the Velebit and Kapela forests, from the border with Venetian Republic (Trstenica St. Magdalena, today's Tribanj Sv. Magdalena) to Mrkoplje (in the northeast, near Požega in Slavonia) and Ogulin (in the north, close to Senj), with a special focus on the forests located in the coastal Dinaric Karst area due to their easier exploitation. His report is the first comprehensive description of the state of forests in the Karlovac Generalate, including the biggest forest management problems and guidelines for improving the situation. He also refers extensively to roads, which he considers the main aggravating circumstance for the potential economic exploitation of forests. Therefore, he repeatedly points

[50] Štefanec (2003), p. 343.

[51] Božičević (2005), pp. 626–627.

out the need to repair existing roads and wood extraction tracks.[52] As for the Velebit foothills in particular, he states that at the end of the seventeenth century, the entire western slope of Velebit was overgrown with dense forest, while at his time, it was almost completely bare. For such conditions, he blames unregulated and arbitrary felling, but also the high intensity storms that destroyed bits of forest which survived the mismanagement of the local population. In the part of the Velebit Podgorje that belongs to the Otočac and Ogulin regiments (from Novi in the north to Prizna in the south), Pierker states that there are 5286 acres (30.41 km^2) of forest, with an average density of 174 trees per acre. There are 1646 acres (9.48 km^2) of forest in the coastal part of Lika regiment with an average density of 163 trees per acre (see Tables 7.1 and 7.2).[53] This shows how much richer in forest areas was the northern part of the Velebit foothills, while the part around Karlobag and further south in the middle of the eighteenth century was already heavily influenced by deforestation.

These differences are also visible on the maps of the Lika and Otočac regiments prepared by Major Pierker as part of his forest descriptions.[54] Pierker's maps are unfortunately not available in high-resolution digital copies, but nearly all the mentioned differences can also be seen in the First Military Survey maps of the Karlovac Generalate (1774–1775). The maps are combined into one larger map that is available in high resolution on Arcanum maps website.[55] Below you may see the map of the coastal karst in Karlovac Generalate, marking the area of the researched Velebit coastline (Fig. 7.2). Interesting to see here is the fact that Wessely does not recognize "meadows" as a distinct category. The reason for that could be because of off around 83,600 acres (480.95 km^2) of meadows, 21,000 acres (120.81 km^2) are forested meadows, 25,000 acres (143.82 km^2) are bare land meadows without vegetation, and only about 36,000 acres (207.11 km^2) are "real" meadows, meaning meadows with grassy vegetation and arable soil.[56] This is why the meadows are not represented on the map as a distinct category, but split between the other represented categories.

As well as forest areas three times larger in the coastal part of the Otočac regiment than those in the coastal part of the Lika regiment, there was also a great difference in the estimated number of trees. According to major Pierker, there were 901,230 trees on 5286 acres (30.41 km^2) in the coastal part of the Otočac regiment compared to only 288,480 trees on 1646 acres (9.48 km^2) in the coastal part of the Lika regiment.

It is also interesting to observe the types of trees in these two areas. In both cases, oak and beech are dominant, and they grow successfully and supply the state and the population with timber and firewood. Although contemporary sources do not make a distinction between different species of a certain genus, by looking at the

[52] Kosović (2015), pp. 11–49.

[53] Ibid., pp. 23–25, 43–45.

[54] HR-HDA-902, B.III.3, Karta Ličke pukovnije (Map of Lika Regiment), 1764–1765; HR-HDA-902, B.III.10, Karta Otočke pukovnije (Map of Otočac Regiment) ca. 1766.

[55] "Arcanum Maps", Maps of Europe, Europe in the XVIII. [!] century: Karlstädter Generalat (1774–1775)—First Military Survey.

[56] Wessely (1876), p. 34.

Table 7.1 Forests of the coastal part of Otočac Regiment by acres, number and type of trees per acre. *Source* Kosović (2015), pp. 43–45

Forests of the coastal part of Otocac regiment																
	Forest name	Senjska Draga		Krmpote		Sv. Juraj		Visibaba		Velika Brisnica		Trolokvc		Karaula North		Total trees
	Acres/Trees per acre	1.088	165	747	165	730	180	338	180	307	185	1.376	170	700	170	
Tree types Total/ per acre	Downy oak (*Quercus pubescens)*	43.520	40	29.880	40	21.900	30	10.140	30	6.140	20	55.040	40	28.000	40	194.620
	Maple *(Acer pseudoplatanus)*	0	0	11.205	15	0	0	0	0	0	0	0	0	0	0	11.205
	Montpellier Maple *{Acer monspessulanum)*	5.440	5	0	0	0	0	0	0	7.675	25	34.400	25	21.000	30	68.515
	Beech *(Fagus sylvatica)*	108.800	100	22.410	30	0	0	3380	10	6.140	20	48.160	35	28.000	40	216.890
	Hornbeam (*Ostrya carpinifolia)*	10.880	10	29.880	40	14.600	20	10.140	30	9.210	30	55.040	40	24.500	35	154.250
	Black pine (*Pinus nigra)*	10.880	10	0	0	73.000	100	33.800	100	18.420	60	0	0	0	0	136.100
	Ash (*Fraxinus excelsior)*	0	0	22.410	30	14.600	20	0	0	4.605	15	41.280	30	14.000	20	96.895
	Fir (*Abies alba)*	0	0	7.470	10	7.300	10	3.380	10	4.605	15	0	0	0	0	22.755
	Linden *(Tilia cordata/Tilia platyphyllos)*	0	0	0	0	0	0	0	0	0	0	0	0	0	0	0
																901.230

Table 7.2 Forests of the coastal part of Lika Regiment by acres, number and type of trees per acre. *Source* Kosović (2015), pp. 23–24

Forests of the coastal part of Lika Regiment																
	Forest name	Karaula South		Karlohag		Lukovo		Pavica Glavica		Sugarje		Olovac		Porobjenik		Total trees
	Acres/Trees per acre	347	130	147	200	133	110	59	140	647,5	200	46	180	266,5	190	
Tree types total/per acre	Downy oak (*Quercus pubescens*)	20.820	60	8.820	60	9.310	70	3.540	60	25.900	40	3.680	80	10.660	40	82.730
	Maple (*Acer pseudoplatanus*)	0	0	2.205	15	0	0	0	0	6.475	10	0	0	2.665	10	11.345
	Montpellier Maple (*Acer monspessulanum*)	6.940	20	4.410	30	2.660	20	2.360	40	19.425	30	1.840	40	7.995	30	45.630
	Beech (*Fagus svlvatica*)	0	0	5.880	40	0	0	0	0	38.850	60	0	0	13.325	50	58.055
	Hornbeam (*Ostrya carpinifolia*)	6.940	20	4.410	30	2.660	20	2.360	40	19.425	30	2.760	60	13.325	50	51.880
	Black pine (*Pinus nigra*)	0	0	0	0	0	0	0	0	0	0	0	0	0	0	0
	Ash (*Fraxinus excelsior*)	10.410	30	3.675	25	0	0	0	0	19.425	30	0	0	5.330	20	38.840
	Fir (*Abies alba*)	0	0	0	0	0	0	0	0	0	0	0	0	0	0	0
	Linden (*Tilia cordata/Tilia platvphvllos*)	0	0	0	0	0	0	0	0	0	0	0	0	2.665	10	2.665
																288.480

Fig. 7.2 Coastal karst under the jurisdiction of the Military Frontier. Map drawn by Joseph Wessely between 1872 and 1875 and published in Wessely (1876). Facsimile (Map legend in English is inserted for better visibility of data; the abbreviation, "W. Kl." designates "Wiener Klafter", i.e., Viennese fathom, 1.895 m)

current situation, we can try to guess which specific species lived in the area of the coastal Dinaric Karst of the Military Frontier. In the lower zone, on the foothills and inside the drywall, it is possible to find better developed thickets and smaller groves in which ash (*Fraxinus excelsior L.*), black hornbeam (*Ostrya carpinifolia Scop.*), white hornbeam (*Carpinus orientalis Mill.*), European nettle tree (*Celtis australis L.*), maple (*Acer monspessulanum L.*) and to a lesser extent downy oak (*Quercus pubescens Willd.*). Above the lowest area, in the smaller steep slopes and the series of grooved karst bays, there are irregularly scattered groves and thickets in which ash, white hornbeam and black hornbeam are present. In locations with a thicker layer of soil, there is also downy oak.[57] It is a degraded mixed forest, i.e., thicket of downy oak and black hornbeam (*As. Ostyro-Quercetum pubescentis*).[58] The wild black pine (*Pinus nigra Arnold*) exists only on the northern and southern part of the coastal slope of Velebit, and since it is easily susceptible to fires, it is possible that it died in the fire in other parts of the slope.[59] Thickets of beech (*Fagus sylvatica*) with only a few trees occur between 800 and 900 m above sea level. The beech tree spreads downwards in the dry valleys, even down to 250 m in the Senj valley. In the zone of 900–1000 m, there are large areas covered with thickets of beech, which in Senjsko bilo, the northernmost branch of Velebit, has the character of a forest.[60] In general, the Dinaric beech–fir forest is widespread on Velebit, where besides those two we can also find mountain maple (*Acer pseudoplatanus*) and spruce (*Picea abies*).[61] In his report, Pierker also mentions linden, but in very small numbers. Since he does not specify the type of linden, according to the current distribution, we can conclude that it is either a small-leaved (*Tilia cordata*) or large-leaved linden (*Tilia platyphyllos*), or possibly a combination of the two.

The biggest difference between the northern and southern part of the Military Frontier's coastal Dinaric Karst is the number of pine trees. While the northern part of the Velebit foothills was rich in pines (15 percent of the total number of woods in the Otočac regiment), in the middle part (the coastal area of the Lika regiment), there is not a single pine. The situation is the same with firs that grow only in the Otočac regiment. On the other hand, in the southern part of the Velebit Podgorje, there is a much higher percentage of Montpelier maple than in the northern part. In the coastal part of the Lika regiment, there was even a small number of linden trees, probably due to a few fertile sinkholes protected from the impact of storms (see Fig. 7.3).

Based on Pierker's report, in 1764 baron Beck drew up a temporary forest code for all the forests of the Karlovac Generalate with twelve points. It was compiled according to the forest code for Goriška as recommended by the court advisor Mygine.[62] Temporary code stipulates that not a single living tree may be cut down,

---

[57] Rogić (1957), pp. 97–98.

[58] Vukelić et al. (2018), p. 102.

[59] Forenbacher (2002); Rogić (1957), p. 99.

[60] Rogić (1975), p. 130.

[61] Vukelić et al. (2018), p. 145.

[62] Kosović (2015), p. 50.

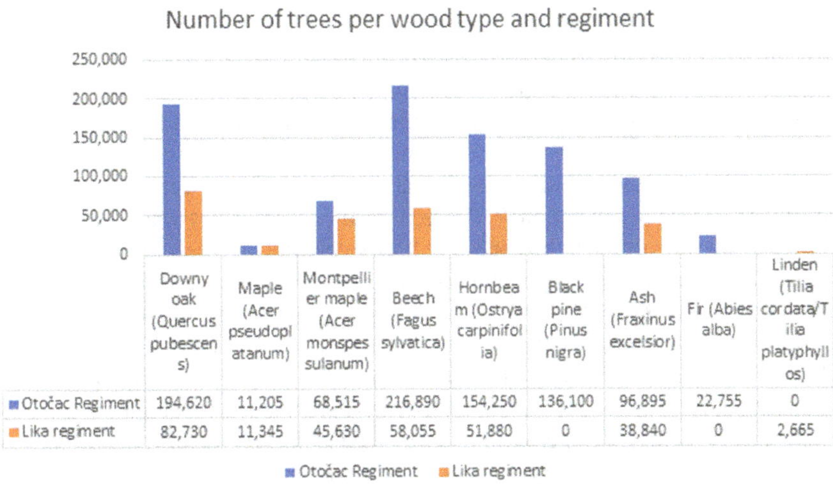

Fig. 7.3 Number of trees per wood type and regiment. *Source* Kosović (2015), pp. 23–45

with only fallen trees being allowed for use, defines a temporary method of monitoring forests, prohibits the keeping of goats with a six-year implementation period and orders the planting of acorns and beech trees and relocation of existing saplings in forests to municipal lands.[63]

7.2.1.4 The First Forest Code for Karlovac Generalate Forests

Based on the temporary forest code, the first proper Forest Code (*Waldordnung*) was created and confirmed for the forests of the Karlovac Generalate in 1765. It retains all the decisions of the temporary forest code and expands them with additional directives. They define which trees can be cut down and when (strict ban on cutting down oak trees without a special permit), an attempt is made to put an end to certain bad habits of the local inhabitants (whitening young oaks, cutting young trees for making tools and dishes, making small cuts in trees to check wood quality, cutting resin, burning fire in the forest, entering a young forest by cart), and fines are determined for the population and forest wardens in case of negligence (with only one witness enough to prove guilt).[64]

At the same time, a decision was made to assign three foresters to oversee the forests of the Karlovac Generalate. One forester was designated to the forests of the Lika Regiment and was based in Oštarije on Karlobag road, and another eight forest wardens were assigned to him (two foreign and six natives).

The forests of the Otočac regiment and the coastal part of the Ogulin regiment shared one forester, who was assigned 12 forest wardens (four foreign and eight

[63] Ibid., pp. 50–52.

[64] Ibid., pp. 76–82.

native), while the third forester oversaw the other forests of the Karlovac Generalate further inland. Instructions have been drawn up for foresters, which must be adhered to and implemented, and which coincide with the Waldordnung.[65] Forest codes have already been implemented within the German lands of the Holy Roman Empire during the sixteenth and seventeenth centuries as instruments of control of forest resources by the local nobility, but also, as in Karlovac Generalate case, by the central court administration.[66] The goal of the forest codes was to ensure the interest of the ruler through the control of the exploitation of forest resources. Since mercantilism was the leading economic policy of the Habsburg Monarchy at that time there, a vibrant trade of wood for lumber and firewood was encouraged. It developed parallel with the expansion of manufactures that used wood as energy (glassworks, steelworks, saltworks, etc.) or as a building material (shipbuilding, construction, carpentry, etc.). At the same time, there was the development of cities as urban and production centers within the general area of Adriatic, due to which the growing need for wood came into conflict with the traditional rights and methods of forest exploitation by the rural population, primarily peasants and serfs. Bernd-Stefan Grewe provides a good overview of the dynamics of forest codes and their impact on the environment and society from an ecohistorical, economic and legal-historical perspective in German historical lands.[67] Forest codes of the eighteenth century for the Military Frontier are therefore not only precursors or foundations of forestry codes from the middle of the nineteenth century, but also a legitimizing and legalizing instrument for the central military authorities of the Habsburg court of a mostly economic nature. Although the above-mentioned codes possess some ecological insights, it can be viewed as such almost exclusively from the position of royal and imperial foresters and the forestry profession. Forest codes therefore consider local traditional self-sustaining agroforestry methods and methods for gathering forest resources bad.

The process of reforestation is primarily tied to the nineteenth century, but it should be noted that even during the eighteenth century, there were attempts to plant new types of wood in the area of the coastal Dinaric Karst. At the request of Maria Theresa, a project to cultivate olives on the western slope of Velebit was launched in 1770. After initial successes (3248 saplings out of 3384 initially planted were successful), the project failed due to the cold weather, the Venetian ban on the olive saplings export and the transfer of the responsible captain Zandonati to Senj. Unfortunately, although it is stated in which areas the olive trees were planted, the exact altitude is not stated and due to the steep western slope of Velebit, it can vary significantly in the coastal part of Military Frontier. Attempts to plant almonds on that slope ended equally unsuccessfully due to a mice infestation which destroyed the saplings.[68] The period of the second half of the eighteenth century was characterized by various attempts to cultivate new plant species in the territory of Hungary and the

[65] Ibid., pp. 64–65.

[66] Sonnlechner (2004), pp. 268–277.

[67] Grewe (2004), pp. 24–41.

[68] Vaniček (1875), pp. 602–604.

Military Frontier. Mulberry silk began to be cultivated in 1765 at the intercession of Maria Theresa, as well as various plants from which dyes are obtained. At the same time, attempts were being made to plant cotton and rice in the flat parts of Slavonia, Serbia and Hungary.[69] Nikola Škrlec Lomnički, one of the most important politicians in Hungary at the end of the eighteenth century, also advocated for the further cultivation of flax, hemp and mulberry silk.[70]

Parallel with the questions of forest exploitation and forest conservation, the authorities had to deal with the question of forest privileges. Although all the forests are located on the territory of the Karlovac Generalate, under the jurisdiction of the military authorities, Senj and Karlobag were put under the Trieste-Commercial-Intendance management in 1754. They have their own statutes, and at the time of the adoption of the Forest Code, they were not under the jurisdiction of the military authorities. The question of forest privileges was raised in the relation between those towns and Intendance on one side, and the military authorities on the other. The dispute was present regarding the division of forest—which forests belong to the mentioned towns to meet their needs, and which are under the administration of the Karlovac Generalate and are to be used by the state? With the placing of these cities under military administration in 1776, the aforementioned disputes did not disappear, as the inhabitants of the cities continued to invoke their old rights to cut forests for their needs.[71]

The most heated topic was the forest at Senjska Draga, with the size of 1088 acres (6.26 km^2) in the immediate vicinity of Senj. In previous centuries, the trees were cut from it for the daily and commercial needs of the citizens of Senj. The entire forest was located on a slope facing the city, which made it suitable for transporting heavy logs to Senj. On the other hand, Senjska Draga could also serve as a warning to the rest of the coastal karst under the administration of the Military Frontier. Since the sixteenth century, due to the cutting of the forest, Senj began to face uncontrolled and dangerous torrents, flash floods and slides of gravel descending into the Senj harbor. In the second half of the seventeenth century, Senj was already at risk of floods and torrents following heavy rains and storms or the melting of snow in the mountains. The strength of the torrents also affected the disappearance of vegetation in the area of the Senjska Draga, and a small gate had to be built into the Senj walls so the water could pass through the town in the safest possible way.[72] In addition to the human factor of cutting the forest cover, the reasons for the mentioned problems can also be sought in the Little Ice Age (fourteenth–nineteenth centuries).[73] It was a global phenomenon which had an impact on the whole northern hemisphere, although its consequences were not the same in all of the areas that have been researched up to now. Generally, in Europe, it was a time with cooler and wetter conditions and general temperature decline per each season and on an annual

69 Škrlec Lomnički (2005), pp. 77–86.

70 Ibid., p. 202.

71 Grüenfelder (2003), p. 371.

72 Štefanec (2003), p. 344.

73 White (2014), p. 327.

basis.[74] Some of the phenomenon that are probably connected to the Little Ice Age include The Maunder minimum (1645–1715), a period of reduced solar activity that potentially led to lower global temperatures,[75] and a negative index of the North Atlantic Oscillation (NAO), which causes a higher amount of precipitation in the Mediterranean area.[76] Ultimately, in 1765, it was agreed, at least temporarily, that Senj would have jurisdiction over forests half an hour's walk from the city, while the rest would belong to the Habsburg Military Frontier.[77]

Another problem of the Military Frontier government is the wood trade. To prevent smuggling, a decision was made that all wood intended for sale must be sold to wood warehouses on the coast. However, it was very difficult to find the right price to give to the Frontier inhabitants. Although initially a price of 24 *kreuzers* (*xr.*) was set for each tree per one foot of circumference, it was not financially viable.[78] For this reason, at the end of 1765, Maria Therese decreed that only oak must be sold in wood stores, while the rest of the wood could be traded by the people of Senj and Karlobag. In order to prevent exploitation of the local people by merchants in Senj and Karlobag, the queen ordered that fixed prices should be defined, which will allow the people to sell the rest of the wood in the warehouse at a fixed price.[79]

All these problems shed light on the first challenges that the authorities had in bringing order to the forests of the Karlovac Generalate. The forest codes created under the influence of similar codes passed in the northern regions (e.g., the Code for Goriška in 1732) were not fully harmonized with the situation in the Karlovac Generalate—the difficult terrain for timber transportation and the local population living conditions, heavily influenced by the extreme weather conditions and poverty. One of the best examples of those conditions is a record by Joseph II made during his trip through the Karlovac Generalate. In his diary, he states that the inhabitants of Karlobag live in severe poverty and that they are hit by frequent violent storms during which it is not possible to travel by land, let alone by sea.[80]

During the cantonal reorganization of the Military Territory in 1787, the existing forest codes were confirmed, with certain details to which foresters and lumberjacks had to pay special attention, such as the explicit prohibition of grazing in forests or the necessity of using felled trees for firewood.[81] There is one interesting recommendation regarding the punishments for the local people—above all, admonitions and education of residents are advocated. Fines were charged only in the case of illegal deforestation when an unknown perpetrator was being searched for and when the entire municipality needed to compensate for the damage.[82] This

[74] Mrgić (2018), p. 140.

[75] Slavin (2016), p. 438.

[76] Scourse et al. (2012), p. 53.

[77] Kosović (2015), p. 69.

[78] Ibid., p. 68.

[79] Ibid., pp. 88–89.

[80] Erceg (1966), p. 248.

[81] Buczynski (2011), pp. 414–416.

[82] Ibid., pp. 177–178.

indicates that the care for the forests was left to the municipalities in a peculiar way—communal responsibility and care. If the unknown perpetrator was found, the compensation for damages was taken away from the already impoverished municipalities. Unfortunately, we have no data on how often such situations occurred.

According to all available sources, the western slope of Velebit in the eighteenth century was already experiencing advanced stages of deforestation. That was the result of the natural processes as well as the impact of the local population's economic activities. In the seventeenth century, the mentioned area was characterized by a sudden increase in the number of inhabitants due to immigration. Lack of arable land and poverty forced people to exploit timber or completely cut down forests in order to get another acre of arable land and generate additional income by selling timber. By reducing the forest cover, the fertile soil of the Velebit foothills lost its natural protection and became exposed to violent blows of the icy north wind and rainfall, not only triggered but accelerated erosion. When the influence of grazing sheep and (especially) goats is added, and on which entire families depend, the forests of the Dinaric Karst of the Velebit Podgorje did not actually ever receive a chance to recover. The constant pressure put on them due to their importance for local communities' survival made their recovery impossible until the second half of the nineteenth century, when the state itself seriously approached the reforestation project. According to McNeill, the resources of the future in mountainous regions have a much lower value than the resources that are necessary in the present, which is a sign of a subsistence economy.[83] The Velebit foothills and its forests in the eighteenth century representthe struggle to survive the difficult present, thus compromising the entire future of that ecosystem. In the end, we can conclude that the processes that took place in this area are very similar to those described by McNeill for the areas of other Mediterranean mountains.[84]

7.2.2 The Nineteenth Century

7.2.2.1 The Beginnings of Modern Forestry on the Dinaric Karst of the Military Frontier

When we talk about statistical data for the karst forests of the Military Frontier coast, we do not yet find completely relevant and reliable data at the beginning of the nineteenth century. In fact, it is difficult to talk about forests on the territory of individual regiments in their entire territorial scope since various territorial changes took place in the regiments between 1777 and 1815 as part of administrative reforms (the Josephine cantonal system) or as a result of political changes (the period of the French administration 1806–1813). The data left to us by Demian (1802–1804)[85] and

[83] McNeill (1992), p. 274.

[84] McNeill (1992).

[85] Demian (1804–1807).

Hietzinger (1816–1817)[86] are mostly aggregated for the areas of entire regiments and imprecise, since the trigonometric measurement of forestland only started in 1807. Hostinek[87] notes that in 1858, triangulation and surveying of the forest area of the Karlovac Generalate was still ongoing. According to the available data for the total forest area of the Karlovac Generalate, we can conclude that in the period between 1777 (642,865 acres or 3699.04 km^2 of forest area) and 1817 (592,954 acres or 3411.85 km^2 of forest area), about 49,911 acres (287.18 km^2) of forest was cut down.[88] This does not necessarily mean the total amount of forest trees cut down, but the territory that was no longer managed as a forest, but as some other type of arable or non-arable land. The amount of wood cut down is clearly greater, but the forest areas were obviously left to the forest to regenerate naturally.

Forest Ordinances of Josephine administration were confirmed with minor changes in the Basic Laws for the Military Frontier (Ger. *Gründgesetze für die Militär Grenze*) in 1807. The year 1801 is particularly important for the area of the Karlovac Generalate, when two "chief foresters" (ger. *Walddirector*) were appointed. In addition to these two chief foresters, there are a total of 96 members of forestry staff (foresters, forest rangers and other forest workers)[89] operating in the area of the Karlovac Generalate. The task of the foresters was to introduce forestry laws and order in the forests of the Military Frontier, according to the established norms of the German–Austrian forestry profession, i.e., according to the model of the imperial foresters in the Austrian part of the Monarchy. In the beginning, forestry in the Military Frontier area was extremely economic in nature, where the emphasis was placed on establishing control over forest resources and their exploitation. From the perspective of the state, i.e., the military authorities and the Habsburg Court in Vienna, it was a question of establishing state control and order in the newly conquered territory (first half of the eighteenth century), which for almost two centuries was a constant battlefield between the Ottoman Empire, the Habsburg Monarchy and the Venetian Republic. Situated on the periphery of all three imperial systems, in some parts, it was almost an anarchic state (e.g., the complete or partial absence of central administration, the dynamics of everyday life in the border war zone, safe havens for bandits and *uskok*s who plundered and lived off an "economy of violence," as a part of the so-called "gray economy," the absence or destruction of the existing transport, residential and economic infrastructure). The state of society was also "mirrored" in the state of forests. For this reason, there was a need to introduce order, supervision and care, not only to ensure the exploitation of the economic potential of the forests of the Military Frontier and to strengthen the Habsburg central authority in the area, but also to preserve this same potential for future generations. That especially related to the Dinaric Karst forests, which were threatened with destruction in certain places. For the purposes of forestry policy, professional development and education of future foresters was necessary, which was contributed to by the establishment of the first

[86] Hietzinger (1817–1823).

[87] Hostinek (1861).

[88] Hietzinger (1817–1823), p. 125.

[89] Hietzinger (1817–1823), pp. 135–136.

Forestry School in Mariabrunn in Vienna (1805/1813), which later became the first Forestry Academy (1867) where future foresters from the entire Monarchy were educated, as well as foresters from the area of the Military Frontier.

With the development of the profession, the expansion of forestry administrations and educational institutions, the forest began to be perceived from an ecological aspect, although the economic aspect was always the primary aspect in forestry practices. Today, the main postulate of forestry, the idea of permanence (sustainability), can be seen because of the economic and ecological aspect of forest management. The introduction of the new General Forest Act (1852), which entered into force in 1860, can be taken as the beginning of permanent forest management in the area of the Military Frontier. According to the new structure, the forests of the Military Frontier were managed by the Forestry directorates in Zagreb and Timişoara, and forestry offices operated at the regimental level, which supervised and administered the company and regimental forest areas.[90] From that law onwards, terms like "durability," "sustainability" and "reforestation" appear more and more in forestry literature, files of forestry offices and "on the field."

7.2.2.2 Deforestation and Reforestation of the Coastal Dinaric Karst of the Military Frontier

Although the previous chapters mentioned the economic and legislative interest of the military authorities in the establishment of forestry orders and administration over the forests of the karst coast of the Military Frontier, in the nineteenth century, it also took on an ecological character. Of course, the ecological character of the reforms from the eighteenth century cannot be ignored, although they primarily emphasized the "danger of a shortage of wood" (Ger. *Holznot*), which often served the rulers as the so-called "ecological reason or excuse" for adopting economic-administrative reforms.[91] However, it is understandable from the perspective of the forestry profession that a certain concern for the environment was created by the glaring whiteness of the bare rocks and karst slopes of the Velebit coast, which still dominate the view today if you travel the road between Senj and Karlobag. For this reason, even during the nineteenth century, there was an idea of resettlement the population of the Velebit coast to the continental part of the Military Frontier, but the famous Austrian forestry expert Joseph Wessely criticized the idea of the complete displacement of the population, because he believed that to recover the environment naturally, hundreds, if not a thousand years would have to pass. The karst soil has no potential for self-restoration, but only through intensive afforestation would it be possible to create a humus substrate and prevent further erosion of the karst soil. For the population, he suggests changing the economic paradigm and the semi-nomadic way of life to a more "cultivated," sedentary way of life, with more

[90] Hostinek (1861), p. 25.

[91] Grewe (2004), p. 34.

organized care for forests.[92] At the same time, this perspective can serve us as an example of the misunderstanding of the local population's traditional sustainable agroforestry practices by the forestry profession, which is guided by the postulates of the German school of sustainable forestry.[93] Although Wessely tries to protect the local population from forced migration, he still proposes changing the economic activities and the way of life of the same population, and therefore changing the environment.

In the 1870s, Wessely was employed to write a study on the forests of the Military Frontier on the karst slopes of Velebit. Very extensive, detailed and analytical study was published in 1876 under the title "The Karst of the Military-Croatia and its rescue and the question of karst in general" (*Das Karstgebiet Militär- Kroatiens und seine Rettung, dann die Karstfrage überhaupt*). This monumental work for Austrian and Croatian forestry on the issue of forests in the karst area informed the profession, the administration and the public about the problem of deforestation in the Velebit forests and the spread of karst. After that, the first measures were taken to solve the problems and challenges that Wessely stated in his study. The first step was the systematic and supervised afforestation (reforestation) of the Dinaric Karst of the Military Frontier. Although the afforestation method was already known at that time and in that area and was even applied in several cases, those afforestation processes were of a very local nature in a small area, since larger afforestation actions required better preparation, logistical capacities and ultimately financial resources.[94] Afforestation became more intensive after the adoption of the General Forestry Act of 1860, so, for example, the Administration of Border Forests in the period 1865–1872 reforested around 4464 acres (25.68 km^2) of common (municipal) pastures in the coastal karst area of the Military Frontier.[95]

The first major reforestation before the establishment of the Inspectorate was carried out in the "Senjska Draga I" protected forestland (ger. *Bannwald*) on a total area of 18.31 ha in 1869. For the need of further afforestation, the so-called "Inspectorate" (Ger. *Karstaufforstung Inspekorat für das Grenz Landesgebiet*) was founded as the main institution for the afforestation of the area between Povila (the northern border of the Military Frontier coast) and Tribanj Sv. Magdalena (southern border of the Military Frontier coast). The Inspectorate was financed from the funds of the Frontier Investment Fund and covered a territory of about 46,000 ha (later expanded to 51777 ha)[96] of completely bare or vegetation-poor karst.[97] The Inspectorate established 95 forest areas, so-called *branjevina-cultures* (karst wasteland on which vegetation has been restored through natural rejuvenation and afforestation, another name for *Bannwald*)[98] on about 10,097 ha, where about 1738 ha have been

[92] Wessely (1876), pp. 21–22.

[93] Panjek (2018), pp. 55–57.

[94] Ivančević (1999), p. 366.

[95] Wessely (1876), pp. 38–39.

[96] Ivančević (1999), p. 361.

[97] Kaser (2003), p. 394.

[98] Ivančević (1999), p. 366.

afforested, about 1441 ha of forests have been filled in, and about 7435 ha of karst have been rejuvenated by natural methods (Data refer to the entire period of activity of the Inspectorate between 1878 and 1942).

In the preface of his book, Wessely mentions that in the order of Emperor and King Franz Joseph I on the abolition of the Military Frontier on June 8, 1871, it was noted that the income from the surplus wood sold from the frontier forests would be directed to the Frontier Investment Fund for reforestation purposes. That means, clearing or deforestation of forests in one area of the Military Frontier (continental part of Velebit, forests of the Slavonian part of the Military Frontier) should finance reforestation or restoration of forests in another area. The Inspectorate founded its own sapling nurseries for the needs of afforestation, and during its activity, over 82 million young trees passed through the nurseries, some grown for local afforestation needs, some for afforestation somewhere else. Afforestation by planting was done in the traditional way, by burying a bare seedling in a dug hole measuring 40×40x40 cm. The Inspectorate is also considered the founder of the first sapling nursery in today's territory of the Republic of Croatia, which is the nursery of St. Mihovil near Senjska Draga, which was founded in 1879.[99] From that moment, the trend of opening sapling nurseries throughout the territory of today's Republic of Croatia slowly developed, primarily in those areas that were threatened by deforestation or that were partially bare. Today, there are 12 active nurseries in Croatia on a total area of about 307 ha.[100] In the first afforestation, it was mostly conifer seedlings (98.5%)—black pine (*Pinus nigra*) was dominantly planted with about 76% of the total seedlings. Spruce (*Picea abies*) had a share of 21%, while the share of other types of conifers was about 3%.[101] According to the average afforestation data gathered during the Inspectorate's 64-year period of activity, about 50 ha of area were afforested annually with about 233,000 pieces of seedlings and about 56.4 kg of seeds from a total of 78 types of seedlings (61 deciduous, 17 conifers).[102] Forester Eduard Malbohan, whose previously collected data from the field were used by Wessely in writing his study, was appointed as the first manager of the Inspectorate. Meta Remec's work in this book provides an interesting analysis and perspective on the methodology of reforestation of karst areas with resistant, non-indigenous cultures (such as black pine), derived from the German school of forestry and the impact of monoculture conifer plantations on ecological and social changes within the karst environment.[103]

The aforementioned study by Joseph Wessely is a rich source of data not only for the forestry profession, but also for a historical review of deforestation processes in the coastal karst area of the Military Frontier. Wessely gives a short socio-historical overview in the first two chapters of the book, where he states that more extensive deforestation processes began in the southern parts of the coastal karst (southern Velebit–northern Dalmatia) as early as the Middle Ages, and in the central and

[99] Ibid., p. 367.

[100] "Hrvatske šume, Rasadnici". https://www.hrsume.hr/sume/rasadnici/.

[101] Ivančević (1999), pp. 368–369.

[102] Ibid., pp. 373–374.

[103] See also Chap. 8, in this volume: Remec (2024).

northern parts no earlier than the seventeenth and eighteenth century, although he does not cite sources or literature from which he would draw such conclusions; therefore, they should not be viewed as indisputable facts.[104] The author's opinion on the causes of deforestation in the mentioned area is interesting, and he believes that the main culprit for the devastation of the karst vegetation is the expansion of pastures through imprudent and thoughtless deforestation and the destruction of the vegetation layer of the soil, which is a phenomenon that is also visible in other livestock areas in the Mediterranean.[105] The general impression of the author's criticism is that municipal land was most often not treated as a "common good" but rather as a "no man's land," free to be exploited by all without prudence. Wessely sees the solution for the preservation of karst forests and the greatest potential for reforestation in communal, municipal land, which is defined as pasture or meadow. In the area between Povile (ca. 20 km north of Senj) and Sv. Magdalena, there is about 83,600 acres (481.03 km^2) of such land, of which 21,000 acres (120.83 km^2) are land with low vegetation or bushes (which, according to Wessely, is mainly intended for forest), 46,000 acres (264, 68 km^2) is land with exclusively grassy (most often poor) vegetation, while the remaining 16,600 acres (92.06 km^2), or 25,000 acres (143.85 km^2) in total, according to Wessely, are bare rocks and bare karst fields.[106] Wessely's expressed views come from the perspective of a forester trained according to the principles of German school of forestry, which often directly clashed with the principles of local and traditional agroforestry techniques and methods of exploiting and shaping the environment of the local population. That is why, for example, the land that Wessely considers to be "poor" in terms of vegetation or where there is no "real" forest (high forest) with only some bushy vegetation (maquis shrubland), local population considers as "forestland" (low forests).

In his work on karst deforestation in Slovenia, Panjek states that the peasants considered a low forest with deciduous trees to be the most useful form of forest, since it was from such a forest that they extracted the most benefits and resources for their daily existence, primarily cattle grazing, firewood and, if necessary, wood for tool material and construction.[107] It is interesting to note that with both mentioned perspectives, the emphasis is actually placed on economic factors and not on ecological ones. Although Wessely and Panjek provide arguments for both perspectives, these arguments mainly refer to the (self-)sustainability of the economic model and the regulation of the use of environmental resources, not necessarily to ecological revitalization and the return of the environment to some sort of "original state" as it was before the anthropization. Since we will not deal with these phenomena in this paper, we will continue with the analysis of the data brought to us by Wessely. However, it was necessary to emphasize the existence of different perspectives and approaches to environmental issues in order to gain a better insight into the forestry policy of that period.

---

[104] Wessely (1876), pp. 3–4.

[105] Ibid., pp. 13–14; see also Mc Neill (1992).

[106] Wessely (1876), p. 34.

[107] Panjek (2018), pp. 59–61.

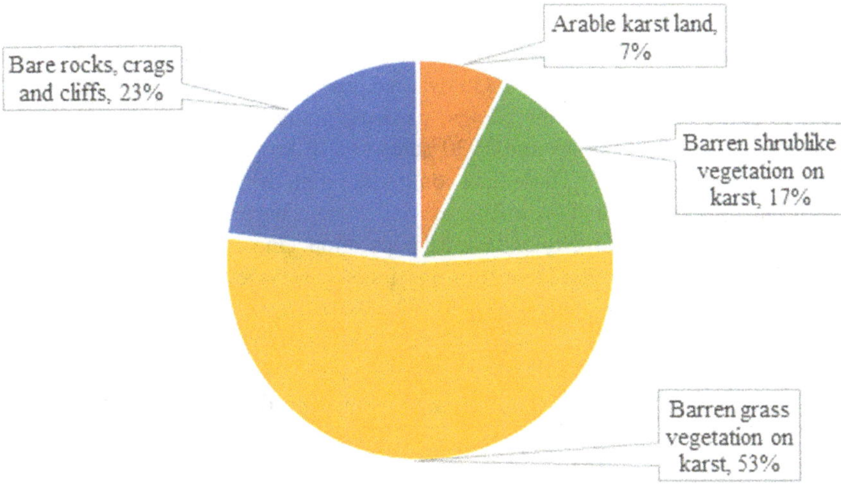

Fig. 7.4 Ratios of land types in the Military Frontier coast. *Source* Calculated according Wessely, (1876), 209

Statistical data from the second half of the nineteenth century (1855–1876) show us a better picture of the state of the environment, degree of deforestation and the spread of karst barrens in the coastal part of the Military Frontier (Fig. 7.4). We will use Austrian square miles (sm) to express the area. On the total area of 9 sm (about 518 km^2), there is only 0.62 sm (about 38.27 km^2) of arable land together with houses and gardens, while the remaining 8.4 sm (about 483.42 km^2) is karst wasteland and uncultivable land. Of the mentioned karst, about 1.5 sm (86.31 km^2) of the surface is covered with poor(er) vegetation and lower shrublike vegetation, while 4.75 sm (273.32 km^2) is covered with poor grassy vegetation, and an area of 2.08 sm (119.68 km^2) are bare rocks without vegetation.[108] Other data provided by Wessely are also interesting, where according to the classification of the environment, the following percentages are shown for the entire karst area of the Military Frontier (around 518 km^2): 21.2% of the territory is arable land (fields, orchards, meadows), 35.3% of the territory is forests, and 42.4% of the territory is pastures and uncultivable land, while the data for the exclusively coastal karst area of the three Frontier regiments (Ogulin, Otočac and Lika) are even worse: Only 7.4% of the territory is arable, and 92.6% is uncultivable soil with poor vegetation and bare karst.[109]

In the karst area of the Military Frontier, there are about 167 settlements, mostly hamlets and smaller, isolated villages, in which a total of 22,411 people live (an average of about 2420 people per 1 square mile). On average, one settlement (village, hamlet) has a total of 40 acres (0.23 km^2) of arable land and 500 acres of wasteland (2.88 km^2), and an average of 134 people live in one settlement.[110] For the sake

[108] Wessely (1876), p. 209.

[109] Ibid., p. 162.

[110] Ibid., p. 210.

of comparison, on the inner continental part of the three mentioned regiments (140 square miles), about 222,000 people live (an average of 1585 people per 1 square mile) in a total of 147 settlements. On average, one settlement has a total of 1470 acres (8.45 km^2) of arable land (fields, gardens, orchards and homesteads), 632 acres (3.63 km^2) of meadows and 1930 acres (11.10 km^2) of pasture, with about 1510 people living there.[111] These data confirm the high degree of deforestation and environmental devastation of the Military Frontier coast, but also the difficult survival conditions and agrarian poverty of the local population. One of the survival strategies was certainly illegal activity. Data from the forestry service show that in 1868, as many as 1209 cubic fathoms (1 $ftm^3 = 6.82\ m^3$) or about 8245 m^3 of wood were illegally cut in the coastal area of the Military Frontier. For example, nine families with about 160 people in the area of Lukovo Šugarje (ca. 20 km south of Karlobag at the seaside) lived and survived exclusively from the illegal cutting and sale of wood![112]

7.3 Conclusion

Different types of historical sources—primarily documentary and legislative, but also narrative, and cartographic, provided a solid base for the research of gathered data, which point to the following conclusions: The process of deforestation and environmental devastation of the Dinaric Karst of the coastal part of the Military Frontier was ubiquitous, in greater or lesser intensity, until the second half of the nineteenth century, and it was caused primarily by economic poverty and preindustrial agro-ecological regime of the local population. In the absence of arable land and mineral wealth for mining, the absence of manufacturing and industrial production on the coastal Dinaric Karst of the Military Frontier, the people lived in constant poverty on substance-based economy, often accompanied by years of hunger and famine. Main provision of food and income was from extensive animal husbandry and forest exploitation (traditional economic activities of many mountain peoples). The Habsburg government's attempt to intervene and improve these living conditions in the second half of the eighteenth century was primarily motivated by the idea of exploiting the timber potential of the area. The ideas of sustainability and concern for the forest environment were well known to the Habsburg government since the release of the *Sylvicultura Oeconomica* (1713). However, the idea of rich forests so close to seaports was too tempting for the court not to try to extract profit from them, as shipbuilding materials and trade items. At that moment, forests are the center of interest of two parties—the state and the population. The logging practice of the local population allowed strong winds and rains to wash away and erode the fertile soil exposed by the disappearance of the trees that preserved it with their underground root system. Under these conditions, a forest once cut down had no possibility of

[111] Ibid., p. 211.

[112] Ibid., p. 57.

regeneration because there was thin and sparse fertile soil which remained. Even when a young forest or some kind of bushy vegetation developed, it perished being eaten by grazing cattle, above all goats, whose cultivation and grazing were tried to be limited and prohibited by strict legal regulations. The regulations were not strictly enforced, since extensive animal husbandry was the most important subsistent economic activity of the local population, and goats, along with sheep, were one of the few animal species that could be raised in the karst area with poor vegetation cover and a lack of watering holes.

Although most of the sources of the time presented the ideas for reforestation in different ways, the state effectively accepted this undertaking only in the second half of the nineteenth century, after state administration implemented infrastructure: Logistical capacities (establishment of the Inspectorate for Afforestation, employment of professional staff, organization in the field, establishment of nurseries and growing saplings) and financial resources (Frontier Investment Fund) for planned afforestation on larger areas were secured. The Forestry Service developed following the German sustainable model of forestry initially had the task of introducing order and authority in state and communal forests and to arrange the same forests according to the principles of the sustainable model. It coped with these challenges until the middle of the nineteenth century, when the establishment and development of educational forestry institutions, together with administrative and legal reforms, enabled the necessary conditions needed to begin work on the ecological aspects of forests and their preservation. Nevertheless, the vision of some kind of "ideal" forest environment by the forestry profession differed greatly from that of the local population, which is also shown in this paper. Representatives of the forestry profession, as well as representatives of the central authorities, did not share the same vision of the environment and the way of exploiting its resources. The application of the German model of permanent forest management would often be opposed to local, traditional ways of exploiting forest resources, which could be categorized as agroforestry (agrosilviculture). On the coastal karst of the Military Frontier, this conflict of opinions and forestry methods in practice was best manifested in the intentions of the forestry profession to greatly limit and strictly regulate Dinaric-type livestock farming and timber trade—the economic base of the local population, in favor of the renewal and expansion of forest areas with high coniferous forests, from which the state itself would benefit economically. Due to the development and work of the forestry service in the area of the Military Frontier, from its beginnings with the Teresian Forest Order (1765) to the establishment of the Karst Reforestation Inspectorate (1878), the process of spreading karst clearings in the coastal Dinaric Karst was partially stopped, although the reforestation process itself contributed to changes in local ecosystems and thus opened up new problems and long-term challenges in the interaction between man and nature.[113]

[113] See also Chapter 8, in this volume: Remec (2024).

References

Sources

Published Sources

"Arcanum Maps", Maps of Europe, Europe in the XIX. [!] century. Accessed on 25 Oct 2022. https://maps.arcanum.com/en/map/europe-19century-secondsurvey/?layers=158%2C164&bbox=1483604.285170887%2C5502028.354307513%2C2084093.5793792314%2C5731339.439163041

"Arcanum Maps", Maps of Europe, Europe in the XVIII. [!] century. Accessed on 25 Oct 2022. https://maps.arcanum.com/en/map/europe-18century-firstsurvey/?layers=163%2C165&bbox=1453482.967312411%2C5487111.036163977%2C2153319.4590046923%2C5754360.205598402ž

Buczynski A (2011). Pa to su samo Hrvati! Građa za povijest kantonske reoganizacije Vojne krajine 1787. godine [They are just Croats!: sources for the history of the cantonal reorganization of the Military Frontier in 1787]. Hrvatski institut za povijest, Zagreb

Danica horvatska, slavonska i dalmatinska, 5 February 1848

Demian JA (1804–1807) Darstellung der Oesterreichischen Monarchie nach den neuesten statistischen Beziehungen: Theil 4 - Militärgrenze in Slavonien und Ungarn. Wien

Demian JA (1806). Darstellung der Oesterreichischen Monarchie nach den neuesten statistischen Beziehungen, Theil 4, Band 1 - Militär-grenze in Kroatien. Wien

Erceg I (1966) Dnevnik Josipa II. o prilikama u Hrvatskoj i na jadranskoj obali god. 1775 [The diary of Joseph II. regarding the conditions in Croatia and adriatic coast during the year 1775]. Starine 53:223–262

Hietzinger CB (1817–1823) Versuch einer Statistik der k. k. Militärgrenze des österrechischen Kaisertums (3 Bände). Carl Gerold Verlag, Wien

Hostinek J (1861) Die K.K. Militär-Grenze: Ihre Organisation und Verfassung, Theil 1. Kaiserlich-königliche Hof. und Staatsdruckerei, Wien

Kosović B (2015) Prvi šumarski stručni opis i nacrt šuma na Velebitu i Velikoj Kapeli od Dalmatinske međe do Mrkoplja i Ogulina [The first forestry professional description and plan of the forests on Velebit and Velika Kapela from the Dalmatian border to Mrkoplje and Ogulin]. Hrvatsko šumarsko društvo, Zagreb

Škrlec Lomnički N (2005) Stanje hrvatskog gospodarstva krajem XVIII. stoljeća [The state of the Croatian economy at the end of 18th century]. Dom i svijet: Ekonomski fakultet Sveučilišta u Zagrebu, Zagreb

Vukasović ID (2005) Zemljopisni i povijesni novi opis Karlovačkog generalata u Kraljevini Hrvatskoj [Geographical and Historical New Description of the Karlovac Generalate in the Kingdom of Croatia]. Holjevac Ž (ed). Državni arhiv, Gospić

Wessely J (1876) Das Karstgebiet Militär-Kroatiens und seine Rettung, dann die Karstfrage überhaupt. Commissions-Verlag der Universitäts-Buchhandlung Albrecht & Fiedler, Agram (Zagreb)

Unpublished Archive Records

HR-HDA-883, Parish Karlobag: Register of deaths 1774–1804

HR-HDA-883, Parish Cesarica: Register of deaths 1795–1839

HR-HDA-902, B.III.3, Map of Lika Regiment, 1764–1765

HR-HDA-902, B.III.10, Map of Otočac Regiment ca. 1766

Literature

Bara M (2012) Ekohistorijski pristup migracijama Bunjevaca [Eco-historical approach to migrations of Bunjevci]. Godišnjak za znanstvena istraživanja 4:11–31

Božičević I (2005) Karlovac, središte šumarstva Hrvatske vojne granice [Karlovac—The Forestry Centre of the Croatian Military Border]. Šumarski list 129(11–12):623–632

Bracewell CW (1992) The Uskoks of Senj. Piracy, Banditry, and Holy War in the Sixteenth-century Adriatic. Cornell University Press, Ithaca, NY

Cvijić J (1893) Das Karstphänomen: Versuch einer morphologischen Monographie. Ed. Hölzel, Wien

Dávid G, Fodor P (eds) (2021) Ottomans, Hungarians, and Habsburgs in Central Europe: The Military Confines in the Era of the Ottoman Conquest. In: Faroqhi S, Inalcik H (eds) The Ottoman Empire and its heritage, politics, society and economy, vol 20. Brill, Leiden–Boston–Köln

Degrassi D (2014) Water, wood, minerals: the resources of Friuli and their use between the XIIIth and XVth centuries. In: Štih P, Zwitter Ž (eds) Man, nature and environment between the Northern Adriatic and the Eastern Alps in premodern times. Ljubljana University Press, Faculty of Arts, and Historical Association of Slovenia, Ljubljana, pp 74–85

Feletar P (2016) Hrvatske povijesne ceste. Karolina, Jozefina i Lujzijana. Prometno-geografska studija o povezivanju kontinentalne i jadranske Hrvatske [Croatian Historical Roads: Karolina, Jozefina and Lujziana: Geotraffical Study of Connecting Continental and Adriatical Croatia]. Meridijani, Zagreb and Samobor

Forenbacher S (2002) Velebit i njegov biljni svijet [Velebit and its flora]. Školska knjiga, Zagreb

Fuerst-Bjeliš B (2018) Deforestacija i reforestacija u hrvatskome dinarskom kršu: odnosi i implikacije [Deforestation and Reforestation in Croatian Dinaric Karst: Human-environmental relations and implications]. Ekonomska i Ekohistorija 14:136–144

Fuerst-Bjeliš B, Lozić S, Perica D (2000) Man and the environment in the Central Velebit Area—Baške Oštarije and surroundings. Acta Geographica Croatica 35(1):111–132

Fuerst-Bjeliš B, Lozić S (2006) Environmental impact and change on the Velebit Mountain, Croatia: an outline of the periodization. In: Armiero M (ed) Views from the south, environmental stories from the Mediterranean World (19th–20th centuries). Consiglio Nazionale delle Ricerche, Istituto di Studi sulle Societa del Mediterraneo, Napoli, pp 127–139

Grewe B-S (2004) "Man sollte sehen und weinen!" Holznotalarm und Waldzerstörung vor der Industrialisierung. In: Uekötter F, Hohensee J (eds) Wird Kassandra heiser? Die geschichte falscher Oekoalarme. Franz Steiner Verlag, Stuttgart, pp 24–41

Grüenfelder AM (1999–2000) Karte kao izvori za proučavanje senjskih uskoka [Maps as a source for the research on the Uskoks of Senj]. Radovi Zavoda za hrvatsku povijest 32/33:355–363

Grüenfelder AM (2002) Senj i Karlobag u doba merkantilističke gospodarske politike [Senj and Karlobag during the mercantilist trade politics]. Senjski zbornik 29(1):125–154

Grüenfelder AM (2003) Velebitske šume od 16. do 18. stoljeća [Velebit Forests from the 16th until the 18th Century]. In: Roksandić D, Štefanec N, Glunčić-Bužančić V, Mimica I (eds) Triplex Confinium (1500–1800): Ekohistorija. Književni krug, Split; Zavod za hrvatsku povijest Filozofskog fakulteta Sveučilišta u Zagrebu, Zagreb, pp 365–377

Holjevac Ž (2014) Karlovački generalat 1802. godine: ustrojstvo, naselja i stanovništvo [The Karlovac Generalate in 1802: Organization, Settlements, and Population] In: Karbić M, Kekez H, Novak A, Horvat Z (eds) Ascendere historiam. Hrvatski institut za povijest, Zagreb, pp 553–571

Hosch R (1875) Fromme's Österreichischer Feuerwehr-Kalender, Zweiter Jahrgang. Carl Fromme Verlag, Wien

"Hrvatske šume, Rasadnici". https://www.hrsume.hr/sume/rasadnici/. Accessed on 12 February 2023

Ingrao C (1994) The Habsburg Monarchy 1618–1815. Cambridge University Press, Cambridge

Ivančević V (1999) 120. obljetnica osnutka 'Kraljevskog nadzorništva za pošumljavanje krasa krajiškog područja - Inspektorata za pošumljavanje krševa, goleti i uređenje bujica u Senju'

(1878–1998) [The 120th anniversary of the foundation of the 'Royal inspectorate for the afforestation of karst in the Military Frontier region—The inspectorate for the afforestation of karst, barren areas and torrent control in Senj (1878–1998]). Senjski zbornik 26:361–400

Kaser K (1997) Freier Bauer und Soldat: Die Militarisierung der agrarischen Gesellschaft an der kroatisch-slawonischen Militärgrenze (1535–1881). Böhlau, Wien

Kaser K (2003) Uništenje šuma na obalnom kraškom području hrvatske Vojne krajine u prvoj polovici 18. stoljeća. Njegovi demografski, privredni i socijalni uzroci. [Deforestation of the Littoral Carstic Area of the Croatian Military Border in the first half of the 18th century. Its demographic, economic and social causes]. In: Roksandić D, Štefanec N, Glunčić-Bužančić V, Mimica I (eds) Triplex Confinium (1500–1800): Ekohistorija. Književni krug, Split; Zavod za hrvatsku povijest Filozofskog fakulteta Sveučilišta u Zagrebu, Zagreb, pp 377–395

Klemenčić M (2017) Poimanje i temeljna zemljopisna obilježja Podgorja. [The understanding and basic geographical features of Podgorje]. Senjski zbornik 44:5–16

Kovačević I (1981) Stoljetni izazov Senja [The century-old challenge of Senj]. Zavod za poljoprivredna ispitivanja, Zagreb

Kužić K (2020) Examples of the Bura wind effects in the Eastern Adriatic area according to chronicles, travelogues, and military reports (15th Century–18th Century). Ekonomska i Ekohistorija 16(1):55–80

Marković M (1980) Narodni život i običaji sezonskih stočara na Velebitu [Folk life and customs of seasonal herders on Velebit]. Zbornik za narodni život i običaje Južnih Slavena 48:5–139

McNeill JR (1992) The mountains of the Mediterranean world. An environmental history. Cambridge University Press, Cambridge

Mioč B et al (2008) Vanjština različitih kategorija hrvatske šarene koze [Exterior characteristics of different categories of croatian coloured goat]. Stočarstvo 62(6):439–447

Mlacović D (2014) The communes in the Northern Istria in the late Middle Ages and wood. In: Štih P, Zwitter Ž (eds) Man, nature and environment between the northern Adriatic and the Eastern Alps in premodern times. Ljubljana University Press, Faculty of Arts, and Historical Association of Slovenia, Ljubljana, pp 86–101

Moačanin F (1991–1992) Organizacijske strukture Vojne krajine do sredine 18. stoljeća [Organisational structures of Military Frontier up to the middle of 18th century]. Arhivski vjesnik 34–35:157–163

Mrgić J (2018) Intemperate weather in violent times—narratives from the Western Balkans during the Little Ice Age (17–18th centuries). Cuadernos de investigacion geografica 44(1):137–169

Panjek A (2018) Against the desert in the Karst. A paradigm-shift from ruined landscape to cultural savannah. Ekonomska i Ekohistorija 14:52–71

Penzar B, Penzar I, Orlić M (2001) Vrijeme i klima hrvatskog Jadrana [Weather and Climate of the Croatian Adriatics]. Nakladna kuća dr. Feletar, Zagreb

Remec M (2024) Afforestation with non-native *Pinus nigra* in the karstic areas of southwestern Slovenia since the mid-nineteenth century: environmental implications. In: Fuerst-Bjeliš B, Mrgić J, Petrić H, Zorn M, Zwitter Ž (eds) Environmental histories of the Dinaric Karst. Springer, Cham, pp 203–237

Rogić P (1965) Naseljenost velebitske primorske padine kroz historiju [Population of the Velebit coastal slope throughout history]. Senjski zbornik 1:35–68

Rogić V (1957) Velebitska primorska padina. Prilog evolucije krškog pejzaža [Velebit coastal slope. Contribution of the evolution of the karst landscape]. Geografski glasnik 19:61–102

Rogić V (1958) Velebitska primorska padina. Prilog evolucije krškog pejzaža (cont.) [Velebit coastal slope. Contribution of the evolution of the karst landscape]. Geografski glasnik 20:53–115

Rogić V (1975) Velebitsko podgorje [Velebit coastal area]. Geografija SR Hrvatske, Sjeverno hrvatsko primorje 5:125–136. Školska knjiga, Zagreb

Roksandić D (1988) Vojna Hrvatska I. [La Croatie Militaire]. Školska knjiga, Zagreb

Rothenberg GE (1960) The Austrian military border in Croatia 1522–1747. The University of Illinois, Chicago

Rukavina A (1970) Privreda ličke i primorske strane Velebita [The Economy of the continental and coastal slope of Velebit]. Senjski zbornik 4:99–124
Rukavina A (1990) Još žive velebitske šume [Forests of Velebit are still alive]. Senjski zbornik 17(1):281–290
Sabljak S (1990) Povijesno-topografska skica Senja [Historical and topographical sketch of Senj]. Senjski zbornik 17(1):267–272
Scourse J, Trouet V, Raible C (2012) The medieval climate anomaly and the little ice age: testing the NAO hypothesis. Global Planet Change 84–85:48–55
Severinski V (1965) Senjske šume i njihova eksploatacija [Forests of Senj and their exploitation]. Senjski zbornik 1:232–245
Slavin P (2016) Climate and famines: a historical reassessment. WIREs Climate Change 7:433–447
Sonnlechner C (2004) Waldordnungen und ergänzende Quellen. In: Pauser JJ Scheutz M, Winkelbauer T (eds) Quellenkunde der Habsburgermonarchie (16.-18. Jahrhundert): Ein exemplarisches Handbuch. Oldenbourg Wissenschaftsverlag, Wien/München, pp 268–277
Štefanec N (2003) Trgovina drvetom na Triplex Confiniumu ili kako izvući novac iz senjskih šuma (1600.–1630.) [Wood Trade on the Triplex Confinium or how to Extract Money from Senj's Woodlands? (1600–1630)]. In: Roksandić D, Štefanec N, Glunčić-Bužančić V, Mimica I (eds) Triplex Confinium (1500–1800): Ekohistorija. Književni krug, Split; Zavod za hrvatsku povijest Filozofskog fakulteta Sveučilišta u Zagrebu, Zagreb, pp. 337–363
Vaniček F (1875) Specialgeschichte der Militärgrenze, aus Originalquellen und Quellenwerken geschöpft (II). K.K. Hof- und Staatsdruckerei, Beč
Vukelić J, Mikac S, Baričević D, Bakšić D, Rosavec R (2018) Šumska staništa i šumske zajednice u Hrvatskoj [Forest sites and forest communities in Croatia]. Državni zavod za zaštitu prirode, Zagreb
White S (2014) The real little ice age. J Interdisc History 44(3):327–352

Filip Pavelić He has a master's degree in history education obtained at the Faculty of Humanities and Social Sciences in Zagreb, Croatia. He was awarded a "Ferdo Šišić" award by *Croatian National Committee for Historical Sciences* and *Croatian History Society* for his graduation thesis "Ecological Factors of Famine in the Middle Lika in the Second Half of the 18th and Early 19th Centuries". He is currently working in the private sector in a communications company, while his academic interests are environmental history and historical demography.

Hrvoje Petrić Professor of history. His research deals with environmental history, economic history and Croatian early modern history. He has written several books and a series of articles (more than 20). He is the editor of "Ekonomska i ekohistorija" [*Economic and Environmental History*] Journal.

Mislav Radošević He graduated Early-modern history and German language and literature at the Faculty of Humanities and Social Sciences, University of Zagreb, Croatia with the dissertation titled "Forests of the Brod Grenzer Regiment in the Second Half of the 18th and the Beginning of the 19th Century". His academic interests are environmental and military history, particularly the environmental and military history of the Habsburg Military Frontier.

Chapter 8
Afforestation with Non-native *Pinus nigra* in the Karstic Areas of Southwestern Slovenia Since the Mid-Nineteenth Century: Environmental Implications

Meta Remec

Abstract The paper deals with the afforestation project and the creation of monocultural forest stands of non-native black pine *(Pinus nigra)* in southwestern Slovenia, mainly on the Kras Plateau and in the Vipava Valley since the middle of the nineteenth century, and its social and environmental implications. It is based on the archival sources kept in the state, regional, and local institutions involved in the afforestation project at various levels and over the time span of three different state entities with their distinctive social orders that shaped the territory under consideration in the twentieth century. The paper analyses the reasons for the afforestation and the consequences of this human intervention in the environment, which has completely changed the characteristics of the landscape and, consequently, the economic situation of the local inhabitants. The paper outlines the various environmental impacts caused by afforestation, such as changes in the soil, vegetation, and microclimate, while focusing primarily on the infestation of the non-native processionary moth *(Thaumetopoea pityocampa)*, which, due to the specific characteristics of the Slovenian karstic region and local climatic conditions, spread rapidly even in the forested areas up to 1200 m above sea level. It describes the methods of controlling this pest, which were initially mechanical, while after WWII, the pest was also aggressively controlled with chemical agents, contaminating the groundwater in the karst areas and the entire local environment.

Keywords Environmental history · Kras · Non-native tree species · *Pinus nigra* · Monoculture plantations · Tree pests · *Thaumetopoea pityocampa* · Pesticides · Groundwater contamination

M. Remec (✉)
Institute of Contemporary History, Ljubljana, Slovenia
e-mail: meta.remec@inz.si

B. Fuerst-Bjeliš et al. (eds.), *Environmental Histories of the Dinaric Karst*, Environmental History 17, https://doi.org/10.1007/978-3-031-56089-7_8

8.1 Dinaric Karst Forests in Slovenia and the Non-existent Forest on the Kras Plateau

The Dinaric Karst is the largest uninterrupted karstic region in Slovenia, stretching to the Alpine foothills and the Ljubljana Marshes in the north, the Gorjanci Mountains in the east, and the river Soča and the Gulf of Trieste in the west.[1] In the distant past, this region used to be almost completely covered by forests, which have either changed their composition due to anthropogenic influences or were converted into open landscape due to several factors, such as farming, mainly sheep and goats grazing, and wood felling starting in prehistory. Even nowadays, the Dinaric Forests still represent a large phytogeographical unit. The geological formations that lie beneath its surface are dominated by limestones and therefore permeable. Under the influence of various factors, especially relatively abundant rainfall that partly compensated for the lack of groundwater, the forests covered the land. However, the environmental conditions do not allow for any major disturbances to their balance: it seems that the path from a primaeval forest to a "desert" is the shortest precisely in the karstic regions.[2]

In terms of location and time, two types of change can be observed that shaped the appearance, structure, and size of forests in the Dinaric Karst. In some parts of the territory, the forest was completely cleared, while elsewhere, its structure was altered. Both share the establishment of large-scale monoculture plantations of non-native tree species in order to achieve higher economic returns. In the Dinaric fir-beech forests, which account for 10% of all Slovenian forests, most areas were not completely cleared, but the tree cover was somewhat reduced in many places. However, the forests' composition was quite different from their natural structure. It was especially the extensive Norway spruce (*Picea abies*) stands that kept spreading, as forests were intensively exploited in connection with industrial development, mining, and trade.[3]

A different and unique phenomenon was the largely deforested area of the carbonate plateau between the Gulf of Trieste, the Vipava Valley, Northern Istria, and Brkini Hills, known as the Kras Plateau (Fig. 8.1).[4]

This "non-existent forest" kept attracting a great deal of attention from the forestry profession as well as other branches of science. The analyses focused on the causes for the disappearance of oak, beech, and hornbeam forests—which, according to the pollen analysis of the sediments[5] and archival sources were prevalent in the Kras Plateau up until the fourteenth century—as well as the afforestation of this area, long considered to be the first successful experiment of its kind on the European

[1] Ravbar (2007), p. 25.

[2] Culiberg (2005) pp. 149–154; Kranjc (2005), pp. 141–148; Panjek (2006), p. 76; Hufnagl (2002), p. 11; Mlinšek (1993), pp. 6–7; Bončina et al. (2003), pp. 227–232.

[3] Perko (1989), pp. 6–7.

[4] Zorn et al. (2015), p. 569.

[5] Šercelj (1996), pp. 71–73.

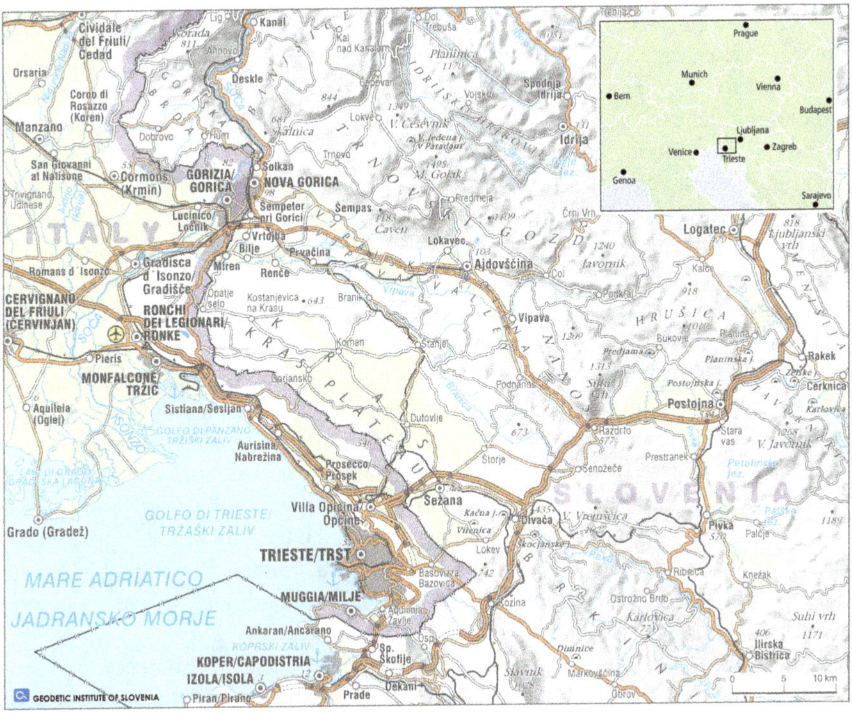

Fig. 8.1 Map of the Kras Plateau and Vipava Valley. *Note* For easier orientation, today's national borders as well as the present road and railroad network, which corresponds or resembles the one of the time in question, are drawn in. *Source* Geodetic Institute of Slovenia, Jamova 2, SI-1000 Ljubljana

scale.[6] The phenomenon of the afforestation of Kras was thus discussed on several occasions, and the afforestation with a non-native species, the black pine (*Pinus nigra*), was most often portrayed as a success, mostly due to environmental and ecological reasons and because it was an improvement compared to the "bare and degraded landscape," supposedly resulting from the poorly controlled exploitation of natural resources, deforestation, and overgrazing in Kras.[7] In recent years, it became clear that the economic reasons behind the afforestation attempts need to be studied as well. The prevailing discourse of the archival and other sources but also of the traditional studies—that the afforestation resulted from an attempt to restore a largely deforested and ruined area—was challenged by the studies that have shown that the economic interests were, in fact, paramount from the very beginning of the process. Previously, it has often been overlooked that extensive areas outside the Dinaric Karst area were reforested with black pine: the Gorizia-Karst Afforestation Commission was active in the predominantly flysch Vipava Valley, and in the Kanal and Tolmin

[6] Grove and Rackham (2001), pp. 187, 324–325; Panjek (2006), p. 76.

[7] Gašperšič and Winkler (1986), pp. 169–184; Perko (2016); Johann (2018), pp. 42–46.

areas, which were clearly not as "degraded" as the Kras Plateau was believed to have been.[8] However, also on the Kras Plateau itself, it became clear that the economic interests of the state and its need for a useful and profitable deciduous forest were considered more crucial than the short-term economic interests of the peasants, the driving force behind the reforestation project.[9]

So far, few analyses have taken into account the local population and its needs or provided a historical analysis of the consequences of the centrally-led and planned environmental change since 1850.[10] The latter coincided with the period of major social reforms and deliberate re-education of the population that often resisted innovation in the name of progress. Convinced that they knew what was good for the people, the authorities started to intervene in all areas of human life. One of the very first aspects that they concerned themselves with was the modification and bonification of the human living environment.[11] Afforestation in the name of restoring the forested landscape corresponded to the idea of the nineteenth- and twentieth-century advances in forestry science and was in fact completely at odds with the traditions of the local population. Since the aim was not only afforestation but also re-education that would ensure a change in values and behaviour, the farmers were supposed to be persuaded to give up their "narrow-mindedness" and "outdated" working methods. Allegedly, landowners needed to be educated, re-educated, and, where necessary, even forced to make changes.[12]

The analyses of social, economic, and environmental history offer a new perspective on this process. In addition to the prevailing discourse that farmers, through their ruthless exploitation of the environment, had eventually degraded this environment to the level of a desert, it was increasingly often pointed out that for centuries, the inhabitants had successfully managed the environment sustainably according to their needs while largely preserving the vital natural environment on which their very lives ultimately depended. The intellectual-history-rooted predominant "degradation theory" is contrary to the more recent interpretations of the Kras Plateau in the past as a "cultural savannah." The bare acknowledgement of the complex consequences of the afforestation project is often perceived as barely acceptable questioning of natural facts and as a means of undermining the achievements of forestry experts. The prevailing "degradation narrative" with the terms such as "erosion," "deforestation," "destruction," "stony desert," "devastation," and "exploitation" still shifts the blame on the inhabitants of the region for first clearing the once richly forested landscape in the name of profit and then hindering afforestation through grazing, littering, and logging. The degradation theory ignores the full spectrum of the consequences associated with afforestation: from the collapse of the existing economic system, social change, and emigration to changes in the ecosystem, such as the emergence

---

[8] *Poročilo deželnemu zboru poknežene grofovine Goriško-Gradiščanske* (1903); *Deželni zbor goriški [Dieta provinciale di Gorizia]* (1902).

[9] Johann (2018), pp. 32–51.

[10] Panjek (2018), pp. 52–71; Panjek (2015), p. 89.

[11] Remec (2015), pp. 40–49.

[12] Rubbia (1912), p. 1; Johann (2018), p. 44.

of non-indigenous species of tree pests, which have emerged as a direct result of the creation of artificial black pine monoculture plantations that had not been traditionally present in the area.[13]

The afforestation project and its different social and environmental implications will be studied based on the expert forestry press and literature of the time, newspapers, and foremost archival sources of the state, regional, and local institutions involved in the afforestation project at various levels and over the time span of three different state entities with their distinctive social orders that shaped the entire region in the twentieth century. The paper outlines the various environmental impacts caused by the afforestation, such as changes in soil, vegetation, and microclimate, while focusing primarily on the spread of non-native tree pests and the environmental consequences of pesticide use in controlling the non-native pine processionary moth, which, due to the specific characteristics of the karstic regions of southwestern Slovenia and the local climatic conditions, spread rapidly even in the forested areas up to 1200 m above sea level.

8.2 Afforestation and Its Environmental and Social Implications

The afforestation of areas once covered by forests and then converted to farmland for a longer or shorter period is interesting from several perspectives. From the socio-historical point of view, these processes are worth studying, as they testify to changes in the economic and social dynamics of a certain society. However, they are also interesting from a natural scientific viewpoint, as they reveal the strategies used, the pitfalls, and the difficulties encountered in doing so. The new forests that were created were usually quite different from the original ones—either because the spontaneous overgrowth of abandoned farmland led to a predominance of plant species other than those that had originally covered the area or because of human interventions aimed at creating the most suitable environment according to needs and values.[14] Afforestation attempts in karstic areas are a special phenomenon. All attempts to save the areas that foresters considered degraded were similar: just like in the Slovenian Dinaric Karst, the foresters elsewhere else (in Croatia, Hungary, Spain, etc.) would also often resort to afforestation with non-native pioneer pine species and the establishment of monoculture forest stands, which, thanks to their resilience, were able to survive in the extremely harsh conditions, but which also led to altered ecological processes and soil acidification. All the areas where monoculture plantations (mostly pine, but also eucalyptus and other non-native tree species) were established in line with the German forestry school share changes in flora and fauna as

[13] Kranjc (2012), p. 82; Kogovšek and Kranjc (1999), p. 150; Williams (1993), pp. 4–6; Panjek (2018), p. 68.

[14] Brock (2014), pp. 156–157.

well as in the microclimatic and soil conditions, ultimately leading to the emergence of new diseases and pests.[15]

The Kras Plateau is one of the areas where the afforestation process was extremely slow and initially failed to produce any particular results. The afforestation of this area is interesting because economic objectives were supposedly not the main motivation for undertaking it. It was clear from the very outset that the economic return would be very modest and that it would take several decades to see any minimum profits. Unlike several other regions, such as certain areas in India, South Africa, Spain, Portugal, and Sardinia, where attempts were made to reduce the shortage of forests by planting fast-growing species like eucalyptuses, pines, and poplars, leading to the creation of tree plantations that provided raw materials for the industry and fuelwood,[16] in Kras, those in charge set the goal of restoring the deciduous forest in what they believed was its original composition. After the first unsuccessful attempts of sowing seeds of tree cultures considered autochthonous, especially oak, the authorities realised that the afforestation could only be achieved by taking a big step in the opposite direction and first getting anything at all to grow there. Only then, it was declared, could any economic goals even be considered. As of the second half of the nineteenth century, deliberate attempts took place to afforest the bare areas of Kras, based on the establishment of monoculture plantations of non-native black pine, leading to profound structural changes in the appearance of the Kras Plateau. The project was not abandoned even after WWII, when forestry experts were already clearly aware that afforestation with non-native cultures and the creation of monocultural forest stands was a mistake that would take its toll sooner or later.[17] The absence of forests on the Kras Plateau was considered so pressing and alarming that any means of rectifying this seemed plausible. Black pine was considered the perfect tree species for the task: modest enough to take root even on the bare karstic rock, yet hardy enough to withstand the severe bora wind, droughts, and lack of humidity.[18]

Environmental conditions such as soil permeability, drought, and bora wind are not only characteristic of the Kras Plateau but also of the other areas of the Dinaric Karst. However, the situation on the Kras Plateau is worth studying due to the unique interconnection of the local environmental and socio-political factors. On the one hand, the constant changes in afforestation laws in three different states (the Habsburg Empire up to 1918, the Kingdom of Italy between the two world wars, and socialist Yugoslavia between 1945 and 1991) and the devastation caused by the two world wars[19] severely impacted the afforestation project, especially during its initial stage. On the other hand, afforestation had numerous environmental and social consequences that were not predictable at the beginning of the project and that altered the landscape and characteristics of the entire region. It has resulted in substantial

[15] Bárány-Kevei (2003), pp. 177–179; Grove and Rackham (2001), pp. 69 and 187–189; Gams (1991).

[16] Bennet (2011); Grove and Rackham (2001), pp. 187–189; Brock (2014), pp. 166–167.

[17] Wraber (1950), pp. 40–46.

[18] Fras (1959), pp. 11–12, 14 and 17; Perko (2016), pp. 213–214.

[19] Perko (2016), pp. 189–197; Masiello (1992), p. 52; Klavora (1997, 2014).

environmental changes in the local climate due to weakened bora wind, greater evapotranspiration, and the associated lower runoff, which, in turn, has eventually led to less water in the karst springs, thus causing water supply problems. Greater ground cover due to the canopy has resulted in local climate changes: more shade means that the ground does not heat up as much and the soil stays moist longer, causing a drop in the average temperature of the region and a delayed harvest.[20]

Social implications were significant as well and originated from the fact that the authorities completely ignored the interests of the local inhabitants, as they were considered ignorant, uncultured, and, above all, short-sighted. The assurances of the authorities at the beginning of the project (stated, for example, in the Karst Afforestation Law of Carniola in 1885)—that the afforestation would not damage the primary agricultural activities on the farms—were soon forgotten, even more so after the territory had become a part of a completely different state: the Kingdom of Italy after WWI.[21] The local farmers unsuccessfully demanded tax reductions for the afforested terrains and aid to ease the transition to stable stock farming, required because it was in fact forbidden to put small cattle to pasture in large areas, which has led to a rapid decline in the number of livestock in the area. The consequences were not only economic but also environmental: the dry grass that was not removed by these animals anymore contributed to the fire hazard that had already increased significantly after the construction of the railroad between Ljubljana and Trieste in 1857. In the territory with a severe shortage of water and dense and hardly accessible monoculture black pine forest stands, numerous fires caused economic and environmental damage. The authorities failed to predict the severe combustibility of black pine trees and pine cones, often described by the locals as "fire projectiles." Numerous deliberate arsons committed by the local inhabitants, who often opposed the afforestation and attempted to restore the previous use of the afforested areas as pastures, were unexpected as well.[22] The landowners were responsible for the fire protection of the tree plantations, which meant that they only had obligations while they were not permitted to use the land for grazing, collect branches, hay, or leaves, and especially cut the trees on their own land. Their tasks included the maintenance of stone walls, paths, and even a constant fire watch during dry seasons, which was a huge burden that they tried to avoid at all costs. In the case of plantations on municipal land, these responsibilities fell on the entire community. The inhabitants were also affected because they were forced to move the fertile soil from the so-called *dolinas*—rare locations with deep soil on the Kras Plateau—where they had previously cultivated potatoes and other vegetables to the areas designated for afforestation.[23] This had an important impact on the landscape as well. The first step of the afforestation project

---

[20] Zorn et al. (2015), p. 63; Grove and Rackham (2001), pp. 324–325.

[21] *Deželni zakonik in vladni list za kranjsko kronovino* 37(7, 21 November 1885): Zakon z dné 9. marca 1885 o pogozdovanji Krasa v vojvodini Kranjski, pp. 37; Panjek (2015), p. 104.

[22] Panjek (2015), pp. 105–106; Rubbia (1912), pp. 41–43; SI PAK KP 638, box 1, archival unit 7; Remec (2021).

[23] SI PAK KP 638, box 1, archival unit 7; SI PAK KP 629, box 3, archival unit 25, no. 9, Johann (2018), p. 48.

was often the removal and relocation of rocks, which triggered additional soil erosion as the rain easily flushed away the exposed soil. The inhabitants were encouraged to use the removed stones to build paths, windshields, and especially additional stone walls, which served as a form of protection against fire and animal grazing. The walls complemented the ones that had already been present all over the region and had been created while adapting the land to agricultural use in an attempt to protect the layer of fertile soil. At the beginning of the afforestation, the highest hopes were in fact invested in the restoration of the humus layer. In reality, in little more than a century after the afforestation had begun, only 2–4 cm of soil was created, which was less than expected. However, most importantly, the planners also failed to predict that the new soil would be acidic. It took a long time until the realisation that the new soil would not be suitable for the tree cultures considered autochthonous. By the time the forestry authorities started to discuss this problem, the local population had already adopted the black pine as a part of their character and regional identity and started to oppose the attempts at removing the pine stands.[24] Ultimately, the black pine was not substituted by tree species of higher quality and greater market value after the first generation, as planned at the beginning. With the collapse of small farms, even the areas that had not been destined for afforestation were overgrown by black pine that became naturalised, so that it started to spread on its own. This resulted in the alteration of the cultural landscape and reduction of biodiversity associated with unforested land. The fact that, since the year 1874, the forest cover on the Kras Plateau has increased from 14 to 60% has affected the entire flora and, of course, fauna, as the number of species associated with extensive forms of agriculture and animal husbandry started to decline. Furthermore, non-native tree pests began to appear. The brown hare (*Lepus europaeus*), grey partridge (*Perdix perdix*), and rock partridge (*Alectoris graeca*), which were the most important game species in the past, have practically disappeared from this area to this day. To preserve the bird species whose natural habitats are open spaces, sparsely populated with trees, and clearings (e.g. Woodlark, *Lullula arborea*), some previously afforested areas in the Vipava Valley had to be clear-cut. At the same time, large game and large carnivores started appearing regularly.[25] The newly emerging monocultural forests were also affected by a non-native tree pest, the processionary moth (*Thaumetopoea pityocampa*), which had major environmental implications.

[24] Gams and Gabrovec (1999), p. 64; Mihevc (2005).

[25] Perko (2016), p. 237; Remec (2021), p. 62; project *Projekt VIPava*, www.projektvipava.si, 26 May 2022.

8.3 The Appearance of the Pine Processionary Moth *(Thaumetopoea pityocampa)* in Black Pine Monoculture Plantations

The interventions in the natural balance of the Slovenian Dinaric Karst forests and the creation of extensive monoculture forest stands led to the spread of various non-native tree pests. Spruce, planted extensively at lower altitudes and outside its natural habitat in some regions of the Dinaric Karst,[26] was most threatened by the six-toothed spruce bark beetle (*Pityogenes chalcographus*) and the eight-toothed or European spruce bark beetle (*Ips typographus*), as well as by the little spruce sawfly (*Pristiphora abietina*).[27] Meanwhile, the monoculture black pine plantations in Kras and Vipava Valley (Fig. 8.2) were most threatened by the pine processionary moth *(Thaumetopoea pityocampa)* and the cockchafer (*Melolontha melolontha*). Artificial monoculture plantations turned out to be completely non-resistant to the predation of pests, which would therefore spread unhindered in the milder climatic conditions at lower altitudes.[28] These pests were extremely difficult to control because foresters had little experience with them, and they spread mainly because the natural balance in the forest had been disturbed.

The pine processionary moth is a typical representative of pests that have appeared in Slovenia due to the planned afforestation and the creation of monoculture forest stands. Several different factors need to be considered when identifying the reasons for its expansion, both the impact of human interventions such as timber trade and the creation of monocultural plantations, as well as the changes in the natural conditions. According to the Intergovernmental Panel on Climate Change (IPCC), the pine processionary moth can serve as an indicator of climate change and global warming, as climate change has allowed it to conquer higher altitudes. The increasingly mild winters and the spreading pine plantations have been recognised as crucial factors, as they have enabled a much larger percentage of the pest population to survive.[29]

As a pest typical of the Mediterranean, the pine processionary moth had not been naturally present in Kras. Foresters first registered it near Komen in 1903 or 1904. Planned eradication began relatively quickly, as early as 1906, but could not prevent its expansion. In a very short time, the pest spread over the entire Kras Plateau, Vipava Valley, the basin of the river Soča, and as far as the river Pivka. In addition to the pine plantations, the pine processionary moth also encountered more than favourable temperature conditions during its oviposition, which usually takes place in the second half of August. Until the pest was still confined to a rather restricted area, the opinions on whether it should be controlled or not differed, as normally, it does not cause fatal damage to trees, which fully regrow within three to four years. In

[26] Bončina et al. (2003).

[27] Knížek and Beaver (2004), pp. 49–51; Grégoire and Evans (2004), pp. 20–23; Zupančič (1980), pp. 14–15.

[28] Jactel et al. (2015), pp. 281–288.

[29] Roques et al. (2015), pp. 82–86 and 152.

Fig. 8.2 Heavily infested plantation near Črniče in Vipava Valley in 1950. *Source* SI AS 675, box 13, Letalska akcija proti borovemu prelcu [Airplane action to suppress pine processionary moth]. Reproduced with permission

addition, the prevailing opinion was that black pine plantations were only transitory and were only supposed to prepare the soil for the regrowth of the autochthonous forests and did not need to be saved at all costs.[30]

In the first years after WWII, the experts closely observed and studied the behaviour of this pest on the Kras Plateau and in the Vipava Valley. At first, they believed that natural processes alone could regulate its expansion. Overpopulation of pine processionary moth was always supposed to be followed by food shortages and then diseases affecting this tree pest, while its natural enemies, which this non-native organism initially did not have, were also expected to multiply.[31] These optimistic projections did not come true, and the extraordinarily rapid advance of this pest that followed in just a few years changed the attitude towards it. In 1948, after several consecutive hot and dry summers, the pine processionary moth caused extremely severe damage, resulting in the stripping of between 25 and 100% of trees and almost destroying the more recent post-WWII afforestation. This revealed the specificity of black pine afforestation in Slovenian Dinaric Karst compared to the southern Mediterranean, North Africa, the Middle East, and southern Europe, where this pest was autochthonous. This specificity mainly stemmed from the higher amount of rainfall in the Dinaric Karst of Slovenia compared to the areas where the

[30] Blažič (1959), pp. 3 and 15–16.

[31] Jošt (1949).

pest was naturally present. Its survival was also facilitated by the proximity of farmland, which made it easier for the pest to burrow in the soil that had been dug over, and by the increasingly mild winters, which allowed it to expand to 600 m above sea level, which had been, in 1950, believed to be the pest's upper altitudinal limit. It is precisely this fact that testifies to the intensity of climate change and the successful adaptation of the pine processionary moth to its new environment, as over several decades, it spread even up to 1200 m above sea level, where it can be found today.[32]

The older plantations were initially considered to be less threatened. Younger trees with fewer branches often dried out completely due to the pest infestation, while older trees were usually expected to survive the infestation and regrow after a year or two. However, it turned out that the pine processionary moth significantly increased the trees' susceptibility to secondary pests, which then caused most of the damage. While the risk was indeed much greater in the case of the more recent afforestation, where the pine processionary moth could also cause complete dieback, the economic damage in both cases was so extensive that it was deemed necessary to implement all the measures that science and the market could offer at the time.[33]

8.3.1 *Pine Processionary Moth Control and the Use of Pesticides in Forestry*

Initially, the pine processionary moth was controlled mechanically: branches with cocoons would be pruned, which was possible as long as the trees were small enough. Caterpillar nests in the treetops would be broken up with a special fork, as otherwise excessive damage could be caused to the trees. Landowners were obliged to control the insects that could damage the new plantations, and they showed a considerable willingness to cooperate, mainly because they took the opportunity to cut off more branches than necessary and thus supply themselves with wood. Insect control did not represent a financial burden for the afforestation commissions, as it was carried out on an (in)voluntary basis by the local peasants. A tradition of work campaigns was established, aimed at cutting away cocoons together with branches. At least one member of each household had to take part in these actions, bringing their own tools and providing their own food.[34]

In the years after WWII, the prevalence of the pine processionary moth was so critical that its physical removal seemed impossible. The first chemical control experiments were carried out as early as 1949 with a Yugoslav-made DDT-based product called Pantakan. In the test area, trees were sprayed with ordinary backpack hand sprayers. Thus, the pest was eradicated up to the height of three meters. Only

---

[32] Jurc et al. (2019), pp. 3–5; SI AS 675, box 13, Letalska akcija proti borovemu prelcu [Airplane action to suppress pine processionary moth]; Roques et al. (2015), p. 114; De Groot and Hauptman (2012), p. 73.

[33] Barič (1951).

[34] SI PAK KP 638, box 1, archival unit 6 and 7; Johann (2018), pp. 45–46.

a detailed analysis of the pine processionary moth's behaviour revealed that control should have started much earlier in the year than originally believed—already in February—and that in the territory of Kras and Vipava Valley, the pest behaved completely differently than described in the expert forestry literature. It would appear at much higher altitudes, leave its cocoons much earlier, and would often completely skip the first feeding phase and burrow into the soil, which limited the window during which the foresters could act.[35] Technological advances also offered other solutions. The Pasteur Institute in Novi Sad (Voivodine, Serbia) was developing viruses and intended to breed parasites that would kill particular pests in a targeted manner without affecting the other species present in the natural environment. The attempts to control the common pine sawfly (*Diprion pini*) raised hopes: a pine plantation in Kras between Črnotiče and Podgorje was infected as a test. However, the Institute did not manage to develop a virus that would attack the pine processionary moth, and it was believed that the final solution would be brought by pesticides, especially DDT, as the cutting and burning of the nests was seen as a sign of under-development unworthy of the new socialist regime.[36]

The pine processionary moth control attained a strong ideological charge. Even in the expert forestry press, it was all about fighting the enemy of the black pine that was there to rescue a ruined region like "an army on a triumphant unstoppable march," "like the Partisans during the national liberation struggle." The new socialist government was on the road to "re-green" the territories that had been most profoundly affected by foreigners, among them especially the Kras Plateau, which had supposedly been devastated already by the exploitation of the Republic of Venice and the fascist Italian state. The pine processionary moth was an obstacle to be overcome by all means.[37] Afforestation on the one hand and the suppression of the pine processionary moth on the other were presented not only as cultivation and technical, but mainly socio-economic problems, whose solution depended on how many resources the community was willing to sacrifice for this purpose and how much it was willing to invest in overcoming the people's economic and cultural backwardness.[38]

The use of pesticides took on militaristic overtones, and the socialist authorities did everything in their power to show that they were not lagging in development and were, in fact, even ahead of the curve, setting the guidelines for others to follow. Supposedly, the efforts of the authorities were constantly undermined by the peasants unfavourably inclined towards them. The catastrophic spread of the pine processionary moth was largely attributed to both the irresponsible behaviour and neglect of this area by the Italian fascist authorities during the inter-war period, as well as to the "deliberate sabotage" of the "progressive" efforts of the new socialist authorities from the Italian side of the border. Despite the large amounts of chemicals they had at their disposal, the Italian authorities were in fact accused of deliberately neglecting

[35] Barič (1951); SI AS 675, box 13, Letalska akcija proti borovemu prelcu [Airplane action to suppress pine processionary moth].

[36] SI PAK KP 607, box 18, archival unit 52.

[37] Beltram (1946), pp. 13–14 and 37.

[38] Jurhar, et al. (1963), pp. 10–12.

the plantations in the border area, which caused the infection to immediately spread back to the Slovenian side.[39]

Despite all the propaganda, even in the 1960s, most control efforts were still based on the manual removal of the pine processionary moth nests. The inventories of fixed assets do not indicate that any significant progress was made during this period in terms of mechanisation. The Secretariat of Agriculture and Forestry promised a helicopter that could be used in the "fight" against harmful plant insects and diseases. However, in reality, the pine processionary moth was still mainly controlled with "scissors and curved knives" and "manual bellows." The situation across the border in Italy, where plenty of DDT-based preparations were always available, could only be observed with envy.[40] The lack of awareness regarding the dangers of physically removing the nests was still evident. Children from nearby villages were often employed for this chore: they would climb the trees and cut the cocoons with scissors. Safety goggles and gloves for the pine processionary moth removal appeared in the inventories of fixed assets in the mid-1960s, but it is clear from the job descriptions that most manual removals were still carried out without any protection. To a significant extent, unprofessional handling, stemming from the ignorance about the correct removal procedures, even contributed to the spreading of the pest. It was only the health consequences that eventually prompted the need to better equip foresters and other individuals who were coming into contact with the pine processionary moth.[41] Skin and eye infections, as well as respiratory problems due to the inhalation of the toxic hairs of caterpillars and contact with both dead and live nests, became a pressing problem, hindering the attempts to present the Kras Plateau as a tourist destination and, in the following decades, reintroduce small cattle pastures as a way of reviving what was considered local and traditional. However, above all, the caterpillars were in the way of the altered lifestyle of the people, for whom forests primarily became a place for recreation and dog walking. Even the authorities seemed to have taken note of the people's changing habits and started to categorise certain areas as "recreational forests" in their land inventories. In the attempt to protect the inhabitants, the decision was made that pine processionary moth nests were to be completely destroyed in public parks and in the vicinity of settlements, which meant an increase in pesticide use, especially in populated areas.[42]

The use of pesticides in forestry never became as widespread as in agriculture, mainly due to the lack of chemical agents and the required machinery but partly also due to the scepticism on the part of those in charge who did not believe in its economic justification. The chemical industry that emerged after WWII devoted most of its products to other agricultural industries that fought the Colorado potato beetle, the cockchafer, and other fruit tree pests. Meanwhile, the forestry industry

[39] SI AS 675, box 13, Pinijev sprevodni prelec [Pine processionary moth], No. 5/31 and No. 5/32; Langston (2014), p. 267.

[40] SI PAK KP 607, box 8, archival unit 14; SI PAK KP 607, box 17, archival unit 45.

[41] De Groot and Hauptman (2012), p. 76; SI PAK KP 607, box 17, archival unit 45, Osnovna sredstva 1961 [Fixed assets 1961].

[42] SI PAK KP 795, box 4, Seznam parcel (Piran) 1961 [List of plots (Piran) 1961].

remained on the sidelines. Forestry focused mainly on protecting the already felled timber or logs from the bark beetle, which caused the most serious economic damage, while protecting the still-standing trees called for huge investments and would be technically difficult.[43]

Considering the general lack of pesticides, the karstic areas in southwestern Slovenia were simply not prioritised. Due to the shortage of materials and irregular deliveries of pesticides, only minor quantities of pesticides were applied in the area of Ilirska Bistrica and Sežana. In all other areas, control was carried out mechanically, employing the traditional removal of the pine processionary moth nests during mass work campaigns which were taking place between September and May. For example, in 1953, chemical treatments of plantations were almost abandoned, as the shortage of insecticides meant that most of the available quantities were diverted for use in tree nurseries, where, in addition to DDT, carbon disulphide, Bordeaux broth,[44] and sulphur and arsenic preparations were also used.[45] The Koper District People's Committee successfully cleared 2999 ha of pine plantations between 1 September 1955 and 10 May 1956, when 155,479 nests were collected and destroyed. During the same period, only nine hectares of forests were treated with chemical agents, mainly to control the common pine sawfly (*Diprion pini*), which proved to be an easier target for insecticide control than the pine processionary moth: in the case of the latter, it was extremely difficult to determine the proper time for control, as this varied even within the same plantation.[46] Forest insect control became a regular budget item of the Section for the Afforestation and Amelioration of Kras, but implementation fell far short of annual plans. In 1960, 1632 kg of chemical preparations for disease and insect control were used on 2810 ha of forests in the entire area under the jurisdiction of the Section.[47] According to the estimates, 200 kg of insecticide (e.g. Gamexan) would be needed for effective control in the area of one hectare, but by no means could the industry meet these needs.[48]

A quick look at the archival sources of the Ministry of Agriculture and Forestry shows a clear increase in requests for various pesticides and technical equipment that would facilitate their use. The requests sent from the local to the ministry level show that it was rather the lack of funds and not the lack of will that restricted the use of

---

[43] SI PAK KP 607, box 1, Zaščita hlodovine pred žuželkami 1955 [Protection of logs from insects 1955]; Kovačević (1953), p. 75.

[44] Bordeaux broth is a mixture of water, copper sulphate, and lime.

[45] SI PAK KP 795, box 4, Kalkulacija stroškov proizvodnje 1952–1953 [Calculation of production costs 1952–1953]; SI PAK KP 607, box 1, Zaščita hlodovine pred žuželkami 1955 [Protection of logs from insects].

[46] SI PAK KP 607, box 25, archival unit 72, Pregled dela uprave za gozdarstvo OLO Koper od 1. 9. 1955 do 10. 5. 1956 [Review of the work of The Forestry Administration at the Regional People's Committee of Koper between 1 September 1955 and 10 May 1956].

[47] SI PAK KP 607, box 27, archival unit 80; SI PAK KP 795, box 3, Zapisnik o delovnem sestanku delovodij objektov, 1958 [Notes of the working meeting of the foremen of the facilities, 1958].

[48] SI AS 675, box 13, Pantakan za emulzije za uničenje borovega prelca [Pantakan for the emulsions to destroy the pine processionary moth], No. 4350/50 and No. 4486/50.

chemical agents in Slovenian forests during the first two decades after WWII.[49] Slovenian foresters were more than aware that successful control was possible by spraying the trees with motorised knapsack sprayers and, over large areas, by spraying with insecticides using suitable aircraft. Foresters faced a shortage of suitable sprayers, as a rough estimate indicates that the entire Slovenian forestry sector only had a hundred motorised knapsack sprayers at its disposal, and almost half of these did not work due to the lack of spare parts. According to the assessments, 900 motorised knapsack sprayers, 1600 transportable motorised sprayers, and 2300 backpack hand sprayers would be needed for smooth operation. However, these would have to be imported, which represented a considerable bureaucratic burden. The sourcing of the required machinery was purely administrative and did not consider the actual needs in the field. Foresters were often left to their own ingenuity. Most often, they simply used vineyard sprayers, exposing themselves to extremely high levels of pesticides, mostly DDT-based products (Pantakan and Pepein) produced in Yugoslavia.[50]

Nevertheless, several reports about experiments carried out to demonstrate the effectiveness and feasibility of pesticides in forestry can be found in the expert forestry press. Already in the early 1950s, DDT and HCH were presented as highly effective agents, and foresters were given some basic information on how to work in the absence of industrial pesticides. Foresters often depended on their own resourcefulness. In their over-zealousness, they would prepare emulsions with five times the recommended dose of the powder and add talc, kerosene derivatives, and fungicides to the basic DDT-based preparation Pantakan to extend the period of effectiveness and reduce the impact of weather, especially wind. In this way, they wanted to maximise the effectiveness but were unaware of the impact on the environment and their own health. Only a single article cautioned that prolonged inhalation could cause dizziness and vomiting, but this did not deter the same author from recommending DDT even for domestic use against cockroaches, flies, mosquitoes, and bedbugs. In the case of moths, it allegedly needed to be sprayed directly on clothing.[51] When DDT was unavailable, foresters would prepare mixtures of the more readily available kerosene and carbolineum[52] in various proportions, most often 1:1.[53] Reports from the field indicated that they mainly experimented with combining 25% of lime sulphur slurry

---

[49] SI PAK KP 607, box 1, Zaščita hlodovine pred žuželkami, 1955 [Protection of logs from insects, 1955].

[50] Jošt (1949), pp. 5–6; SI AS 244, box 13, archival unit 19; SI AS 675, box 13, Gozdna škoda po mrčesih [Forest damage due to insect], No. 86/1-45; SI AS 675, box 13, Gozdarski inštitut Slovenije, Ljubljana 1950, Podatki o škodi v gozdovih, [Forestry Institute of Slovenia, Ljubljana 1950, Data on forest damage], No. 420; SI AS 675, box 13, Škropilnice in razpršilci za zatiranje bor. spr. prelca [Sprayers for the control of pine processionary moth], No. 363/2–49.

[51] Šebenik (1950); Kovačević (1953), p. 75.

[52] Carbolineum and kerosene are mineral insecticides. Carbolineum is an oily liquid produced by the distillation of coal tar. The exact formula cannot be given, but it contains anthracene and phenol, among other compounds. Kerosene or petroleum ether is the distillation product of petroleum. It also consists of a combination of C5-C6 hydrocarbons. Kerosene was once used for lice removal. Hutter (2007), p. 196.

[53] SI PAK KP 607, box 1, Zaščita hlodovine pred žuželkami 1955 [Protection of logs from insects, 1955].

with liquid DDT and other preparations, as they were trying to control various pests as well as fungi and diseases in a single application.[54] Considerable enthusiasm about various other chemical preparations—notably a mixture of carbolineum and kerosene,[55] Gesarol and Gamexane, diesel oil, lime arsenate, and Hexa[56]—was also evident from the professional forestry press. Chemical preparations were also used to fight against mice, which caused a lot of damage by gnawing on tree bark and tree seeds, posing a major threat to afforestation. However, poisoning mice with chemicals over large areas was difficult. To this end, seeds were soaked in lead oxide. Phosphorus preparations and strychnine were also used, which was expensive as well as devastating to other species, especially birds, which fed on the poisoned mice. Attempts were also made to kill mice with the typhoid bacillus, but the side effects were once again ignored.[57]

8.3.1.1 Aerial Pest Control on the Kras Plateau and in the Vipava Valley

In 1950, the authorities decided to use aerial pest control, i.e. spraying with insecticides using suitable aircraft, against the pine processionary moth—a method that was widespread in the nearby countries. In Germany, it started to be used in 1925, in Poland in 1928, and in Italy in 1930. It was also known in Yugoslavia, having been used in Bosnia as early as 1932; while in the period from 1947 to 1949, it was employed regularly against the gypsy moth in the Croatian region of Slavonia. In the territory of present-day Slovenia, airplanes were first used against the little spruce sawfly in 1936, in the vicinity of Brežice.[58] After this one-off event, aircraft were not used in Slovenia for tree pest control for quite some time. The pine processionary moth control in 1950 thus represented the first major attempt to eradicate forest pests in the territory of today's Slovenia. Due to doubts about the justification of procuring special airplanes and helicopters at the state level, it was considered more efficient to establish a special unit in the framework of JAT Yugoslav Airlines, which was named PA ("Privredna avijacija" or "Agricultural Aviation"). This unit was supposed to cover the needs of the entire territory of Yugoslavia. The Ministries of Forestry at the level of the republics had until 10 November to apply for pest control in 1952. The price for forestry applications was set at 400 dinars/ha—cheaper than for farming, viticulture, and fruit cultivation, where it was set at 600 dinars/ha. The fuel costs were covered by JAT Airlines, while the appropriate insecticides had to be provided by the clients. The concerns of Slovenian foresters that this aerial unit would be stationed too far away to meet Slovenian requirements proved to be justified. With weather conditions changing from hour to hour, it seemed impossible to

54 SI AS 244, box 13, archival unit 19.

55 Šlander (1953), p. 68.

56 Šlander (1952), p. 7.

57 Kordiš (1952).

58 Virnik (1951), p. 209.

plan any pest control months in advance. Furthermore, it soon became apparent that agriculture and forestry would not be the priority of this unit, which then spent the rest of its existence mainly on antimalarial campaigns, where DDT was also used.[59]

Aerial spraying represented progress and symbolised the onset of a new era. Its effectiveness was particularly important for conifers, as DDT was successfully retained in the foliage, providing the trees with long-term protection against pests that caused damage by biting the needles.[60] While there were occasional concerns that such a system could be too indiscriminate and potentially dangerous even for harmless and beneficial organisms, the benefits often prevailed.[61] As initially, the use of DDT would be mostly limited to uninhabited areas, the impact on humans was thought to be so negligible that it was not even considered at first.[62] In fact, it seems that the main reason for the delay in following the Western trends was mainly technical, and the modest amounts of DDT used in comparison with other countries did not result from greater environmental awareness but rather from the lack of technical capacities to keep up with such rapid progress.[63]

Not only the areas of the Kras Plateau, but also the plantations in the Vipava Valley were chosen for testing the new method, mainly for ideological and political reasons. To prove the superiority of the new socialist ideology, the authorities wanted to encourage the development of the territory that bordered the former Italian occupiers and was, at the same time, the main area with a high abundance of the pine processionary moth. However, forestry professionals were concerned that Kras Plateau and Vipava Valley would be the precise regions where it would be most difficult to fly due to the severe bora winds and the fragmentation of the ownership structure, and that it would be better to make this attempt elsewhere, in areas with a more favourable terrain configuration and less adverse climate conditions. Some of the forestry experts who took an analytical approach to the options for controlling the pine processionary moth and made their recommendations based on field observations and their own analyses strongly advocated an approach that was contrary to the established practices in the neighbouring countries: they argued in favour of controlling the pest in the autumn and combining aerial control with cutting cocoons, installing sticky tape, and using motorised knapsack and hand-held sprayers.[64]

The Ministry of Agriculture and Forestry nevertheless decided to ignore the local knowledge and warnings of the local foresters and rely on the information about this pest that came from abroad. The first aerial campaign took place from 26 February to 18 March 1950. It was conducted by a team consisting of forestry experts, a flight crew, two meteorologists, delegates from the Federal Ministry of Agriculture

[59] SI AS 675, box 13, Gozdarski institut Slovenije, Ljubljana 1950, Podatki o škodi v gozdovih [Forestry Institute of Slovenia, Ljubljana 1950, Data on forest damage], No. 3750/1.

[60] Mellanby (1992), p. 39.

[61] Kinkela (2011), pp. 30–31.

[62] Dunlap (1982), p. 59; Russell (2001), p. 159.

[63] Dunlap (1982), p. 38.

[64] SI AS 675, box 13, Biološka opažanja o borovem sprevodnem prelcu [Biological observations of the pine processionary moth], No. 5/10.

Table 8.1 Distribution of black pine plantations by altitude in the districts of Gorica and Sežana

District	Altitude		
Gorica	Up to 200 m	200–400 m	400–600 m
	169 ha 13 plantations	164 ha 11 plantations	108 ha 14 plantations
Sežana	/	1693 ha 69 plantations	1118 ha 40 plantations

Source SI AS 675, box 13, Letalska akcija proti borovemu prelcu [Airplane action to suppress pine processionary moth]

Table 8.2 Black pine plantation damage in the districts of Gorica and Sežana

District	Non-affected area	Affected area			
		10%	10–25%	25–50%	50–100%
Gorica (ha)	0	269	117	30	25
Sežana (ha)	363	1859	385	160	44
Total (ha)	363	2128	502	190	69

Source SI AS 675, box 13, Letalska akcija proti borovemu prelcu [Airplane action to suppress pine processionary moth]. Author's own elaboration

and Forestry, a biologist, field and biological observation assistants, an economist, and two drivers, and it cost 573 dinars per hectare.[65] In the months leading up to the campaign, field observers conducted a rough survey of the situation in the two districts that were to be included: Gorica and Sežana.

The plantations smaller than one hectare were not included in the survey, although their final share in the total area was very high. The foresters observed that while the worst infestations were taking place at the lowest altitudes, plantations at higher altitudes were not immune either (Table 8.1). According to the estimates of the local foresters, ten cocoons on a single tree meant that the caterpillars would completely strip the tree within a year, and many trees with 25 or more cocoons were observed.[66]

Control was sensible in the areas exhibiting more than 25% damage (Table 8.2) as at that point, the vitality of the trees was already affected, but pointless in the areas with the most serious damage, were more that 50% of the plantations were infested (Table 8.2), as these trees were already beyond salvation. While the foresters reported that some trees recovered even after being completely stripped, the problem was that the infestations kept repeating year after year, and the pine processionary moth attacks were followed by secondary pests.[67]

65 SI AS 675, box 13, Letalska akcija proti borovemu prelcu [Airplane action to suppress pine processionary moth].

66 SI AS 675, box 13, Letalska akcija proti borovemu prelcu [Airplane action to suppress pine processionary moth].

67 SI AS 675, box 13, Letalska akcija proti borovemu prelcu [Airplane action to suppress pine processionary moth].

Already in the days around 20 February, there were warnings from the Gorica district that more than 70% of the caterpillars had already left their nests and that spraying at that point would therefore be futile. The leading staff at the Ministry received warnings that it was already too late to carry out the action and that the airplanes should have been nearby and on standby to ensure a timely response. The district forester urged the Ministry to call off the campaign, as otherwise, they would be spraying empty cocoons. However, the Ministry did not decide to cancel the action: too many press teams had already been informed, the insecticide had been ordered and paid for, and the aircrew had been assigned as well.[68]

The start of the operation was delayed several times, mainly due to bora wind and bad weather, and when it finally took place, it was obstructed by the many thermal currents during the day that made it impossible to fly safely. Ultimately, 62 flights were carried out (40 h of flying time altogether).[69]

The planes that were used had been specially adapted and modified for the occasion. Special tanks for the pesticide, which was filled in the Vipava airfield before every flight, were installed to the airplanes (Fig. 8.3). The airplanes used in this airplane action were not able to carry huge additional weight which meant that the pilot had to return to Vipava airfield repeatedly, losing precious time, and resources. The plan was to spray the gaseous and sticky pesticide in a low overflight, about 5 m above the surface (Fig. 8.4), which would place Slovenia alongside technologically developed and advanced countries that also applied modern technology in agriculture and forestry.[70]

Out of the planned 2,500 ha of pine plantations in Vipava Valley and Kras, aerial control was carried out on 1,502 ha, using an average of only 3 kg of the Pantakan pesticide per hectare. As shown by the field observations of forest experts, who laid out several rather simple square observation plots with sticky surfaces (Fig. 8.5), the substance was effective immediately and during the next day, after which the bora wind completely nullified its effectiveness. Complaints were also voiced about the flight crew, which was apparently not sufficiently trained to perform low overflights over rough terrain typical of the Branica Valley, an area near the village of Branik (Fig. 8.6) between Kras Plateau and the Vipava Valley that was the first to be chosen for treatment with DDT for several reasons, but primarily because the plantings in that area are denser than elsewhere and because the area is sparsely populated. Trials with different concentrations of the insecticide could also not be carried out. As the campaign was cut short, only plantations up to 400 m of altitude were successfully treated. There, up to 60–70% of the pests were eradicated, making it necessary to keep repeating the campaigns. To ensure successful extermination, the action would need to last at least two months, but this was impossible due to the lack of funds, planes, and flight personnel. Furthermore, smaller plantations would simultaneously

---

[68] SI AS 675, box 13, Gozdarski institut Slovenije, Ljubljana 1950, Podatki o škodi v gozdovih [Forestry Institute of Slovenia, Ljubljana 1950, Data on forest damage], No. 1068/29.

[69] SI AS 675, box 13, Letalska akcija proti borovemu prelcu [Airplane action to suppress pine processionary moth].

[70] *V obrambi gozdnega bogastva Slovenije [In defence of Slovenia's forest wealth]* (1950).

Fig. 8.3 Filling the aircraft with pesticide in Vipava airfield in 1950. *Source* SI AS 675, box 13, Letalska akcija proti borovemu prelcu [Airplane action to suppress pine processionary moth]. Reproduced with permission

Fig. 8.4 Aircraft spraying DDT during a challenging low flight over rugged terrain in southwestern Slovenia in 1950. *Source* SI AS 675, box 13, Letalska akcija proti borovemu prelcu [Airplane action to suppress pine processionary moth]. Reproduced with permission

Fig. 8.5 Forester inspects the outcomes after the aerial pine processionary moth control action, southwestern Slovenia, 1950. *Source* SI AS 675, box 13, Letalska akcija proti borovemu prelcu [Airplane action to suppress pine processionary moth]. Reproduced with permission

need to be sprayed with motorised knapsack sprayers. The experts suggested that an observation station be set up in Kras, headed by a biologist or a forestry engineer, and that the aerial campaigns should take place in October and November rather than in the spring.[71]

Therefore, already the first use of aircraft in 1950 largely confirmed the scepticism of those experts who doubted the applicability of the aerial method in the area of Kras and Vipava Valley, even though it had proved successful elsewhere. It turned out that the pine plantations were too scattered and fragmented for serious aerial work, that the effectiveness of such control was greatly reduced by the wind, and that the costs involved in such campaigns, which would need to be repeated periodically, simply could not be justified. It was hoped that a single systematic pesticide treatment would weaken the pine processionary moth population to the point where it would be attacked by natural enemies and that afterwards, it would be sufficient to continue

[71] SI AS 675, box 13, Letalska akcija proti borovemu prelcu [Airplane action to suppress pine processionary moth].

Fig. 8.6 Aircraft during the pine processionary moth control action in the proximity of the village of Branik. *Source* SI AS 675, box 13, Letalska akcija proti borovemu prelcu [Airplane action to suppress pine processionary moth]. Reproduced with permission

the treatment with hand sprayers. During this period, ecological factors were not yet among the concerns regarding the use of aircraft for dispersing DDT agents.[72]

8.3.1.2 Responses to the Aerial Campaign and the Shifting of Responsibility to Landowners

The responses to the air campaign were rather mixed. It has revealed the lack of insecticide availability and has led the Ministries of Agriculture and Forestry of the individual republics to find chlorine on foreign markets and supply it to the chemical factories in order to secure the insecticide they needed. It became clear that too much coordination was called for: between meteorologists, foresters, the local authorities, the chemical industry, and, finally, JAT Yugoslav Airlines, which provided the planes, pilots, and mechanics. Moreover, pest control had to be carried out in cooperation with neighbouring countries, which was not always easy since, after WWII, the border line over the territory was not definitively drawn for a long time. The Allied Military Administration (until 1947) and the Free Territory of Trieste (1947–1954) were naturally not overly concerned with forest pest control. The situation improved only after 1954 when the demarcation of the territories was recognised de facto

[72] Barič (1951); *Več pozornosti zaščiti gozdov [More attention to forest protection]* (1950).

(*de iure* only in 1975 with the Treaty of Osimo) as the border between Italy and Yugoslavia. Until then, the Slovenian foresters had communication problems with their Italian colleagues on the one hand, as it was often unclear with whom they should coordinate their activities, but also with their Croatian colleagues within Yugoslavia on the other, with whom it was also necessary to establish cooperation, lest all activities prove unsuccessful. The aerial campaigns turned out to be an extraordinary bureaucratic feat that ultimately produced only partial results, far from those reported in the USA.[73] So, it is not an overstatement that The Ministry of Agriculture and Forestry used the air campaign mainly for promotional purposes. The entire action was monitored and filmed by a Triglav Film crew, and the Ministry tried to attract as many journalists and photographers as possible to cover it in the media, especially in Slovenian newspapers that were published in Trieste and Gorizia.[74] However, it fuelled considerable enthusiasm for the use of chemical preparations, which is evident from the reports published in the professional forestry press (the *Gozdarski vestnik* magazine). While centrally controlled aerial campaigns were discontinued, this does not mean that the use of pesticides was abandoned. Backpack hand and motorised sprayers were used in smaller areas to apply solutions in much stronger concentrations than during the aerial campaign, while the use of special protective equipment was mostly not even mentioned.[75]

In the autumn, a few months after the aerial control, an extremely strong outbreak of the pest was observed. This was considered proof that the procedure had not taken enough account of the pest's biology and, in particular, of the unstable weather conditions. It was obvious that the aerial campaign was based on assumptions rather than on actual observation. The foresters in the field complained that the Ministry was too busy focusing on the propaganda part of the campaign, while it did not even provide them with batteries for the pocket torches needed for nocturnal observations.[76] The foul mood even intensified during the field observations in the autumn of 1950, which revealed that the pest had reached such proportions that the pine plantations were in danger of being destroyed. The Ministry failed to react properly and provide either machinery or insecticide for the potential eradication in the autumn and instead passed the initiative on to private owners, which proved to be a terrible decision. Some of them did not take any action at all, while others used insecticides as they saw fit, making their own homemade mixtures and uncontrollably poisoning the surroundings of their homes.[77] Without any regulation, the authorities allowed

[73] SI AS 675, box 13, Pantakan za emulzije za uničenje borovega prelca, [Pantakan for the emulsions to destroy the pine processionary moth] No. 4480, Troha (2006), pp. 908–913.

[74] SI AS 675, box 13, Letalska akcija proti borovemu prelcu [Airplane action to suppress pine processionary moth]; V obrambi gozdnega bogastva Slovenije [In defence of Slovenia's forest wealth] (1950).

[75] Kofol (1951).

[76] SI AS 675, box 13, Letalska akcija proti borovemu prelcu [Airplane action to suppress pine processionary moth].

[77] SI AS 675, box 13, Pinijev sprevodni prelec [Pine processionary moth], No. 4717/50; SI AS 675, box 13, Uničevanje gosenic pinijevega sprevodnega prelca [Destruction of caterpillars of the pine processionary moth], No. 5/30.

the sale of even the most toxic chemical agents to individuals who had not been trained in their use. Consequently, people would get poisoned by fruit, especially in springtime, as the landowners used pesticides and other chemical agents at their own discretion and utilised the substances intended for use in the forest on fruit trees instead. Forestry experts and health inspectors strongly advocated the idea that the use of the most toxic chemical substances should be allowed only to the personnel of agricultural cooperatives, but the regulation by the Federal Sanitary Inspector that would restrict sales to private individuals was implemented too late.[78]

The production and marketing of poisons remained completely unregulated. Field inspections revealed that toxic products were stored and sold at the same premises as animal feed, that the rooms containing chemicals were not locked, and that insecticides were often repackaged in inappropriate containers without any labels, which people brought from home. Even the people in charge of the cooperatives and the vendors were not sufficiently informed about the toxicity of the materials they were selling, and they put themselves and their own health at risk by not wearing any masks and gloves. In other Slovenian territories, the factor that regulated the market was the obsolescence of the chemical industry, which was unable to keep up with the considerable market demand. In the area of Kras Plateau and Vipava Valley, the situation was different: with the proximity of the border with Italy, pesticides soon became one of the most frequently smuggled goods, and the authorities had no control whatsoever over the situation. Efforts were made to inform people of at least the minimal preventive measures during insecticide application. They were warned not to inhale the spray, put a bandana over their nose, not to spray while the wind was blowing, not to ingest the plants, and prevent animals from grazing near the trees treated with insecticides for at least ten days after spraying. Ideally, they should wear special protective clothing. However, as the text stated rather dramatically, most people only had a single shirt and a pair of trousers at their disposal, so they were advised to remove their clothing immediately after spraying and wash well. Gauze bags or old socks were used for dusting small areas, which, in the case of the pine processionary moth control, meant that village children were recruited to climb trees and dust the nests.[79]

Allegedly, most landowners were completely careless, openly mocking the demands to note the number of moth nests on each tree. They saw no real point in taking any action, as this property was "dead" to them: they were not allowed to exploit the plantations on their own land. The demands made by the authorities were unrealistic, and not even the threat of actions under Article 39 of the General Forest Act had any effect. Warnings about the dangers of the improper handling of harvested nests further discouraged people from any attempts to control the pest and risk exposing their own health, even if it was for the greater good.[80]

---

[78] SI AS 1127, box 20, archival unit 121.

[79] SI AS 1127, box 20, archival unit 121, Letno poročilo o varstvu rastlin za leto 1959 [Annual report on plant protection for the year 1959].

[80] SI AS 675, box 13, Pinijev sprevodni prelec [Pine processionary moth], No. 4717/8.

Towards the end of the 1950s, the interest in controlling the pine processionary moth gradually started to wane due to various reasons that had nothing to do with the characteristics of the pest itself. Thus, it also continued to spread in the following decades and is currently present even as far from the original sites on the Kras Plateau as the area around Log pod Mangartom, in the northwestern part of Slovenia, and in the Triglav National Park at over 1200 m above sea level.[81]

Apart from the scepticism of certain forestry experts, already evident in the 1950s, who opposed saving the pine plantations at all costs, in the 1960s, the pine processionary moth was increasingly often seen as a factor that would facilitate the transition to a more profitable forest composition. In the same period, the political authorities made the decision to redirect the very limited funds and resources at their disposal towards controlling the pests that were easier to suppress and where greater success was expected. Last but not least, the situation on the Kras Plateau was no longer considered alarming—quite the reverse. At first glance, the objective had been achieved: at the beginning of the 1970s, the forest cover on the Kras Plateau reached 37%, already exceeding the target of 35%, which the authorities at the time believed would ensure an ecological balance.[82] More and more measures were implemented to control the cockchafer, which had also not been completely ignored earlier. The struggle against it was mainly based on the tradition during this insect's time of swarming: each family had to collect a certain number of beetles, which were received and then destroyed by the village mayors, who also gave rewards to the most diligent individuals. As of 1955, this pest was controlled with DDT, but as the substance had to be buried in 10 cm of soil, this treatment was only feasible on new plantations.[83] The pine moth seems to have been taken less and less seriously as a pest, and in the Goriška region, it was practically no longer suppressed.[84] In fact, during the period of the most intensive use of chemical preparations, an extremely broad range of insecticides with various active ingredients were used in the area of the Kras Plateau and Vipava Valley—xyloline, dendroline, and most often DDT-based preparations, but no longer only against the pine processionary moth.[85]

[81] Jurc et al. (2019), pp. 3–5.

[82] Nicod (2003), pp. 28–29; Perko (2016), pp. 236–237.

[83] SI AS 1127, box 1, archival unit 123, Pogozdovanje Krasa in uničevanje majskega hrošča [Afforestation of the Kras Plateau and the control of the cockchafer], No. 10/101.

[84] SI PAK KP 607, box 25, archival unit 72.

[85] Titovšek (1984), p. 5.

8.3.2 The Use of Pesticides and Their Impacts on the Non-target Elements of Karstic Environment, Especially Groundwater

There is a general conviction among Slovenian forestry experts that in Slovenian forests, pesticides were only used occasionally and not systematically—as experiments in rather restricted areas, and that the quantities of pesticides used were relatively modest compared to western countries like the USA, Sweden, Switzerland, and Germany.[86] So far, no actual analyses have been carried out, as in the 1950s and 1960s, pesticide use was not monitored and information was scarce. Whether it was due to environmental awareness or lack of resources, forestry was considered less of a threat to the ecosystem than the other branches of agriculture, while the wood-processing industry has had a more severe impact on the environment than forestry itself. However, the impact of pesticide use in Slovenian forests cannot be simply ignored, even if the quantities of pesticides and other chemical products used in forests were minor compared to the environmental impact of the other agricultural industries that used to consume an estimated 80 railroad cars of DDT each year alongside other insecticides, herbicides, and artificial fertilisers.[87]

Undoubtedly, there was very little awareness of the specific nature of the karstic environment and its water sources, and the fact that the exposure of the water sources to sunlight was very brief or non-existent. The use of hazardous substances such as solvents and pesticides, which could lead to the contamination of groundwater sources, was not regulated. Moreover, due to the nature of the karstic surface, physical filtration was generally absent as well.[88] In the times when people would spray DDT on their clothes and gardens and use it to control all sorts of insects, there was, of course, no awareness that any use of pesticides, herbicides, and artificial fertilisers in agriculture and forestry could represent a source of major pollution in the karst. Due to the rapid distribution of water, consequences became evident over large areas and threatened the already scarce sources of drinking water.[89] DDT's saintly halo started to fade, but no one was willing to give it up that easily.[90] Quite the opposite: its widespread use in agriculture, forestry, and public health supposedly confirmed its complete harmlessness to humans. People had been exposed to DDT in almost every conceivable way over several decades, so it seemed completely unnecessary to ban its use in forestry, as in that case, it was used in uninhabited areas and in much lower concentrations than in the past.[91] During the afforestation project on the Kras Plateau and in Vipava Valley, DDT was not the only nor the most problematic among

[86] Jurc et al. (2017), p. 97; Kinkela (2011), pp. 104–105; Mrema et al. (2013), pp. 1–21; Christian (1999), pp. 95–99; *Uradni list RS 30/93*: Zakon o gozdovih; De Groot and Hauptman (2012), p. 76.

[87] Kogovšek and Kranjc (1999), p. 150. See also SI AS 244, box 13, folder 19; and Kranjc (2012), p. 82.

[88] Williams (1993), p. 7.

[89] Coxon (1999), pp. 37 and 46; Williams (1993), p. 90.

[90] Clarke (2012), p. 136.

[91] Roberts (2010), p. 188.

the pesticides that were being used. Lindane[92] was also in extremely high demand among foresters, as it was added as a contact insecticide to the roots at the time of afforestation with very approximate dosing, often in accordance with the "better more than less" principle. The poison thus ended up directly in the soil, unaffected by UV light and wind. Therefore, its effectiveness was prolonged, but, unfortunately, its penetration into groundwater was also greatly facilitated. In some places, Lindane use increased the efficiency of afforestation to as much as 98%, and the foresters greatly regretted that they had not been able to obtain it in sufficient quantities.[93]

During the period under consideration, karstic water sources, which cover more than 50% of Slovenia's drinking water needs, were believed to be protected by their forest cover, sparse population, and low and predominantly extensive agricultural activity. The danger stemming from inadequate sewerage systems of settlements, traffic, and industrial plants—which, in line with post-war policy, were also established in the previously industrially undeveloped environments—as well as from the unregulated use of pesticides and herbicides in agriculture and forestry was neglected for a long time.[94] In the first two decades after WWII, the public still believed that any preparation used in agriculture or forestry was safe if its use did not lead directly to the poisoning of the person who administered it. In the post-war years, very little was known about the long-term consequences.[95]

Water protection in Slovenia was already regulated, at least at the basic level, by the Water Protection Act of 1957, which highlighted the special importance of water for the economy, public health, and climatic conditions, and therefore placed water under special social protection.[96] Furthermore, the danger of groundwater contamination was discussed as early as the 1960s. However, it seems that there was little consideration of the specific situation of the karstic environment in this respect. There was certainly little awareness among the general population in the 1950s and 1960s—during the period of the most intensive use of DDT and Lindane—that anthropogenic activities in the impact zone of karstic watercourses could have long-term and very far-reaching consequences not only in the immediate environment where the activities took place but also in the form of multiplier effects over extremely large areas.[97] Moreover, the karstic waters conveniently subsided, taking with them the traces of human activities. It is also necessary to consider the fact that the organised water supply of the Slovene Dinaric Karst started as late as the 1970s and 1980s and that the basic infrastructure was therefore established late, although the water from the springs of Rižana and those at Malni at Planina polje and Hubelj in the Vipava

[92] Lindane is a gamma-isomer of benzene hexachloride, a colourless to white coloured, synthetic, crystalline solid neurotoxin with a slight musty odour that emits toxic fumes of hydrochloric acid and other chlorinated compounds when heated to decomposition. Lindane is used as an insecticide for hardwood logs, lumber, and crops.

[93] SI PAK KP 795, box 3, Zapisnik o delovnem sestanku delovodij objektov, 1958 [Notes of the working meeting of the foremen of the facilities, 1958].

[94] Ravbar (2007), pp. 29–30.

[95] Langston (2014), pp. 262–263.

[96] Stariha (2000), p. 22.

[97] Ravbar (2007), pp. 11–12; Kovačič (2003), p. 221.

Valley had already been used earlier, of course. For example, it was not until 1976 that the groundwater protection methodology was first established and the protection zones and belts were defined, while the appropriate legislation was adopted as late as 1981.[98]

Also in this case, the karstic areas proved to be specific. Unlike other areas of Slovenia, where the water self-purification capacity was usually higher and surface water is much more common, the water supply in the karstic terrain represented a particular challenge. In the absence of special cleaning plants, the water was often murky after heavy rainfall or microbiologically contaminated, and excessive concentrations of certain chemical compounds kept occurring as well.[99] The enforcement of the legislative provisions often depended on the will of the local authorities, which were frequently unwilling to agree to any self-restraint in agricultural and industrial activities in the name of protecting some "invisible" groundwater. The Ordinance on the Protection and Buffer Zones of the Source of the River Rižana, which was, in any case, adopted late—in 1987—was enforced only in one-third of the entire water catchment area because the Kras municipalities refused to implement the Ordinance and activities simply continued in the rest of the area.[100] DDT was not deliberately searched for in karstic springs until as late as 1990, when its use had been banned for almost two decades. For this year, data are available and show that DDT was not detected in any of the tested karstic springs in the wider Slovenian Dinaric Karst region (the Rižana spring, The Malenščica spring in Malni at Planina polje, the Tominčev studenec spring in the Krka basin, the Dobličica spring in Bela krajina, the Bistrica spring near Ilirska Bistrica, Podroteja near Idrija, and Bužini springs near Bužin in Istria) nor at Hubelj and Mrzlek springs in Vipava Valley.[101] However, significant contamination of certain stretches of the individual karstic watercourses such as Hubelj, Reka, and Rižana was proved, making the drinking water supply increasingly unreliable.[102]

8.4 Conclusions

For a long time, the phenomenon of the afforestation of Kras with a non-native black pine was perceived as a success and an improvement compared to the bare and degraded landscape. However, in the afforestation plans, the interests of the local population were not considered, nor were the consequences that the interventions would have on the natural environment and the rest of the economic and social system. Especially, the regime's post-WWII propaganda glorified the restoration of

---

[98] Breznik (1976); Uradni list SFRJ 33/1981; Uradni list SRS 35/1981.

[99] Macarol-Hitti (1990), p. 58.

[100] Hočevar et al. (2010), p. 151.

[101] ARSO, Baza podatkov državnega monitoringa kakovosti voda, DDT in DDD v vodnih izvirih 1990.

[102] Kokolj (1990), pp. 44–45.

the landscape, which was claimed to have been exploited by foreigners for centuries to the point of almost total destruction. The afforestation became a myth that could not (or should not) be doubted. A firmly rooted image of the Karst plateau as a completely barren and infertile landscape, beset by fierce winds that kept carrying away the last remnants of fertile soil, which needed to be saved by all means and at all costs, was created. The project was glorified for its supposed predominant ecological rather than economic reasons and its supposedly sincere desire to restore and revitalise the natural environment. It is only in the more recent period that the researchers of different scientific fields started to challenge the degradation theory, take into consideration the social and economic implications, and acknowledge that the attempt to regrow the forest by planting large-scale monoculture black pine plantations has not only permanently altered the character of the landscape and the environment but also the way of life in the area in general.

The afforestation with non-native black pine has led to long-term structural consequences. It has clearly caused changes not only in soil and water balance but also in biodiversity with the changes in the flora in fauna. The formation of vast monoculture plantations has led to the spread of new, non-native pests and, consequently, to the rather uncontrolled use of chemical pesticides in the post-WWII period in an attempt to control them. The use of pesticides in the Kras Plateau and the Vipava Valley can be examined as a case study of the combination of environmental and socio-political factors. As the research has shown, the most prominent and high-profile attempt to control the pine processionary moth by spraying DDT from airplanes in 1950 primarily served the political goals of the socialist regime while overriding the opinions of the local forestry experts. Furthermore, the fact that such large-scale actions were not repeated and that the control of the pine processionary moth remained mostly mechanical cannot be interpreted as a sign of increased environmental awareness. In fact, it is not an overstatement that it was only the lack of pesticides and other chemical agents on the market that prevented the use of larger quantities of pesticides and greater environmental damage in the 1950s and 1960s.

Pine processionary moth, "the external enemy" of the socialist regime, turned out to be less dangerous for black pine trees than the authorities feared. Black pine, which was given the honorary title of "a pioneer of the bare karst," survived, and the afforestation with this species was clearly not as transitional as it had been planned. The expectations that the black pine would be removed from the karst after one generation were not fulfilled. It has survived the attack of the non-native pests and started spreading by itself, strongly encouraged by the abandonment of agricultural land. Only a few decades after the extraordinary efforts to afforest the karst areas, the first initiatives to eliminate the overgrowth appeared in order to contribute to the restoration of the species-rich habitats in artificially forested and overgrown areas. In recent years, the removal of black pine has become an important part of the strategy to save endangered species, preserve the cultural landscape, and perhaps even revive agricultural activity in these areas. Nonetheless, black pine remains an integral part of the karst landscape and its identity. The plan to largely restore a natural environment has clearly failed. The afforestation project proved to be a profound and extensive human intervention in the environment, resulting in rapid and drastic changes in land use patterns with far-reaching consequences for the entire ecosystem and society as a whole.

Acknowledgements The research was carried out within the framework of the research programme Economic, Social and Environmental History of Slovenia, No. P6-0280, financed by the Slovenian Research and Innovation Agency.

Abbreviations

DDT Dichlorodiphenyltrichloroethane
HCH Hexachlorocyclohexane
DDD Dichlorodiphenyldichloroethane

References

Sources

Published Sources

ARSO Agencija Republike Slovenije za okolje, Baza podatkov državnega monitoringa kakovosti voda, DDT in DDD v vodnih izvirih 1990 [Slovenian Environment Agency, Database of the national monitoring of water quality, DDT and DDD in water springs 1990]

Deželni zakonik in vladni list za kranjsko kronovino 37(7, 21 November 1885): Zakon z dné 9. marca 1885 o pogozdovanji Krasa v vojvodini Kranjski [Landes-Gesetz- und Regierungs- Blatt für das Kronland Krain 37(7, 21 November 1885): Gesetz vom 9. März 1885 betreffend die Karstaufforstung im Herzogthume Krain]. https://www.dlib.si/details/URN:NBN:SI:DOC-0P2UNRBU

Jošt J (1949) Borov sprevodni prelec na Krasu [Pine processionary moth on the Kras Plateau], 3. https://www.zdravgozd.si/dat/pdp/19491223.pdf

Jurc M, Hauptman T, Pavlin R, Borkovič D, Zavrtanik Z, Papler-Lampe V (2019) Pinijev sprevodni prelec (*Thaumetopoea pityocampa*) v Sloveniji [Pine processionary moth (*Thaumetopoea pityocampa*) in Slovenia] Novice iz varstva gozdov (12):3–5. https://www.zdravgozd.si/nvg/prispevek.aspx?idzapis=12-2

Poročilo deželnemu zboru poknežene grofovine Goriško - Gradiščanske o delovanju deželnega odbora od 1. januarja do konca decembra 1902 [Pertrattazioni della Dieta provinciale della Contea principesca di Gorizia e Gradisca dal 1° gennaio alla fine del dicembre del 1902] (1903). Deželni odbor, Gorica. https://www.dlib.si/details/URN:NBN:SI:spr-VKMPFCRL

Projekt VIPava [Project VIPava], www.projektvipava.si

Titovšek J et al (1984) Sušenje črnega bora na Krasu [Drying of black pine on the Kras Plateau]. https://www.zdravgozd.si/dat/pdp/19840312.pdf

Uradni list Republike Slovenije 30/93, Zakon o gozdovih [The Official Gazette of the Republic of Slovenia 30/93, Act on Forests]

Uradni list SFRJ 33/1981 [The Official Gazette of the Socialist Federal Republic of Yugoslavia 33/1981]

Uradni list SRS 35/1981 [The Official Gazette of the Socialist Republic of Slovenia 35/1981]

Newspaper Sources

Barič J (1951) O razvoju in zatiranju pinijevega sprevodnega prelca na Krasu [About the development and suppression of the pine processionary moth on the Kras Plateau]. Gozdarski vestnik 9:144–148

Beltram V (1946) Pogozdovanje Krasa ni problem [Afforestation of the Kras Plateau is not a problem]. Gozdarski vestnik, 5:9–18 and 29–39

Deželni zbor goriški [Dieta provinciale di Gorizia]. Gorica 4(57, 19 July 1902):1

Gašperšič F, Winkler I (1986) Ponovna ozelenitev in gozdnogospodarsko aktiviranje slovenskega krasa [Re-greening and forest-management activation of the Slovenian karst]. Gozdarski vestnik 44(5):169–184

Kofol A (1951) Iz zgodovine pinijevega sprevodnega prelca [From the history of the pine processionary moth]. Gozdarski vestnik 9:243–246

Kordiš F (1952) Miši, nevarni gozdni škodljivci [Mice, dangerous forest pests]. Gozdarski vestnik 10:174–175

Kovačević Ž (1953) Ne lovna debla – temveč lovna drevesa! [Trap trees—not just trap logs!]. Gozdarski vestnik 11:71–75

Šebenik M (1950) Gozdarska in lesna posvetovalnica: DDT ali Pantakan [Forestry and wood consultation column: DDT or Pantakan]. Gozdarski vestnik 8:115–116

Šlander J (1952) Ne lovna debla – temveč lovna drevesa! [Trap trees—not just trap logs!]. Gozdarski vestnik 10:1–7

Šlander J (1953) Kako zavarujemo les pred žuželkami [How to protect wood from insects]. Gozdarski vestnik 11:65–71

V obrambi gozdnega bogastva Slovenije. Z letali nad uničevalca borovih gozdov Krasa [In defence of Slovenia's forest wealth. Airplanes against the destroyer of the pine forests on the Kras Plateau]. Primorski dnevnik 6(47=1438, 11 March 1950):4

Več pozornosti zaščiti gozdov [More attention to forest protection]. Ljudska pravica 11(235, 2 October 1950):1

Virnik F (1951) Smrekova grizlica in naše izkušnje z njo [Small spruce fly and our experience with it]. Gozdarski vestnik 9:209–220

Wraber M (1950) O pomenu fitosociologije za gozdarstvo [On the importance of phytosociology for forestry]. Gozdarski vestnik 8:40–46

Archival Sources

SI PAK KP—Provincial Archives of Koper

SI PAK KP 607, Uprava za gozdarstvo pri OLO Koper, 1951–1963 [The Forestry Administration at the Regional People's Committee of Koper, 1951–1963]

SI PAK KP 629, Okrajni urad Komen [District office of Komen]

SI PAK KP 638, Občina Štorje [Municipality of Štorje]

SI PAK KP 795, Zavod za pogozdovanje in melioracijo Krasa, Ajševica–Gorica [Institute for Afforestation and Melioration of the Kras, Ajševica–Gorica]

SI AS—Archives of the Republic of Slovenia

SI AS 244, Svet za kmetijstvo in gozdarstvo LRS [Council for Agriculture and Forestry of the People's Republic of Slovenia]

SI AS 675, Ministrstvo za gozdarstvo Ljudske republike Slovenije, 1945–1951 [Ministry of Forestry of the People's Republic of Slovenia, 1945–1951]

SI AS 1127, Republiški sekretariat za kmetijstvo in gozdarstvo Socialistične republike Slovenije [Republic Secretariat for Agriculture and Forestry of the Socialist Republic of Slovenia]

Literature

Bárány-Kevei I (2003) Human impact on Hungarian karst terrains, with special regard to silviculture in Hungary. Acta Carsologica 32(2):175–185

Bennet BM (2011) A Global History of Australian Trees. J Hist Biol 44(1):125–145

Blažič M (1959) Borov sprevodni prelec [The pine processionary moth]. [self-published], Ljubljana

Bončina A, Gašperšič F, Diaci J (2003) Long term changes in tree species composition in the Dinaric mountain forests of Slovenia. For Chronicle 79(2):227–232

Breznik M (1976) Metodologija za določitev varstvenih območij in pasov pitne vode v podzemnih vodnih virih [Methodology for determining protection areas and drinking water stripes in groundwater sources]. FAGG, Hidrotehnična smer, Ljubljana

Brock EK (2014) New patterns in old places. Forest history for the global present. In: Isenberg AC (ed) The Oxford Handbook of Environmental History. Oxford University Press, New York, pp 154–177

Christian S (1999) DDT: Kulturgeschichte einer chemischen Verbindung. Christoph Merian Verlag, Basel

Clarke S (2012) Rethinking the Post-War Hegemony of DDT: Insecticides Research and the British Colonial Empire. In: Berridge V, Gorsky M (eds) Environment, Health and History. Palgrave Macmillan, Basingstoke, pp 133–153

Coxon C (1999) The nature of human impacts on karst waters. Agriculturally induced impacts. In: Drew D, Hötzl H (eds) Karst Hydrogeology and Human Activities. Impacts, Consequences and Implications. Balkema, Rotterdam, pp 37–53

Culiberg M (2005) Paleobotanične raziskave na Krasu [Paleobotanical research on the Kras]. In: Mihevc A (ed.) Kras: voda in življenje v kamniti pokrajini. Založba ZRC, ZRC SAZU, Ljubljana, pp 149–154

De Groot M, Hauptman T (2012) Pinijev sprevodni prelec—*Thaumetopoea pityocampa* [The processionary moth—*Thaumetopoea pityocampa*]. In: Navodila za preprečevanje in zatiranje škodljivcev in bolezni gozdnega drevja v Sloveniji: priročnik za javno gozdarsko službo. Gozdarski inštitut Slovenije, Založba Silva Slovenica, Ljubljana, pp 73–77

Dunlap TR (1982) DDT: scientists, citizens, and public policy. Princeton University Press, Princeton

Fras S (1959) Pomen črnega bora za pogozdovanje Krasa in njegovi abiotični ter biotični škodljivi spremljevalci [The importance of black pine for Karst afforestation and its abiotic and biotic harmful companions]. [self-published], Ljubljana

Gams I (1991) Sistemi prilagoditve primorskega dinarskega krasa na kmetijsko rabo tal [Systems of adapting the littoral dinaric karst to agrarian land use]. Geografski zbornik 31:5–106

Gams I, Gabrovec M (1999) Land use and human impact in the Dinarc karst. Int J Speleol 28:55–70

Grégoire JC, Evans HF (2004) Damage and control of BAWBILT organisms: an overview. In: Lieutier F, Day KR, Battisti A, Grégoire JC, Evans HF (eds) Bark and Wood Boring Insects in Living Trees in Europe: A Synthesis. Kluwer, Dordrecht [etc.] pp 19–37

Grove AT, Rackham O (2001) The Nature of Mediterranean Europe. An Ecological History. Yale University Press, New Haven, London

Hočevar Z, Knez K, Krbavčič S, Križman D, Sau S, Valentič D, Žigon I (2010) Rižanski vodovod Koper – 75 [Rižana river water supply network – 75]. Rižanski vodovod, Koper

Hufnagl L (2002) Splošni del gospodarskih načrtov za kočevsko gospostvo 1894 [General part of economic plans for the Kočevje estate 1894]. Biotehniška fakulteta, Oddelek za gozdarstvo in obnovljive gozdne vire, Ljubljana

Hutter H-P, Moshammer H, Wallner P, Damberger B, Tappler P, Schleritzko K, Kundi M (2007) Belastungen durch Carbolineum in der Wiener Kaisergruft. Umweltmed Forsch Prax 12(4):194–198

Jactel H, Barbaro L, Battisti A, Bosc A, Branco M, Brockerhoff E, Castagneyrol B, Dulaurent A-M, Hódar JA, Jacquet J-S, Mateus E, Paiva M-R, Roques A, Samalens J-C, Santos H, Schlyter F (2015) Insect—Tree Interactions in *Thaumetopoea pityocampa*. In: Roques A (ed) Processionary Moths and Climate Change: An Update. Springer, Cham, pp 265–310

Johann E (2018) The interrelation of wood requirements of the Austrian navy and the shaping of the cultural landscape in the northern Adriatic region. Ekonomska i ekohistorija 14:32–51

Jurc M, Pavlin R, Andreja K, De Groot M, Hauptman T (2017) Priporočila za uporabo različnih biotehniških metod in kemičnih sredstev za obvladovanje podlubnikov (Curculionidae: *Scolytinae*) [Recommendations for the Use of Various Biotechnical Methods and Chemical Agents for Bark Beetle Control (Curculionidae: *Scolytinae*)]. Gozdarski vestnik 75(2):94–111

Jurhar F, Miklavžič J, Sevnik F, Žagar B (1963) Gozd na krasu Slovenskega primorja [Il bosco sul carso del Litorale Sloveno]. Tehniški muzej Slovenije, Ljubljana

Kinkela D (2011) DDT and the American century. Global health, environmental politics, and the pesticide that changed the world. The University of North Carolina Press, Chapel Hill

Klavora V (1997) Škabrijel. Mohorjeva založba, Celovec-Ljubljana-Dunaj

Klavora V (2014) Fajti hrib: bojišče na Komenskem Krasu 1916–1917 [Der Fajti hrib – Berg des Todes]. Mohorjeva družba, Celovec

Knížek M, Beaver R (2004) Taxonomy and systematics of bark and ambrosia beetles. In: Lieutier F, Day KR, Batisti A, Grégoire JC, Evans HF (eds) Bark and Wood Boring Insects in Living Trees in Europe: A Synthesis. Kluwer, Dordrecht [etc.], pp 41–54

Kogovšek J, Kranjc A (1999) Pollution by surface water: The Reka River. In: Drew D, Hötzl H (eds) Karst Hydrogeology and Human Activities. Impacts, Consequences and Implications. Balkema, Rotterdam, pp 146–152

Kokolj J (1990) Zaloge, raba in varovanje pitne vode v Sloveniji [Supplies, use and protection of drinking water in Slovenia]. In: Voda – v prehrani in živilstvu. Društvo živilskih in prehranskih strokovnih delavcev Slovenije, Ljubljana, pp 39–56

Kovačič G (2003) The protection of karst aquifers: the example of the Bistrica Karst Spring (SW Slovenia). Acta Carsologica 32(2):219–234

Kranjc A (2005) Kratka zgodovina gozda na Krasu [A Short History of the Forest on Kras]. In: Mihevc A (ed) Kras: voda in življenje v kamniti pokrajini. Založba ZRC, ZRC SAZU, Ljubljana, pp 141–148

Kranjc A (2012) Dinaric Karst: an example of deforestation and desertification of limestone terrain. In: Moutinho P (ed) Deforestation Around the World. InTech, Rijeka, pp 73–94

Langston N (2014) New chemical bodies. Synthetic chemicals, regulation and human health. In: Isenberg AC (ed) The Oxford Handbook of Environmental History. Oxford University Press, New York, pp 259–281

Macarol-Hitti M (1990) Nadzor nad pitno vodo in problematika v Sloveniji 1989 [Drinking water control and problems in Slovenia 1989]. In: Voda – v prehrani in živilstvu. Društvo živilskih in prehranskih strokovnih delavcev Slovenije, Ljubljana, pp 57–60

Masiello D (1992) Cenni di storia sui boschi carsici con particolare riferimento al comune di Trieste. In: Barocchi R (ed) L'imboschimento del Carso: cenni di storia e previsioni sul futuro dei boschi carsici e, in appendice Piano naturalistico forestale del Carso Triestino e Goriziano redatto dall'Università di Trieste per conto dell'Amministrazione regionale. Regione autonoma FVG, Direzione regionale delle foreste e dei parchi, Udine, pp 43–54

Mellanby K (1992) The DDT story. British Crop Protection Council, Surrey

Mihevc A (2005) Suhi zidovi in delane vrtače – antropogena preoblikovanost kraškega površja na območju Račic, Divače in Volčjega Gradu ["Dry walls and transformed dolines - anthropogenic influence on the surface of the Kras in the area of Divača, Račice and Volčji Grad"]. In: Kras: voda in življenje v kamniti pokrajini, edited by Andrej Mihevc. Založba ZRC, ZRC SAZU, Ljubljana, pp 57–71

Mlinšek D (1993) Beseda o knjigi [Foreword]. In: Kordiš, Franjo. Dinarski jelovo bukovi gozdovi v Sloveniji [Dinaric Fir-Beech Forests in the Republic of Slovenia]. Univerza v Ljubljani, Biotehniška fakulteta, Oddelek za gozdarstvo in gozdna gospodarstva, Ljubljana, pp 6–7

Mrema EJ, Rubino FM, Colosio C (2013) Obsolete Pesticides—A Threat to Environment, Biodiversity and Human Health. In: Simeonov L, Macaev F, Simeonova B (eds) Environmental Security Assessment and Management of Obsolete Pesticides in Southeast Europe, NATO Science for

Peace and Security Series C: Environmental Security. Springer, Dordrecht, pp 1–21. https://doi.org/10.1007/978-94-007-6461-3_1

Nicod J (2003) Understanding environmental problems in Dinaric Karst. In: Natek K (ed) Fizična geografija pred novimi izzivi [Physical geography facing new challenges]. Znanstveni simpozij ob 80-letnici akademika prof. dr. Ivana Gamsa. Oddelek za geografijo Filozofske fakultete, Ljubljana, pp 27–41

Panjek A (2006) Človek, zemlja, kamen in burja. Zgodovina kulturne krajine Krasa (oris od 16. do 20. stoletja) [Man, earth, stone and bora wind. History of the cultural landscape of the karst (outline from the 16th to the 20th century)]. Univerza na Primorskem, Znanstveno-raziskovalno središče, Zal. Annales, Koper

Panjek A (2015) Kulturna krajina in okolje Krasa: o rabi naravnih virov v novem veku [Paesaggio culturale e ambiente del Carso: l'uso delle risorse naturali in età moderna]. Založba Univerze na Primorskem, Koper

Panjek A (2018) Against the desert in the Karst: a paradigm-shift from ruined landscape to cultural savannah. Ekonomska i ekohistorija 14:52–71

Perko F (1989) Ekološka niša in gospodarski pomen smreke na jelovo-bukovih rastiščih visokega Krasa [Ecological niche and economic importance of spruce in fir-beech habitats of the High Karst]. [self-published], Postojna

Perko F (2016) Od ogolelega do gozdnatega krasa: pogozdovanje krasa [From bare to forested karst: karst afforestation]. Zveza gozdarskih društev Slovenije – Gozdarska založba, and Jutro, Ljubljana

Ravbar N (2007) The protection of karst waters: a comprehensive Slovene approach to vulnerability and contamination risk mapping. Inštitut za raziskovanje krasa ZRC SAZU, Postojna; Založba ZRC, Ljubljana

Remec M (2015) Podrgni, očedi, živali otrebi: higiena in snaga v dobi meščanstva [Rub, tidy, tend to animals: Hygiene and sanitation in the bourgeois period]. Inštitut za novejšo zgodovino, Ljubljana

Remec M (2021) Črni bor kot črna kuga: odnos lokalnega prebivalstva do pogozdovanja s črnim borom na Krasu v 19. in 20. stoletju [Black Pine as the Black Death: The Local People's Attitudes towards Black Pine Afforestation in the Slovenian Karst in the 19th and 20th Century]. Prispevki za novejšo zgodovino 61(2):43–66

Roberts D (2010) The Excellent Powder: DDT's political and scientific history. Dog Ear Publishing, Indianapolis

Roques A, Rousselet J, Avci M, Robinet C (2015) Climate warming and past and present distribution of the processionary moths (*Thaumetopoea* spp.) in Europe, Asia Minor and North Africa. In: Roques A (ed) Processionary Moths and Climate Change: An Update. Springer, Dordrecht, pp 81–161. http://dx.doi.org/10.1007/978-94-017-9340-7_3

Rubbia K (1912) Petindvajset let pogozdovanja Krasa na Kranjskem: poročilo komisije za pogozdovanje Krasa v vojvodini Kranjski o delovanju, od leta 1886 do konca leta 1911: z dvema podobama [Fünfundzwanzig Jahre Karstaufforstung in Krain: Bericht der Aufforstungskommission für das Karstgebiet des Herzogtums Krain über die Tätigkeit vom Jahre 1886 bis Ende 1911]. Pogozdovalna komisija, Ljubljana

Russell E (2001) War and nature. Fighting humans and insects with chemicals from World War I to Silent Spring. Cambridge University Press, Cambridge

Šercelj A (1996) Začetki in razvoj gozdov v Sloveniji [The origins and development of forests in Slovenia]. SAZU, Ljubljana

Stariha G (2000) Vodno pravo in vodne pravice [The water legislation and water rights]. Arhivi 22(2):11–24

Troha N (2006) Urejanje mejnih vprašanj [Regulation of boundary issues]. In: Čepič Z, Borak N, Fischer J (eds) Slovenska novejša zgodovina: od programa Zedinjena Slovenija do mednarodnega priznanja Republike Slovenije: 1848–1992 [Recent Slovenian history: from the United Slovenia program to the international recognition of the Republic of Slovenia: 1848–1992]. Mladinska knjiga and Inštitut za novejšo zgodovino, Ljubljana, pp 908–915

Williams PW (ed) (1993) Karst terrains: environmental changes and human impact. Catena, Cremlingen

Zorn M, Kumer P, Ferk M (2015) Od gozda do gozda ali kje je je goli, kamniti Kras? [From forest to forest or where is the bare rocky Karst?]. Kronika 63(3):561–574

Zupančič M (1980) Smrekovi gozdovi v mraziščih Dinarskega gorstva Slovenije [Spruce forests in the frost hollows of the Dinaric Mountains of Slovenia]. SAZU, Ljubljana

Meta Remec Research fellow, Ph.D. After her B.A. studies in History and Italian language and literature, she obtained her Ph.D. in History in 2015. She authored two monographs and several scientific articles. She has worked on issues regarding the history of everyday life, agricultural development, and social and economic conditions in the Slovenian rural areas. She has also explored various questions concerning the effects of the rapid industrialisation since World War II and the role of mechanisation in the development of agriculture during the socialist era in Slovenia.

Chapter 9
Traditional Management of Forest Resources in Mountain Sites of the Dinaric Transhumance in Montenegro

Ivan Laković and Jelena Lazarević

Abstract The paper provides valuable insight into the traditional approach to the management and use of forest resources, related to seasonal summer settlements—*katun*s in the mountains in the Dinaric Karst areas of Montenegro from the nineteenth until the mid-twentieth century. It lays the focus on the experiences of the population living within the case study of the Žijevo mountain massif, situated in the south-eastern part of Montenegro. This region was chosen as its characteristics encompass all the main features of a high-mountain area within the Dinaric Karst and has a developed transhumance tradition. The results of other research activities, conducted on Sinjajevina, Prokletije, and Durmitor mountains, have also contributed to the picture presented in this paper. The results originate from a series of research activities carried out within the framework of several projects funded by the Montenegrin Ministry of Science from 2015 until 2019. The paper presents the main categories of katun dwellers' perception of forests based on relevant historical sources, literature, and findings of intensive fieldwork and communication with the local population. These categories include the relationship between the position of katuns in relation to the forests; forests as a source of construction wood and firewood along with the traditional use of herbs and forest fruits in nutrition; forests as a domain of protection or danger; and traditional forest management procedures and modalities. Particular attention is paid to protection against wolves, one of the most widespread dangers commonly assumed to have lurked from the nearby forests, threatening the sheep and cattle on the mountain pastures. Such data are generally missing from the existing literature on Montenegrin mountain settlements and the exploitation of forest resources.

I. Laković (✉)
Historical Institute of the University of Montenegro, Podgorica, Montenegro
e-mail: ivan.lakovic@ucg.ac.me

J. Lazarević
Biotechnical Faculty of the University of Montenegro, Podgorica, Montenegro
e-mail: enalazarevic@ucg.ac.me

B. Fuerst-Bjeliš et al. (eds.), *Environmental Histories of the Dinaric Karst*, Environmental History 17, https://doi.org/10.1007/978-3-031-56089-7_9

Keywords Montenegro · Transhumance · Seasonal mountain settlements · Mountain forests · Beech · Bosnian pine · Forest food

9.1 Introduction

Montenegro is situated in the western part of the Balkan Peninsula, at the junction of two important geographical domains: the Dinarides and the central Mediterranean. While the southern part of the country follows the Adriatic coast, the central and northern regions are hilly and mountainous areas with distinctive relief composed of mountain chains and massifs reaching heights of over 2500 m, intersected by deep and narrow river valleys. Only 16% of its territory is covered by terrain with an altitude of up to 500 m. The climate in Montenegro varies greatly within short distances, from a moderately warm Mediterranean climate, through sub-continental and continental to a moderately cold boreal climate.[1] Although Montenegro's territory is composed of different types of sedimentary, igneous, and metamorphic rocks, the karst terrains are dominantly shaping its landscapes as over two-thirds lie on carbonates.[2]

The historical lifestyle of the population inhabiting the Dinaric Karst areas of Montenegro was strongly impacted by the lack of sufficient arable land, which was present and available only in lowlands and river valleys that directed most of the people toward livestock rearing as an activity better suited to rugged, mountainous terrain. Furthermore, the climate conditions, infamous for the long, dry, and hot summers limited the context for this activity further, allowing grazing in lower areas only during colder and more humid periods in late fall, winter, and early spring. After that, the grass was mowed and stored for winter reserves, while the flocks and herds were transferred to mountain pastures for summer grazing. The migration, widely known as transhumance, or nomadic pastoralism, involved almost all family members and resulted in the complete displacement of all activities from villages to *katuns*—seasonal mountain settlements founded within and around mountain pastures. Such practice enabled the continuation and complementing of life in villages, not only through surviving the dry period but also through enlarging the livestock population, making food reserves for winter, and providing surpluses for marketing. Households stayed in katuns with livestock, mostly from June to October.[3]

There were strict rules for the use of mountain pastures throughout the country. All the tribes, clans, and villages that formed the structure of traditional Montenegrin society had precisely defined areas for establishing katuns. The size and quality of the relevant pastures, water sources, and forests resulted in a strict territorial demarcation of mountain areas belonging to a particular katun. Those who acquired the right to use a katun in a specific location in the mountains were entitled only to that particular place, which was regarded as their common or, to a lesser extent, private

[1] See more in: Burić et al. (2014).

[2] See more in: Bešić (1975); Fuštić and Đuretić (2000).

[3] See more in: Laković (2017).

property.[4] Throughout Montenegro's history, transhumance was the primary manner of livestock rearing. Even the colloquial use of the name "Montenegro" was derived from the "katuns of Montenegro," an area on the western slopes of Lovćen Mountain, where the local population was bringing their livestock for summer grazing as early as the fourteenth century.[5]

Transhumance has been practiced in the Dinaric Karst areas, as well as in most of the Mediterranean basin, since the distant past.[6] Such development pattern in Montenegro persisted as the predominant agricultural activity until the mid-twentieth century, when producing milk and meat products was the main source of income for its population.[7] Based on the generational cycle of movement from the village to katun and vice versa, the local population established systems for managing relevant natural resources, organizing their lives according to their quantity, quality, vicinity, and available tools and means for their use.

The Dinaric Karst area is naturally covered with forest vegetation and various grassland types. It is known that the forests on the lowland and hilly terrains of the Dinaric Karst were devastated leading to deforestation of large areas under human pressure such as urbanization, development of transport infrastructure, and agriculture.[8] Even the forests in high mountains, close to and on the upper forest line, have been affected by high-altitude livestock farming.[9] Deforestation has led to the formation of new agricultural landscapes, which now have a great economic, cultural, and biological importance,[10] and their protection is recognized as a strategic goal.[11] However, in the Montenegrin Dinaric Karst, it seems that in the zone, roughly speaking, between 700 and 1100 m asl., the forest remained preserved to a significant extent throughout the long historical period. Traditionally, its use was limited and, according to contemporary terminology, sustainable. A major shift in forest resource exploitation and the resulting major disruptions in forest ecosystems occurred only at the end of the nineteenth and during the twentieth century.[12]

This paper aims to demonstrate the local population's relationship with the surrounding forests and their resources in the period before the change in the exploitation mode and extent, which coincided with the abandonment of the traditional transhumant cattle rearing and the gradual depopulation of mountain katuns. The study mainly focuses on the experience within the exemplary region of the Žijevo mountain massif, a part of a wider transhumant area called *Kučka planina (The Mountain of Kuči),* which contains all the significant attributes of the Dinaric Karst and has a developed transhumance tradition. The very nature of the mountain settlements built

[4] Kapetanović et al. (2017), 92–95; Laković et al. (2018), 209–210.

[5] Božić (1979), 150–155.

[6] See more in: Braudel (1995).

[7] See more in: Bogetić-Ćirić (1966).

[8] Stevanović and Vasić (1995), 22–23; Fuerst-Bjeliš (2018), 136, 139–140.

[9] Janković (1991), 63–65.

[10] Liechti and Biber (2016), 569–572; Opperman et al. (2012), 16–24.

[11] Marković et al. (2012), 302–310.

[12] See more in: Vučković (1972).

for summer grazing implies the reliance of the local population on forests mainly as the source of construction and firewood. As well, these forests were understood as domains of lurking dangers that were to be thwarted through a relevant system of measures. Traditional ecological knowledge developed in such a way enabled a sustainable coexistence of shepherds' communities and larger forest complexes, which came to an end with fundamental social and economic changes in the country after WWII.

9.2 Material and Methodology

Most of the results presented here are from a series of research studies conducted between 2015 and 2019. These studies involved field surveys and interviews with katun dwellers on general katun-related issues and those related to forests, wood harvesting, non-timber forest products, and ecosystem services provided by the forests. Additionally, a pilot survey using an online Google questionnaire was conducted in the summer on similar topics. Among other forestry-related topics, we were interested in how local communities have taken care of natural resources in the past, including their known management practices. About 400 respondents took part in the survey. Fifteen respondents from the territory of Kuči (the Montenegrin tribe traditionally using the katuns in the case study area) were interviewed to prepare this article during the winter of 2023.

Although some relevant available literature was consulted, it was apparent that a significant amount of data was missing. Compared with areas of the Dinaric Karst, historically under the rule of the Venetian Republic and Austrian/Austro-Hungarian Monarchy, there is scarcity of historical sources and related research on forest themes in the regions that experienced centuries-long Ottoman rule, as was the case with Montenegro. This does not mean that such documents do not exist in relevant archives, but no significant research effort has been taken in this direction yet.

9.2.1 Case Study Area

The Žijevo mountain massif, chosen for the case study area, covers an area of up to 150 km^2 and is situated northeast of Podgorica (Fig. 9.1). The terrain of Žijevo Mt. is made of solid Cretaceous limestone and Triassic limestone mixed with dolomites and dolomitic limestone, exhibiting all the characteristics of a pronounced karst relief. It is rugged and rocky, with around 20 peaks of 2000 m and above. Typical dark soil (calcomelanosol) is formed on the limestone. Locally, on flattened terrains and

Fig. 9.1 Map of Montenegro with the case study area

at lower elevations (below 1400 m), it develops into a browned dark soil (calcocambisol).[13] The climate is characterized by dry summers and cold, snowy winters. The period of summer drought is harsh, with frequent month-long periods without precipitation. Heavy precipitations occur in the late autumn and early spring period. Total annual precipitation is high, ranging from 2000 to 3000 mm, and sometimes exceeding 3500 mm. The humid, warm temperate climate type (moderately Cf) is represented by subtypes Cfb and Cfws "bx".[14]

[13] Fuštić and Đuretić (2000), 156–172.

[14] See more in: Burić et al. (2012).

Fig. 9.2 Katun Radan on the Žijevo Mountain, with the forest of Bosnian pine in the background. *Photo* I. Laković (2015)

The forests at Mt. Žijevo are mostly deciduous beech (*Fagus sylvatica* L.) forests, occurring at altitudes between 700 and 1600 m. In the upper belt, the forested areas are interspersed with mountain pastures and meadows. Bosnian pine forests (*Pinus heldreichii* H. Christ; Serb. *munika*) are present at the altitudinal belt between 1200 and 2000 m asl., usually on rocky and sloped terrains (Fig. 9.2). These forests have rich ground flora, dendroflora, and herbaceous flora, with a large number of fungi and animals. The forests in the Dinaric Karst mountain regions are important for the protection of soil and water basins, as well as due to the wealth of biodiversity.[15]

9.2.2 Katuns Position in Relation to Forests

The mountain shepherd settlements in the Žijevo area, along with other parts of Montenegrin Dinaric Karst, blend naturally into the landscape, following the logic of the terrain, utilizing favorable environmental factors, and construction material available in situ. The establishment and organization of these katun-type settlements were based on three main premises: pastures, as a source of food for animals, water as a precondition of sustaining life, and forests, both the domain of indispensable wood resources and habitat for wild animals that posed constant hazard.[16] The katuns were built according to traditional knowledge of planning, architecture, construction, protection, and use. The domain of protection was not limited to inclement weather only, but was also supposed to provide security against intruders and natural hazards (flooding, avalanches, forest fires, etc.). The primary function of katuns was to provide infrastructure for organizing cattle grazing in mountain pastures, which included many significant aspects. They were built within or next to pastures to provide short and direct access to grazing. At the same time, their position had to enable a permanent water supply, be it from springs, lakes, ponds, or snow caves.

[15] Lazarević (2017), 64–65.

[16] See more in: McNeill (1992); Berkes (1993), 1–9; Laković et al. (2020), 1–26.

They also had to be within reasonable reach of sources of construction materials, namely stone and wood, for building housing and auxiliary facilities, and enable the collecting and transportation of firewood for households.

The altitudinal dispersion of the katuns in Montenegro ranges (roughly) between 1100 and 1900 m. Depending on the area, forest type, density, and spatial distribution, the katuns were mostly built in this zone, enabling the communal use of pastures above, and the forests below or above the settlements. As exceptions from the general rule, there were also katuns built within the forests, without direct exit to the pastures below, above, or around it, but all these cases had a special, particular rationale behind them. Around 180 katuns have been recorded in the case study area, whose characteristics fully fit into the described context.[17]

As mentioned, the very location of the katun was further envisioned to provide defense, against both wild animals (wolves and bears) and humans. The whole architectural organization of katuns was meant to disable, repel, and mitigate such assaults.[18] The protective character was emphasized through the settlements' locations, layout, built material and construction characteristics of housing and auxiliary facilities, as well as the warning and defensive measures taken in this regard.[19] The hazard of an attack by other people varied depending on many different factors, from the vicinity of the border with other tribes or states, warfare, social and political turmoil, to the ever-present outlaws of all kinds. While most attacks of this type were motivated by the desire to expel the current or newly arrived katuns' residents or seize and take away their animals, the ones with the roots in private, social, religious, or political animosities were not uncommon. In this sphere, the forest was treated ambivalently, both as a domain of hazard and protection. It could provide cover for the attackers, hiding their preparations and presence, but it was also the first refuge where the attacked would have sought shelter when overrun, which explains why the majority of traditional katun locations tended to be close to, but not in, or at the very edge of the forests, to be far enough away in case of attack, but not too far if retreat and hiding became necessary.

9.2.3 Coniferous Forests on Žijevo and Their Use

Coniferous forests on Mt. Žijevo appear at an altitudinal range between 1100 and 1900 m. They are composed primarily of Bosnian pine as the dominant and usually the only tree species present (Ass. *Pinetum heldreichii mediteraneo-montanum.* Bleč et Lak 69.).[20] Bosnian pine is a tertiary period relict and is sub-endemic to the Balkans and the southern parts of the Apennine peninsula (Fig. 9.3). It is well adapted to the

---

[17] "Valorization of Montenegrin katuns through sustainable development of agriculture and tourism—KATUN", (2015–2017).

[18] Laković (2017), 14; Laković et al. (2020), 2–3.

[19] Krstić (1979), 130.

[20] Lakušić (1987), 394.

Fig. 9.3 Map of distribution of the Bosnian pine. *Author* J. Lazarević (2020)

short vegetative period of the high-mountain zone, a dry and warm mountain summer with very intense sun radiation, but also to severe winters.[21] Its forests (*Pinion heldreichii* Horv.50, em Bleč 59) represent the climax, potential vegetation on the upper forest line on southeastern Dinaric Karst, on the high mountains under the influence of the Mediterranean climate (Fig. 9.4). Bosnian pine, along with Macedonian pine (*Pinus peuce* Griseb., Serb. molika), which grow in the conditions of greater humidity and colder climate, once formed a massive belt of the high-mountain forest vegetation of the Balkan peninsula. Under the pressure of animal husbandry, for the purpose of forming pasturelands, this vegetation has been severely degraded and high mountain areas have been deforested.[22]

Bosnian pine is classified as nearly threatened by the International Union for Conservation of Nature Red List of Threatened Species (IUCN), and it is protected in the Balkan countries (in Montenegro since 1977).[23] With reduced grazing, in the second half of the twentieth century, the Bosnian pine area was successfully regenerated and expanded on many Montenegrin mountains.[24] It is expected to adapt well to climate change and could be a promising tree species for forestry and nature conservation in high-altitude mountain areas.[25]

---

[21] Janković (1960), 149–153; Lazarević and Menkis (2020), 172.

[22] Stevanović and Vasić (1995), 18–20; Janković (1991), 65.

[23] "Rješenje o stavljanju pod zaštitu pojedinih biljnih i životinjskih vrsta [Decision on putting certain plant and animal species under the protection]" (2020).

[24] Lazarević and Menkis (2018), 189–197.

[25] Jovanović (1967), 123–127; Vendramin et al. (2008).

Fig. 9.4 Bosnian pine (*Pinus heldreichii*) above the Kučka Korita, southeastern slopes of Žijevo. *Photo* I. Laković (2018)

9.2.4 Bosnian Pine Wood in Construction of Facilities

Bosnian pine is known for its hard and durable wood. In the katuns of Žijevo, it was used in the construction of housing and auxiliary facilities. Life in katuns was organized in different types of cottages and auxiliary facilities, whose evolution advanced in the late nineteenth and throughout the twentieth century. The oldest constructions were made using a circular or rectangular plan, with a simple construction of logs and different kinds of cover—*krovina* (roofing) from available materials. The most widespread were the ones of a circular plan, with a hearth in the middle and a cone-shaped roof. In the Kuči Mountain, they were called *dubirog(s)*, while the name *savardak* was used in the northern regions (Fig. 9.5). The original cottages were covered with cloth made of goat hair or straw. More complex types of constructions followed, with drystone walls, wooden roof construction, and a roof cover, locally called *glada(s)* (Fig. 9.6).[26]

It is believed that since the late nineteenth century, the walls of housing and auxiliary facilities in Mt. Žijevo katuns were made of stone, while the roof construction, cover, and furniture were made of wood. Bosnian pine wood was primarily used for the roof cover, although beech wood was an option in some cases. The wood was brought to the katuns from the surrounding or nearby forests, considered as an essential part of the katuns' domain together with the pastures.

This type of construction was also common throughout the Montenegrin part of the Dinarides, where both high-quality stone and knowledge of drywall masonry were available. In areas where quality stone for construction was not readily available in situ due to the prevalence of shales, the cottage walls were made of a wooden skeleton, with stones stacked around it. Over time, as cottages became dilapidated,

[26] Laković et al. (2020), 4.

Fig. 9.5 Modern incarnation of *dubirog,* covered with beech planks. Kučka Korita katun area. *Photo* I. Laković (2016)

Fig. 9.6 Traditional *glada* with drystone walls and wooden (beech) roof cover. Gornji Rikavac katun area. *Photo* I. Laković (2017)

these wooden skeletons often outlived the stone constructions they had supported (Fig. 9.7).[27]

The third type of construction for both housing and auxiliary facilities within the Dinaric Karst domain in Montenegro can be found in the northern and north-eastern mountain areas, such as Durmitor, Sinjajevina, Bjelasica, Komovi, Hajla, and Prokletije. Due to different climate conditions, the boreal forests of spruce (*Picea abies* Karst.) and fir (*Abies alba* Mill.) were spread out along these massifs. As a result, the entire building process of cottages (both exterior and interior), fences, corrals, etc., relied on wood as the primary material. Wood was used for walls and roofs, both for construction and cover (Fig. 9.8). However, even within the Durmitor mountain massif, known for its tradition of wooden cottages, when the katuns were established at higher altitudes and above the forest zone, the cottages were made of stone and only covered by wood.

In any case, the local population used the readily available material for construction, both stone and wood. Stone was the preferred material for wall building when

[27] Laković (2017), 23–40.

Fig. 9.7 Cottage built around the pinewood frame. Sumor, Lještanski katun. *Photo* I. Laković (2017)

Fig. 9.8 Traditional wooden (fir wood) cottage. Katun Ćićerača. *Photo* I. Laković (2017)

the needed quantity, and sufficient quality was available; otherwise, people turned to wood. It is worth noting that the needed carpentry tools were not readily available as often as it could be expected—even basic items, such as nails, were only widely used in the twentieth century.

Bosnian pine wood was commonly used to make food containers, such as buckets, pails, firkins, pitchers, and similar household items.[28] It was also used to make water troughs for livestock and ladders for descending into caves with snow (Fig. 9.9). Its resin was poured into candles and containers with wicks for lighting. It was commonly believed to accelerate the healing of sprains and broken bones and alleviate asthmatic attacks by inhaling the vapors of pre-heated resin.

[28] Dučić (1998), 74.

Fig. 9.9 Water trough carved from Bosnian pine. Katun Gladišta, Žijevo. *Photo* I. Laković (2016)

9.2.5 Deciduous Forests on Žijevo

The dominant, prevailing forests in Žijevo Mt. are beech (alliance *Fagion moesiaca* Bleč. et Lak. 1970, suballiance *Fagenion moesiacae montanum* B. Jov.,1976).[29] Attitudinally, it occupies the central belt of mountains (700–1600 m), which is characterized by more favorable conditions for survival due to a larger amount of precipitation, higher relative air and soil humidity, and lower temperatures compared to the warm belt of oaks at lower elevations and the Bosnian pine at higher ones. The forests develop on slopes of all aspects, and are interspersed with mountain pastures and meadows, which are often of secondary nature. They occupy analog habitats to related variants of forests from the alliance *Fagion illiricum* Horv. 38, in the northwestern Dinarides.[30]

Throughout its distribution in Dinaric Karst's mountains, the beech forms monodominant forests in the lower highland belt, starting at an altitude of 500 m (Fig. 9.10). On warmer slope aspects (S, SE, SW) and all inclinations over 30°, the beech forests developed on shallow and very skeletal soils, spreading up to 1400 m alt. These habitats represent marginal environmental conditions for beech and are sensitive to temperature extremes, soil dryness, and all disturbances caused by human activity (logging, canopy opening, etc.). The forests on gentler, northern, and shady slopes have a better canopy, more favorable microclimatic conditions, and greater stability. The beech forest belt on such terrains spreads out between 800 and 1100 m, while at higher altitudes (1000–1400 m), mixed fir-beech forests are present (*Abieti-Fagetum illiricum*, Fuk. Et Stef., 1958, similar as *Abieti -Fagetum* "*dinaricum*").[31] Natural virgin forests of beech and fir are preserved even today throughout the Dinaric Karst in a few locations, best known of which are Perućica in Bosnia and Herzegovina and around Plitvice lakes in Croatia. The existence of still preserved virgin forests suggests that significant anthropogenic impact has been absent from these areas for an extended period of time.

[29] Lazarević (2017), 65.

[30] Lakušić (1987), 391; Stevanović (1977), 166–170.

[31] Stevanović (1977), 176–177.

Fig. 9.10 Beech forest in the Jezera katun area. *Photo* I. Laković (2017)

9.2.6 Use of Beech as Firewood

The firewood used in Žijevo katuns almost exclusively originated from beech forests due to their altitudinal dispersion. Though its prevailing use was firewood, beech wood was used also in construction. Like the Bosnian pine wood, beech wood was primarily used in roof construction and cover, with rare examples of entire cottages made from it. However, in the past, wood harvesting was limited due to the lack of appropriate tools and transportation facilities. The firewood was acquired mostly by collecting fallen and dry branches, which were then cut and brought to the katuns, usually on one's back.[32] A pile of branches that a single person could carry was called *breme*. For larger quantities, when available, horses, donkeys, and mules were used, a practice still present in remote areas and rugged terrains.

The design of the cottages and the way fire was burned within them also impacted the tradition of firewood collection. Until the mid-twentieth century, the fire in the cottages was mainly made at the hearth, placed at two-thirds of the longitudinal extension of the floor (Fig. 9.11). There were no chimneys or smokestacks, so the smoke escaped through the cracks in the roof construction and the small apertures on the gable-sided roofs. In such conditions, only the dry and the least smoky wood could be used for burning.[33]

The firewood collection was, generally, performed by women and children.[34] The practice relied on the preservation of as much forest area as possible, since only the large forests could provide enough firewood for more populated katuns, or those intended for longer stays. Collecting fallen, dead, and dry branches from the forest also had an environmental protection purpose, since removing this flammable material from the forest floor prevented and decreased the possibility of naturally or anthropogenically caused forest fires escaping human control. Ownership rights

[32] Statements of katun users from the area of Kuči, given to authors in February of 2023.

[33] Kapetanović et al. (2017), 102.

[34] Statements of katun users from the area of Kuči, given to authors in February of 2023.

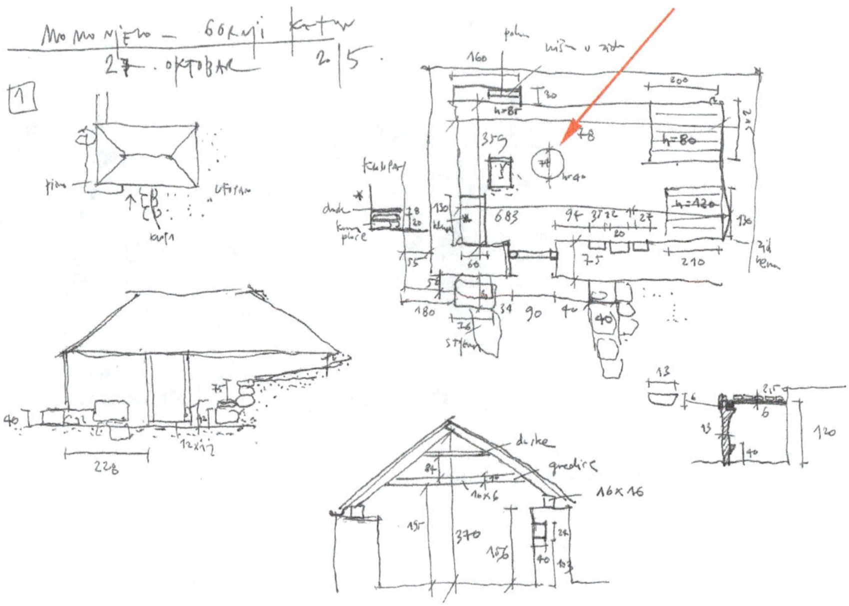

Fig. 9.11 Drawing of *glada* and its elements. Red arrow marks the position of the hearth. *Author* A. Kapetanović (2016). Reproduced with permission

were respected in regard to firewood collection. This practice not only provided the balance in the communal use of mountains, but also enabled the preservation of both the pastures and forests.[35]

The practice of cutting down standing trees for firewood became common in the later part of the twentieth century, when the availability of sawing tools, road infrastructure, and the use of lorries allowed for the transportation of larger quantities of heavier and more voluminous chunks of wood,[36] which fundamentally altered the centuries-long relationship between the people at katuns and the surrounding forests. Until then, only the timber needed for construction purposes was cut in forests, which, due to the number of objects requiring it as materiel, available tools and transportation means, was not a significant quantity. Even with the expansion of industrial forest exploitation in Montenegro in the early twentieth century, the forests surrounding the katuns remained mostly out of reach, due to their inaccessibility, leaving those near the rivers to stand the first blow.[37] With road infrastructure development, almost every mountain forest became subject to some form of harvest regime, with the intensive exploitation of timber, both for industrial and firewood use, reaching the climax in the 1960s and 1970s.[38]

35 See more in: Čubrilović (1976); Krstić (1956).

36 See more in: Vučković (1972).

37 See more in: Vučković (1965).

38 Marović (2018), 592–599.

In the second half of the twentieth century, the development of road infrastructure also allowed for the transportation of bulkier and heavier household appliances, such as metal stoves and smokestacks to the majority of katuns. As a result, the rudimentary practice of collecting dry wood waned, and larger quantities of cut wood became readily available for domestic use. As of the 1970s, collecting firewood has become a disappearing practice, pursued only in the most remote katun areas without road access.[39]

9.2.7 Traditional Use of Plants in Diet, Forest Fruits, and Medicinal Herbs

During the stay in the katun, meals were primarily made up of cereals, plant, and milk-based products. Bread was typically made of maize (*frumetin*), barley, or buckwheat, which was rarely brought from the villages more than once per season.[40] Along with cereals, the diet in the katun primarily consisted of different types of dairy products. According to Rovinsky's observations at the end of the nineteenth century, the katun community members almost never ate meat, except on special occasions, and relied mostly on dairy and plant-based foods.[41] Consequently, locally collectible plant-based food was of utmost importance. A kind of porridge (starch) was made from redroot pigweed (*Amaranthus retroflexus* L.) and nettle (*Urtica dioica* L.), while cabbage and potatoes were also used in the diet. Our most recent interviews with locals and existing practices confirm that stews, pies, broths, and salads were made from the leaves of common nettle, redroot pigweed, different dock species (*Rumex* spp.), as well as the leaves of dandelion (*Taraxacum officinale* (L.) Wener ex F.H. Wigg.) and broadleaf plantain (*Plantago major L.*). All these species grew within the katuns, in nutrient-rich (nitrogen) areas, usually where the livestock were kept for longer periods.

It is interesting to note that redroot pigweed is not an autochthonous plant, present in the region since ancient times. It is an invasive weed species, brought to Europe from North America which spread throughout the Balkans in the early nineteenth century. By sharing its habitat and nutritional role with autochthonous species like common nettle and the patience dock, redroot pigweed significantly enriched the number of edible plant species in the mountains. The increase in its significance for nutrition in the katuns temporarily coincided with the trend of constructing more solidly built, stone-walled cottages. Elder (*Sambucus nigra* L.), juniper (*Juniperus communis* L.), dog rose (*Rosa* sp.), raspberry (*Rubus idaeus* L.), strawberry (*Fragaria vesca* L.), and other forest fruits growing on the edges of beech forests and forest clearings, such as rowan (*Sorbus aucuparia* L.), were used to make refreshing juices, water extracts, and spreads.

---

[39] Statements of katun users from the area of Kuči, given to authors in February of 2023.

[40] Dučić (1998), 103.

[41] Rovinski (1998), 370.

The use of herbs for healing is a part of the local population's historical culture. Studying traditional medicine and pharmacy, Jovan Tucakov concluded that "there may be no other place with as many remnants of ancient medicine as here, which has been preserved and handed down from generation to generation for centuries."[42] In Montenegro, about 300 species of medicinal herbs are traditionally widely used, of which 133 species are considered pharmacopeias.[43]

Numerous species of medicinal herbs have been traditionally used on Kuči Mountain and in other mountainous areas for various purposes. Interviewed people from the Žijevo katuns, as well as population in the wider area covered by the questionnaire, nowadays recognize and frequently use a dozen of medicinal herb species, while the rest of their biodiversity is recognized and used only by professionals (including folk herbalists). These species include Perforate St John's wort (*Hypericum perforatum* L.), yarrow (*Achillea millefolium* L.), creeping thyme (*Thymus* spp.), winter savory (*Satureja montana* L.), wild marjoram (*Origanum vulgare*), raspberry (*Rubus* sp.), dog rose (*Rosa canina* L.), *as well as* juniper (*Juniperus sp.*), rosebay willowherb (*Epilobium angustifolium* L.), and primrose (*Primula vulgaris* Huds.). From a large number of species, medicinal teas are prepared for the prevention and treatment of certain illnesses. Additionally, these herbs are used as spices to enhance the flavor and preserve food (additives to cheeses, meat, and cured meat). Although almost all forest fruits and berries (raspberry, strawberry, blueberry, apple, gooseberry, dogwood, currant, hazelnuts, etc.) are commonly used, there is almost no significant tradition in the collection and use of fungi. Despite being present in forests and pastures around the katuns of Žijevo, the lack of knowledge, as well as the fear of unwanted consequences, have decisively affected the general restraint from their inclusion in the traditional diet.

9.2.8 *Perception of Danger from Wolves*

Another important issue regarding the relationship between the katun communities and the surrounding forests were the domain of lurking dangers and coping with them. The katuns were established around the core idea of bringing numerous animals to the vast mountain areas, and the protection of livestock, shepherds, and their families was a crucial task. The people living in katuns were primarily afraid of violent attacks by wolves and robbers.

Millennial experience of living under the constant threat of wolves placed them near the very top of the hierarchy of human nemeses.[44] Myths and legends about their supernatural powers were easily formed on the real examples of their cleverness, bloodlust, and the harm inflicted on humans and their insufficiently or inadequately protected animals. Traditionally considered one of the most dangerous human

[42] Tucakov (1990), 9–10.

[43] See more in: Lazarević and Bojović (2017).

[44] Frank and Sjöström (2007).

enemies, wolves were the most feared source of danger for the katun population. In many cases, the fear of wolves reached the level of colloquial treatment reserved for the most dreadful personifications of evil, such as devil and demons, whose name should not be publicly and loudly spoken to avoid invoking their actions.[45] For instance, in northern Montenegro, the term *strašivoća* (something that causes or brings fear) was used when referring to the potential presence of wolves in a particular area. Although this term could encompass anything from mythical creatures to wild animals and notorious outlaws, it was predominantly used with wolves in mind.

On the other hand, that millennial experience also resulted in a significant accumulation of traditional knowledge on how to cope with, live next to, and be efficiently protected against them. The katuns dwellers developed a deep insight into the psychology, behavioral patterns, and reasons behind the wolves' actions. Consequently, they developed procedures to monitor the animals in the open pastures and protect them while gathered for overnight rest.

The first step was to use shepherd dogs to follow the flocks, although the expectation was not that they could, generally, overpower the wolves in a direct encounter. In the summer months, attacks on flocks protected by shepherd dogs were deadly dangerous for wolves themselves. Such attempts would result in serious wounds from the dogs' teeth, which, though not lethal, would lead to infections with the same outcome. Killing the shepherd dog (or more of them) to get the animals it protected was meaningless if it would result in death from the consequences of such a victory. In desperate situations caused by prolonged hunger, wolf packs were easily taking this risk, but it was not the regular practice, i.e., why a couple of dogs were mostly enough to protect the flock, together with the shepherd, throughout the entire summer grazing season. When faced with the possibility of a clash with loose dogs in daylight, wolves almost always preferred to wait for an opportunity during the night's sleep.[46]

The people in katuns were aware that the wolves' posed the greatest threat during the night, when the sheep and cattle were gathered within resting places. These were stone and wooden walled corrals (locally called *torina* or *tor*), stables or natural shelters adjusted for this purpose. Defense strategies varied depending on the area, available material, and location of the corrals. In the katuns of Žijevo, the corrals were placed right next to the housing facilities with at least one watchdog tied to it. It could have been the same dog that guarded the flock during the day, but here the size, courage, and strength of the shepherd dogs were not that important. Their purpose was to raise the alarm barking when sensing danger and alerting people to come, not to directly confront the attackers. For this reason, the role of watchdogs was often entrusted to smaller ones, easier and cheaper to acquire, and more expendable than standard, large shepherd dogs. In the open pasture, it was common to have only shepherds supervising the flocks without dogs, but when in pens, the presence of dogs was mandatory. In larger katuns, there was a tendency to place more pens near

[45] Čajkanović (1994), 183–202.

[46] Kaartinen et al. (2009); Ambarlı (2019), 1–9.

Fig. 9.12 Stone corral. Kastrat, western slopes of Žijevo. *Photo* I. Laković (2015)

Fig. 9.13 Wooden corral Okrugljak, Sinjajevina. *Photo* A. Kapetanović (2022). Reproduced with permission

each other to further enhance the effects of watchdogs, i.e., to have the same number of dogs protecting more flocks.[47]

9.2.9 Passive Protection—Corrals and Pens

Corrals were, generally, made of stone or wood (Figs. 9.12 and 9.13), but the stone ones prevail in the Žijevo massif. Varying in design, size, and shape, these corrals followed the shape of the terrain, encircling small depressions in the ground or connecting the existing rock formations, which made keeping the animals inside easier. The stone corrals were around 1–1.5 m in height, 50–70 cm in width and additionally covered with thorny branches or, in later periods, barbed wire.[48]

[47] Statements of katun users from the area of Kuči, given to authors in February of 2023.

[48] Picuno et al. (2017), 1–15.

It seems that the idea behind the corral structure and location was not so much to prevent the wolves from getting in, but rather to make their exit with the prey impossible. Generational experience taught that those clever animals would eventually find a way to sneak or break in any denied area, but only if they would expect to get out alive and with prey. The stone corrals with thorny branches placed on top were hardly insurmountable barriers if wolves really wanted to jump over them, while the wooden ones, placed on the softer pasture ground, could not prevent them from digging holes underneath to squeeze through from below. In any case, returning with prey would have been impossible, while noise raised by the sheep or watchdogs would have certainly attracted the attention of humans. Although not significant hurdles for wolves to overcome while getting in, these structures would have certainly become deadly traps. The people at katuns knew that the wolves feared that too, so the corrals have continued to serve as the dominant type of shelters for sheep in Montenegrin katuns to this day. Their design emerged as a compromise between available materiel, ease of construction and, most importantly, understanding of the nature of the threat they should provide protection against.[49]

Unlike the stone corals in the study area, wooden ones were much more prevalent in the northern parts of the country. More mobile and faster to repair, suited to the areas without enough stone available, they served only as temporary and short-term solutions. Their biggest comparative advantage was the possibility to be partially dismantled, removed, and repositioned, allowing for flock protection both within the katuns and on different parts of pastures. This practice was followed by placing mobile, sled-mounted wooden cabins for shepherds next to them (Fig. 9.14), enabling prolonged grazing outside of katuns. Such cottages were small in size and usually provided a tight sleeping place for two shepherds, resembling enlarged doghouses. Mounted on wooden skis, they could be towed by horses or oxen from one position to another, so that the shepherds could sleep close to their livestock and quickly react in case of wolves' attacks. The combination of light, wooden corrals, and such cottages increased the grazing range, which was important in the vast pastures of Sinjajevina and Durmitor mountains, while maintaining the necessary level of protection against wolves. Additionally, the practice of frequent repositioning such corrals all over the pastures provided better soil fertilization, especially in areas intended for future potato crops.[50]

9.2.10 *Forest Fires as Remnants of Past Practices*

As the overall social, economic, and demographic landscape changed after WWII, the traditional approach to natural resources and the perception of their importance was drastically disrupted. It became especially evident in relation to forest fires. Although the sheep and cattle population in the mountains was decimated,

[49] Vujošević (2010), 48–50; Laković (2017), 40–41.

[50] Kapetanović et al. (2019), 96.

Fig. 9.14 Sled-mounted wooden cabins for shepherds next to the corral. Katun Dobri Do, Durmitor. *Photo* A. Kapetanović (2018). Reproduced with permission

many manmade forest fires occur under the pretext of securing more grazing areas. One may wonder how the existing pasturelands once accommodated a much larger number of animals, while the current, drastically smaller one, desperately needs more grazing area prompting their owners to resort to setting fires in the surrounding forests. The answer cannot be easily found in the animals' needs, but rather in the significant degradation of the common understanding of the importance of resource preservation.[51]

9.3 Conclusion

This research has only scratched the surface of the vast topic of traditional forest resource management in seasonal mountain settlements in the Dinaric Karst areas of Montenegro. As this subject has not received much research attention previously, the inquiries, interviews, surveys, and fieldwork conducted have provided some insight into the matter, revealing a strong and multifaceted connection between the local population and the surrounding forests. The presented overview of recognized and determined approaches, practices, and tendencies, although broad in nature, is the first deliberate attempt to decipher the layers of forest importance in the context of transhumance between the winter and summer rural dwellings in Montenegrin history. The exemplary case study highlights the relevance of the conclusions drawn to other similar areas in the country, which are more important due to the lack of dedicated research in this field.

The Žijevo mountain massif, with approximately 180 katuns scattered throughout its slopes, beech, and Bosnian pine forests, and a centuries-long tradition of transhumance, provided all the needed categories for this research venture. The effort

[51] *Informacija o šumskim požarima* [Information on the forest fires] (2014); Kalač and Rudović (2021).

was made to depict the most important and recognizable aspects of the practices and activities related to and relevant to them, presenting a domain of their strong interdependence, which greatly impacted the lives of the local population in such circumstances. Although varying among the localities, there has always been a direct connection and causality between the availability of quality forest and the successful development of given katun, since all the major ones have been situated next to or in the immediate vicinity of forests. Comparing this practice with other similar areas in Montenegro confirms such an attitude.

Altogether, life in seasonal mountain settlements in the Montenegrin Dinaric mountains was impossible without appropriate forest resources. While pastures provided food for cattle, becoming the main reason for such a way of life, the surrounding forests were crucial for its sustainability. They provided wood for construction and fire, fruits, and berries for food as well as shelter and refuge during hard times. As such, they were treated as valuable parts of common property, but also as sources of dangers lurking upon shepherds, their families, and their property. This danger was primarily related to attacks by robbers or wolves that could use the forests for cover before the sudden strike. Defense measures were taken through the development of behavioral habits and patterns that encompassed every sphere of lifestyle during the grazing season, from shepherding practices to the architectural layout of settlements, housing cottages, and adjusted auxiliary facilities. Through generational cycles of transhumance between villages and katuns, the local population established a sustainable system of reliance and dependence on mountain pastures and forests, based on knowledge and understanding of the genuine nature of the challenges and dangers they were facing. Therefore, the reflections of such accumulated traditional ecological knowledge are still clearly visible in the cultural and national identity of the contemporary population, although the current context of life there barely resembles the one of that era.

References

Literature

Ambarlı H (2019) Analysis of wolf-human conflicts: implications for damage mitigation measures. Eur J Wildl Res 65. https://doi.org/10.1007/s10344-019-1320-4

Berkes F (1993) Traditional ecological knowledge in perspective. In: Inglis JT (ed) Traditional ecological knowledge—concepts and case. International Program on Traditional Ecological Knowledge and International Development Research Centre, Ottawa, pp 1–9

Bešić Z (1975) Geologija Crne Gore [The Geology of Montenegro], vol 1. CANU, Titograd

Bogetić-Ćirić L (1966) Komunice u Crnoj Gori u XIX i početkom XX veka [The commons in Montenegro in the 19th and in the beginning of 20th century]. Istorijski institut Crne Gore, Titograd

Božić I (1979) Nemirno pomorje XV veka [The restless seaside of the XV century]. Srpska književna zadruga, Beograd

Braudel F (1995) The Mediterranean and the Mediterranean world in the age of Philip II 1&2. University of California Press, Berkeley-Los Angeles-London

Burić D, Ducić V, Mihajlović J (2014) The climate of Montenegro: modificators and types-part two. Bulletin of the Serbian Geographical Society 94:73–90

Burić M, Micev B, Mitrović L (2012) Atlas klime Crne Gore [The Atlas of Montenegrin Climate]. CANU, Leksikografski centar, Podgorica

Čajkanović V (1994) Religija i kult životinja [Religion and the cult of anmals]. In: Stara srpska religija i mitologija [The old Serbian religion and mythology], Sabrana dela iz srpske religije i mitologije, knjiga 5. In: Đurić V (ed). Srpska književna zadruga, BIGZ, Prosveta, Partenon M.A.M, Beograd, pp 183–202

Čubrilović V (ed) (1976) Odredbe pozitivnog zakonodavstva i običajnog prava o sezonskim kretanjima stočara u Jugoistočnoj Evropi kroz vekove [Provisions of positive legislation and customary law on the seasonal movements of herders in Southeast Europe throughout the centuries]. In: Zbornik radova međunarodnog naučnog skupa. Balkanološki institut, Beograd

Dučić S (1998) Pleme Kuči, život i običaji [The tribe of Kuči, life and customs]. CID, Podgorica

Frank J, Sjöström M (2007) Human attitudes towards wolves, a matter of distance. Biol Cons 137(4):610–616. https://doi.org/10.1016/j.biocon.2007.03.023

Fuerst-Bjeliš B (2018) Deforestacija i reforestacija u hrvatskome dinarskom kršu: odnosi i implikacije [Deforestation and reforestation in Croatian Dinaric Karst: Human-environmental relations and implications]. Ekonomska i ekohistorija 14:136–144

Fuštić B, Đuretić G (2000) Zemljišta Crne Gore [The soils of Montenegro]. Biotehnički institut, Univerzitet Crne Gore, Podgorica

Informacija o šumskim požarima [Information on the forest fires] (2014) Ministarstvo poljoprivrede i ruralnog razvoja, Podgorica. https://www.gov.me/dokumenta/93f504a1-1577-437e-ac7e-b4ac11b61e0c

Janković MM (1960) Razmatranje o uzajamnim odnosima molike (*Pius peuce*) i munike (*Pinus heldreichii),* kao i njihovim ekološkim osobinama, posebno u odnosu na podlogu [Consideration of the mutual relationship between Macedonian pine (*Pinus peuce*) and Bosnian pine (*Pinus heldreichii*), as well as their ecological characteristics, particularly in relation to the substrate]. Glasnik Botaničkog zavoda i Bašte Univerziteta u Beogradu 1(2):141–181

Janković MM (1991) O veoma značajnoj potrebi uspostavljanja kontinuiranog vegetacijskog pojasa endemoreliktnih balkanskih borova *Pinus heldreichii* i *Pinus peuce* u planinama SR Srbije i Balkanskog poluostrva [On the very significant need for establishing a continuous vegetation belt of endemorelict Balkan pines *Pinus heldreichii* and *Pinus peuce* in the mountains of the Republic of Serbia and the Balkan Peninsula]. Ekologija 26:61–67

Jovanović B (1967) Dendrologija sa osnvima fitocenologije [Dendrology with the bases of phytocenology]. Šumarski fakultet Univerziteta u Beogradu, Naučna knjiga, Beograd

Kaartinen S, Luoto M, Kojola I (2009) Carnivore–livestock conflicts: determinants of wolf (*Canis lupus*) depredation on sheep farms in Finland. Biodivers Conserv 18:3503–3517. https://doi.org/10.1007/s10531-009-9657-8

Kalač D, Rudović M (2021) Šume, vazduh i novac nestaju u dimu [Forests, air and money disappear in smoke]. Vijesti 18.8.2021. https://www.vijesti.me/vijesti/drustvo/562807/sume-vazduh-i-novac-nestaju-u-dimu

Kapetanović A, Laković I, Koprivica T, Pelcer-Vujačić O (2017) Graditeljska baština katuna na Kučkoj planini [The Architectural heritage of the katuns at the Kuči Mountain]. Istorijski Zapisi 1–2:92–95

Kapetanović A, Gligorić B, Rajić T (2019) Studija tradicionalne arhitekture na području opština Berane, Petnjica, Andrijevica, Plav i Gusinje [The study of traditional architecture in the area of municipalities of Berane, Petnjica, Andrijevica, Plav and Gusinje]. Polimski muzej, Regionalna razvojna agencija Bjelasica, Komovi i Prokletije, Berane

Krstić Đ (1956) Planinski i šumski pašnjaci Jugoslavije [Mountain and forest pastures of Yugoslavia]. Minerva, Subotica

Krstić Đ (1979) Pravni običaji kod Kuča [Legal customs of Kuči tribe]. SANU, Balkanološki institut, Beograd

Laković I (ed) (2017) Katuni Kučke planine: monografija [The katuns of the Kuči Mountain: monograph]. Univerzitet Crne Gore, Podgorica

Laković I, Kapetanović A, Koprivica T, Pelcer-Vujačić O (2018) Katuni Kučke planine [The katuns of the Kuči Mountain]. In: Održivi razvoj planinskih područja – iskustva, izazovi i perspektive, CANU book 148. Crnogorska akademija nauka i umjetnosti, Podgorica, pp 209–210

Laković I, Kapetanović A, Pelcer-Vujačić O, Koprivica T (2020) Endangered Mediterranean heritage—case study of katuns at the Kuči Mountains in Montenegro. Land 9(8). https://doi.org/10.3390/land9080248

Lakušić R (1987) Šumske zajednice Jugoslavije—SR Crna Gora [The forest communities of Yugoslavia—SR Montenegro]. In: Šumarska enciklopedija. JLZ "Miroslav Krleža", Zagreb, pp 388–395

Lazarević J (2017) Šume i biodiverzitet Kučke planine [Forests and biodiversity of the Kučka mountains]. In: Laković I (ed) Katuni Kučke planine. Istorijski institut, Biotehnički fakultet, Univerzitet Crne Gore, Podgorica, pp 64–78

Lazarević J, Bojović D (2017) Hrana i ljekovi iz prirode, priručnik o jestivim gljivama i ljekovitom bilju planinskih područja [The food and medicine from the nature, handbook on edible fungi and medicinal herbs of the mountain regions]. Univerzitet Crne Gore, Podgorica

Lazarević J, Menkis A (2018) Fungi inhabiting fine roots of *Pinus heldreichii* in the Montenegrin montane forests. Symbiosis 74:189–197. https://doi.org/10.21203/rs.3.rs-669235/v1

Lazarević J, Menkis A (2020) Fungal diversity in the Phyllosphere of *Pinus heldreichii* H. Christ—an endemic and high-altitude Pine of the Mediterranean Region. Diversity 12(5). https://doi.org/10.3390/d12050172

Liechti K, Biber J-P (2016) Pastoralism in Europe: characteristics and challenges of highland-lowland transhumance. Rev Sci Tech 35(2):561–575. https://doi.org/10.20506/rst.35.2.2541. PMID: 27917971

Marović B (2018) Ekonomska istorija Crne Gore [Economic history of Montenegro]. Privredna komora Crne Gore, Podgorica

McNeill JR (1992) The mountains of the Mediterranean world. An environmental history. Cambridge University Press, New York

Oppermann R, Beaufoy G, Jones G (eds) (2012) High nature value farming in Europe. Verlag Regionalkultur

Picuno CA, Laković I, Roubis D, Picuno P, Kapetanović A (2017) Analysis of the characteristics of traditional rural constructions for animal corrals in the Adriatic-Ionian area. Sustainability 9(8):1–15. https://doi.org/10.3390/su9081441

Republički zavod za zaštitu prirode (2020) Rješenje o stavljanju pod zaštitu pojedinih biljnih i životinjskih vrsta [Decision on putting certain plant and animal species under the protection]. Službeni list RCG, broj 76 od 12. decembra 2006. Podgorica

Rovinski PA (1998) Etnografija Crne Gore, Tom I [The ethnography of Montenegro, Volume I]. CID, Podgorica

Stevanović V, Vasić V (1995) Pregled antropogenih faktora koji ugrožavaju biodiverzitet Jugoslavije [Overview of anthropogenic factors threatening the biodiversity of Yugoslavia]. In: Stevanović V, Vasić V (eds) Biodiverzitet Jugoslavije sa pregledom vrsta od međunarodnog značaja. Biološki fakultet and Ecolibri, Beograd, pp 19–37

Stevanović V (1977) Fitocenologija sa pregledom šumskih fitocenoza Jugoslavije [Phytocoenology with an overview of forest phytocoenoses of Yugoslavia]. IGKRO "Svjetlost", OOUR Zavod za udžbenike, Sarajevo

Tucakov J (1990) Lečenje biljem. Fitoterapija [Healing by herbs. The phytotherapy]. Rad, Beograd

Valorization of Montenegrin katuns through sustainable development of agriculture and tourism—KATUN, No. 01-648, realized by the University of Montenegro through the INVO-HERIC program of the Ministry of Science (2015–2017)

Vendramin GG, Fineschi S, Fady B (2008) Technical Guidelines for genetic conservation and use for Bosnian pine (*Pinus heldreichii*). EUFORGEN Bioversity International, Rome

Vučković D (1965) Kapitalističko iskoriščavanje šuma u Crnoj Gori [The capitalist exploitation of the forests in Montenegro]. Istorijski institut Crne Gore, Titograd
Vučković D (1972) Šumarstvo i lovstvo u Crnoj Gori u drugoj polovini XIX i početkom XX vijeka [Forestry and hunting in Montenegro in the second half of 19th and in the beginning of 20th century]. Istorijski institut Crne Gore, Titograd
Vujošević N (2010) Kučki katuni, nekad i sad [The katuns of Kuči, then and now]. NVO „Budućnost koju želimo", Podgorica

Ivan Laković Historian, senior scientific associate, Ph.D. He is dealing with the Cold War history and the cultural heritage of the seasonal mountain settlements. He took part in several projects that have introduced environmental history in the mainstream scientific tendencies in Montenegro. His main contribution to the topic was editing and co-authoring the monograph on the katuns (traditional mountain settlements built for summer grazing) at Kuči Mountain, Montenegro.

Jelena Lazarević Senior researcher in forestry, Ph.D. Her engagement in research projects covers the area of forest plant pathology, mycology, landscape architecture, nursery production, and seedling mycorrhisation. She authored and co-authored more than fifty scientific articles, and participated in numerous international conferences, workshops, and professional trainings. She represents Montenegro in the projects of European cooperation in science and technology, in the domain of forest protection and biodiversity.

Part III
Water and Climate

Chapter 10
Scarcity Versus Abundance: A Case Study of the Hardship and Imminent Starvation in the District of Senožeče in 1864 and 1865

Andrej Studen

Abstract The unusually rainy and cold weather of 1864 caused a disastrous harvest failure in the Senožeče district and its surroundings. By studying the archives of the Imperial Royal Presidency of the province of Carniola and newspaper reports, the present study deals with the broader socio-economic and environmental setting, assembles data on the weather in 1864 placed in a macroregional context, and focuses on the interplay between the environmental and socio-economic dimensions of the vulnerability of farm economy in the area under consideration. The author also draws attention to the diversity of the environments within the analysed Dinaric Karst and the diversity of nineteenth-century farm economies. The study outlines the plight of many farmers during the severe agrarian crisis and the farmers' and political strategies for overcoming the subsistence crisis.

Keywords Environmental history · Climate history · Economic history · Dinaric Karst · Subsistence crisis · Vulnerability · Natural disasters · Nineteenth century

After Studen's death, his original manuscript was improved by Žiga Zwitter; Dragica Čeč checked and improved the table and data for graphs, and Meta Remec helped.

A. Studen (Deceased) (✉)
Institute of Contemporary History, Ljubljana, Slovenia
e-mail: inz@inz.si

B. Fuerst-Bjeliš et al. (eds.), *Environmental Histories of the Dinaric Karst*, Environmental History 17, https://doi.org/10.1007/978-3-031-56089-7_10

10.1 Introduction

After the last evidently climate-driven great subsistence crises in the Western World in the second half of the 1810s,[1] poor harvests due to hail, excessive rainfall or drought, cold, frost, or hoarfrost were occasional phenomena in the countryside of what is now Slovenia in the nineteenth century and later. Of course, the poor among smallholder peasantry suffered the most in such cases: these people would "*normally face food shortage already during the winter, and hunger would drag on far into the springtime months*".[2] By studying the newspaper reports and archives of the Imperial Royal Presidency of Carniola (the province where the district of Senožeče was located) and by providing the broader economic history setting, the present case study shows that severe weather still played a crucial role among the different dimensions of the vulnerability of the farm economy[3] at the local to regional scale when the railroad arrived in the area under consideration. This was a consequence of mutual interactions between the environmental and socio-economic conditions. The scale of the subsistence crisis in the microregion was, however, not comparable with the one in 1810s.[4]

At the transition from present-day continental Slovenia to the hinterlands of the Gulf of Trieste/Trst and Kvarner in the Adriatic Sea, the Senožeče district (Fig. 10.1: settlements from Senadole in the southwest and Rakulik in southeast to Bukovje in the northeast) in Inner Carniola was one of the regions that suffered several poor harvests in the middle of the nineteenth century. Let us mention two examples: (a) the dry and hot year of 1859, when, in addition to the prolonged drought, Inner Carniola was also plagued by cockchafer larvae, which "*destroyed the roots of the spring cereals as well as all kinds of legumes and root vegetables*" and "*were responsible for so little being produced that the peasants, with their measly crops, could barely survive on the most basic of food until Christmas*"[5]; (b) after a few years with a medium harvest,[6] the unusually rainy, wet, and cold year of 1864 followed, which will be examined closely in the present case study. We mainly chose this year because the severe hardship and imminent starvation in the above-mentioned territory in the mid-1860s are well-documented in the preserved sources. Moreover, the area under consideration is located in the northern Dinaric Karst, the region dealt with in this volume. The same subsistence crisis has already been explored in the Lož/Laas district, located in the Dinaric Karst some 25 km east of the analysed area (district

---

[1] Pfister and White (2018) and further literature cited there; White et al. (2018), p. 342; for recent, often neglected findings on the dimensions of the crisis in Eastern Europe see Rus (2020). For the dimensions of the crisis in the broader study region see Mal (1993), pp. 504–513; recently, Dragica Čeč has carried out in-depth research into this subsistence crisis in Slovenian territories, see Čeč (2015), pp. 243–275; Čeč (2017), pp. 3–16.

[2] Makarovič (1988–1990), p. 158; Pirc and Baš (1938), pp. 102–103; White et al. (2018), p. 346.

[3] For more details about the concept of vulnerability, see Füssel (2007), pp. 155–167.

[4] Cf. *Postojinsko okrajno glavarstvo* (1889), p. 133.

[5] *Kmetijske in rokodelske novice*, 11 April 1860; *Postojinsko okrajno glavarstvo* (1889), p. 62.

[6] Studen (2020), p. 178; Studen (2018), p. 47.

centre: Fig. 10.1), and in the Postojna/Adelsberg district just east of the area studied here (district centre: Fig. 10.1).[7]

10.1.1 Socio-Economic Setting

The vast majority of the population in the analysed area consisted of farmers. Nonetheless, the integrated peasant economy involved various economic activities that far exceeded farming.[8] Until the late 1850s, the area under consideration was traversed by a busy thoroughfare, the main commercial road that connected Vienna and Trieste, the city boasting the most important port in the Habsburg Monarchy; Vienna and the port of Rijeka/Fiume; and Vienna and the Apennine Peninsula through Gorizia/Görz/Gorica. Thus, medium and long-distance transport provided many opportunities for even the poorest inhabitants to earn some money and contributed considerably to the integrated farm economy. As of the eighteenth century, Trieste also played an important role in the agrarian and non-agrarian market economy of the district of Senožeče, especially after its declaration as a free port in 1719, *"either as a direct market for goods and services or as a promoter of economic activities in its immediate and broader hinterlands"*.[9] *"With their draught animals and undersized carts, smallholder farmers could not keep up with the big foreign freight forwarders that boasted heavy carts and good horses. Nevertheless, more than enough work was still left for small transporters. Numerous innkeepers and tradesmen who accommodated and fed the carriers as well as serviced their vehicles could also make a lot of money from these operations (blacksmiths, wheelwrights)"*.[10] The market town of Senožeče was one of the settlements that benefited from this profit in particular. The example of wood trade shows that large quantities of this trade item were brought to Senožeče from the broader mesoregion (Planina, Logatec, Vrhnika, Unec, Cerknica, and other places—Fig. 10.1) and sold there every Friday.[11] This was an era when *"some money could be found even in the poorest houses in the remotest of villages"*.[12] People from the analysed area were repeatedly accused of neglecting farms for the sake of road transport.[13] Thus, the inhabitants of the settlements under consideration could usually secure relatively decent and steadiest profits from transportation, as well as some other types of non-agrarian income, which encouraged the fragmentation of the existing farms into smallholdings and the creation of new smallholdings. These processes were characteristic of the Senožeče district (Figs. 10.2 and 10.3) and

[7] Studen (2018); Studen (2020).

[8] Beguš (2017); Blaznik (1970), pp. 168–175; Panjek (2017a); Panjek (2017b); Kalc (2017); Voje (1952–1953), pp. 650–662.

[9] Kalc (2018) (quote from page 155); Klemenčič (1959), pp. 135–140.

[10] Klemenčič (1959), p. 133; Pančur (2009), p. 64 (quote).

[11] *Kmetijske in rokodelske novice*, 14 March 1866; Melik (1960), pp. 238, 320.

[12] Pančur (2009), p. 66.

[13] *Kmetijske in rokodelske novice*, 6 June 1860, p. 177.

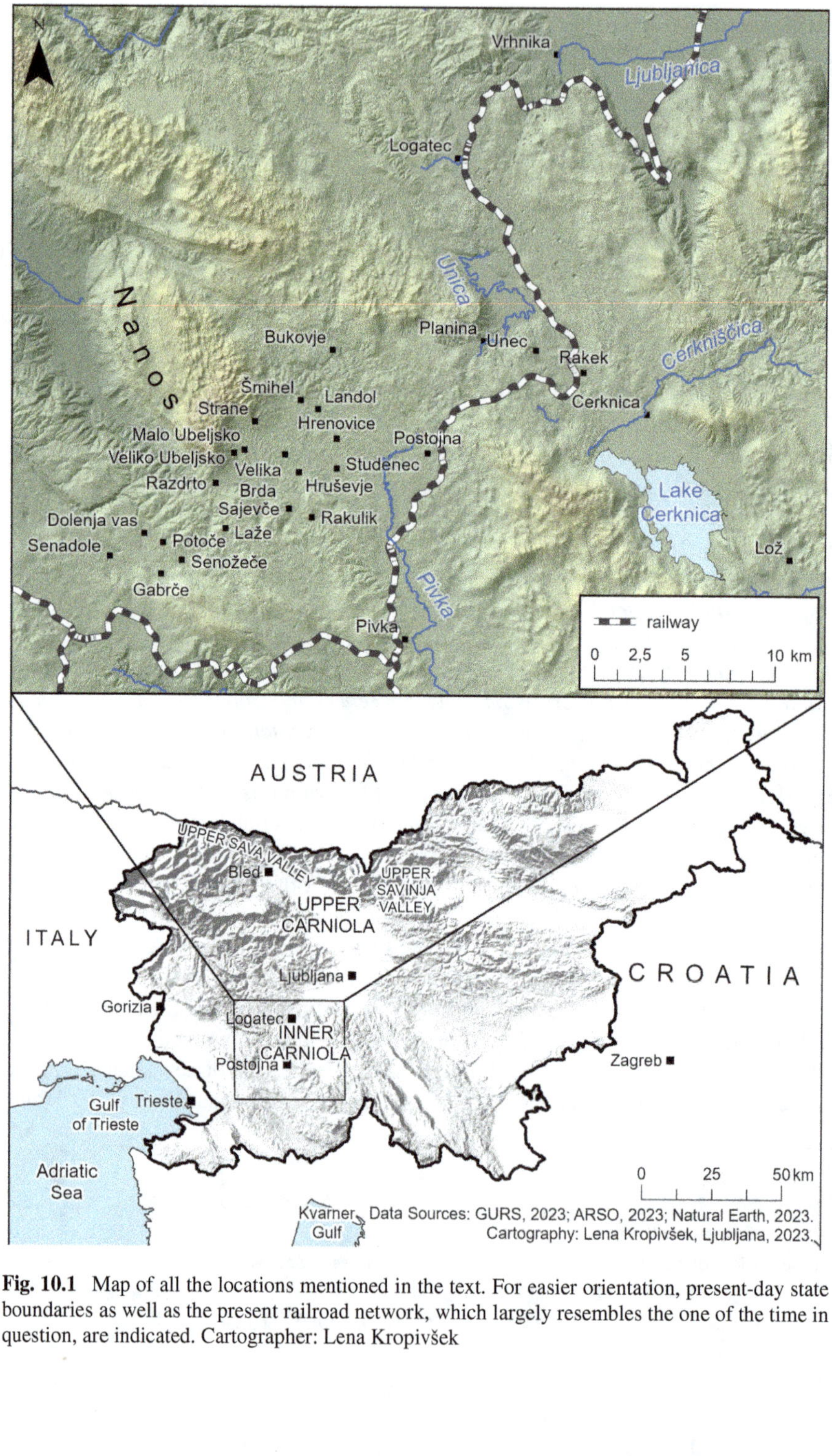

Fig. 10.1 Map of all the locations mentioned in the text. For easier orientation, present-day state boundaries as well as the present railroad network, which largely resembles the one of the time in question, are indicated. Cartographer: Lena Kropivšek

the broader area, including the districts of Postojna and Lož, which are occasionally mentioned in this chapter.[14] According to the data available for January 1865, most of the Senožeče district's holdings consisted of cottagers (24% of households), and there were relatively many lodgers or lodger family heads (7%). The latter were a markedly heterogeneous, flexible, and dynamic population group without a home or land of their own.[15] It is not insignificant that in 1865, the census enumerators recorded the largest number of them in the cadastral municipalities of Razdrto and Senožeče, i.e. in the settlements along the main commercial road. As the analysis of the ownership structure in the Razdrto cadastral municipality indicates, the cottagers there amounted to as much as 40% of holdings, while 7% of lodgers were registered as well (Figs. 10.2 and 10.3). The cadastral municipality of Senožeče had 26% of cottage holdings and 8% of lodgers (Fig. 10.2). In both cadastral municipalities, these people obviously assisted the larger farmers by working in the fields, while during the winter, they would survive on small artisanal works (e.g. as tailors, shoemakers, and carpenters) or other side businesses, e.g. those related to transit trade and wood transport, and work in urban settlements or forests in the Pannonian Basin, in Posavina (see Sect. 10.3).

In the mid-1860s, however—in the times of a severe farm economy crisis—such fragmentation of farm holdings was an anachronism. As the general economic situation decreased significantly, such a distinctly unfavourable land structure further aggravated the agrarian crisis, characteristic of many places in the broader mesoregion of Inner Carniola.[16]

Apart from the small farm size, various further socio-economic factors significantly contributed to the agrarian crisis, strengthening the environmental vulnerability of farm economies in the Senožeče district. The economic situation of the peasants deteriorated with their considerable indebtedness after the 1848 abolishment of serfdom when the land had been allocated to the peasants in return for partial monetary compensation.[17] In addition, especially small farmers and cottagers, as well as, to a degree, medium-sized farmers, were impaired by the abolishment of advantages stemming from peasant *profits-à-prendre* (in forests, for example, the right of grazing, wood provisioning, and litter raking) or their redemption, which once again represented considerable monetary obligations.[18] Aside from the changes in the agrarian sector after 1848, the *"abolishment, redemption, and regulation of* profits-à-prendre *were among the changes that had the strongest economic impact on the lives of the majority of the population"*.[19] Even the abolishment of serfdom in itself failed to encourage the development of the agrarian sector and did not result in

[14] Compare Klemenčič (1959), pp. 59–60; Kačičnik Gabrič (2004), pp. 143–144; Voje (1952–1953), pp. 650–662.

[15] On lodgers, see Sovič (2005), p. 161.

[16] Fischer (2005), pp. 72–73; *Kmetijske in rokodelske novice*, 6 June 1860, pp. 177–178, 186.

[17] For more information about agricultural debt, see for example: Lazarević (1994); Kačičnik Gabrič (2004).

[18] For more information, see: Lazarević (1994), pp. 14–19.

[19] Kačičnik Gabrič (2014), p. 382 (quote), p. 384.

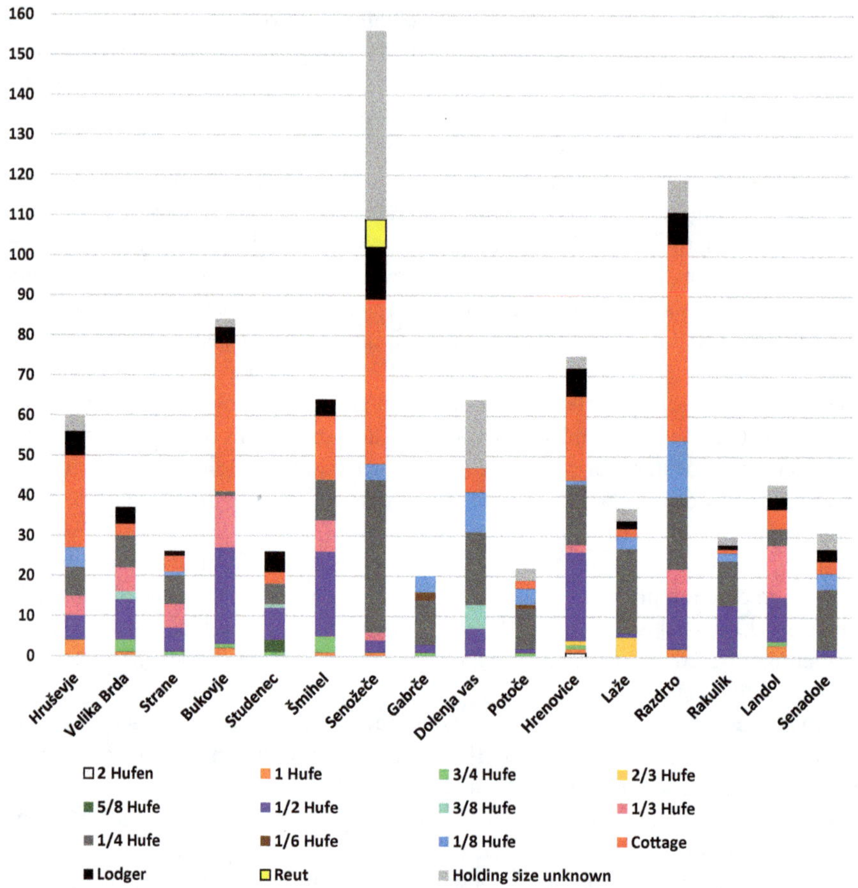

Fig. 10.2 Land ownership structure in the cadastral municipalities of the district of Senožeče in 1865. A *Hufe* is a hide or family farm, lodgers (single lodgers and family heads are given in Fig. 10.2) were a population group without a home or land of their own, and *Reut* is a farm with a shorter history. Author's own elaboration. A last check of a selected example in the archives showed that holdings of unknown size are counted rather generously in some cases, but this does not have any significant impact on the results. *Sources:* SI AS 16, box 10, bundle 23: Ausweis über den Vorrath an Lebensmitteln in der Ortsgemeinde Bründl (18 January 1865), Bukuje (12 January 1865), Gross-Berdu (13 January 1865), Hrenovitz (10 January 1865), Hrušuje (9 January 1865), Landoll (11 January 1865), Niederdorf (11 January 1865), Prävald (12 January 1865), Laže (14 January 1865), Rakulik (17 January 1865), Senožeč (19 January 1865), Sinadolle (13 January 1865), St Michael (17 January 1865), Strane (10 January 1865); Ausweis über den Vorrath an Lebensmitteln in der zur Ortsgemeinde Niederdorf gehörigen Steuergemeinde Potoče (14 January 1865); and Ausweis über den Vorrath an Lebensmitteln in der zur Ortsgemeinde Senosetsch gehörigen Steuergemeinde Gaberče, 14 January 1865

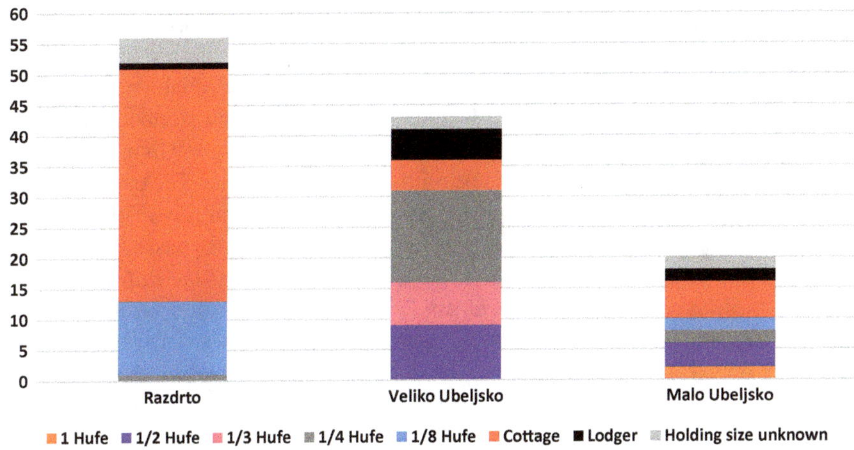

Fig. 10.3 Land ownership structure in the cadastral municipality of Razdrto. Half of the cadastral municipalities in Fig. 10.2 consisted of a single settlement, while the other half combined a few settlements.[20] An example of the latter is the cadastral municipality of Razdrto, whose land ownership structure at the village level is represented in Fig. 10.3. Author's own elaboration. *Source* SI AS 16, box 10, bundle 23: Ausweis über den Vorrath an Lebensmitteln in der Ortsgemeinde Prävald, 12 January 1865

increased productivity. Farm debt, resulting from the general crisis and the structural disparities in agriculture, contributed to the proletarianisation of the countryside or the *flight from the countryside* to the urban areas. In addition to the above-mentioned instalments for the abolition of serfdom and the redemption of *profits-à-prendre*, the payment of taxes also contributed to the oppressive and detrimental indebtedness of the peasants. In the broader context of the Habsburg Monarchy, taxes were high in Carniola. Therefore, most of the local peasant population constantly struggled to make ends meet. When harvests were bad, most farmers even had to go into further debt just to pay the taxes. In the middle of the nineteenth century, rinderpest also affected the economy of Inner Carniola, indirectly, at least. Over-indebtedness as a cumulative result of many factors led to many farms going bankrupt or being auctioned off.[21] Due to the severe indebtedness, some farmers were, according to the reports on the situation in the district of Senožeče, forced to sell the best or some plots of their land.[22]

In addition, the Southern Railway connection between Vienna and Trieste through Ljubljana/Laibach (Fig. 10.1) became operational in 1857. The railroad did not cross the area under consideration. Thus, the medium and long-range traffic on the main commercial road in the analysed area started to decrease drastically. The redirection

[20] "Arcanum Maps", Cadastral maps, Habsburg Empire—Cadastral maps (XIX. [!] century).

[21] Fischer (2005), pp. 72–73; Granda (2003); Lazarević (1994), especially page 15; *Stenographischer Bericht*, pp. 16–17; Studen (2020), pp. 176, 178; Studen (2018), p. 47.

[22] E.g. SI AS 16, box 10, bundle 23, Ausweis über den Vorrath an Lebensmitteln in der Ortsgemeinde Präwald, 12 January 1865.

of freight transport to the railway caused a significant stagnation of the transportation business in the district under consideration, severely hampering the possibilities for additional earnings. Indeed, after the opening of the final section of the Southern Railway in 1857, Razdrto, located on the main road, started to decline. Many innkeepers, who had previously made more than merely a good living, went out of business, as did artisans such as blacksmiths and wheelwrights. The once-thriving district centre of Senožeče lost its remarkable traffic role entirely. Timber trade in Senožeče completely disappeared, and because the railway bypassed Senožeče, the annual timber fairs *"moved to Rakek* (Fig. 10.1)*, which was the railway station nearest to the* [large] *Inner Carniolan forests".*[23] In 1866, a newspaper wrote that *"Six trade companies where goods could be submitted* [freight forwarders in Senožeče] *and several innkeepers closed down immediately. Now we have four shops and too many taverns left. Since that time, even the wealthiest houses have been starving. With the exception of some seven houses, which still earn extra income, not a single farmer in our parish* [Senožeče] *can live on the produce of their land. Everyone is in debt! People face hunger. Last year, the harvest was even worse than the year before".*[24] This report refers to the subsistence crises in the years following the severe weather of 1864 and early 1865.

10.1.2 A Brief Outline of the Environmental Conditions for Farming in the Analysed Area with an Emphasis on Climate

According to the usual regionalisations, the entire analysed area of the district of Senožeče (Fig. 10.1) belongs to the Dinaric Karst.[25] However, the territory under consideration is geologically diverse. On the one hand, karst prevails in the landscape around Senožeče, Laže, Potoče, Dolenja vas, Senadole, and Gabrče. On the other hand, the Lower Pivka (Fig. 10.1: the microregion between Razdrto in the west, Postojna in the east, Šmihel in the north, and Hruševje in the south) is not a typical karstic landscape but rather a microregion of non-karst with flysch and alluvial deposits as well as some (contact) karst. Most arable plots of land and meadows in Lower Pivka were located on flysch and alluvial deposits, while some pastures and forest were in fact located in the karstic landscape.[26]

Most of the settlements in the district of Senožeče are located at an altitude between 500 and 600 m, while the villages of Laže (681 m), Strane (658 m), and Bukovje (650 m) are located somewhat higher and Senadole (474 m) lower. This area belongs

23 Dolenc (1994), p. 59 (quote); Klemenčič (1959), p. 133; *Kmetijske in rokodelske novice*, 14 March 1866; Melik (1960), pp. 238, 320, 322; *Stenographischer Bericht*, p. 16; Studen (2018), p. 65.

24 *Kmetijske in rokodelske novice*, 14 March 1866.

25 Mihevc and Prelovšek (2010), p. 6; Perko (1998), especially pp. 20–23, 30–31.

26 Compare Gams (2004), pp. 290–291 and "Arcanum Maps", Cadastral maps, Habsburg Empire—Cadastral maps (XIX. [!] century); see also Melik (1960), pp. 234–238, 310, 312.

to the temperate continental climate type of western and southern Slovenia, characterised by the sub-Mediterranean precipitation regime and October temperatures exceeding those in April.[27] The estimated average annual precipitation in Postojna between 1851 and 1880 (based largely on the measurements taken during only two years, 1852 and 1853) was 1338 mm[28] (an estimate that is probably lower than actual annual precipitation between 1851 and 1880). The measurements in Postojna determined an average annual precipitation of 1832 mm between 1925 and 1940,[29] 1703 mm in Postojna between 1919 and 1939, and 1728 mm at Razdrto from 1919 to 1939.[30] The estimated average annual temperature in Postojna between 1851 and 1880 (based largely on the measurements during 52 months of the years 1854–1856, 1871, 1872, and 1879) was 8.7 °C, in July 18.8 °C, and in January −1.2 °C.[31] *"The bora wind, which is most frequent in January and most violent in the village of Razdrto"* often causes cold spells in the area.[32] In times of anticyclone windless winter weather, temperature inversion also results in severe cold.[33] During the warm part of the year, particularly in the summer, droughts represent a much more serious hazard in the karst around Senožeče than in the Lower Pivka.[34] The analysed area is characterised by frequent storms, sometimes accompanied by hail, early frost or early snow, which can come as a surprise as soon as at the end of September or in October, and often by late snow or frost in May.[35]

Relatively extensive sections of the flatland in the analysed area of Lower Pivka are flood-prone, experiencing flooding that does not occur regularly, though. The wet and flood-prone parcels were usually used as hay meadows. Hay represented one of the main crops in the Senožeče district, providing not only feed for the local livestock but also a marketable product that could be sold in Trieste. Namely, the demand for hay for the livestock kept by Trieste's inhabitants and other merchants operating in that port city was considerable at the time. To meet it, many farmers from the Lower Pivka region usually sold hundreds of metric tonnes of their hay crops during the mowing season or even in advance. Sometimes, floods taking place during the wrong time (from the farmers' point of view), caused by the abundant precipitation during the summer, would destroy the hay harvest.[36] Large quantities of hay were also sold to Trieste from Senožeče and its close surroundings, where

[27] Ogrin (1996), pp. 47, 49.

[28] Melik (1960) p. 309; Seidl (1894), pp. 8, 11.

[29] *Padavine u Jugoslaviji* (1957), p. 13.

[30] Reya (1940), p. 27.

[31] Seidl (1890), pp. 308–309.

[32] *Krajevni leksikon Slovenije* (1968), p. 245; Melik (1960), p. 307.

[33] Melik (1960), pp. 307–308; Plut (2002), p. 177.

[34] Klemenčič (1959), pp. 30–34, 36, 90.

[35] Klemenčič (1959), pp. 34, 36, 69; *Krajevni leksikon Slovenije* (1968), p. 245.

[36] Klemenčič (1959), p. 38; *Kmetijske in rokodelske novice*, 28 December 1864; Melik (1960), p. 312; *Postojinsko okrajno glavarstvo* (1889), pp. 118–120, 131, 133.

many hay meadows were located on the rocky limestone soils.[37] This is reflected in nineteenth-century percentages of land use categories: at Senožeče, Laže, Gabrče, Potoče, Dolenja vas, and Senadole, open hay meadows and hay meadows with trees covered 19.7 km^2, i.e. 47% of all agricultural land or 27% of all productive land.[38]

The environmental conditions also co-influenced the decision which crops to grow on the arable land. Potatoes usually performed well, at least in Lower Pivka. They represented one of the main crops in the settlements of that region in the 1860s, while maize was less adapted to the local environmental conditions and not commonly grown. Similarly, due to strong and stormy bora wind, farmers around Senožeče avoided sowing millet (Table 10.1).[39] While most cereal types of the temperate continental climate were grown in the area, yields were highly diverse. In the central Lower Pivka region, for example around Hrenovice, all cereals were grown. However, with the exception of oats (and probably rye, as suggested by Table 10.1), farmers were rarely happy with the yield.[40] Vegetables such as cabbage, turnips, carrots, beetroots, and pulses were also frequently grown.[41]

10.2 1864: An Exceptional Year of Severe Weather Conditions

The weather anomalies in 1864 and early 1865 were probably a part of the last weather deviations of the Little Ice Age in what is now Slovenia,[42] which appears to have been coming to an end in the 1860s. As revealed by data from the area under consideration and sources from present-day Germany, in 1864, most of Central Europe experienced a cold spring followed by a cool summer.[43] In the analysed area, weather was particularly severe in 1864—the scarcity in the districts of Senožeče and Lož was the most serious in the entire province (Sect. 10.3). The average annual temperature was one of the four coldest in Ljubljana and Zagreb since the beginning of instrumental measurements, i.e. since 1851 in Ljubljana and since 1862 in Zagreb. Similarly, it was one of the five coldest years in Trieste since 1841.[44] In 1864, newspapers in Carniola reported that the unusual weather had already begun on New

[37] ARSO (2023a), Atlas okolja [Environmental altas of Slovenia], Land and Soil, Soil map 1:25000; "Arcanum Maps", Cadastral maps, Habsburg Empire—Cadastral maps (XIX. [!] century); Melik (1960), pp. 234–238; *Postojinsko okrajno glavarstvo* (1889), p. 109.

[38] Calculated from *Postojinsko okrajno glavarstvo* (1889), p. 109.

[39] *Kmetijske in rokodelske novice*, 28 December 1864; Melik 1960, pp. 311–312; *Postojinsko okrajno glavarstvo* (1889), pp. 109, 118, 131, 133; Studen (2020), pp. 173–174, 180.

[40] Melik (1960), pp. 311–312; *Postojinsko okrajno glavarstvo* (1889), pp. 109, 118.

[41] *Postojinsko okrajno glavarstvo* (1889), pp. 109, 118, 133.

[42] On the Little Ice Age, see, for instance, Glaser (2001); White (2014). More information about the Little Ice Age in what is now Slovenia, see, for instance, in: Zwitter (2013); Zwitter (2014); Zwitter (2015), pp. 662–675; Čufar et al. (2008).

[43] Glaser (2001), pp. 176, 178 (on his study area, see pp. 11–12, 17, 38, and 57, for instance).

[44] Ogrin (2015), pp. 17–19, 21–23.

Table 10.1 Overview of the quantity and approximate sufficiency of food available in the cadastral municipalities of the Senožeče district, threatened by hunger

Cadastral municipality	A	B	C	D	E	F	G	H	I	J	K	L	M	N	O
Hruševje	60	317	26 1/2	5 1/2	31	25 1/2	1	2	15	67	110	198	12 1/2	1/10	3–4 weeks
Velika Brda	37	231	12	6 1/2	21 1/2	43	1	2	4 1/2	50	116	101	14 1/2	•	4–5 weeks
Strane	25	194	6	2	43	86	2	1 1/2	5 1/2	61	153	174	88	4.90	**8 weeks**
Bukovje	84	557	2 1/2	6	8 1/2	1	1	•	•	43	192	65 1/2	142	0.03	2–3 weeks
Studenec	26	171	1/2	1/2	5 1/2	4	•	•	•	13	30 1/2	79	•	•	2–3 weeks
Šmihel	64	449	15	1 1/2	17 1/2	54	3 1/2	1 1/2	2 1/2	40	73	162	36	0.30	4 weeks
Senožeče	156	594	•	16 1/2	34 1/2	•	•	•	8 1/4	21 1/4	36	96 1/2	34 3/4	•	2–3 weeks
Gabrče	20	138	•	27	21	5 1/2	•	•	2	6 3/4	62	54 1/2	54	•	4 weeks
Dolenja vas	64	277	•	82 1/2	66 1/4	18 1/2	2	•	42 1/2	27 5/8	100	124	14 1/2	1/2	3–4 weeks
Potoče	22	120	•	8 3/4	21	1 3/4	•	•	8 3/4	5 1/4	34 1/2	45 1/2	32 1/2	•	1 week
Hrenovice	75	418	46 3/8	68 5/8	20 3/4	49 3/4	1/4	12 3/4	11 1/4	147 1/18	346 1/2	340 1/4	43	•	2–3 weeks
Laže	30	239	1 1/2	5 1/2	49 1/2	15 1/2	•	•	2	18 1/4	76 1/2	92 1/2	80 3/4	•	1 week

(continued)

Table 10.1 (continued)

Cadastral municipality	A	B	C	D	E	F	G	H	I	J	K	L	M	N	O
Razdrto	122	605	1 3/8	3 1/2	41 1/2	28 7/8	•	•	1/2	42 3/4	102 3/4	106 3/4	65 1/2	1 1/2	**3 days**
Rakulik	30	188	1 1/2	3	13	29	•	•	26 3/4	27	126 1/2	59 1/2	53 1/4	1 1/2	1 week
Landol	43	256	4 1/2	•	6	15 1/2	1	1	•	22	84 1/2	102	7	•	2–3 weeks
Senadole	31	169	5 1/2	52	25	7	1/2	•	26 1/2	23	48 1/2	48	37	•	5 weeks
SUM	**901**	**4923**	**123**	**227** 3/8	**425 1/2**	**384** 7/8	**12 1/4**	**20 3/4**	**156**	**615**	**1692 1/4**	**1849**	**715 1/4**	**8. 83**	

Legend: A = number of houses; B = number of inhabitants; C = wheat; D = rye; E = barley; F = oats; G = maize; H = millet; I = buckwheat; J = beans and other legumes; K = potatoes; L = turnips, carrots and/or beetroots; M = cabbage; N = fruit; O = approximate average duration of food sufficiency, • = 0. Food stocks are measured in *Metzen* (61.5 L), except for fruit, which is reported in *Zentner* (56 kg). Note that the indicated stocks also include the quantity of required seeds, and if this is deducted from the total amount of stocks, the latter is reduced by almost half

Source: SI AS 16, box 10, bundle 23, Z 328 Übersicht über die Menge der vorhandenen Lebensmittel und die aproximative Zulänglichkeit derselben in den vom Notstande bedrohten Ortschaften des Bezirkes Senosetsch, 19 January 1865

The table was signed by the Head of the Senožeče District Office Klančič

Year's Eve. A series of daily reports referred to "*fear and terror*" while describing extreme bora wind and snow falling "*as if someone were tearing rags apart*". Due to the bitter cold, even the roads in the coastal city of Trieste were frozen over, while the fierce wind uncovered roofs and knocked down chimneys. The bora wind raged in Inner Carniola as well, causing snowdrifts. When it subsided, the bitter cold set in.[45] An icy winter with plenty of snow came over the land.[46] The cold and rainy weather persisted in Carniola until as late as the end of May. At that point, it even snowed in some of the lowlands.[47] The analysis of oak tree rings in the southeast of present-day Slovenia identified June 1864 as one of the twenty-eight wettest and coldest Junes in the period between 1497 and 2003,[48] with a certain level of diversity between the individual "wettest and coldest" Junes.[49] Bad weather also characterised the summer, resulting in a delayed and meagre harvest. Towards the end of December 1864, the head of the District of Lož reported to the Imperial Royal Provincial Presidency in Ljubljana that the large intermittent "*Lake Cerknica failed to dry up due to the constant rain*", that the villagers' meadows and fields around the lake remained flooded, that "*people lost the majority of fodder*" otherwise cultivated on what was now still the bottom of the lake, and that "*only a little was left in the unflooded places*".[50] Cool and rainy weather must have also been a severe nuisance for hay harvest elsewhere. For instance, the *Novice* newspaper, published during the 1864 mowing season, advised farmers to turn over the wet grass lying in heaps or scattered on hay fields during prolonged rainfall "*even while it rains so that it does not turn yellow or become tainted*".[51] A cold autumn with early frost and an extreme winter followed.[52] We can get some idea of how harsh the winter and early spring in Inner Carniola were by reading an excerpt from the report of the district office in Lož of 3 April 1865, in which the District Head lamented that "*a harsh winter, the likes of which have not been seen for eighty years, has dragged on into the spring and still persists. In March, when the soggy ground was already supposed to start drying up, it continued to snow, resulting in four to seven or eight feet of snow (4 × 0.32 m = 1.28 m to 2.24 or 2.56 m). This meant that communication was incessantly down, while the bora wind made unblocking the roads impossible despite all the efforts, as it would instantly blow in more snow before it could even be cleared. This, of course, has drastically exacerbated the situation of the distressed people all around the district. From then on, the suffering has worsened not only in terms of intensity but also scope, as people were deprived of any possibility of earning money. They*

---

[45] *Kmetijske in rokodelske novice*, 13 January 1864.

[46] *Kmetijske in rokodelske novice*, 27 January 1864.

[47] *Kmetijske in rokodelske novice*, 25 May 1864; 8 June 1864; 22 June 1864.

[48] Čufar et al. (2008), pp. 608–614.

[49] Compare Pfister et al. (2015).

[50] Studen (2018), pp. 56–57, 60, 62; *Kmetijske in rokodelske novice*, 28 December 1864.

[51] *Kmetijske in rokodelske novice*, 22 June 1864.

[52] Studen (2018), pp. 56, 60.

hope that the situation will improve and that aid in the form of food and seeds will arrive once the fierce bora wind stops."[53]

Many parallels from the south-easternmost Alps add further pieces to the puzzle of how severe the weather in 1864 was. In the Upper Savinja Valley (Fig. 10.1), northern winds carrying wet air and precipitation in the form of rain and snow were common in spring, summer and autumn. Each month, fresh snow covered the alpine peaks; the highest mountains in the area reach the altitudes of around 2000–2500 m. Due to cold weather, ripening was delayed and cereals, especially oats, had to be cut green on the high-altitude farms; the highest farms in that area are located at about 1100–1300 m, depending on the microregion.[54] Similarly, snow was recorded in the Upper Sava Valley mountains (Fig. 10.1) as late as early July. *"Our lush Alpine pastures were covered with thick layers of snow. Hundreds of small livestock, as well as most cattle and horses, died of hunger and cold." "It keeps raining incessantly, and it almost seems like nature no longer follows its old ways."*[55] The cool July also upset the holidaymakers—the bourgeoisie that wanted to enjoy the Bled resort in the Julian Alps (Fig. 10.1). Due to the unpleasant weather, the local spa guests and bathers *"were cold and completely bored"*. The ladies were severely disappointed and deprived by the cold and rainy weather, as they were unable to show off the splendid newest fashion in public.[56]

In addition, in the summer of 1846, devastating hailstorms occurred in many parts of Inner Carniola. In August, a hailstorm hit the villages below the Nanos Plateau (Fig. 10.1).[57] In the district of Lož, hailstorms caused severe destruction and immense damage in August 1864. Fifteen out of twenty-two of the district's cadastral municipalities were severely affected.[58] Similarly, in August 1864, a hailstorm in the Postojna district almost completely ruined the harvest and resulted in severe hardship, as people faced starvation due to the resulting shortages.[59]

10.3 Impacts of the Severe Weather in 1864 on the Farm Economy in the Senožeče District

In 1864, the extreme weather was the main reason for the disastrous harvest failure in the Senožeče district and some regions in its broader surroundings. Within the entire province of Carniola—encompassing what is now southeastern, southern, central, and most of southwestern Slovenia—scarcity was the strongest precisely in the districts of Senožeče and Lož. In addition to the low temperatures, constant

[53] Studen (2018), p. 70 (quote); Studen (2020), p. 180.
[54] Zwitter (2015), p. 151.
[55] *Laibacher Zeitung*, 8 July 1864.
[56] *Laibacher Zeitung*, 8 July 1864.
[57] *Kmetijske in rokodelske novice*, 28 December 1864.
[58] Studen (2018), pp. 55–57, 59–60.
[59] Studen (2020), pp. 167–183.

rain, and locally flooded agricultural land, premature frost in the autumn destroyed the harvest even further.[60] "*… not only the individual weather factors but also their diverse mutual interactions matter. The detrimental effect of a particular weather phenomenon depends on the plant's developmental stage and varies from species to species. However, certain phenomena affect almost all crops at particular times of the year, increasing or decreasing the yields.*"[61] Cold springs and wet midsummers were exactly the types of seasonal weather patterns that agriculture in Central Europe was particularly vulnerable to, as they "*affected most sources of food and animal feed, resulting in [serious] … crop failures*".[62]

The heavy snow and cold caused concerns that the winter would be too long and that other mishaps might spoil the spring sowing. The *Novice* newspaper took the opportunity to advise farmers to sow less grain and produce more livestock fodder, thus contributing to the "*development of livestock farming.*"[63] Indeed, the cold and rainy spring and summer months (Sect. 10.2) caused many problems for farmers already during the spring sowing and while harvesting or storing the staple crops, resulting in a shortage of food and seeds. In addition, a severe hailstorm "*precisely at harvest time knocked down all the fruit and destroyed the crops*" in the villages below Nanos in August 1864. According to the report, the fields there appeared "*as if they had been burnt.*"[64] That summer, the farmers also suffered due to the rainy weather during mowing.[65] In the non-karst Lower Pivka region, the floods washed away half of the hay from the wet meadows, while the rest was muddy and possibly mouldy. As such, it was unsuitable for fodder. Therefore, not only did the local livestock suffer: what was usually a lucrative deal with the hay for Trieste was also off the table.[66] Moreover, in the autumn, "*much further damage was caused by premature frost.*"[67] In October 1864, many people "*grumbled about the weather this year*" because the harvest was "*mostly bad*".[68] In the district of Senožeče, potato yield, crucial for the local agriculture, was reduced as the tubers rotted.[69] Like in the excessively dry year of 1859, severe hardship and deprivation already began around Christmas, while in January 1860 and January 1865, respectively, the population already faced a severe food shortage. The daily experience of deprivation, hardship, and looming hunger, especially by the poorest inhabitants, was in fact ubiquitous in the area.[70]

---

[60] Studen (2018).

[61] Pfister (1985), p. 34.

[62] White et al. (2018), p. 339.

[63] *Kmetijske in rokodelske novice*, 24 February 1864.

[64] *Kmetijske in rokodelske novice*, 28 December 1864.

[65] *Kmetijske in rokodelske novice*, 6 July 1864.

[66] *Kmetijske in rokodelske novice*, 28 December 1864.

[67] Studen (2018), p. 60.

[68] *Kmetijske in rokodelske novice*, 12 October 1864.

[69] *Kmetijske in rokodelske novice*, 28 December 1864; Studen (2018), p. 65.

[70] SI AS 16, box 10, bundle 23, Ad Nr. 4/Pr Hohes k.k. Landes-Präsidium, 19 January 1865; see also introduction to this chapter.

In December 1864, when the hardships in the Senožeče district were getting worse by the day and the hunger already threatened the poor, the Head of the Imperial Royal Senožeče District informed the Provincial Presidency in Ljubljana of the extremely difficult conditions and poverty, and asked for urgent assistance, as the district treasury did not have the necessary funds to mitigate the crisis. At the beginning of January, the Provincial Presidency instructed the Head of the District Office to draw up and submit statements on the quantities of food and seeds available in the deprived cadastral municipalities of the Senožeče district by 15 January 1865.[71] The district head, assisted by the census enumerators who were familiar with the local situation, immediately set to work. In the Senožeče district, sixteen cadastral municipalities were surveyed to establish their stocks of food and seeds needed for the spring sowing. The enumerators visited 901 houses inhabited by 4923 people. They drew up tabular statements with the following headings: house number, surname and name, place of residence, number of family members, and property or livelihood status. This was followed by an inventory of food items in *Metzen* (61.5 L or about 46 kg of potatoes[72]): wheat; rye; barley; oats; maize; millet; buckwheat; beans and other legumes; potatoes; turnips, carrots and beetroots ("Rüben", including also "gelbe Rüben", i.e. carrots); and cabbage. Fruits were measured in *Zentner* (100 *Pfunds* = 56 kg) and pounds (*Pfund* = 0.56 kg). An approximate estimate of how long, on average, the inventoried food stocks should last was noted (Table 10.1). The recorded comments are also very relevant, as they add precious information to the situation in the individual cadastral municipalities.[73]

Although the enumerators in the field hurried to take stock and prepare *"statements of the available food items and their approximate sufficiency in the cadastral municipalities in need"*, the investigation of the situation *"could not proceed at the desired pace because of the awful, partly snow-covered and partly icy roads, as well as due to the shortness of the day"*. The district office was therefore unable to complete and submit the investigation report by the required deadline of 15 January. Thus, the Head of the District Office contacted the Provincial Presidency in Ljubljana and asked for a four-day delay for the submission of the accounts.[74] The district office then sent these statements to Ljubljana on 19 January 1865. On this occasion, the Head of the Senožeče District Office Štefan Klančič also addressed the following letter to the Imperial Royal Provincial Presidency:

> With regard to the High Decree No. 2265/Pr of 3 January of this year, sixteen summary statements on the quantity of available food items and their approximate sufficiency in the cadastral municipalities of the district of Senožeče, which are threatened by starvation, are

[71] Studen (2020), p. 173; Studen (2018), pp. 48–57, 62; SI AS 16, box 10, bundle 23, Nr. 4/Pr Hohes k.k. Landes-Präsidium, 15 January 1865.

[72] Britovšek (1958–1959), p. 123.

[73] SI AS 16, box 10, bundle 23, Ad Nr. 4/Pr Hohes k.k. Landes-Präsidium, 15 January 1865; SI AS 16, box 10, bundle 23, Nr. 4/Pr Hohes k.k. Landes-Präsidium, 19 January 1865; SI AS 16, box 10, bundle 23, Z 328 Übersicht über die Menge der vorhandenen Lebensmittel und die approximative Zulänglichkeit derselben in den vom Notstande bedrohten Ortschaften des Bezirkes Senosetsch, 19 January 1865, signed by the Head of the Senožeče District Office Klančič.

[74] SI AS 16, box 10, bundle 23, Nr. 4/Pr Hohes k.k. Landes-Präsidium, 15 January 1865.

> annexed hereto. I should humbly note that the statements on the established quantities of cereals and potatoes also take into account the quantities that the affected farmers need to preserve as seeds and which they have set aside regardless of the current hardship in order not to be left without seeds at sowing time or forced to purchase them at any price. If the required seeds are deducted from the identified and inventoried stocks, the latter are reduced by almost half and are thus sufficient to feed the households for only five to six weeks, even on the wealthier farms. Meanwhile, the less prosperous landowners as well as cottagers can survive on the available food items for only one to two weeks, after which they will be exposed to enormous hardship. There is therefore no doubt whatsoever that they are in dire need of support and assistance.
>
> The misery that has descended upon this area can only be imagined by those who witness it with their own eyes. Following the house-to-house search, we are saddened to find that the owners of a quarter and even half a Hufe have no choice but to eat unsalted food without any dressing, while bread can barely be found anywhere. The poor usually have many children who now suffer deprivation and sit half-naked around the hearth, nibbling on half-baked carrots for the lack of bread. In many houses, even carrots can no longer be found.
>
> In such circumstances, the District Office humbly asks the Provincial Presidency for at least 2000 to 3000 Gulden for emergency aid.[75]

The methodology of the census enumerators led to precise estimations in some cases and poor results in others. Namely, the approximate durations of food sufficiency, based on the amounts of inventoried stocks, reflect the census enumerators' subjective judgements. However, significant differences in the calorific value of the food stocks per person per day must have been partly influenced by the differences in the age and gender structure of the households. In addition to certain inconsistencies in the work of the census takers, (some) people probably concealed a significant percentage of their stocks. Nonetheless, these data provide a relevant insight into the region's hardship. Here are some examples from the census in the village of Razdrto:

- Cottager and blacksmith Jožef Krajnc, the head of a family of four, stocked a *Metzen* of potatoes, which was enough for **three days**. Calculated, each household member would supposedly consume 11.5 kg of potatoes, which amounts to about 3.8 kg per person daily (or less, as the conversion 61.5 l of potatoes = 46 kg[76] is somewhat optimistic and could in fact easily mean as little as 31 kg[77]).[78] This means about 3300 kilocalories per person daily.[79] According to today's findings, this should be sufficient, as a physically active adult woman should consume about 2250–4000 kilocalories per day (depending on the sort of work and the environment), while around 3100–5000 kilocalories per day are enough for an adult man (in fact, somewhat less, as the average body height was lower in the

[75] SI AS 16, box 10, bundle 23, Ad Nr. 4/Pr Hohes k.k. Landes-Präsidium, 19 January 1865.

[76] Britovšek (1958–1959), p. 123.

[77] Cf. Makarovič (1988–1990), p. 151.

[78] SI AS 16, box 10, bundle 23, Ausweis über den Vorrath an Lebensmitteln in der Ortsgemeinde Prävald, 12 January 1865.

[79] The calculation is based on Makarovič (1988–1990), p. 156.

nineteenth century[80]). Potatoes provided the feeling of satiety sooner than cereals despite being less nutritious.[81]

- The owner of an eighth of a *Hufe* Jožef Blažek, the head of a family of six, stocked a single *Metzen* of turnips, carrots and/or beetroots (61.5 l), enough for **two days**. Calculated, this meant 5.1 l of turnips, carrots and/or beetroots per household member daily.
- Cottager Anton Osana, the head of a family of twelve, had a *Metzen* of cabbage, which was supposed to suffice for **two days**. Calculated, this meant 2.6 l of cabbage per household member daily.
- The owner of an eighth of a *Hufe* Jožef Osana, the head of a family of four, had half a *Metzen* of beans (or other legumes), and two *Metzen* of turnips, carrots and/or beetroots, supposedly enough for **a week**. Calculated, this meant 1.1 l of beans (or other legumes) and 4.4 l of turnips, carrots and/or beetroots per person daily.[82]

Turnips, carrots and/or beetroots, and potatoes, in several cases also cabbage and beans or other legumes, were the main crops stocked in the district of Senožeče in early 1865 (Table 10.1). In several cadastral municipalities in Lower Pivka, Senožeče, and Laže, barley was the leading cereal harvested in 1864 (Table 10.1). Barley's advantage is that it can perform well in relatively harsh climatic conditions: "*Except for the beginning of the growing season, it prefers cool weather. However, its yields are lower than those of other main cereals*[,]" and weather that is excessively cold or wet (especially in the summer), as well as inappropriate cultivation, can cause the greatest damage to it of all the main cereals.[83] Barley is harvested sooner than any other of the most usual cereals, so its exposure to weather is shorter.[84] Oats were the cereal kept in the largest quantities in the five cadastral municipalities in the analysed area in January 1865 (Table 10.1). This is completely understandable, as oats harvest does not decline during exceedingly wet summers.[85]

Significant differences existed between cadastral municipalities, both in terms of the approximate duration of food sufficiency and the amount of inventoried stocks of individual crops. In the cadastral municipality of Hrenovice, for example, where the census takers estimated that the supplies still sufficed for about two to three weeks, the quantities of cereals were particularly interesting. The farmers only stocked somewhat more than twenty *Metzen* of barley, the most common cereal in the district. Instead, however, they had almost 30% of all the recorded rye, 13% of all the inventoried oats, and 37% of all the recorded wheat in comparison with the entire district. Their stocks also included 12 3/4 *Metzen* of millet, which represented almost 62% of the total millet stocks inventoried in the district. Potato and turnip, carrot and/

[80] Remec (2015), p. 169; Studen (2020), pp. 172, 182.

[81] Makarovič (1988–1990), p. 155.

[82] SI AS 16, box 10, bundle 23, Ausweis über den Vorrath an Lebensmitteln in der Ortsgemeinde Prävald, 12 January 1865.

[83] Zwitter (2015), p. 152.

[84] Zwitter (2015), p. 152.

[85] Zwitter (2015), p. 151.

or beetroot stocks in that cadastral municipality also stood out in terms of quantity (Table 10.1). Note, however, that the stocks kept by the families not threatened by the subsistence crisis (roughly corresponding with the light grey bar sections in Figs. 10.2 and 10.3) were often not recorded.

The situation in the cadastral municipality of Strane is also interesting regarding cereal; potato; turnip, carrot and/or beetroot; cabbage; and fruit stocks. Considering the number of households and people in the cadastral municipality, the census enumerators estimated that the supplies in Strane would suffice for as long as eight weeks. There, more than half of the district's total fruit stocks were recorded, among other food items. The hail that had severely damaged fruit trees in the area under consideration in August 1864 was apparently less severe in the cadastral municipality of Strane. The following examples from Strane will contribute to our understanding of the circumstances in the village. The cottager Jakob Tominc and his family of six owned one and a half *Metzen* of oats, the same amount of beans (or other legumes), a *Metzen* of potatoes as many as 12 *Metzen* of turnips, carrots and/or beetroots, and 20 pounds (*Pfund*) of bacon, which was estimated to suffice for three months.[86] Comparatively, the owner of half a *Hufe* and the head of a family of twelve, Janez Tominc, was even better off with about four more months of supplies in stock: 3 *Metzen* of barley, 5 *Metzen* of oats, and one and a half *Metzen* of millet and buckwheat each, 5 *Metzen* of beans (or other legumes), as much as 25 *Metzen* of potatoes, 17 *Metzen* of turnips, carrots and/or beetroots, 7 *Metzen* of cabbage, and half a *Zentner* (28 kg) of bacon. Matija Brezec, who also owned half a *Hufe* and was the head of a family of seven, was also doing relatively well. He had 2.5 *Metzen* of wheat, 2 *Metzen* of rye, 8 *Metzen* of barley, 6 *Metzen* of oats, and one *Metzen* of buckwheat. He also stocked 4 *Metzen* of beans (or other legumes), 7 *Metzen* of potatoes, 12 *Metzen* of turnips, carrots and/or beetroots, 6 *Metzen* of cabbage, 1.5 *Zentner* (84 kg) of fruit, and one *Zentner* (56 kg) of bacon.[87]

However, the reliability of these estimations is clearly uncertain. For instance, if we compare the cadastral municipalities of Šmihel and Hrenovice, the number of inhabitants of the former was allegedly higher, while the amount of calories reportedly stocked at Šmihel was significantly lower. However, the stocks were estimated to suffice for almost twice as long at Šmihel than at Hrenovice (Table 10.1). What is clear, however, is that there was a general lack of food and seeds and that food availability per inhabitant differed significantly from one cadastral municipality to the other, from village to village, from holding to holding. The present case study will focus on the cadastral municipality of Razdrto. Figure 10.3 shows the property structure in that microregion, which, during the severe distress in 1865, included the villages of Razdrto, Veliko Ubeljsko, and Malo Ubeljsko. The village of Razdrto had a few more than 242 inhabitants. Their landholdings were fragmented and tiny: the census takers only registered a single owner of a quarter of a *Hufe*, and no farm whose

[86] SI AS 16, box 10, bundle 23, Ausweis über den Vorrath an Lebensmitteln in der Ortsgemeinde Strane, 10 January 1865.

[87] SI AS 16, box 10, bundle 23, Ausweis über den Vorrath an Lebensmitteln in der Ortsgemeinde Strane, 10 January 1865.

size category was indicated in that village was larger than that. Note, however, that the economic situation of the holdings whose size was not recorded in the sources as the basis for Figs. 10.2 and 10.3 (light grey bar sections), was better, they were not threatened by the subsistence crisis as a general rule. The village of Razdrto consisted mostly of cottagers, who represented about 68% of the total number of households. In the cadastral municipality, some larger farms could only be found in Veliko and Malo Ubeljsko, but not at the formerly busy thoroughfare at Razdrto, with a possible exception of the households represented by the light grey column section in Fig. 10.3. Many people in Razdrto were left completely without food or with stocks that would only suffice for a day or two, as recorded in January 1865. Nonetheless, the crisis threatened the villages of Veliko and Malo Ubeljsko as well; in January 1865, more than one fifth of the heads of the farm, cottage, and lodger economies in Veliko and Malo Ubeljsko begged, mainly some lodgers and cottagers (in one case, children partly begged and partly worked for others). In the case of individual farmers, Veliko and Malo Ubeljsko stood out with regard to bacon and pork stocks,[88] which were nothing unusual for January 1865, i.e. for the festive season after the slaughtering of pigs in the context of autarkic pig farming.[89] In Razdrto, there was no mention of any bacon or pork stocks, which could be added to the dishes to enrich them.[90]

In the village of Razdrto, extremely modest stocks of cereals were particularly striking, as the entire settlement only had 1 3/8 *Metzen* of wheat, half a *Metzen* of rye, and four and a half *Metzen* of barley. Those who still had anything at all to eat mainly consumed turnips, carrots and/or beetroots; potatoes; and beans of other legumes. However, even the stocks of these food items were extremely scarce. According to official data, households with inventoried food stocks in the entire village kept 47 *Metzen* of turnips, carrots and/or beetroots, 18.5 *Metzen* of potatoes, and 9.5 *Metzen* of beans or other legumes. Only 3 *Metzen* of cabbage were inventoried. Today's observers can barely imagine the scope of hardship in Razdrto in January 1865. Compared with the other villages and cadastral municipalities in the district, the situation in Razdrto was extremely critical.[91]

A third of the cottagers from Razdrto mentioned an additional source of income, often still related to road transport. For example, three of them stated that they were cottagers and blacksmiths, while one of them was a roadman (*Strasseneinräumer*), i.e. in charge of ensuring smooth traffic and maintaining the commercial road to Trieste. The remaining cottagers listed the following as secondary sources of income: saddler, leatherworker, shoemaker, hatter, field guard (*Feldhüter*), and, in two cases, day labourer (*Tagelöhner*). Roughly a quarter (or a bit more) of 56 inventoried households in the village did not have any food items in stock. For a few

[88] SI AS 16, box 10, bundle 23, Ausweis über den Vorrath an Lebensmitteln in der Ortsgemeinde Präwald, 12 January 1865.

[89] Compare Pančur (2009), p. 59.

[90] SI AS 16, box 10, bundle 23, Ausweis über den Vorrath an Lebensmitteln in der Ortsgemeinde Präwald, 12 January 1865.

[91] SI AS 16, box 10, bundle 23, Ausweis über den Vorrath an Lebensmitteln in der Ortsgemeinde Präwald, 12 January 1865.

heads of households, we can only find a brief note stating that they owned nothing and earned their living as day labourers. The lodger Jernej Sila, the head of a two-member household, was registered as a beggar.[92]

Unlike the beggar Sila, the Razdrto miller Anton Bole, the parish priest Franc Milavc, the road master's wife Apolonija Toman, and the innkeeper Andrej Premru certainly did not suffer any hunger and deprivation. The census enumerators did not indicate their food supplies but only noted that they did not need any assistance.[93] For comparison and as an illustration, we should mention the example of an innkeeper in the nearby village of Sajevče in the cadastral municipality of Rakulik. There, the inventory of the quarter-*Hufe* owner, innkeeper, and head of a family of ten Matej Gorjanc stated that he did not require assistance. In his case, an abundant amount of supplies were inventoried in comparison with the other farmers in the cadastral municipality of Rakulik: one *Metzen* (61.5 l) of wheat, two *Metzen* of rye, three *Metzen* of barley, six *Metzen* of oats, five *Metzen* of buckwheat, two and a half *Metzen* of beans or other legumes, twenty *Metzen* (1230 l or around 920 kg) of potatoes, twelve *Metzen* (738 l) of turnips, carrots and/or beetroots, ten and a half *Metzen* (646 l) of cabbage, and 140 *Pfund* (78.5 kg) of bacon and pork.[94]

Although food was extremely scarce in many households, famine did not occur or many people from the area under consideration would have died.[95] This was a combined result of an extensive charitable action, coordinated at the level of the province, and the local tactics of coping with the disaster. Fundamental changes in the broader economic conditions played a prominent role as well: railways represented a turning point in cheap long-distance inland food transport in places located far away from navigable waterways,[96] such as the district under consideration. Thus, in the nineteenth century, the railway also put an end to severe starvation triggered by the environmental conditions in this area—an observation reported already in 1889.[97]

Regarding the extensive charitable action, its initial steps, involving the cooperation between the District Office in Senožeče and the Provincial Presidency in December 1864 and January 1865, were already presented. The census results from January 1865, dealt with in this chapter, persuaded the president of the province of Carniola that this was a natural disaster. In February, he asked the state minister in Vienna for financial aid from the state treasury. In late February, however, the petition was rejected in Vienna with the argument that the state treasury cannot help in case of local disasters. Nonetheless, the president of the province initiated an extensive alms campaign for the people from Inner Carniolan districts, who suffered severe distress

[92] SI AS 16, box 10, bundle 23, Ausweis über den Vorrath an Lebensmitteln in der Ortsgemeinde Präwald, 12 January 1865.

[93] SI AS 16, box 10, bundle 23, Ausweis über den Vorrath an Lebensmitteln in der Ortsgemeinde Präwald, 12 January 1865.

[94] SI AS 16, box 10, bundle 23, Ausweis über den Vorrath an Lebensmitteln in der Ortsgemeinde Rakulik, 17 January 1865.

[95] Pančur (2009), p. 55.

[96] White et al. (2018), p. 340 (quote), p. 344; Krausmann et al., (2016), pp. 73, 77.

[97] *Postojinsko okrajno glavarstvo* (1889), p. 62.

and poverty. His appeal for charitable donations was sent to the district offices in Carniola, the Ljubljana Town Hall (in the province capital), and the archbishop's office in Ljubljana, which spread it further, as well as to three newspapers published in Ljubljana. The appeal was read in newspapers in late February and early March 1865, and the president of the province asked the editorships to publish the lists of donors and thus increase their public reputation. A diverse group of people and legal entities responded favourably: the most generous included the 7,200 *Gulden* donated by some members of the Vienna Imperial Court (including Emperor Franz Joseph, his wife Karolina Augusta, and the Archdukes Franz Karl and Albrecht), a donation of 1,000 *Gulden* by the Provincial Committee (*Landesausschuss*), 500 *Gulden* by the Carniolan provincial fund for agriculture, and 500 *Gulden* by the Ljubljana Town Hall. These were joined by donations from many individuals, charitable social and cultural events, etc. The district office at Senožeče was asked to inform the Provincial Presidency of the quantities of different types of missing seeds and seed potatoes by 10 April 1865; it was emphasised that the ratio between different seed types should take account of the fact that the forthcoming sowing would take place in unfavourable weather. By early April 1865, the first 1,600 *Gulden* of emergency aid were allocated to the disaster-stricken Inner Carniolan districts, moreover a total of 14,208 *Gulden* were donated. On 8 April, 13,400 *Gulden* from that sum were distributed among the districts of Senožeče, Lož, and Postojna. The district of Senožeče received 4000 *Gulden.* As the total donated increased to 17,675.8 *Gulden* by mid-June, more money was available for the poor in the area under consideration: the Senožeče and Lož districts were awarded at least an additional 442 *Gulden.* A comparison between the donated 17,675.8 *Gulden* and the 22,000 *Gulden* as the total value of victuals and seeds needed to prevent starvation and lack of seeds in the districts of Senožeče, Lož, and Postojna, based on the data from April, suggests that people in distress received most of what the census enumerators asked for, but not the total aid presumably needed.[98] The monetary assistance for the district of Senožeče was invested in purchases of seeds and seed potatoes (most of the aid) and victuals (a lower percentage of the aid) in Ljubljana. Seeds and victuals were then transported by train, free of charge, from Ljubljana to the closest railway station and then on the roads to the relevant area; some funds for the transport costs were available. The following data from the same charity campaign for parts of Inner Carniola illustrate what the aid in the amount of more than 4000 *Gulden* meant for the district of Senožeče: 3762.65 *Gulden* were used during that campaign to purchase the following victuals: 1100 *Metzen* (about 68,000 l) of maize, 110 *Zentner* (about 6.2 metric tonnes) of flour, 75 *Metzen* (about 4,600 l) of beans, 6 *Zentner* (336 kg) of salt, and 4 *Zentner* and 14 *Pfund* (232 kg) of lard (this illustration only includes the prices of victuals, which differed from seed prices). In the district of Senožeče, the aid provided in this manner was distributed by a so-called *"hardship commission"*; such commissions typically consisted of farmers' representatives and a priest.[99]

[98] Studen (2020), pp. 179–181; Studen (2018), pp. 62–68, 70–72; *Laibacher Zeitung*, 3 July 1865.

[99] Studen (2020), pp. 179–181; Studen (2018), pp. 67–68, 71–72; *Laibacher Zeitung*, 3 July 1865.

However, in the spring of 1865, the shortage of seeds represented an even more complex issue than it appears; namely, the serious problem of seed availability was not the only challenge. Reports of the district heads from Lož and Postojna, addressed to the Imperial Royal Presidency of Carniola, emphasised that due to the harsh environmental conditions in the region, the seeds and seed potatoes should originate from Upper Carniola, where the climate is similarly harsh—from "*the areas where the soil resembles the local conditions*", as plants from regions with a significantly milder climate would not thrive in Inner Carniola.[100] A newspaper report claiming that the 1865 harvest around Senožeče was even worse than 1864,[101] although the weather was not as severe as in 1864,[102] suggests that the challenge of the lack of (appropriate) seeds was solved only partly in 1865. Possibly, the unfavourable weather in the early spring of 1865 and the potential lack of manure (see below) contributed to the poor harvest,[103] or the newspaper report on the 1865 harvest was too pessimistic.

In addition to the charitable donations, the deficient food and seeds were also partly bought with money earned by the locals. Food purchases are mentioned several times in the census reports from January 1865.[104] Such activities had started before the aid in victuals and seeds bought with charitable donations arrived. The severe poverty and lack of other means of earning money had forced many people to look elsewhere for earnings.[105] For instance, the cold early spring prevented the usual work in the fields "*making it impossible for the poor to earn anything*", as reported by the Head of the Senožeče District Office in late March 1865.[106] By the mid-nineteenth century, the inhabitants of the analysed area had already participated in seasonal work as carpenters in forests in Slavonia (Posavina) or even northern Ottoman Bosnia.[107] For instance, the owner of an eighth of a *Hufe* at Rakulik, whose wife was begging in the village, was reportedly working as a day labourer in Croatia in January 1865.[108] People from the studied settlements also worked as seasonal or day labourers in the busy port cities of Trieste and Rijeka[109] or, to a much more limited degree, in

---

[100] Pančur (2009), p. 55; Studen (2020), pp. 176, 178–180 (quote from page 178).

[101] *Kmetijske in rokodelske novice*, 14 March 1866.

[102] E.g., Čufar et al. (2008), pp. 608–614; Ogrin (2015).

[103] Studen (2018), pp. 48, 62, 64–72.

[104] SI AS 16, bundle: 23, box 10, e.g. Ausweis über den Vorrath an Lebensmitteln in der Ortsgemeinde Prävald, 12 January 1865.

[105] Studen (2020), p. 177.

[106] Studen (2020), p. 180.

[107] Klemenčič (1959), pp. 60, 150–152, 155, 159; *Postojinsko okrajno glavarstvo* (1889), pp. 60, 62; SI AS 16, bundle 23, box 10, e.g. Ausweis über den Vorrath an Lebensmitteln in der Ortsgemeinde Prävald, 12 January 1865 (Malo Ubeljsko 8).

[108] SI AS 16, box 10, bundle 23, Ausweis über den Vorrath an Lebensmitteln in der Ortsgemeinde Rakulik, 17 January 1865.

[109] Klemenčič (1959), p. 150; *Postojinsko okrajno glavarstvo* (1889), pp. 109, 118, 131; see, for instance, a couple of day labourers in SI AS 16, bundle 234, box 10, Ausweis über den Vorrath an Lebensmitteln in der Ortsgemeinde Präwald, 12 January 1865.

the market town of Postojna.[110] The manners of making a living in times of crisis differed significantly within the microregion under consideration: the village of Laže was surrounded by exceptionally extensive forests, as mapped in 1822,[111] and its inhabitants were selling firewood—beech bundles—to Trieste.[112] The inhabitants of villages in the Lower Pivka region, close to the Nanoščica and smaller streams (at Hrenovice, for example), earned quite a lot by selling crayfish in Trieste, as recorded in the 1880s. According to the data from the 1880s, the inhabitants of Lower Pivka also produced large quantities of bricks for the market and sold a lot of charcoal to Trieste[113]; the question of whether brick and charcoal production had already been a relevant income source in the 1860s remains open. Some villagers were artisans, as demonstrated by the examples from Razdrto. The following examples from the villages of Malo Ubeljsko and Veliko Ubeljsko confirm this as well: in Malo Ubeljsko, Janez Simčič was a cottager and innkeeper, while his neighbour Jakob Debevc was a cottager and shoemaker. In Veliko Ubeljsko, the owner of a quarter of a *Hufe* Peter Simšič was a bread baker, while the third-*Hufe* owner Anton Slivar was a blacksmith. In Veliko Ubeljsko, the miller Blaž Žuža (and the chaplain Martin Počkaj) required no assistance in January 1865. Further examples from Veliko and Malo Ubeljsko also demonstrate that some farmers sold or leased a part of their land to overcome the contemporaneous crisis, though this represented a detriment to their agricultural economy in the long run.[114] The destruction and scarcity of hay encouraged the sales of livestock, which might have reduced the quantities of manure available for the fields next year.[115] Most likely, loans (unavailable to the poorest) and, certainly, begging also played a role in some households.[116] The role of permanent emigration to Trieste or Rijeka or the potential early overseas emigration from the study area at the time is not yet clear.[117]

10.4 Conclusions

"*The effects of weather on agriculture and food supplies cannot be understood apart from the fragility and resilience of societies, states, and economies.*"[118] The case study deals with an example of a Little Ice Age-type year-to-year weather variability at the

---

[110] *Postojinsko okrajno glavarstvo* (1889), p. 60.

[111] "Arcanum Maps", Cadastral maps, Habsburg Empire—Cadastral maps (XIX. [!] century).

[112] *Postojinsko okrajno glavarstvo* (1889), p. 109; Studen (2018), p. 65.

[113] *Postojinsko okrajno glavarstvo* (1889), p. 119; Studen (2018), p. 65.

[114] SI AS 16, bundle 23, box 10, Ausweis über den Vorrath an Lebensmitteln in der Ortsgemeinde Prävald, 12 January 1865.

[115] Studen (2018), pp. 62–63, 65.

[116] Studen (2020), p. 177; Studen (2018), pp. 55, 69; on beggars, see above in Sect. 10.3.

[117] Compare Klemenčič (1959), pp. 157–160; Melik (1960), pp. 320–322; Studen (2020), p. 181.

[118] White et al. (2018), p. 331.

very end of the Little Ice Age rather than with a long-term climate cooling. Nonetheless, this weather variability caused a natural disaster in parts of Inner Carniola in times of severe economic depression of the integrated farm economy. More precisely, the analysis of sources kept in the archives of the Imperial Royal Presidency of Carniola and newspapers revealed the existence of extensive diversity in the resilience of farm economies within the same settlements and between the various settlements in the analysed area as a consequence of the interplay between the human and environmental factors. Focusing on the latter, the considerable environmental diversity of Dinaric Karst played a prominent role. Hay farming played a significant role in the karst of Senožeče, the karst polje with the intermittent Lake Cerknica, and the predominately non-karstic meadows in the Lower Pivka region. While the environmental conditions usually enabled much higher hay yields in Lower Pivka and at the bottom of the intermittent Lake Cerknica (after the karst water retreated and the lake disappeared) than in the dry and drought-prone calcareous grasslands around Senožeče, the 1864 hay harvest in the calcareous grasslands around Senožeče was not swept away or smeared by mud like in many meadows in Lower Pivka, or prevented by flooding like at Lake Cerknica. Nonetheless, due to the challenges posed by rainy weather in times of haymaking, the 1864 quantities of hay stored around Senožeče were likely well below average.

The railway played an ambiguous role in the local economy during the 1860s. On the one hand, it aggravated the crisis of the integrated farm economy in the area under consideration, as many former sources of non-agricultural income along the busy thoroughfare virtually disappeared. On the other, it is clear that the railway, enabling cheap long-distance inland cargo transport, contributed significantly to the ability of states and economies "to cope with a growing population and subsistence crises".[119]

Several questions deserve future research attention, such as (1) how the locals coped with the analysed subsistence crisis between January 1865 and the arrival of (most of) the victuals, seeds, and seed potatoes purchased with charitable donations, as the aid arrived significantly later (with the exception of the relatively low sum of 1,600 *Gulden* for all the disaster-stricken districts,[120] allocated to the affected areas at an unknown time before 7 April, the funds were distributed among these districts in April) than the stocks inventoried in January 1865 had been expected to last in most cases; in late March 1865, the head of the nearby district of Postojna reported that "*he and the priest keep coming across large groups of people begging for alms*"[121]; (2) whether this subsistence crisis caused a severe increase in the use of natural resources, like a potentially heavier crayfish hunting in the Lower Pivka region or increased (fire)wood cutting around Laže and elsewhere (to make charcoal, for instance)[122]; and (3) whether this subsistence crisis had any significant demographic impacts in the area under consideration.

[119] Compare White et al. (2018), pp. 340, 342 (quote).

[120] See also Studen (2018); Studen (2020).

[121] Studen (2020), p. 180.

[122] See Studen (2018), p. 55; Studen (2020), p. 177.

Acknowledgements The research was carried out within the framework of the research programme Economic, Social and Environmental History of Slovenia, No. P6-0280, financed by the Slovenian Research and Innovation Agency.

References

Sources

Published Sources

"Arcanum Maps", Cadastral maps, Habsburg Empire—Cadastral maps (XIX. [!] century). Accessed 17 July 2023. https://maps.arcanum.com/en/map/cadastral/?layers=3%2C4&bbox=1794877.5643639148%2C6125569.198257975%2C1867760.27083383%2C6149035.3659415245

ARSO (2023a) Atlas okolja [Environmental Atlas of Slovenia], Land and Soil, Soil map 1:25000. Accessed 11 July 2023. http://gis.arso.gov.si/atlasokolja/profile.aspx?id=Atlas_Okolja_AXL@Arso&culture=en-US

ARSO (2023b) Hidrografija [Hydrography]. Agencija Republike Slovenije za okolje, Ljubljana

ARSO (2023c) Lidar. Agencija Republike Slovenije za okolje, Ljubljana

GURS (2023a) Državna meja [State border]. Geodetska uprava Republike Slovenije, Ljubljana

GURS (2023b) Železnice [Railways]. Geodetska uprava Republike Slovenije, Ljubljana

Kmetijske in rokodelske novice [Agricultural and Artisan News] 1860, 1864, 1866

Laibacher Zeitung 1864, 1865

Natural earth. Accessed 9 July 2023. https://naturalearthdata.com

Stenographischer Bericht der zwanzigsten Sitzung des krain. Landtages zu Laibach am 14. April 1864. Laibach: Verlag des krainischen Landes-Ausschusses. s.d.

Archival Sources

Archives of the Republic of Slovenia: SI AS 16, Deželno predsedstvo za Kranjsko (Presidency of the province of Carniola): box 10, bundle 23

Literature

Beguš I (2017) Peasant income integration in early modern Slovenia: A historiographical review. In: Panjek A, Larsson J, Mocarelli L (eds) Integrated Peasant Economy in a Comparative Perspective: Alps, Scandinavia and Beyond. University of Primorska Press, Koper, pp 51–81

Blaznik P (1970) Enote individualne posesti [Einheiten des individuellen Besitzes]. In: Blaznik P, Grafenauer B, Kos M, Vilfan S, Zwitter F, Žontar J (eds) Agrarno gospodarstvo, Vol 1 of Gospodarska in družbena zgodovina Slovencev: zgodovina agrarnih panog. Slovenska akademija znanosti in umetnosti and Državna založba Slovenije, Ljubljana, pp 161–184

Britovšek M (1958–1959) Uvajanje novih kultur na Kranjskem v drugi polovici 18. in v prvi polovici 19. stoletja [Die Einführung der neueren Kulturpflanzen in Krain in der zweiten Hälfte des 18. und in der ersten Hälfte des 19. Jahrhunderts]. Zgodovinski časopis 12–13:111–150

Čeč D (2015) Družbena percepcija in pravna obravnava zločinov iz stiske: primer eksistenčne krize 1816–1817 na Slovenskem [Social perception and legal treatment of crimes committed due to hardship. The example of the 1816–1817 existential crisis in the Slovenian territories]. In: Bianco F, Panjek A (eds) Upor, nasilje in preživetje: slovenski in evropski primeri iz srednjega in novega veka. Založba Univerze na Primorskem, Koper, pp 243–275

Čeč D (2017) Gospodarska in družbena kriza zaradi izbruha vulkana Tambora v Indoneziji aprila 1815 [Economic and social crisis brought on by the eruption of the Tambora volcano in Indonesia in April 1815]. Zgodovina v šoli 25(1):3–16

Čufar K, De Luis M, Eckstein D, Kajfež-Bogataj L (2008) Reconstructing dry and wet summers in SE Slovenia from oak tree-ring series. Int J Biometeorol 52:607–615

Dolenc E (1994) Senožeče: skupnost na prepihu [Senožeče: A community in the draught]. Rubic Trade, Koper

Fischer J (2005) Zagate v kmetijstvu [Farming conundrums]. In: Fischer J, Lazarević Ž, Dolenc E, Perovšek J, Godeša B, Čepič Z, Gabrič A (eds) Slovenska novejša zgodovina: od programa Zedinjena Slovenija do mednarodnega priznanja Republike Slovenije: 1848–1992, Vol 1. Mladinska knjiga and Inštitut za novejšo zgodovino, Ljubljana, pp 72–74

Füssel H-M (2007) Vulnerability: A generally applicable conceptual framework for climate change research. Glob Environ Chang 17(2):155–167

Gams I (2004) Kras v Sloveniji v prostoru in času [Karst in Slovenia in space and time]. Založba ZRC, Ljubljana

Glaser R (2001) Klimageschichte Mitteleuropas: 1000 Jahre Wetter, Klima, Katastrophen. Wissenschaftliche Buchgesellschaft, Darmstadt

Granda S (2003) Davki, davki, davki [Taxes, taxes, taxes]. In: Cvirn J, Granda S, Grdina I, Komelj M, Studen A, Vinkler J (eds) Slovenska kronika XIX. stoletja (1861–1883). Nova revija, Ljubljana, pp 198–199

Kačičnik Gabrič A (2004) O kmečkih dolgovih nekoliko drugače: problem servitutnih pravic na posestvu Snežnik [Bäuerliche Schulden etwas anders gesehen: Das Problem der Servituten auf der Hersschaft Snežnik (Schneeberg)]. Zveza zgodovinskih društev Slovenije and Arhiv Republike Slovenije, Ljubljana

Kačičnik Gabrič A (2014) "To smemo že tako dolgo": kmečke služnosti in njihova odprava ["We have been able to do that for such a long time": peasant easement and its abolishment]. Inštitut za novejšo zgodovino, Ljubljana

Kalc A (2017) Peasant population and income integration: The case of the Trieste port-town hinterland (Karst, 16th-20th Centuries). In: Panjek A, Larsson J, Mocarelli L (eds) Integrated Peasant Economy in a Comparative Perspective: Alps, Scandinavia and Beyond. University of Primorska Press, Koper, pp 205–227

Kalc A (2018) Prebivalstvo in gospodarski viri v tržaškem kraškem zaledju (16.–20. stoletje) [Population and economic resources in the Trieste hinterland of Kras (16th–20th century)]. In: Panjek A, Lazarević Ž (eds) Preživetje in podjetnost: integrirana kmečka ekonomija na Slovenskem od srednjega veka do danes. Založba Univerze na Primorskem, Koper, pp 155–188

Klemenčič V (1959) Pokrajina med Snežnikom in Slavnikom: gospodarska geografija [La région entre le Snežnik et le Slavnik: géographie économique]. Slovenska akademija znanosti in umetnosti, Ljubljana

Krajevni leksikon Slovenije (1968) [Lexicon of settlements of Slovenia]. Vol. 1, Zahodni del Slovenije. Državna založba Slovenije, Ljubljana

Krausmann F, Weisz H, Eisenmenger N (2016) Transitions in sociometabolic regimes throughout human history. In: Haberl H, Fischer-Kowalski M, Krausmann F, Winiwarter V (eds) Social Ecology: Society-Nature Relations across Time and Space. Springer, Cham, pp 63–92. https://doi.org/10.1007/978-3-319-33326-7_3

Lazarević Ž (1994) Kmečki dolgovi na Slovenskem: socialno-ekonomski vidiki zadolženosti slovenskih kmetov 1848–1948 [Agricultural debt in Slovenia: Socio-economic aspects of the indebtedness of Slovenian farmers, 1848–1948]. Znanstveno in publicistično središče, Ljubljana

Makarovič G (1988–1990) Prehrana v 19. stoletju na Slovenskem [Nutrition in the Slovenian territories in the nineteenth century]. Slovenski etnograf 33–34:127–205

Mal J (1993) Zgodovina slovenskega naroda [History of the Slovenian nation]. Part 2. Mohorjeva družba, Celje

Melik A (1960) Slovensko Primorje [The Slovenian Littoral]. Slovenska matica, Ljubljana

Mihevc A, Prelovšek M (2010) Geographical position and general overview. In: Mihevc A, Prelovšek M, Zupan Hajna N (eds) Introduction to the Dinaric Karst. Karst Research Institute at ZRC SAZU, Postojna, pp 6–8

Ogrin D (1996) Podnebni tipi v Sloveniji [The climate types in Slovenia]. Geografski vestnik 68(1):39–68

Ogrin D (2015) Long-term air temperature changes in Ljubljana (Slovenia) in comparison to Trieste (Italy) and Zagreb (Croatia). Moravian Geograph Rep 23(3):17–26

Padavine u Jugoslaviji: rezultati osmatranja za period 1925–1940 (1957) [Precipitations en Yougoslavie: resultats des observations pour la periode 1925–1940]. Institut hydrometeorologique federal, Beograd

Pančur A (2009) Porajanje modernega gospodarstva v 18. in 19. stoletju [The emergence of the modern economy in the 18th and 19th centuries]. In: Fikfak J (ed) Postojna: upravno in gospodarsko središče. Galerija 2, Postojna; Inštitut za novejšo zgodovino, Ljubljana, pp 51–66

Panjek A (2017a) Intangible and material evidence on the Slovenian peasant economy: Custom and land market in the Karst highland (17th–18th Centuries). In: Panjek A, Larsson J, Mocarelli L (eds) Integrated Peasant Economy in a Comparative Perspective: Alps, Scandinavia and Beyond. University of Primorska Press, Koper, pp 229–257

Panjek A (2017b) The integrated peasant economy as a concept in progress. In: Panjek A, Larsson J, Mocarelli L (eds) Integrated Peasant Economy in a Comparative Perspective: Alps, Scandinavia and Beyond. University of Primorska Press, Koper, pp 11–49

Perko D (1998) The regionalization of Slovenia. Geografski zbornik 38:11–57

Pfister C (1985) Das Klima der Schweiz von 1525–1860 und seine Bedeutung in der Geschichte von Bevölkerung und Landwirtschaft. Band 2, Bevölkerung, Klima und Agrarmodernisierung 1525–1860. Verlag Paul Haupt, Bern and Stuttgart

Pfister C, White S (2018) A year without a summer, 1816. In: White S, Pfister C, Mauelshagen F (eds), The Palgrave Handbook of Climate History. Palgrave Macmillan, London, pp 551–561

Pfister C, Wetter O, Brázdil R, Dobrovolný P, Glaser R, Luterbacher J, Seneviratne SI et al (2015) Tree-rings and people—different views on the 1540 megadrought: Reply to Büntgen et al. 2015. Climatic Change 131:191–198

Pirc I, Baš F (1938) Socialni problemi slovenske vasi [Social challenges of Slovenian rural areas], Vol 1. Socialno ekonomski institut, Ljubljana

Plut D (2002) Okoljevarstveni vidiki prostorskega razvoja Slovenije [Environmental-protection aspects of the spatial development in Slovenia]. Znanstveni inštitut Filozofske fakultete, Ljubljana

Postojinsko okrajno glavarstvo: zemljepisni in zgodovinski opis (1889) [The district of Postojna/Adelsberg: A geographical and historical description]. R. Šeber, Postojna

Remec M (2015) Podrgni, očedi, živali otrebi: higiena in snaga v dobi meščanstva [Rub, tidy, tend to animals: Hygiene and sanitation in the bourgeois period]. Inštitut za novejšo zgodovino, Ljubljana

Reya O (1940) Padavine na Slovenskem v dobi 1919–1939 [La répartition de la précipitation en Slovénie]. Geografski vestnik 16(1–4):25–40

Rus D-I (2020) Weather anomalies in Transylvania, the Banat and Partium from 1813 to 1818, as reflected in contemporary sources. Econ and Ecohist: J Econ Hist Environ Hist 16:146–166

Seidl F (1890) Ueber das Klima des Karstes. Mittheilungen des Musealvereines für Krain 3:307–340

Seidl F (1894) Das Klima von Krain, IV. Theil. Mittheilungen des Musealvereines für Krain. Zweite Abtheilung: Naturkundlicher Theil 7:1–68

Sovič S (2005) Families and households of the poor: The 19th-century Slovenian gostači. History Family 10:161–182

Studen A (2018) Lakota na Notranjskem 1865: primer okraja Lož [Starvation in Inner Carniola in 1865: the case of the Lož district]. In: Šorn M (ed) Lakote in pomanjkanje: slovenski primer. Inštitut za novejšo zgodovino, Ljubljana, pp 43–72

Studen A (2020) It is still almost half a year until harvest: The raging hail in the Postojna district in august 1864 and its consequences. Econ and Ecohist: J Econ Hist Environ Hist 16:167–183

Voje I (1952–1953) Kajžarstvo v logaškem gospostvu [Cotters in the manorial estate of Logatec]. Zgodovinski časopis 6–7:650–662

White S (2014) The real Little Ice Age. J Interdiscipl Hist 44(3):327–352

White S, Brook J, Pfister C (2018) Climate, weather, agriculture, and food. In: White S, Pfister C, Mauelshagen F (eds) The Palgrave Handbook of Climate History. Palgrave Macmillan, London, pp 331–353

Zwitter Ž (2013) Vremenska in klimatska zgodovina v koledarjih in podložniških dnevnikih ljubljanskega škofa Tomaža Hrena (1597–1630) [History of weather and climate in calendars and manorial minutes of Tomaž Hren, bishop of Ljubljana (1597–1630)]. Zgodovinski časopis 67(3–4):306–389

Zwitter Ž (2014) Okolje na Kranjskem v 17. stoletju po Slavi vojvodine Kranjske [Die Umwelt in Krain im 17. Jahrhundert in der Ehre Deß Hertzogthums Crain]. In: Weiss J (ed) Studia Valvasoriana: Zbornik spremnih študij ob prvem integralnem prevodu Die Ehre Deß Hertzogthums Crain v slovenski jezik. Zavod Dežela Kranjska, Ljubljana, pp 611–702

Zwitter Ž (2015) Subsistence, prosperity and abandonment of Alpine isolated farms in the dynamic 17th century environment: Case study from the Upper Savinja Valley with special emphasis on tenants' inventories. Econ Ecohist: J Econ Hist Environ Hist 11:139–181

Andrej Studen Assistant professor and senior research associate, Ph.D. He lectured on the history of everyday life and on the subject of historical anthropology at the Department of History of the Faculty of Arts, University of Ljubljana. Studen's research interests included residential culture, the history of modernisation, urbanisation and the process of hygienisation, the bourgeoisie, the arrival of the railway, tourism, crime, punishment and capital punishment, alcoholism, and the reception of labour and self-responsibility. He published numerous scientific papers and several monographs.

Chapter 11
Water Infrastructure in Kras in the Nineteenth Century Between Continuity and Change

Aneja Rože Kravanja

Abstract The article discusses the still poorly researched development of the construction of water infrastructure on the Kras Plateau in the nineteenth century, with special emphasis on the cadastral municipalities of Lokev and Dutovlje. The increasing role of the state (beginning with the Habsburg monarchy at the time of the Theresian reforms with final implementation during the time of Austria-Hungary) in supplying the inhabitants on the Kras Plateau and its livestock with drinking water proves to reach a turning point in the middle of the nineteenth century. The research is based on detailed early nineteenth-century land cadastre, on documents from the second half of the long nineteenth century and kept in the Sežana district board fonds in the Provincial Archives of Koper, on unstructured interviews, and field observations.

Keywords Water supply · Water infrastructure · Water holes · Plumbing · Kras Plateau · Nineteenth century

11.1 Introduction

The Kras Plateau is located on the northwestern edge of the Dinaric Karst, in the Slovenian-Italian borderland. Despite the relatively high amount of precipitation on the entire Kras Plateau (somewhere between 1500 and 1800 mm[1]), rainwater quickly seeps into the karst aquifer. The Kras Plateau has two basic geological compositions. The fringes of the northern, southern, and eastern parts consist of bedded tertiary limestone. The rest of the plateau consists of Cretaceous limestone and partly dolomite. The units are more than 1000 m thick. Tertiary and Cretaceous limestones are karstified, while dolomite layers are less permeable and play the role of a hydrogeological

[1] Slovenian Environment Agency, "Atlas okolja" [Environmental atlas of Slovenia], Climate, Precipitation, Mean annual corrected precipitation 1971–2000.

A. Rože Kravanja (✉)
Municipality of Ilirska Bistrica, Ilirska Bistrica, Slovenia
e-mail: aneja.roze@hotmail.com

B. Fuerst-Bjeliš et al. (eds.), *Environmental Histories of the Dinaric Karst*, Environmental History 17, https://doi.org/10.1007/978-3-031-56089-7_11

isolator. The carbonate massif is surrounded by poorly permeable Eocene flysch which has the important function of an impermeable barrier. In the contact zone between the limestone and flysch, springs appear. In the northwestern boundary of Kras Plateau, the karst aquifer is in contact with the porous aquifer of alluvial sediments deposited along the Soča and Vipava Rivers.[2]

The outlined geological structure resulted in the difficult water supply on the Kras Plateau for centuries.[3] To keep the water from going underground and to be able to use it, they built water reservoirs. Such interventions in the natural environment have varied according to the type of reservoir, relief, soil characteristics, material, and changing construction techniques over time.[4] This article deals with the state and developments of water infrastructure and the growing role of the state in supplying the people on the Kras Plateau and their livestock with drinking water through the "long nineteenth century", starting in the last quarter of the eighteenth century and lasting until the outbreak of the World War I. The period of chronic lack of drinking water in the Kras Plateau ended around 1980 with the successful hydrogeological research of the Geological Institute of Ljubljana in Klariči near Brestovica where large amounts of underground water were discovered (research on groundwater extraction in the Kras Plateau began in 1953).[5] The paper focuses mainly on the results of the microhistory analysis of the cadastral municipalities of Dutovlje and Lokev as two different Kras Plateau microregions (see Fig. 11.1). Even on this small surface, there are significant differences in types of water infrastructure, the time of their construction, and their number. These varieties are excluded from general narratives about the cultural landscape of Kras Plateau, which tend to present the Kras cultural landscape uniformly.[6] The research results here prove that there was a turning point in the construction of water infrastructure in the middle of the nineteenth century when it was regulated under the guidance of the Austro-Hungarian government through the authorities at the state level, the government in Trieste, and the Sežana district board. Certain construction and hygiene standards had to be fulfilled, therefore the construction of *lokevs* and open-type *štirnas* (see Sect. 11.2.2 for explanations) was abandoned.

There are nearly a dozen historical reviews of the Kras Plateau water infrastructure[7] while in-depth studies of specific aspects are sorely missing. In the past

[2] About the depth of the underground water table see in: Petrič (2005), pp. 22–24.

[3] Petrič (2005), p. 25.

[4] Panjek (2015), pp. 457–474.

[5] Kristan and Skrinjar (1998), p. 10–18.

[6] Panjek (2006); Panjek (2015); Belingar (2011a), pp. 71–78; Renčelj and Lah (2008).

[7] Babij et al. (2005), pp. 83–99; Čelik et al. (2005), pp. 72–82; Belingar (2007a), pp. 116–128; Belingar (2011a), pp. 71–78; Belingar (2012), pp. 285–298; Panjek (2006), pp. 44–45; Panjek (2015), pp. 39–59; Renčelj and Lah (2008), pp. 161–164; Zobec (2007), pp. 193–207; Zobec (2008), pp. 495–504.

Fig. 11.1 Study area. Author of the map Filip Viler. *Topographic base source.* Environmental Systems Research Institute (ESRI); Surveying and Mapping Authority of the Republic of Slovenia (GURS) (2023); Ministry of Agriculture, Forestry and Food (MKGP) (2023). With permission

two centuries, researchers of the Kras Plateau (especially speleologist and geologists, like A. C. Morlot, A. Schimdl, A. Hanke)[8] have been particularly interested in the discovery and exploration of its underground terrain. Others focused on the history of knowledge about underground water reservoirs from the early modern period: Father Ferrante Imperato (*Historia Naturale*, 1599) and the Jesuit Athanasius Kircher (*Mundus subterraneus*, 1678). The discoveries of the underground terrain became popular due to tourism, especially to the geotourism,[9] and became more widely known (e.g., the Vilenica cave[10]). While there was much enthusiasm for the Kras Plateau underground cave systems, its surface was overlooked and much less valued.[11] Today, due to the study of biodiversity, biologists[12] pay more attention to the past infrastructure or preserved water bodies. These include water holes (*kals, lokevs*), and open-type *štirnas*, which are evidence of anthropogenically transformed karstic depressions, ranging in diameter from a couple of meters to tens of meters. The study of water infrastructure and water supply on the Kras Plateau seems to be neglected by historians and ethnologist, with only few exceptions, such as Eda Belingar,[13] Miha Zobec,[14] Žiga Zwitter,[15] and Aneja Rože.[16] Furthermore, the holistic concept of studying, preserving, and presenting heritage is also missing when discussing the past water infrastructure or preserved water bodies.[17]

As the middle of the nineteenth century turned out to be pivotal in the development of water infrastructure construction in the period considered, each half of the examined period is presented in its section. For the period up to the first half of the nineteenth century, I mostly relied my research on the Josephine land survey map and the Franciscan Cadaster as sources for studying the water supply on the Kras Plateau.[18] The primary source of information for the second half of the nineteenth century was the Sežana district board's fonds, which are kept in the Provincial Archives of Koper.[19] In addition to historical and archival sources, I have conducted field investigations of the preserved water infrastructure and interviewed the local population in an unstructured manner. Newspaper reports (such as from Trieste's

[8] Kranjc (1994), p. 133. Especially about A. C. Morlot see in: Pavlovec (1997), pp. 167–173. In addition to article by Kranjc for a general overview of the history of exploration of Kras Plateau see also: Kranjc (1997), pp. 15–17; Gams, (2003), pp. 11–28.

[9] Panjek (2015), p. 15; Kranjc (1994), pp. 131–134; Shaw (1998), pp. 235–264; Zorn et al. (2009), pp. 1–10; Zorn et al. (2012), pp. 348–355.

[10] Kavrečič (2015), pp. 549–553.

[11] Panjek (2015), p. 15.

[12] "1001 kal—1001 zgodba o življenju, 2005–2007;" "ENGREEN."

[13] Belingar (2007a), pp. 116–128; Belingar (2007b), pp. 433–452; Belingar (2011a), pp. 71–78; Belingar (2011b), pp. 75–90; Belingar (2012), pp. 285–298.

[14] Zobec (2007), pp. 193–207; Zobec (2008), pp. 495–504.

[15] Zwitter (2015), pp. 457–474.

[16] Rože (2018b), pp. 223–240; Rože (2020a), pp. 99–116; Rože (2020b), pp. 115–140.

[17] Perko (2015), p. 718.

[18] Rože (2020a), pp. 99–116.

[19] SI PAK KP 633—Okrajno glavarstvo Sežana [District board of Sežana].

newspaper *Edinost* and the Gorica's newspapers *Soča*, *Gorica* and *Glas*) from the period in question were also a part of the research.

11.2 The State of Water Infrastructure from the Second Half of the Eighteenth Century to the Middle of the Nineteenth Century: From Methodological Challenges to Historical Insights

The settlement pattern of villages and their spatial distribution on the Kras Plateau indicates that probably ever since the Middle Age, each of them had to have an easy access to water reservoirs, providing the people and livestock with fresh water. Until the nineteenth century, there was scarce information about their exact number and sanitary quality but the report from years 1770 and 1771 on the social and economic situation of Goriška and Gradiščanska, created as an addition to the Reports of Court Military Council, the highest military administrative authority of the Habsburg monarchy (I use this name for the state that was known by different names between 1770 and 1918. Between 1804 and 1867, it was known as the Austrian Empire, and after the agreement of 1867, the monarchy was renamed Austria-Hungary) stipulates that the inadequacy of the water infrastructure was one of the reasons for the population decline on the Kras Plateau.[20] However, it can be assumed that the number of water reservoirs was slightly higher along busy traffic routes that ran along the Karst Plateau in the direction of Trieste–Ljubljana (part of the connection Trieste–Vienna), Trieste–Gorizia (Sln. Gorica)/Vipava, and Trieste–Gradisca d'Isonzo (Sln. Gradišče ob Soči). Therefore, we rightly assume that water reservoirs that served not only the local population but also travelers and their draft animals also existed in the area of the villages of Lokev and Dutovlje.[21]

11.2.1 Water Infrastructure on Early Maps—Critical Examination in the Field

The general lack of water infrastructure can be observed in the comments to the Josephine land survey map from the 1780s. For the Kras Plateau, only *kals* are mentioned in most cases in the textual part, but less in the graphical part (see Fig. 11.2). In the textual part it is also written that they dry up in summer. The only water well which is mentioned is the one from Štanjel.[22] This does not mean, however, that there were no wells elsewhere. They were few in number and smaller

[20] Štih (2009/2010), p. 326.

[21] Zobec (2008), p. 497; Kosi (1998); Holz (2015), pp. 495–512.

[22] Rajšp and Ficko (1997), p. 127.

Fig. 11.2 Kals of the village of Veliki dol near Sežana, which are drawn on the Josephine land survey map. *Source* Facsimile from: Rajšp and Ficko (1997), Sektion 208

and, therefore, most likely not worth mentioning for the needs of the military. The preserved wellheads with inscribed dates confirm the existence of the wells at that time. In Lokev, for example, a preserved head of a municipal well with the year 1787 (adapted in 1887) can be found in the direct vicinity of St. Michael's Church. Moreover, if the Lokevian oral tradition is to be believed, the municipal well Stanka was already there before 1787. Three private wells from the second half of the eighteenth century have also been preserved from Lokev, with the years 1752 (the wellhead of this one has been relocated in the 1970s to Ilirska Bistrica, a town 45 km southeast from Lokev since the 1970s; see Fig. 11.3), 1778, and 1796 engraved on the heads.[23] As early as 1580, a peasant cistern was documented in nearby Lipica in the assessment of the value of peasant holdings in the context of the founding of the Lipica Stud Farm.[24]

[23] Rože (2018a), pp. 9–10; Rože (2018b), pp. 230–232.

[24] Zwitter (2015), pp. 459, 464.

Fig. 11.3 Wellhead of private well of the family *Ruščevi from Lokev*, which is nowadays in Ilirska Bistrica. In the middle of the wellhead, it is written that G. B. (Gregor Božeglav) ordered this cistern to be built in 1752 ("1752/G B/HA FARE QUESTA/CISTEBNA"). *Photo* Rože (2018)

Access to water was a relevant factor in measuring the social and economic power of local inhabitants. Consequently, the authors of the first highly precise early nineteenth century land cadastre, the Franciscan Cadaster,[25] marked the water infrastructure according to required instructions (1819). Conventional graphic signs as symbols were drawn to indicate certain features of water infrastructure (for example see Fig. 11.4), such as for the direction of the water flow of rivers and streams, and also for water wells, which were divided into three groups. For the cistern, a pale blue cross on a white background and a red ring was specified, for the stone shaft well a pale blue circle surrounded by a red ring, and for the wooden shaft well only a pale blue circle surrounded by a black border. The state of the water supply of individual municipalities was also recorded in the cadastral evaluation (1829–1830) belonging to the abovementioned cadaster. The Franciscan Cadaster is, therefore, an indispensable source for the history of water supply and the study of the water infrastructure of the time.

Field research has shown that the marking of the wells' locations is more or less accurate. Conventional signs, however, were used mostly arbitrarily and cannot be relied upon. Microhistorical approach yielded new results, since my field research with questioning informants further revealed that the recorded condition of water infrastructure in both cadastral municipalities is not entirely accurate because, in addition to what has been recorded, there was another water infrastructure that was not foreseen by the instructions. It turned out that the water infrastructure, drawn in the manner of a wooden shaft well in Kreplje, was an open-type *štirna* which is now buried. The water infrastructure in Prelože, drawn in the same way, was most likely a stone drainage canal for water. Field research also revealed that for some unknown reason the abovementioned private wells from the second half of the

[25] Ribnikar (1982), pp. 327–333; Lisec and Ferlan (2017), pp. 80–88.

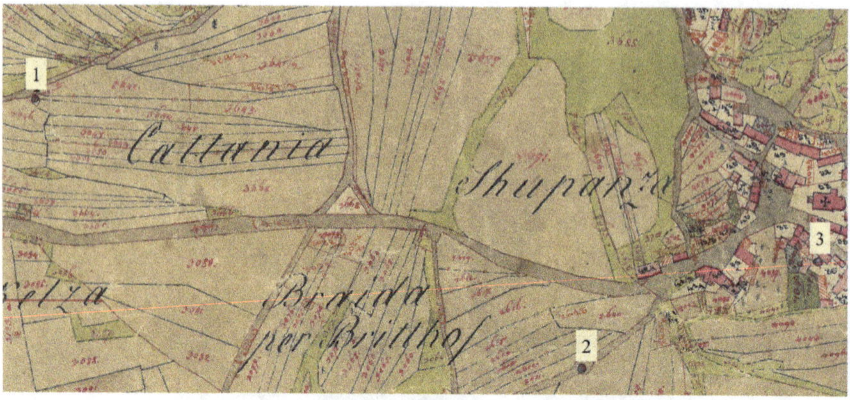

Fig. 11.4 Part of Lokev on the Franciscan Cadaster. Wells are marked with numbers next to blue circles: 1 Municipal well Stanka, 2 Municipal wells near Štoka (or Bekonovi), 3 Private wells. *Source* Koper Regional Geodetic Administration, Geodetic office Sežana; Table 13.3: Condition of the water infrastructure in municipalities Lokev and Dutovlje in the years 1902, 1905, and 1908. Reproduced with permission

eighteenth century were not drawn on the map of the cadastral municipality of Lokev. Table 11.1 shows the overall condition of the water infrastructure in the cadastral municipalities of Lokev and Dutovlje in the first quarter of the nineteenth century. The results were obtained through fieldwork, recording oral tradition, using written historical sources, including cartographic evidence. The two *lokevs* in Dutovlje are not included because they were created after the middle of the nineteenth century.[26] The research shows that the municipality of Lokev was wealthier, evidenced by the number of municipal (2) and private wells (10).

11.2.2 Types and Construction of Water Supply Infrastructure Until the Middle of the Nineteenth Century

The type, location, method of construction, and adaptation of water reservoirs were determined by several interrelated factors, such as the relief, hydrogeological and pedological conditions, the needs of the local population and its economy, the available building materials, knowledge, techniques, and hygiene standards. Until the wider availability of modern Portland cement, patented in 1824 by Joseph Aspdin,[27] only natural local material was used in the construction of water reservoirs in traditional local technique, which was also supported by the ruling classes.[28]

[26] Rože (2020a), pp. 99–116.

[27] Gligić (2011), p. 3.

[28] Zwitter (2015), p. 465.

Table 11.1 The state of the water infrastructure in the cadastral municipalities of Lokev and Dutovlje in the first quarter of the nineteenth century

Municipal wells					Private wells					*Kals*					*Lokevs*					Open-type *štirnas*					Stone drainage canal				
Lokev		Dutovlje			Lokev		Dutovlje			Lokev		Dutovlje			Lokev		Dutovlje			Lokev		Dutovlje			Lokev		Dutovlje		
L	P	D	K	G	L	P	D	K	G	L	P	D	K	G	L	P	D	K	G	L	P	D	K	G	L	P	D	K	G
2	0	0	0	0	10	0	3	0	0	8	2	3	2	0	0	0	0	0	0	0	0	0	1	0	0	1	0	0	0

cadastral municipalities: L = Lokev, P = Prelože; D = Dutovlje, K = Kreplje, G = Godnje

One of the most common and simplest ways to collect water was using an anthropogenically transformed karstic depression (*doline*) with flat edges, which was known as kal on the Kras Plateau (see Fig. 11.5). Kals were primarily intended to provide water for the livestock. The local population first removed a layer of the upper, more permeable layer of soil from the bottom of the doline. This is how they got to the second layer of soil (B or Brz horizon). As much as one-half to three-quarters of this soil layer, which is the accumulated material washed from the overlying soil, consists of clay, a naturally occurring material forms the basis for achieving waterproofing. They then had to add some water to the clay from another already existing water infrastructure. By adding the appropriate amount of water, the clay became plastic and ready for processing. To make it as impermeable as possible, they had to beat it properly.[29] For this purpose, they used about a half-meter-long timber log on which they attached about meter-long timber sticks which were used as holders to hit the ground. They also led the livestock across the doline. Another option was to remove the top of the doline's soil layer and then wait for rainfall. After the rains ended and before the surface dried, they went to the future kal and began to trample the clay. If there was no clay in the area of the future kal, they had to remove the soil by getting to the rocky base. Then gravel and soil were applied. Clay brought from elsewhere was applied to it. Then, in the presence of a suitable amount of water, it was pounded to make it waterproof. In some places, stone shards were placed on the surface of the doline and the spaces in between were filled with clay.[30] Once the kal had already been established, the impermeability was helped by livestock wading through the water when drinking, which resulted in trampling the clay base of the kal (squeezing air and water out of the pores, reducing their size, and filling cracks in the bedrock). The impermeability of the mud was also helped by organic substances in the form of leaves and animal droppings which played the role of a (sticky) binder in the trampled soil.[31]

A reservoir similar to a *kal*, but with steeply built walls, was called *lokev* (e.g., the big *lokev* in Brce near Dutovlje (Fig. 11.6), small Križman's *lokev* in Dutovlje). It was intended to provide drinking water to the inhabitants and was, similarly to a *kal*, made impermeable by using clay. Impermeability was achieved by grouting with a clay layer between the surrounding soil and cut stones of steeply built walls. Lokev's ground was stumped only with timber logs because the livestock could not enter inside due to the steep inclination of the walls. The *štirnas* were the same in terms of construction but they were usually smaller (e.g., the former *štirna* in Kreplje). In this article, I use the term open-type *štirna* for this water hole type because the term *štirna* on the Kras Plateau also refers to a water well and its cistern. A *lokev* and an

[29] Zwitter (2015), p. 465. About B horizon of soil see more in: Grčman et al. (2015).

[30] Belingar (2011a), p. 74.

[31] I want to thank Dr. Žiga Zwitter for the shared piece of information from pedogeographer Dr. Blaž Repe, Geography Department at the Faculty of Arts, University of Ljubljana, on the influence of livestock's feces on *kals* impermeability.

Fig. 11.5 *Stari kal* [Old kal] in Prelože near Lokev. Kal is an anthropogenically transformed karstic depression (doline) with flat edges. *Photo* Rože

Fig. 11.6 Lokev in Brce near Dutovlje. *Photo* Rože (2019)

open-type *štirna* both had stone steps built around the rim to make it easier to capture water.[32]

Folk terminology for water hole types was local to regional specific. Contrary to the statement that *lokev*s had steep walls, inhabitants of the village of Lokev perceive the term *lokev* as an older expression for a *kal* and use it as such to explain the origin of the village name.[33] Similarly, in some other parts of Dinaric Karst of Slovenia (e.g.,

---

[32] Belingar (2012), pp. 286–287; Rože (2020b), p. 122; SI PAK KP 633, box 473, Preskrba z vodo 1871–1914, letter from the municipality of Dutovlje to the district board of Sežana, no. 494, December 20, 1902; Zobec (2008), p. 499.

[33] Rože (2018b), p. 224.

lokev in Bezuljak near Cerknica)[34] and in Croatian Istria, the term *lokev* is also used to describe a *kal* but in Chakavian Croatian "*kal*" means water infrastructure which is defined as *lokev* on the Kras Plateau and in this paper.[35] With the construction of *kals, lokevs,* and open-type *štirnas*, a network of surface freshwater ecosystems was established on the Kras Plateau, which would not otherwise exist.[36]

To supply people with drinking water they built not only *lokevs* and open-type *štirnas* but also wells that were more expensive and required special skilled labor to be constructed. In contrast to the already described water reservoirs, some wells were private. Public wells were usually built on village squares, private ones in courtyards next to homestead houses. Wells were also owned by some institutions, for example, the parishes and after the mid-nineteenth century, especially village schools because the school network expanded significantly at that time.[37] Most of the wells on the Kras Plateau were built in two parts, the wellhead above ground and the underground cistern. First, it was necessary to dig a pit for the cistern whereby the depth of the cistern was conditioned by the terrain due to the hard rocky base. The walls were built of local stone, crushed and dressed, followed by a layer of clay about half a meter thick for waterproofing and then a layer of a homogeneous mass of fine sand and clay.[38]

The cistern was usually filled with rainwater through the opening of the wellhead (only a few wells were also filled with groundwater). For some public wells, canals have been built along field paths or at the foot of hills to supply water to the cistern. From the considered area, such cases are known from Lokev where three of the four canal routes are still noticeable. The oldest public wells in Lokev had an inflow canal (see Fig. 11.8) made along a field path and the younger ones—along the foot of the hills. From the inflow canal of the well from 1848, a stone channel has been preserved, as I have observed in my field trip. It was placed along with others at the foot of Klemenka hill (as far as it is known, the other three public wells did not have them) but the canal route is no longer visible on the slope because now there are houses atop of the slope. There is no known written or material evidence that the public wells of the cadastral municipalities of Dutovlje and Lokev had rainwater led into the cistern through stone gutters, e.g., in Štanjel or Volčji Grad. An opening was usually left on the upper part of the cistern to drain the water out of it in case it overfilled.

Water reservoirs were usually cleaned in groups during droughts when the mud was scraped from the bottom, which—at least in the case of *lokevs*, open-type *štirnas,* and *kals*—affected the established ecosystem in the water reservoir.[39] For the cadastral municipality of Lokev, it is certified for July 1834 that the scraped mud from the

[34] Hren (1942).

[35] Kostelac (2009), p. 271.

[36] Babij (2005), p. 18.

[37] Gabrič (2009), pp. 19–22.

[38] Belingar (2012), p. 288.

[39] Babij et al. (2005), p. 90.

four Lokev's *kals* was sold at auction.[40] As I have recorded in the interviews, mud was first transported from the *kals* as fertilizer to the fields and then to the meadows. At least at the end of the nineteenth and into the twentieth century, the water reservoirs were cleaned in groups by those who used them. Individuals also cleaned them when they were out pasturing their livestock or after bad weather. Cleaning the wells was a special challenge for the villagers because only the bravest men were supposed to descend into the cisterns. Group works of villagers were called *robotas*.[41]

11.3 Limiting Water Usage

Despite progress, there was still not enough water in water reservoirs for everyday needs and it had to be used sparingly. That is why the usage of water was limited both in time and quantity. The codified legal basis for the temporal and quantitative usage of water was provided by the Water Act of 1912, in which Article 55 stated that *"where there is a permanent shortage of water needed for drinking, cooking, washing, and watering or for extinguishing fire or for other public purposes, and where the ability to provide water supply exceeds the power of individual citizens, there it becomes a matter of municipalities or towns according to the provisions of municipal law."*[42] Before the adoption of this law, they most likely entered into agreements under customary law. The control of water reservoirs was usually exercised by one of the villagers. This is shown, for example, by the preserved tradition from Prelože of controlling the use of the public well and the Vroče spring (cadastral municipality of Grozzana/Gročana)[43] and it can be linked to the preserved Vrhovlje tradition of the late nineteenth or the beginning of twentieth century. "*According to the elders, the village open-type štirna was locked. It was unlocked only at certain hours of the day. Its use was supervised by one of the villagers who had the keys. The water from the open-type štirna is only allowed to be used up to a height of one meter, leaving some water as a preventive measure in the case of fire.*"[44] Testimonies about the most important watering hole of the Lipica Stud Farm from 1585 and 1586 show that the water infrastructure was locked as early as the sixteenth century, which was exceptional in a broader sense because the protection of the water infrastructure would only be expected during a drought.[45] The amount of water was most likely

40 SI PAK KP 633, Občinske zadeve [Municipal affairs], box 148, letter on the sale of mud from *kals* in Lokev, July 22, 1834.

41 Author interviews with: Ivanka Stopar (September 3, 2017; January 29, 2023), Milena Stopar (September 2, 2017), Cecilija Sila (September 3, 2017), and Tončka Vojska (December 29, 2017). On the cleaning of *kals*, see also in Zwitter (2015), p. 466.

42 "Vodopravni zakon," 1.

43 Author's interview with Joško Stopar, November 12, 2017; oral traditions collected by students of the Lokev branch school of the Srečko Kosovel Sežana Primary School as part of a school project in 1995.

44 Author's interview with Guštin Jordan, March 5, 2019.

45 Zwitter (2015), p. 465. See also Zwitter (2014), p. 649; Valvasor (2009), p. 325.

determined by the number of family members and the amount of water in the well's cisterns. The schedule for the opening of public water infrastructure also influenced the daily routine of women because women in particular were usually in charge of the water supply. *"My mother* [was] *going to Trieste where she was carrying milk, and if she didn't return home by eleven, we were left without water that day. That was after 1900."*[46]

Similarities regarding the control of the use of public water infrastructure can also be found elsewhere where there is karst terrain and surface water is not abundant. In Sodji Vrh in White Carniola (Sln. Bela Krajina, southeast of Slovenia), the key to the walled spring was held by the village elder who unlocked the door twice a day. Based on the number of family members and heads of livestock, he determined how much water an individual family could draw from a spring.[47] It was similar to the San Felice spring in the karst of Volterra, central Apennine peninsula. In 1834, the commune hired the merchant Pietro Lazzeri for this purpose. The spring was opened every day between six and eight o'clock in the morning and between eleven and midnight. Only then could the people of Volterra draw water from the spring for free, any other time they had to pay for it.[48] In the first half of the twentieth century, in the town of Lovran in Istria (northwest of Croatia), water was distributed by two locals. From a higher-situated reservoir in Tuliševica, they brought water in a *brenta* (container with straps) which they carried on their backs and delivered around the town. For a few liters of water, they received some change.[49]

11.3.1 Water Provisioning in Times of Drought and Winter Cold

The amount of distributed water had to be measured even more carefully during drought events. At that time, the distance to accessible drinking water also increased in many cases. When there was a shortage of water in the municipal wells at villages Lokev in Prelože, the access to the Vroče spring[50] above the village of Prelože was opened under the supervision.[51] The Vroče spring is an anthropogenically modified water cave on the contact karst where alveolinid-nummulitic limestone turns into flysch and to which leads a few meters long arched accessible tunnel that ends with

[46] Krpan (1976), p. 15. I want to thank Srečko Rože from the Military Museum Tabor Lokev for calling my attention to the cited text.

[47] Dichlberger (2020a).

[48] Furiesi (1999), p. 159.

[49] Kostelac (2009), p. 272.

[50] SI PAK KP 633, box 473, Spisi okrajnega glavarstva Sežana, 1871–1914 [Documents of district board of Sežana, 1871–1914], Občina Lokev: Izkaz o načinu kako se preskrbuje pitna voda in o napajališčih za živino [Municipality of Lokev: Statement on the way drinking water is supplied and on watering holes for livestock], April 25, 1905.

[51] Jurkovšek et al. (1996).

a few steps making the access to water easier. It is not known exactly when it was built. There is an oral tradition that an Austro-Hungarian engineer came with his team and started digging the tunnel, and the locals helped.[52] In the area of Žekanc near Dutovlje is also water cave, which lies on the bedded granular dolomite of the Brje Formation.[53] There was another spring near Dutovlje which is mentioned in the appendix to the census on municipal water infrastructure from 1905 but its location remains unknown.[54] Water caves arranged in this way can also be found at several other places in the Dinaric Karst, in White Carniola, for instance, the Pečina spring in Hrast near Vinica and Škalva in Zapudje near Dragatuš from 1885.[55] Valvasor also mentions the importance of water caves with an access tunnel, as was the case at the Bločice.[56] In this case, there were also *lokevs* and open-type *štirnas* in the area. Water from them was also used for drinking during droughts.[57]

Before most of the *kals* dried up and especially after the water level in the reservoir in Vroče decreased and water quality deteriorated, the responsibility to provide water for families and livestock daily fell on men because they had to leave their home settlements and bring back larger quantities of water by cart. According to the evaluation of the Franciscan Cadaster, the inhabitants of the villages went to get water all the way to Dane pri Divači and to the Reka River, a two-hour drive by a small farm wagon away.[58] In the studied written sources from the late nineteenth and early twentieth century, Dane pri Divači is no longer mentioned but the name of the nearby village of Rodik appears in the appendix to the 1905 census.[59] During the severe drought in the summer of 1890, people from Lokev and Prelože went to get water for their livestock to the Reka River (two hours away)[60] and for themselves from time to time to the *Furmanov studenec* spring on the south side of the Veliko Gradišče slope (one and a half hour away).[61] In September 1890, water

[52] Author's interview with Matjaž Stopar, November 1, 2017.

[53] Jurkovšek et al. (1996).

[54] SI PAK KP 633, box 473, Spisi okrajnega glavarstva Sežana, 1871–1914 [Documents of district board of Sežana, 1871–1914], letter of the municipality of Dutovlje to the district board of Sežana, April 20, 1905.

[55] Dichlberger (2020b).

[56] Zwitter (2014), p. 650.

[57] SI PAK KP 633, box 473, Spisi okrajnega glavarstva Sežana, 1871–1914 [Documents of district board of Sežana, 1871–1914], letter of the municipality of Dutovlje to the district board of Sežana, no. 494, December 20, 1902; SI PAK KP 633, box 473, letter of the municipality of Komen to the district board of Sežana, no. 1485, December 1, 1902.

[58] Rože (2020a), p. 102.

[59] SI PAK KP 633, box 473, Spisi okrajnega glavarstva Sežana, 1871–1914 [Documents of district board of Sežana, 1871–1914], Občina Lokev: Izkaz o načinu kako se preskrbuje pitna voda in o napajališčih za živino [Municipality of Lokev: Statement on the way drinking water is supplied and on watering holes for livestock], April 25, 1905.

[60] Rože (2020a), p. 102.

[61] The estimate of time is given based on today's walking length.

was brought from Nasirec[62] (two hours away).[63] The most common water source in the event of a hydrological drought throughout the nineteenth century remained the Reka River. In the first half of the nineteenth century, men from the cadastral municipality of Dutovlje went to Doberdo (Sln. Doberdob) Lake and the Vipava River to get larger amounts of water which were both five hours away.[64] In the second half of the century, according to the locals, their ancestors went to get water from the rivers Branica (three hours away) and Raša (two hours away), which are not mentioned in the censuses.[65]

Cold and harsh winters were exceptionally hard for the people on the Kras Plateau. Men had to break the ice on top of the *kals* to be able to provide water for the livestock. Those who had private wells drew water for livestock from them. They had to be careful, however, to keep enough water for their own needs. During the winter droughts, women from Lokev and Prelože visited the water cave in Vroče where the path was made even more difficult by the cold and an icy cart track. Most likely, they also melted snow and ice (they probably boiled the water for drinking and for preparing the food). For some residents of both municipalities, selling ice from public water bodies was a profitable business because they cut it specifically for this purpose and stored it in ice stores (i.e., underground ice storages built like wells' cisterns with gable roofs or in the case of a round plan with pyramid roofs) and transported it by wagons to inns, breweries, and fish markets.[66]

11.4 Public Health as the Responsibility of the State Government and Its Citizens

The condition of water infrastructure started to improve significantly after the middle of the nineteenth century when the state—the Habsburg monarchy—strengthened its economic, fiscal, and population policies. In the field of economy and politics, the administration made questionnaires for gathering statistical data for providing facts about birth rates, life expectancy, public health, housing, migration, and so on.[67] Natural sciences, such as biology, medicine, and construction, were developing, and their achievements affected the quality of people's lives.[68]

---

62 ŽAL, Šček, *Lokavske starine I*, pp. 35–36.

63 The estimate of time is given based on today's walking length.

64 Rože (2020a), p. 102.

65 Rože (2020b), p. 127. The estimate of time is given based on today's walking length.

66 Author's interviews with Ladislav Škabar, January 7, 2019, and Milena Stopar, Juny 1, 2019; Zwitter (2015), pp. 464–465; Belingar (2011a), pp. 76–78.

67 Foucault (1978), p. 25.

68 Remec (2015), pp. 24–26.

11.4.1 Data Acquisition on Water Infrastructure from the Middle of the Nineteenth Century to 1908

To prepare a strategy for improving the water supply on the Kras Plateau, the state had to collect data on water infrastructure and analyze, assess, and monitor it. We presume that the data collection was initially carried out by commissions that walked around the Kras Plateau with the aim of *"judging justly where lies the greatest need to build a municipal well"* because *"there is a shortage of water almost every year, and the water we do have is mostly in the water holes."*[69] In this way, for example, in 1872, a public well was built in the village center of Dutovlje with the support of the provincial government.[70]

At the beginning of the twentieth century, the recording of the condition of water infrastructure became more systematic because the state established records of water infrastructure through censuses. By order of the government in Trieste, the highest state administrative body in the Austrian Littoral, the Sežana district board, sent questionnaires to its municipalities, and these are valuable historical documents for my research. The questionnaires had to be completed within a specified period, and then were sent by the Sežana district board to the Trieste government. The data were to be examined at the land reclamation office. Records of water infrastructure were used both to monitor the situation and to make decisions to support the construction of new water infrastructure or the adaptation of the existing one.[71]

On October 13, 1902, the district board sent an order to the associated municipalities for the first time to record detailed data on the state of water resources in all associated places. The municipalities' officials answered a questionnaire. The first point requested information on the water supply from water holes (*kals*). They had to reply how many *kals* there were in the municipality and whether those were *"walled up or made by an excavation of land."* The second point referred to the condition of the municipal wells. The state was interested in how many municipal wells there were in the municipality, how much water they held, and whether the amount of water was enough for the residents of the municipality in normal conditions. Further on, they had to reply if the wells had *strainers* (filters) and if the latter were regularly

69 "S Krasa, 13. avg.," 3.

70 Rože (2020b), pp. 118–121.

71 Zobec (2008), p. 498.

inspected. They also had to state whether the location of the municipal wells was in accordance with the health demands due to the chance of dirt leaking into the cistern. The third point questioned the existence of municipal springs. They were required to provide information on their number, depth, amount of water, whether they met needs under normal conditions, and whether they had an appropriate location concerning health demands. The penultimate point referred to springs and if they were drying up. At the same time, sub-questions were raised again about meeting the needs in normal conditions and a health-appropriate location. The last point asked about the existence of a water distribution system. It was followed by sub-questions about the water pumping station (spring or groundwater), the material used, the existence and maintenance of the filter, meeting the needs in normal conditions, and the health-appropriate location of the water pumping station (for an example of the census of the water infrastructure of the municipality of Lokev from the year 1902; see Figs. 11.7, 11.9, and 11.10).[72]

Fig. 11.7 Municipal well in Brje pri Komnu (near Sežana). In the wellhead year of making, the crafstman, customers, Jesus Christ and the Samaritan woman are depicted. The wellhead originally belonged to a private well.[73] *Photo* Rože

[72] SI PAK KP 633, box 473, Preskrba z vodo 1871–1914 [Water supply 1871–1914], letter from the district board of Sežana to the municipalities of the same political district, no. 11810, October 13, 1902.

[73] Čeh (2023).

Fig. 11.8 Public well near *Štoka* (also near *Bekonovi, i.e., house names)* in Lokev with an inflow canal. *Photo* Rože (2018)

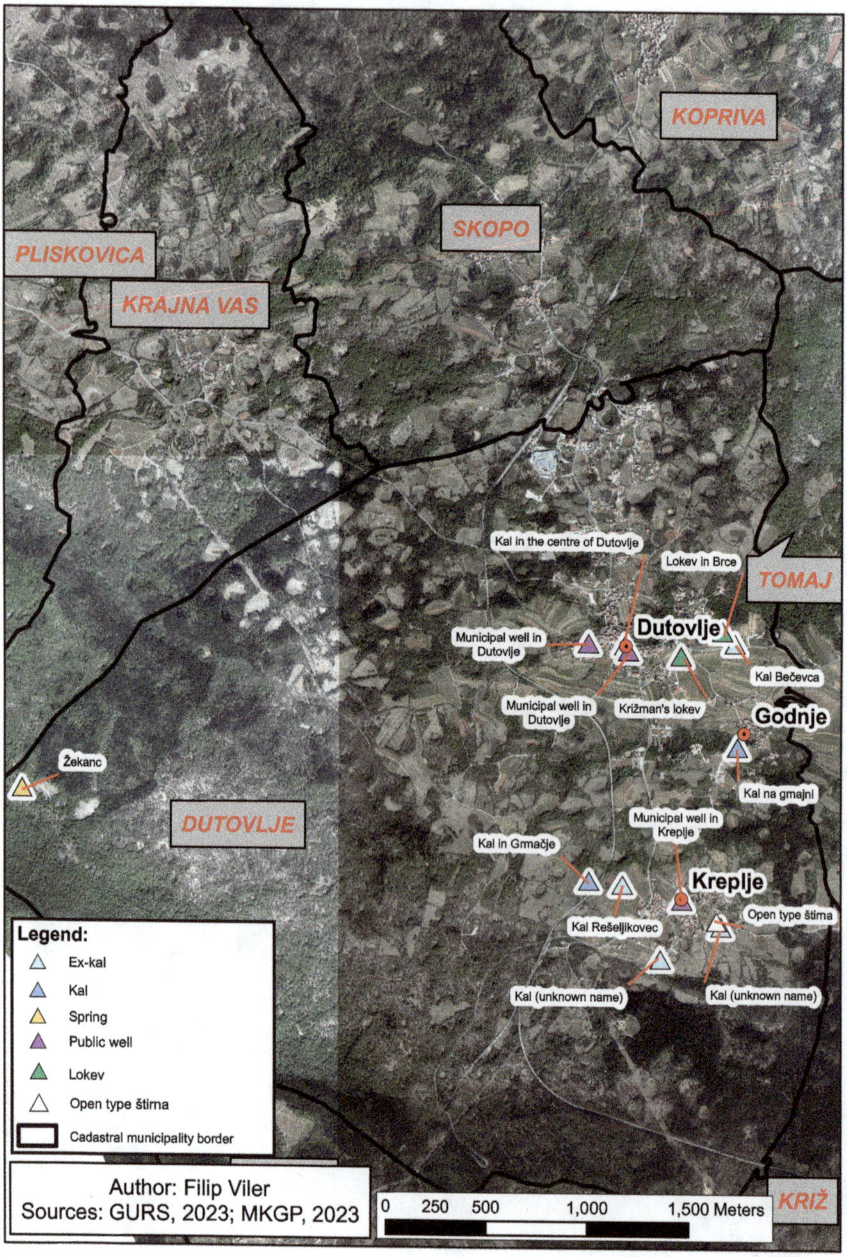
KOPRIVA
SKOPO
PLISKOVICA
KRAJNA VAS
TOMAJ
DUTOVLJE
KRIŽ
Kal in the centre of Dutovlje
Lokev in Brce
Municipal well in Dutovlje
Dutovlje
Kal Bečevca
Municipal well in Dutovlje
Križman's lokev
Godnje
Žekanc
Kal na gmajni
Municipal well in Kreplje
Kal in Grmačje
Kreplje
Open type štirna
Kal Rešeljikovec
Kal (unknown name)
Kal (unknown name)
Legend:
Ex-kal
Kal
Spring
Public well
Lokev
Open type štirna
Cadastral municipality border
Author: Filip Viler
Sources: GURS, 2023; MKGP, 2023
0 250 500 1,000 1,500 Meters

◂**Fig. 11.9** Overview map of the public water infrastructure in the cadastral municipality of Dutovlje in the long nineteenth century. In the case of kals, those which have not been preserved to this day are specially marked. The kal in the area with the fallow name Rupe (northeast of the village of Dutovlje) is not marked on the map, because knowledge about it has not been preserved to this day.[74] Author of the map Filip Viler. *Topographic base sources,* Surveying and Mapping Authority of the Republic of Slovenia (GURS) (2023)

In 1905 and 1908, the Sežana district board sent new decrees which turned out to be clearer and more transparent than the previous document. They also avoided ambiguities and additional work occurring due to the carelessness of some mayors' responses.[75] To get a more complete insight into the state of water infrastructure on the Kras Plateau, the number of private wells and the number of citizens in individual villages as well as the entire municipality had to be recorded in 1905 and 1908.[76]

[74] Data sources for names and locations of public water infrastructure: AST, Catasto Franceschino (CFr), Mappe del Catasto Franceschino (M), Distretto di Sesana (DS), Comune di Duttogliano, 175 b. Available also at: https://www.archiviodistatotrieste.it/AriannaWeb/main.htm?eads=*#35305_archivio (September 30, 2019). Since the Archive of the Republic of Slovenia (Sln. *Arhiv Republike Slovenije*) does not store all the cadastral maps (and elaborations) of the Franciscan Cadaster for the selected tax district, the research of locations was created after an online publication by the Trieste state archive (It. *Archivo di Stato di Trieste*). Available also at: SI AS 179. http://arsq.gov.si/Query/detail.aspx?ID=23256 (October 2, 2019).

[75] Example from Naklo: SI PAK KP 633, box 473, Preskrba z vodo 1871–1914 [Water supply 1871–1914], letter form municipality of Naklo to the district board of Sežana, no. 1105, November 24, 1902; SI PAK KP 633, box 473, Preskrba z vodo 1871–1914 [Water supply 1871–1914], letter from district board of Sežana to the municipality of Naklo, no. 11810, November 26, 1902.

[76] SI PAK KP 633, box 473, Preskrba z vodo 1871–1914 [Water supply 1871–1914], Izkaz o načinu kako se preskrbuje pitna voda in o napajališčih za živino [A report on how drinking water is supplied and on watering holes for livestock], letters from municipalities to the district board of Sežana in 1905; SI PAK KP 633, box 473, Preskrba z vodo 1871–1914 [Water supply 1871–1914], Izkaz pitnih vod in napajališč [List of drinking water and watering holes], letters from municipalities to the district board of Sežana in 1908.

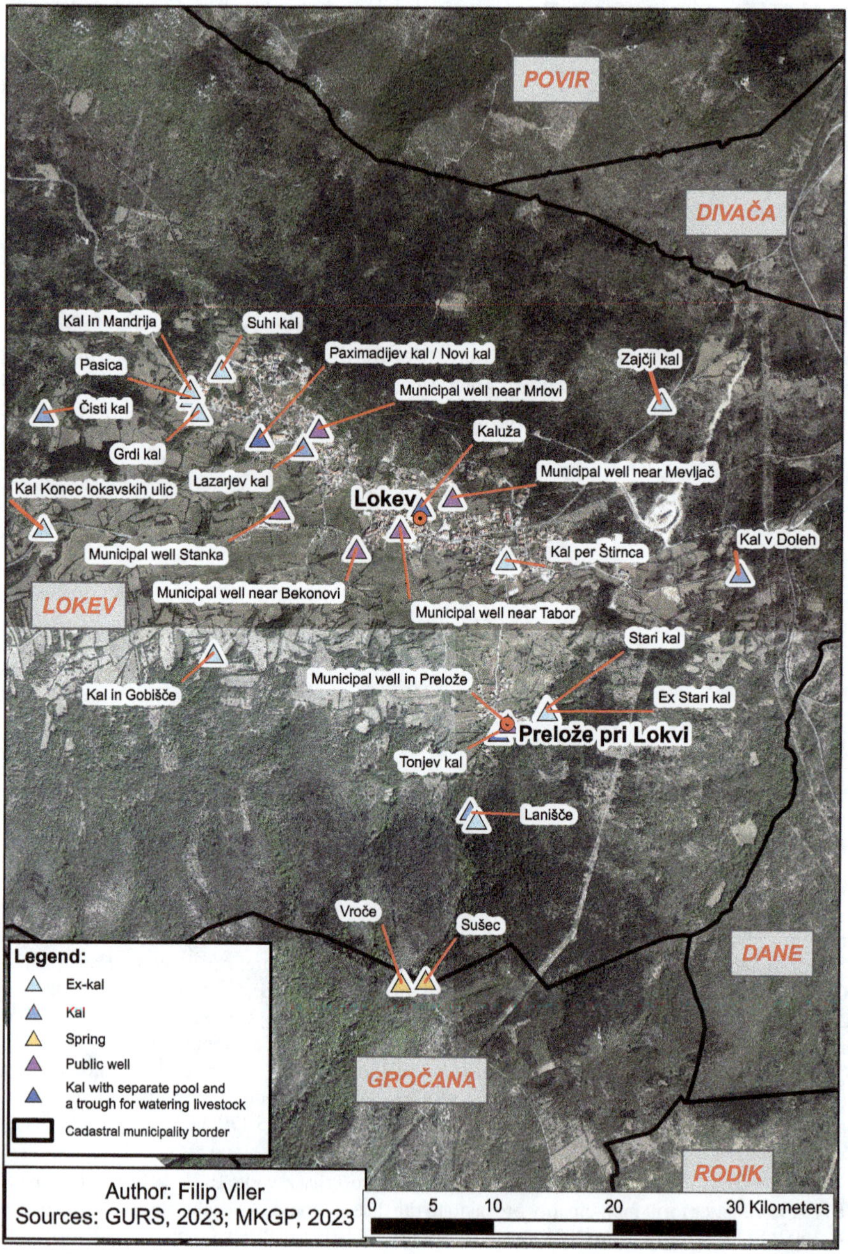
POVIR
DIVAČA
Kal in Mandrija
Suhi kal
Pasica
Čisti kal
Grdi kal
Paximadijev kal / Novi kal
Municipal well near Mrlovi
Zajčji kal
Kaluža
Municipal well near Mevljač
Lazarjev kal
Kal Konec lokavskih ulic
Lokev
Municipal well Stanka
Kal per Štirnca
Kal v Doleh
LOKEV
Municipal well near Bekonovi
Municipal well near Tabor
Stari kal
Kal in Gobišče
Municipal well in Prelože
Ex Stari kal
Prelože pri Lokvi
Tonjev kal
Lanišče
Vroče
Sušec
DANE
Legend:
Ex-kal
Kal
Spring
Public well
Kal with separate pool and a trough for watering livestock
Cadastral municipality border
GROČANA
RODIK
Author: Filip Viler
Sources: GURS, 2023; MKGP, 2023
0 5 10 20 30 Kilometers

◂**Fig. 11.10** Overview map of the public water infrastructure in the cadastral municipality of Lokev in the long nineteenth century. In the case of kals, those which have not been preserved to this day are specially marked. Due to the loss of knowledge in the collective memory of the local population, the location of *Suhi kal* is only approximate. Its basin is no longer perceptible in nature.[77] Author of the map Filip Viler. *Topographic base sources,* Surveying and Mapping Authority of the Republic of Slovenia (GURS) (2023); Ministry of Agriculture, Forestry and Food (MKGP) (2023)

Public water infrastructure for the needs of military exercises was recorded similarly because it put a strain on the water infrastructure and sufficient water availability in the village where the army made a camp.[78]

11.4.2 The Analysis of Water Infrastructure from Early Twentieth Century Census Record

According to the data from the municipal censuses from 1902[79] and 1905,[80] the condition of the water infrastructure in both municipalities improved from the first third of the nineteenth century (compare Tables 11.1 and 11.2). Then the condition of the water infrastructure in the municipality of Dutovlje improved according to the data from the 1908 census. However, the condition of the water infrastructure in the municipality of Lokev deteriorated slightly between 1905 and 1908 (see Table 11.2).[81]

---

[77] Data sources for names and locations of public water infrastructure: AST, CFr, M, DS, Comune di Corgnale, 132 b. Available also at: at: www.archiviodistatotrieste.it/AriannaWeb/main.htm?eads=*#34970_archivio (September 30, 2019). I would like also to thank Joško Stopar (August 28, 2017), Marija Bernetič (December 1, 2017), Frančiška (Francka) Mljač (March 4, 2018), Ivanka Stopar (October 14, 2017), Tončka Vojska (December 29, 2017), and Joško Frankovič (August 31, 2017) for the names of the water infrastructure and their locations on the field.

[78] SI PAK KP 633, box 473, Preskrba z vodo 1871–1914 [Water supply 1871–1914], letter from district board of Sežana to the municipalities, no. 11067, September 10, 1908; "Pomanjkanje vode na Krasu," p. 2; "Posledice pomanjkljivemu preskrbljevanju", p. 1; "Vojaške vaje na Krasu," p. 2.

[79] SI PAK KP 633, box 473, Spisi okrajnega glavarstva Sežana, 1871–1914 [Documents of district board of Sežana, 1871–1914], letter from municipality of Lokev to the district board of Sežana, no. 649, December 19, 1902; SI PAK KP 633, box 473, Spisi okrajnega glavarstva Sežana 1871–1914 [Documents of district board of Sežana, 1871–1914], Občina Lokev: Izkaz o načinu kako se preskrbuje pitna voda in o napajališčih za živino [Municipality of Lokev: Statement on the way drinking water is supplied and on watering holes for livestock], April 25, 1905; SI PAK KP 633, box 473, Spisi okrajnega glavarstva Sežana 1871–1914 [Documents of district board of Sežana, 1871–1914], letter from municipality of Dutovlje to the district board of Sežana, no. 494, December 20, 1902.

[80] SI PAK KP 633, box 473, Spisi okrajnega glavarstva Sežana, 1871–1914 [Documents of district board of Sežana, 1871–1914], Občina Dutovlje: Izkaz o načinu kako se preskrbuje pitna voda in o napajališčih za živino [Municipality of Lokev: Statement on the way drinking water is supplied and on watering holes for livestock], April 20, 1905.

[81] SI PAK KP 633, box 473, Spisi okrajnega glavarstva Sežana, 1871–1914 [Documents of district board of Sežana, 1871–1914], report of drinking water and watering holes in 1908 in the municipality of Dutovlje; SI PAK KP 633, box 473, Spisi okrajnega glavarstva Sežana, 1871–1914 [Documents

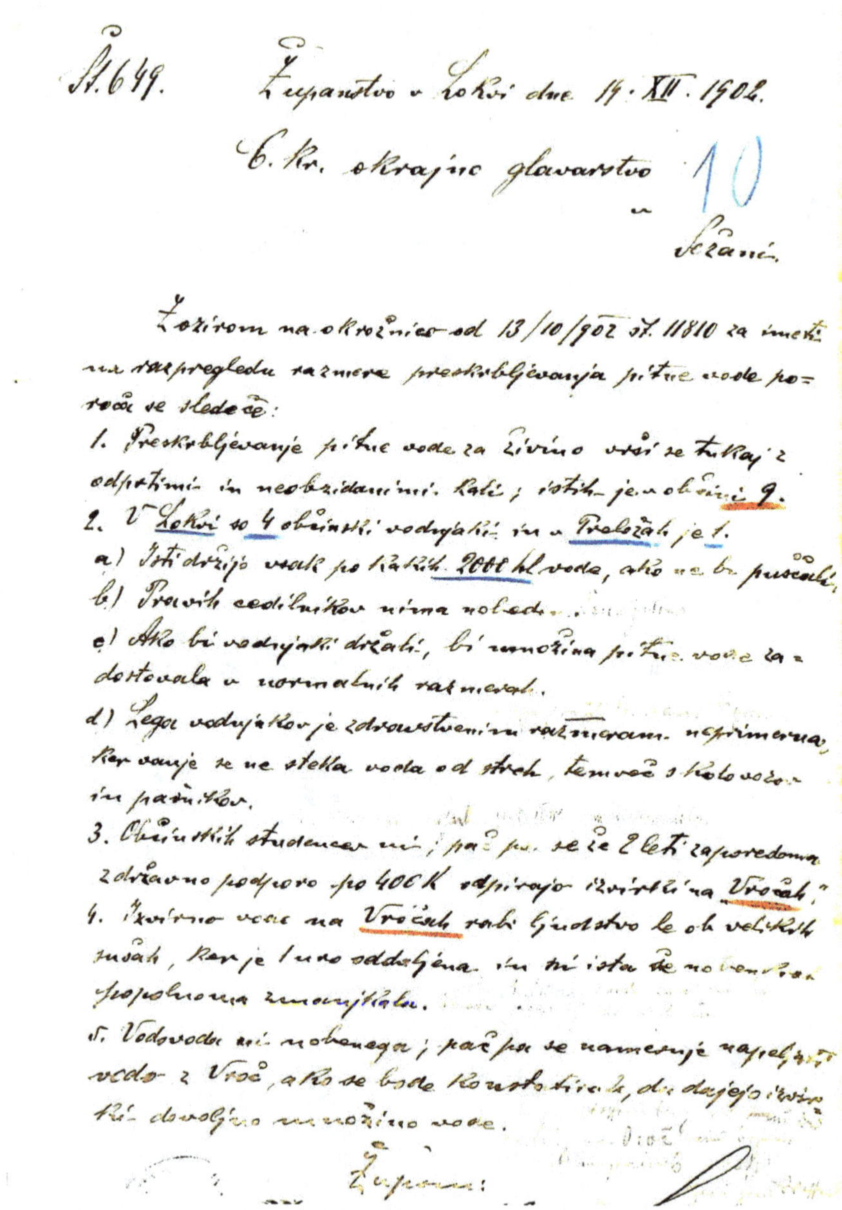

Št. 649. Županstvo v Lokvi dne 19. XII. 1902.

C. kr. okrajno glavarstvo v Sežani.

Z ozirom na okrožnico od 13/10/902 št. 11810 za imeti na razpregledu razmere preskrbljevanja pitne vode poroča se sledeče:

1. Preskrbljevanje pitne vode za živino vrši se tukaj z odprtimi in neobzidanimi kali; istih je v občini 9.
2. V Lokvi so 4 občinski vodnjaki in v Preložah je 1.
a) Isti držijo vsak po kakih 2000 hl vode, ako ne bi puščali,
b) Pravih cedilnikov nima nobeden.
c) Ako bi vodnjaki držali, bi množina pitne vode zadostovala v normalnih razmerah.
d) Lega vodnjakov je zdravstvenim razmeram neprimerna, ker vanje se ne stekla voda od streh, temveč s kolovozov in pašnikov.
3. Občinskih studencev ni; pač pa se že 2 leti zaporedoma z državno podporo po 400 K odpirajo izvirki na Vročah.
4. Izvirno vodo na Vročah rabi ljudstvo le ob velikih sušah, ker je luno oddaljena in bi ista se ne vendar popolnoma zmanjkala.
5. Vodovoda ni nobenega; pač pa se namerava napeljati vodo z Vroč, ako se bode konstruiralo, da dajejo izvirki dovoljno množino vode.

Župan:

Fig. 11.11 Municipal censuses from 1902 of municipality of Lokev. *Source* SI PAK KP 633, box 473, Spisi okrajnega glavarstva Sežana, 1871–1914 [Documents of district board of Sežana, 1871–1914], letter from municipality of Lokev to the district board of Sežana, no. 649, December 19, 1902. Reproduced with permission

Table 11.2 Condition of the water infrastructure in municipalities Lokev and Dutovlje in the years 1902, 1905, and 1908

	Municipal wells					Private wells					*Kals*				
	Lokev		Dutovlje			Lokev		Dutovlje			Lokev		Dutovlje		
	L	P	D	K	G	L	P	D	K	G	L	P	D	K	G
1902	4	1	1	1	0	The census from 1902 does not include private wells					6	3	3	2	1
1905	4	1	1	1	0	74	1	21	7	6	6	3	3	2	1
1908	3	1	1	1	0	53	10	34	9	7	5	1	4	1	1

Lokevs					Open-type *štirnas*					Stone drainage canal				
Lokev		Dutovlje			Lokev		Dutovlje			Lokev		Dutovlje		
L	P	D	K	G	L	P	D	K	G	L	P	D	K	G
0	0	2	0	0	0	0	0	1	0	0	1	0	0	0
0	0	2	0	0	0	0	0	1	0	0	1	0	0	0
0	0	2	0	0	0	0	0	1	0	0	1	0	0	0

* cadastral municipalities: L = Lokev, P = Preložje; D = Dutovlje, K = Kreplje, G = Godnje

The cause of the deterioration can presumably be attributed to the damage to the infrastructure as a result of which it no longer held water or because it was ordered to be filled up due to the contamination of water or because there was no longer a need for it. In 1892, for example, the Sežana district board ordered the municipality of Lokev to fill the Kaluža *kal* in the immediate vicinity of the village center for sanitary reasons, and the following year the municipality asked the district board to rebuild the *kal*. From their request, we also find out that the recently renovated municipal well had started leaking due to extreme cold. Clay at two *kals* started to crack.[82] In contrast, in 1890, the municipality of Dutovlje sent a request to the district board to have the *kal* by the main road filled up.[83] One of the reasons for the reduced number of private wells is also, as already proven in the case of the Franciscan Cadaster, the inconsistency when recording data.

Censuses, especially those in the form of tables, put some municipalities in difficult positions because the tables did not stipulate as *lokevs* and open-type *štirnas*. In 1902, the municipality of Dutovlje solved the problem by recording both *lokevs* in Dutovlje and open-type *štirna* in the answer to the question asking about *kals*. In 1905, their volumes were most likely added to the municipal wells in Dutovlje and Kreplje because in the 1908 census volume of both municipal wells was recorded as in the 1902 census.[84]

A comparison between the censuses from 1902, 1905, and 1908 of the municipality of Dutovlje and field research revealed another discrepancy. The village of Dutovlje has another municipal well in Gornji konec with the wellhead dated 1848. The only municipal well recorded in the mentioned census is undoubtedly the well in the center of the village because it was built and adapted with the financial support of the provincial government. The municipal well with the wellhead dated 1848 was supposedly built after 1908, and its older wellhead was brought from elsewhere.[85] If they had wanted to purposefully omit the existence of the well and thus showcase a poorer condition of the water infrastructure then *lokevs* recorded at Dutovlje in the early nineteenth century (Table 11.1) would probably not have been added to the

of district board of Sežana, 1871–1914], report of drinking water and watering holes in 1908 in the municipality of Lokev.

[82] Rože (2020d).

[83] SI PAK KP 633, box 249, Spisi okrajnega glavarstva Sežana, 1893 [Documents of district board of Sežana, 1893], letter from municipality of Dutovlje to the district board of Sežana, no. 418, October 20, 1890.

[84] SI PAK KP 633, box 473, Spisi okrajnega glavarstva Sežana, 1871–1914 [Documents of district board of Sežana, 1871–1914], letter from municipality of Dutovlje to the district board of Sežana, no. 494, December 20, 1902; SI PAK KP 633, box 473, Spisi okrajnega glavarstva Sežana, 1871–1914 [Documents of district board of Sežana, 1871–1914], Občina Dutovlje: Izkaz o načinu kako se preskrbuje pitna voda in o napajališčih za živino [Municipality of Dutovlje: Statement on the way drinking water is supplied and on watering holes for livestock], April 20, 1905; SI PAK KP 633, box 473, Spisi okrajnega glavarstva Sežana, 1871–1914 [Documents of district board of Sežana, 1871–1914], report of drinking water and watering holes in 1908 in the municipality of Dutovlje.

[85] Rože (2018a), pp. 9–10; Belingar (2012), p. 291.

municipal well in 1905 (and the volume of the open-type *štirna* to the Kreplje well). Instead, they would have recorded the same as in 1902.

11.4.3 Measuring Water Quality

An important piece of information during the inspections was the sanitary quality of the water. Before the usage of advanced microscopes in the nineteenth century, people assessed water by its external criteria, such as color, clarity, odor, and taste. In the second half of the eighteenth century, the quality of water in Princely County of Gorizia and Gradisca was studied also by A. Muznik (1726–1803), a provincial physician and protomedic. "*It is generally believed that plain water is better the purer it is. The purity of water is known especially by its lightness. It depends, firstly, on a smaller number of heterogeneous particles and secondly, on a larger, water-mixed amount of air and ether.*"[86] To find out in greater detail which heterogeneous particles the water from springs and wells (probably from the vicinity of Gorizia) contained, Muznik added different substances in seven experiments: gall powder, gall concoction, some "purple syrup," tartaric oil, nitric acid and finally lead acid, and then concluded: "*Thus it may be rightfully concluded that our water is pure, that it consists of air or etheric vapors, alkaline salts, and of little soil substance, and that it is therefore very healthy.*"[87] Nevertheless, Muznik's findings cannot be taken as representative of the water quality in many cases of water infrastructure in the cadastral municipalities of Lokev and Dutovlje, since in these areas the availability of water in stagnant water bodies gradually decreased with high summer temperatures, while its quality declined, so that the inhabitants obtained water from distant watercourses (see Sect. 11.4.1), which was cleaner than the previous one.

Between 1870 and 1880, the notion of what is good, pristine water changed in natural sciences. Although the idea of microorganisms as pathogens appeared before the nineteenth century, it was not until Pasteur that the postulate *generatio spontanea* (the possibility of the spontaneous generation of visible animals and plants from nonliving matter) was disproved. Instead, Pasteur discovered that a multitude of tiny, unicellular organisms can be found in the human environment (air, water, etc.).[88] The discovery of the world of bacteria changed the notion of washing. Water began to clean also invisible dirt that could not even be smelled.[89] It started to be considered important for the water not to contain microorganisms. "*Good water that will be used for drinking and other needs. It should be colorless and pure. All substances that can be noticed in sediments under the microscope, and especially those that are organic in nature, are all the more suspicious when more floating and sedimentary substances belong to flora and fauna. It must be assumed that water which contains,*

[86] Muznik (2000), p. 58.

[87] *Ibid.*, p. 59.

[88] Borisov (2009), pp. 292–296.

[89] Vigarello (1999), pp. 252–253.

in addition to Crenothrix and other fungal filaments, as well as infusoria, many diatoms, has had an inflow from mostly plant sources [most probably meaning that the water passed mainly through densely vegetated areas, a. n.]. *That type of water must definitely be thought of as unclean. If it turns out that there are different fungal filaments, zoogleals, bacteria, infusoria, and radiolarians of various kinds or even waste from the kitchen, household, residues of animal feces and other types of mud,* etc. *in the water besides chronic signs of decay, then the water must be discarded immediately*," explained the headmaster E. Kramer, from the laboratory of the agro-chemical testing facility for Carniola in Ljubljana at the beginning of the twentieth century.[90]

With the discovery of microorganisms, awareness of the importance of location and construction of the water infrastructure rose (see Sect. 3.1). The method of water collecting in well cisterns[91] and the method of retrieving water[92] from them also became subjects of supervision with water pumps being considered the most hygienic. However, they were likely to freeze in winter. Therefore, they had to be insulated—covered with straw or something similar.[93] Modern standards were mostly met by the wells which were built according to the state's plans and their construction was supervised by the technical body of the land melioration office.[94]

11.4.4 Safe Drinking Water for Livestock Farming

In the second half of the nineteenth century, the awareness rose that adequate water quality also mattered in livestock farming. Due to the influence of the rapidly developing Trieste, sheep farming, once the main livestock farming branch, began to be replaced by cattle farming which better met the demand for milk and dairy products.[95] In addition to archival material, I have used newspapers as historical sources, because they repeatedly published advice on cattle farming and the sale of milk. Important conditions for healthy cattle, and thus cow's milk, were good fodder and clean water. If the entry of livestock into a *kal* was still desirable after the middle of the nineteenth century because they made the bottom of the hollow impermeable by trampling the clay, it was already considered inappropriate toward the end of the nineteenth century. "*Many people ask: Why are the cattle in Kras, especially in the lower reaches, so weak and dull?* […]," wrote Andrej Zavadlav in *Soča* newspaper

---

[90] Kramer (1905), p. 107.

[91] SI PAK KP 633, box 473, Dodatki k izkazu t. u. štev. 3973 [Addendum to statement t. u. number. 3973], 1905.

[92] SI PAK KP 633, box 473, the order district board of the Sežana on the reporting of the method of collecting water in municipal wells, no. 1628, Februar 5, 1903.

[93] SI PAK KP 633, box 472, Preskrba z vodo 1871–1914 [Water supply 1871–1914], ordinance on the construction of wells, no. 10305, August 8, 1904.

[94] Rože (2020c), pp. 40–41.

[95] Panjek (2015), p. 70. See also Vilfan (1957), pp. 78, 86.

Fig. 11.12 *Novi kal* [New kal] in Lokev. *Photo* Rože. Reproduced with permission

in 1899. He considered the water quality highly relevant in this respect. "*The poor cattle never drank immediately. They entered the puddle so that you could see how thirsty they were but wandered around the puddle looking for clean, cold water. In vain. They had to drink what they got and that was that muddy, dirty puddle. And this poor-quality drinking water is therefore the main reason why livestock, especially in summer, are so feeble in Kras.*"[96] The challenge of providing water to the livestock from open, anthropogenically transformed karstic depressions was presented even more drastically (and possibly in a partly exaggerated manner) by the municipality of Brestovica na Krasu, which most likely wanted to obtain co-financing for the construction of a more advanced water infrastructure for the livestock. "*The kals are used as watering places into which water from the slopes flows, as well as from yards and manure pits. Livestock enters the water and leaves dirt and urine there. In summer, the water is half-liquid manure, smelly, and full of animals. Livestock suffers from thirst because they cannot drink enough stinky water and because it is harmful to them. Bad water can lead to serious illnesses and sick animals can further poison water with their excrements and spread the disease.*"[97] Thus, it is not surprising that data collection on water infrastructure at the beginning of the twentieth century also checked the technical details of *kals* and whether livestock could enter the water. It was increasingly advised to build fenced *kals*, especially those consisting of a pool to which the livestock did not have access and a trough that was intended exclusively to provide water flowing from the pool to the livestock without livestock wading in their drinking water (see Fig. 11.12).

The transition to cattle farming is now analyzed from the environmental history perspective since it led to a significant increase in demand for water in this area. Namely, a dairy cow requires 50–100 l while a sheep drinks significantly less, i.e.,

[96] Zavadlav (1899), p. 3.

[97] SI PAK KP 633, box 473, Preskrba z vodo 1871–1914 [Water supply 1871–1914]. Letter from municipality of Brestavica na Krasu to the district board of the Sežana, additions to the statement no. 3973, 1905.

10–20 l water per day.[98] As a result, more and more *kals* began to be built. Among them, however, there were only a handful of advanced ones, i.e., those with a separate pool and a trough for watering livestock because their construction required more material, workers, and finances. According to oral sources, during the winters until the 1970's, livestock was also supplied with water from the wells.[99]

11.4.5 Construction of Water Infrastructure After the Middle of the Nineteenth Century

Initiatives to build or adapt public water infrastructure, partly funded by the state and/or the province, usually came from municipalities where mayors or other prominent villagers recognized such a need. The state encouraged individuals to show initiative at the local level and awarded the most deserving with the Order of Franz Joseph or the Cross of Merit.[100] The municipalities sent the initiatives to the Sežana district board which then forwarded them to the Trieste government. When planning the construction or adaptation of public water infrastructure, municipalities cooperated with civil engineers (a term that appeared in a ministerial decree of December 8, 1860) (for example of the construction plan see, Fig. 11.13).[101] According to the archival document from 1894, civil engineer Karel (Karol) Oberst was appointed to the municipalities of the Sežana district board that also helped municipalities prepare budgets for the construction of the water infrastructure.[102] This was the basis for state and/or province co-financing. The rest of the cost had to be covered by the municipalities.

On July 26, 1906, the mayor of Lokev Janez (Ivan) Obersnel announced to the imperial and royal district board in Sežana an intention to build a public well. In the late summer of 1906, engineer Oberst visited Lokev where he chose a site for a new well that would hold up to 2000 hl of water and suitable roofs to collect water. He determined that the well would be built next to the chaplaincy garden opposite the defensive tower of Tabor in which the seat of the municipality was located. The Sežana district board then had to send plans for the well in question to the Lokev municipality and the municipality had to calculate the construction costs and make a sketch of how water would be collected from the chosen roofs. The municipal elders initially calculated that the new well would cost 13,640 krons and 52 hellers (for this amount, it was possible to buy for example 110 tons of wheat in the Austrian part

[98] Slatnar (2009).

[99] Author's interviews with Peter Skok (September 3, 2017), Franc Čok (August 22, 2018), Ivanka Stopar (January 15, 2023).

[100] Car (2017), pp. 248–250, 290–291.

[101] Serše (1994), p. 74.

[102] SI PAK KP 633, box 251, circular to the all municipalities of the district board of Sežana, November 14, 1894, no. 9872.

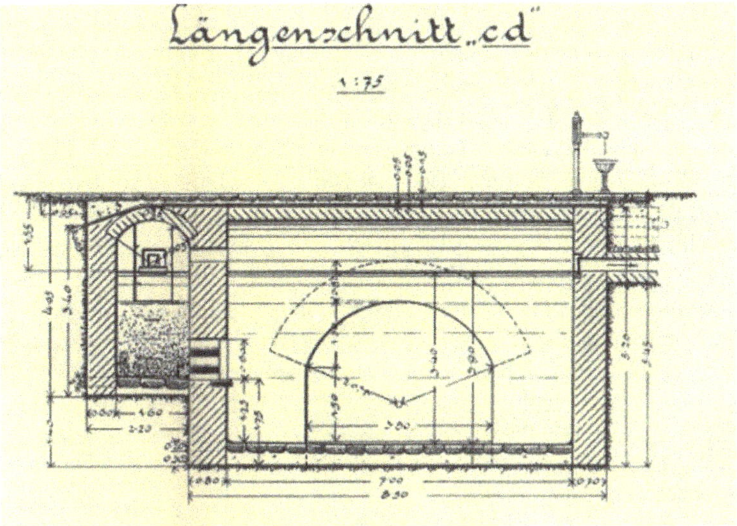

Fig. 11.13 General longitudinal section of the cistern for 2000 hectoliters of water. *Source* SI PAK KP 633, a. e. 472, Cisterne von 2000 hect.a Inhalt. Reproduced with permission

of the Habsburg monarchy in the year 1902)[103] but the Ministry of Agriculture later calculated that all construction costs would be significantly lower, namely 12,500 krons. The municipality received 4300 krons from the province and 5000 krons from state support for the construction of the well. The latter was paid in two installments of 2500 krons each. The conditions for obtaining state support were to remove the manure pit which was located near the planned location of the well, to move the cistern at least 10 m away from three homesteads, and to start building in 1911.[104]

To save part of their money, the municipalities organized *rabotas*. Due to the required physical strength to perform the work, the presence of men was desirable. However, a transcript of the minutes of the meeting of elders in Veliki Dol near Sežana in December 1896, which discussed the construction of the Veliki Dol's municipal well, speaks also of the possible presence of women. As parents, due to farm work, often sent children who could not yet handle the heavy physical work required to these *rabotas*, at a municipal session in Veliki Dol, it was decided not to accept children under the age of 18.[105] At municipal meetings, the time, location, the way it would be done, and the type of *rabota* (manual or transport) were also determined. Workers usually dug a pit, broke the stone, and transported material. The first phase of work had to be completed according to a timetable set by the land reclamation office at the Trieste unit of the Ministry of Agriculture. Otherwise, the

[103] "Izpremembe," p. 154.

[104] Rože (2020c), p. 40.

[105] SI PAK KP 633, box 251, Spisi okrajnega glavarstva Sežana, 1875–1893 [Documents of district board of Sežana, 1875–1893], transcript of the minutes of the municipality meeting in Veliki dol, no. 383, December 10, 1896.

municipality was threatened with the expiry of financial support.[106] To encourage villagers to work, some municipalities proposed monetary compensation.[107]

Despite the *rabota*s, the biggest share of work was done by people who were paid for their work, which also included the village craftsmen.[108] Tenders (they were called auctions) were usually held for the selection of mason's workshops, and applicants were allowed to appeal the final selection. If no one submitted for the tender or if no one met the criteria, the municipality selected a mason's workshop and signed a contract with the leader of the workshop. Other craftsmen, especially tinsmiths and stonemasons, also collaborated. The masons had to follow specific plans drawn up by civil engineers. The construction or adaptation of public water infrastructure had to be monitored by an engineer who was also the designer, and the construction was supervised by the technical body of the land reclamation office if financial support was granted. For example, the municipality of Lokev asked the Sežana district board to submit a request to the Trieste deputy governor to inspect the excavated pit of the public well in the village center so that masonry work could begin.[109]

Following the instructions given by engineer Oberst at the turn of the century to the Sežana district board, the pit had to be excavated as vertically as possible and possible support for the excavation walls had to be thought of. *"The terrain is very hard and we always have to work also with gunpowder, which is why the work is only slowly progressing,"* the Povir municipality reported on the construction work in January 1892.[110] *"The work was also delayed by bad weather and more would be excavated if it had not caused troubles."*[111] Such an excuse was real or there were other problems but the weather was an elegant excuse for municipalities to ask for an extension of time to complete the construction work that had started because they would otherwise lose co-financing benefits.

After the approval of the excavated pit, its walling began. As in the first half of the nineteenth century, the cistern was usually walled with local crushed stone and a novelty was the use of cement mortar which completely replaced clay. Before the mass opening of cement plants (the first in Slovenia was founded in Trbovlje in

[106] SI PAK KP 633, box 251, Spisi okrajnega glavarstva Sežana, 1875–1893 [Documents of district board of Sežana, 1875–1893], letter from municipality of Pliskovica to the district board of Sežana, no. 449, December 7, 1897.

[107] SI PAK KP 633, box 251, Spisi okrajnega glavarstva Sežana 1875–1893 [Documents of district board of Sežana, 1875–1893], letter from municipality of the Veliki dol to the district board of the Sežana, no. 383, December 10, 1896; SI PAK KP 633, box 251, Spisi okrajnega glavarstva Sežana 1875–1893 [Documents of district board of Sežana, 1875–1893], letter from the municipality of Veliki dol to the district board of the Sežana, no. 495, October 7, 1897, Kalc (1999), p. 284.

[108] Kalc (1999); Rože (2020b), p. 122; Rože (2020d).

[109] Rože (2020b), pp. 124–126; Rože (2020c), pp. 40–41.

[110] SI PAK KP 633, box 251, Spisi okrajnega glavarstva Sežana, 1875–1893 [Documents of district board of Sežana, 1875–1893], letter of the economic council of Povir to the district board of Sežana, no. 7, January 10, 1892.

[111] SI PAK KP 633, box 251, Spisi okrajnega glavarstva Sežana, 1875–1893 [Documents of district board of Sežana, 1875–1893], letter of the economic council of Povir to the district board of Sežana, no. 7, January 10, 1892.

1876)[112] and the wider availability of modern Portland cement, they possibly used hydraulic lime mortar (instead of cement mortar) made of marly and clay-limestone burnt at 900 °C while a raw meal for cement clinker is burnt at higher temperatures (first up to 1300 °C and then between 1300 °C and 1500 °C).[113] In line with the instructions of the engineer Oberst, the interior of the pit had to be encompassed by cement mortar from modern Portland cement following the standards accepted by the Association of Architects on January 3, 1889, and fine sand in a ratio of 1:3. For the mixture to become moldable, enough clean water had to be added. Both the sand and the water had to contain as little organic matter as possible. Otherwise, the hydration of the clinker minerals could not run properly. The mortar had to be applied immediately before the cement hydration process began. The crushed and treated local stones were then horizontally embedded and the gaps between them were filled with cement mortar with a binder-aggregate ratio of 1:1. To improve hygienic conditions, the built-in stones could also be plastered with cement mortar in the ratio of ingredients 1:3. With the approval of the supervisory authority, a masonry scaffolding was erected (and removed at the end of the work) to build the upper part of the cistern. The circular-plan cisterns were completed in the upper part with the construction of a dome while the square-plan ones were completed with a cross or groin vault.[114] Both methods of finishing the cistern were not new. Only the use of cement mortar with the abovementioned binder and aggregate in a ratio of 1:1 was new. To enable the installation of the wellhead above the ground, an opening was left in the middle in both cases. This also enabled the free water inflow during precipitation.[115] Consideration of these two different ratios was important precisely because of the two different functions of the mortar. If the mortar made in the ratio of ingredients 1:1 served as a kind of glue between the stones, the mortar in the ratio of ingredients 1:3 together with the built-in stones had the task of holding the weight of the entire structure. Both types of mortar also made the structure impermeable.[116]

Engineer Oberst listed fired brick as another suitable material for walling excavated pits. Bricks had to be soaked in water first.[117] That prevented cracks to appear in the cement mortar and, at the same time, enabled its curing and direct hydration of the cement on the bricks. Oberst listed mineral concrete as the last suitable material. It is made from a dry mixture of modern Portland cement, gravel, and fine sand in

[112] Gligić (2011), p. 13.

[113] Žarnić (2005), pp. 81, 84. I thank Dr. Violeta Bokan Bosiljkov from the University of Ljubljana's Faculty of Civil and Geodetic Engineering for her explanation. See also the document on building a watering place in Skopo: SI PAK KP 633, box 473, Spisi okrajnega glavarstva Sežana, 1875–1893 [Documents of district board of Sežana, 1871–1914], letter from municipality of Skopo to the district board of the Sežana, no. 666, January 1, 1901.

[114] See more in: Vodeb (2018).

[115] SI PAK KP 633, box 472, Preskrba z vodo 1871–1914 [Water supply 1871–1914], Specielle Baubedingnisse; SI PAK KP 633, box 472, Preskrba z vodo 1871–1914 [Water supply 1871–1914], letter from district board of Sežana to all his municipalities, no. 10305, August 8, 1904.

[116] Žarnić (2005), pp. 103–107.

[117] SI PAK KP 633, box 472, Preskrba z vodo 1871–1914 [Water supply 1871–1914], Specielle Baubedingnisse.

the proportions prescribed for individual cases and with such an amount of water that the mixture becomes moldable. Like mortar, it had to be applied immediately before the cement hydration process began.[118] When using reinforced concrete, the upper part of the cistern could be finished flat.

After that, masonry, stonemasonry, and metalwork were performed. Following the construction of the upper part of the cistern, regardless of the building material, a covering layer made of cement mortar followed. They checked whether the cistern was impermeable by filling it with water and the water level had to remain the same for fourteen days. When building cisterns, they had to take into account also water collected through the gutters and drainpipes made of galvanized sheet metal and the installation of a hand-held water pump which had to supply at least 2500 l of water per hour. Under the water pump had to be a funnel with a drainage canal made of metal pipes. To enter the cistern—for example, when it had to be cleaned—a wrought iron ladder was installed. Protective nets made of galvanized iron had to be attached to the opening of the wellhead. All metal parts had to be protected with a double oil coating. From the same material as the cistern was built, a filter chamber was also made. It contained smaller crushed stones, coarse and fine gravel, and coarse and fine sand. It was necessary to clean it and take care of it regularly. Drainage and inflow channels into the cistern were also planned. They were built of crushed and treated local stone which was held together with cement mortar in a 1:1 ratio of ingredients.[119]

Above the ground, they then made the wellhead which was, as before, an excellent piece of stonemason's work. With the development of the building trade, the plans, and instructions for how things were constructed also changed.[120] The wellheads made of concrete began to appear on a larger scale after World War I and prevailed during the second half of the twentieth century. Cement mortar and crushed local stone were also used when building fenced *kals* consisting of the pool and trough. Examples of such *kals* from the period considered are the *kals* Kaluža and Novi kal, both located in Lokev. In the interwar period of the twentieth century, Tonjev *kal* (or a *kal* in Mirišče) in Prelože was adapted similarly. Due to the high cost and demanding construction technique, such *kals* are rare. They are also often exempted from recent general overviews of *kals* on the Kras Plateau because there is an emphasis on

---

[118] SI PAK KP 633, box 472, Preskrba z vodo 1871–1914 [Water supply 1871–1914], Specielle Baubedingnisse.

[119] SI PAK KP 633, box 472, Preskrba z vodo 1871–1914 [Water supply 1871–1914], Specielle Baubedingnisse. See the examples of the plans of the wells: SI PAK KP 633, box 472, Preskrba z vodo 1871–1914 [Water supply 1871–1914], Cisterne von 1200 hect. Inhalt; SI PAK KP 633, box 472, Preskrba z vodo 1871–1914 [Water supply 1871–1914], Cisterne von 2000 hect. Inhalt. I would like to thank MEng Sebastjan Kravanja for his help in understanding the use of cement and reading plans, and to the retired bricklayer Joško Stopar for the help in understanding the construction of the upper part of the cistern.

[120] SI PAK KP 633, box 472, Preskrba z vodo 1871–1914 [Water supply 1871–1914], Odredba o gradnji vodnjakov [Ordinance on the construction of wells], no. 10305, August 8, 1904.

biodiversity in these overviews. However, these *kals* in function are usually not as species-rich as the older historical *kal* type.[121]

11.4.6 Plumbing

Until the advent of a water distribution system, the inhabitants on the Kras Plateau were faced with modest water consumption, which affected various segments of their lives (cooking, the amount of water they drank, washing clothes, and personal hygiene). The underground pipe system was already a part of everyday life in London in the first third of the nineteenth century. Engineers had to follow hygienic laws and calculate the quantities of water to regulate and direct its supply. With the advent of modern plumbing, people's habits also changed. They urged people even more not to spare water and to wash because this was crucial for them to stay healthy. The change of attitude toward quality water which washes the dirt from the body and a change in the image of cities with hidden water and sewage infrastructure was largely due also to cholera epidemics.[122]

In the middle of the nineteenth century when the, e.g., Parisians built the water distribution system,[123] Carl Ghega, the chief designer of the Austrian Southern Railway, studied how to successfully construct the Vienna-Trieste (Southern) Railway across the Kras Plateau where not only the composition and configuration of the terrain, bora wind, but also the lack of water for steam engines were a challenge.[124]

The water supply of steam locomotives was successfully solved by capturing the Reka River by building two water reservoirs that could each hold 850 m^3 of water close to Gornje Ležeče. From there, the water flowed through a 38-km-long pipeline which consisted of 12.7 cm wide cast iron pipes to the railway stations of Divača, Sežana, and Prosécco (Sln. Prosek; modern-day Italy). They were called the imperial *štirnas*. These reservoirs were large pools of water fully enclosed by a high wall forming an arch. Entering the reservoir was possible through a metal gate, while around the pool there was a walking surface. One such building has been preserved in Divača (see Fig. 11.14). Even though this was the first modern water distribution system on the Kras Plateau, it was meant merely to supply water to the locomotives.[125] On the other hand, the growing city of Trieste (modern-day Italy) got its first modern water distributing system already in 1751 (according to the edict of Maria Theresa on November 19, 1749) but it provided water only to a part of the city.[126] In 1916, the Hubelj water distribution system was constructed.

121 Babij et al. (2005), pp. 83–99; *Opredelitev krajinske pestrosti* (2015), p. 176.

122 Vigarello (1999), pp. 225–226.

123 Vigarello (1999), pp. 226–228.

124 Bogić (1998), p. 7.

125 Kristan and Skrinjar (1998), pp. 4–5.

126 Giulia (2009), p. 2.

Fig. 11.14 Exterior of a preserved example of an imperial štirna in Divača. *Photo* Rože

It was named after a spring near Ajdovščina and was used only for the needs of the army on the southern part of the Soča/Isonzo Front. Only some villages, e.g., Dane pri Divači, had local water systems built.[127] Despite numerous initiatives to build both a water distribution system (including one for the construction of a water supply system from the aforementioned Vroče spring)[128] and a public sewer, regular water supply infrastructure started to be built on the Kras Plateau only during the Kingdom of Italy in 1923 when the Company of Kras aqueduct Nanos (*Consorzio dell' Acquedotto carsico Monte Re*) was formed. The company supplied water only to those municipalities that invested capital in it.[129] Among them was also the municipality of Dutovlje. Although these were public waterworks to the village troughs, but not individual households,[130] such water distribution meant great progress because at least some Kras Plateau villages received the long-awaited water supply. In the cadastral municipality of Lokev, the water supply system was built only between 1974 and 1979.[131]

11.5 Conclusion

Considering the used historical sources of the long nineteenth century (mainly the Josephine land survey map, the Franciscan Cadaster, and censuses of 1902, 1905, and 1908), we can conclude that the state—the Habsburg monarchy—tended to unify the types of water infrastructure and that local differences on the Karst Plateau

[127] Belingar (2011b), pp. 75–90.

[128] Author's interviews with Matjaž Stopar, November 1, 2017, and with Joško Stopar, November 22, 2017.

[129] Kristan and Skrinjar (1998), pp. 5–8.

[130] Kristan and Skrinjar (1998), pp. 7–9.

[131] Rože (2019), pp. 16–17.

were not known or taken into account. This is here testified by the instructions and conventional signs of the Franciscan Cadaster or the censuses of water infrastructure, especially from 1905 and 1908, which do not include open-type *štirnas* and *lokevs* that supplied the population in the cadastral municipality of Dutovlje. These two types of water infrastructure are considered either *kal*s or wells. The huge diversity of the types of water infrastructure is evidenced primarily during field research, obtaining photographs of the existing examples, and in interviews with the local population. We also found out that the inhabitants of the cadastral municipality of Lokev do not know open-type *štirna*s and *lokev*s in their area and that they interpret the term *lokev* as an older term for *kal* from which the name of the village is said to have originated. We also note that the local population in the cadastral municipality of Lokev had more water infrastructure at their disposal, which was also of better quality for people's health. This can be attributed to its favorable location on the route of the then Ljubljana (Vienna)-Trieste state road which enabled better economic development than the road that led from Trieste through Dutovlje to the Gorizia.

Despite these differences, we can point to some common features in the construction and use of water infrastructure between the cadastral municipalities of Dutovlje and Lokev, which can be generalized to the entire Karst Plateau. The construction of the water infrastructure on the Kras Plateau followed—on the one hand, the traditional knowledge, techniques, natural resources (building materials), the availability of workers and livestock, and on the other hand, the administrative and political changes in the country. With the expansion of the state apparatus from 1747/48,[132] the assumption of complete authority over the landed estates, and a changed attitude toward population, the care of water supply became more and more the domain of the state—Habsburg monarchy, which is especially evident after the middle of the nineteenth century. To improve the water infrastructure, it took various experts under its wings. They implemented the findings from the natural sciences in the construction of new or the adaptation of the old water infrastructure. To encourage municipalities to make improvements, the Habsburg monarchy also allocated funds subject to compliance with the established rules (funds for the improvement of water infrastructure were also allocated by the province) and individuals of special merit were rewarded with certain decorations. Experts continued to support local techniques to some extent but abandoned those that were too outdated or no longer acceptable (e.g., livestock entering the kals) and replaced them with modern ones. Due to unsanitary microbiological findings, the construction of *lokevs* and open-type *štirnas* also came to an end. Despite visible progress, there was still not enough water, and it was necessary to save it. During droughts, the population had to use older water infrastructure that no longer met modern standards or set out on a journey to the nearest accessible drinking water. In the 1920s a public water supply system that was accessible to the local population for daily needs was constructed in the cadastral municipality of Dutovlje, while the modern water supply system in the cadastral municipality of Lokev was built only in the 1970s. With the construction of the water supply system, access to drinking water improved significantly. The period

[132] Žontar (1980).

of chronic lack of drinking water in the Kras Plateau ended around 1980 with the successful speleological and especially hydrogeological research of the Geological Institute of Ljubljana in Klariči near Brestovica where large amounts of underground water reservoirs were discovered.

Acknowledgements I thank Žiga Zwitter, Ph.D., from the University of Ljubljana's Faculty of Arts, and Sebastjan Kravanja, MEng, for giving me a lot of advice when writing the article. I would also like to thank all the interviewees who selflessly shared information about the water infrastructure with me and showed me its locations. I would like also to thank David Tavčar (October 22, 2019), Srečko Rože (November 19, 2019 and November 29, 2029), Emil Žvab (November 19, 2019), Slavko Lavrenčič (November 19, 2019), Dušan Štok (November 29, 2019), Marta Ščuka (November 29, 2019), and Neda Tavčar Lah (December 12, 2019) for the names of the water infrastructure and their locations on the field.

References

Sources

Published Sources

Newspapers

Izpremembe v trgovinskih in zadružnih registrih na Kranjskem [Changes in trade and cooperative registers in Carniola]. Narodnogospodarski vestnik, 10 Jan 1902

Kramer E (1905) Preiskovanje voda za pitje in domačo vporabo na Kranjskem [Investigation of drinking water and domestic use in Carniola]. Izvestja Muzejskega društva za Kranjsko 15(5–6):101–153

Krpan V (1976) Konec sušnih dni [The end of dry days]. Primorske novice, 6 June 1976

Pomanjkanje vode na Krasu [Water scarcity in the Kras]. Edinost, 31 Aug 1906

Posledice pomanjkljivemu preskrbljevanju z vodo ob vojaških vajah v kraških krajih [Consequences of insufficient water supply during military exercises in Karst places]. Edinost, 29 Sept 1910

S Krasa, 13. avg. [From Kras, August 13]. Glas, 16 Aug 1872

Vodopravni zakon (Dalje.): O uporabi voda [Water law (continued): on the use of water]. Soča, 18 June 1912

Vojaške vaje na Krasu [Military exercises on the Kras]. Gorica, 19 Aug 1911

Zavadlav A (1899) O živinoreji s Krasa [On livestock farming from the Kras"]. Soča, 13 Nov 1899

Cartographic Sources and Database

Environmental Systems Research Institute (ESRI)

Grčman H et al (ed) (2015) Tla Slovenije s pedološko karto v merilu 1:250000 = Soils of Slovenia with soil map 1:250000. Evropska komisija, Skupni raziskovalni center (JRC) = European Commission Joint Reaearch Centre (JRC): Publications Office of the European Union, Luxembourg

Koper Regional Geodetic Administration, Geodetic office Sežana

Ministry of Agriculture, Forestry and Food (2023)

Rajšp V, Ficko M (1997) Slovenija na vojaškem zemljevidu 1763–1787. Zvezek 3, Kartografsko gradivo: karte = Josephinische Landesaufnahme 1763–1787 (1804) für das Gebiet der Republik

Slowenien, Bd. 3, Karten. Znanstvenoraziskovalni center SAZU, Archives of the Republic of Slovenia, Ljubljana
Surveying and Mapping Authority of the Republic of Slovenia (2023)

Websites

Center for the cartography of fauna and flora (2021) 1001 kal—1001 zgodba o življenju, 2005–2007 [1001 kals—1001 stories about life, 2005–2007]. Accessed 20 Aug 2021. https://www.ckff.si/projekt.php?pid=5
Čeh J (2023) Relief na vaškem vodnjaku, "štirni", Brje pri Komnu [Relief on the village well, "štirna", Brje pri Komnu]. Accessed 22 June 2023. https://www.kamra.si/mm-elementi/detajl-na-vodnjaku-stirni-brje-pri-komnu/
Dichlberger P (2020a) Belokranjci in voda: tradicionalni vodni viri Bele krajine: Boj za vodo [People of White Carniola: traditional water resources of White Carniola: Fight for water]. Accessed 12 Dec 2020. https://www.kamra.si/digitalne-zbirke/item/boj-za-vodo.html
Dichlberger P (2020b) Belokranjci in voda: tradicionalni vodni viri Bele krajine: Izviri in vodne jame [People of White Carniola: traditional water resources of White Carniola: Sprigs and water caves]. Accessed 12 Dec 2020. https://www.kamra.si/digitalne-zbirke/item/izviri-in-vodne-jame.html
Interreg Italia-Slovenia (2021) ENGREEN: strengthening of green infrastructures in IT-SI cross-border environment. Accessed 20 Aug 2021. https://www.ita-slo.eu/en/engreen

Archival Sources

AST—Archivio di Stato di Trieste
AST, Catasto franceschino (CFr), Mappe del Catasto franceschino (M), Distretto di Sessana (DS). Available also at: www.archiviodistatotrieste.it/AriannaWeb/main.htm?eads=*#34970_archivio, 30 Sept 2019
SI AS—Arhiv Republike Slovenije [Archives of the Republic of Slovenia]
SI AS 179 (1811–1869) Franciscejski kataster za Primorsko [Franciscan Cadaster for Slovene Littoral]. Available also at: http://arsq.gov.si/Query/detail.aspx?ID=23256, 2 Oct 2019
SI PAK—Pokrajinski arhiv Koper (Archivio Regionale di Capodistria)
SI PAK KP 633—Okrajno glavarstvo Sežana [District board of Sežana]
ŽAL—Župnijski arhiv Lokev [Parish archive of Lokev]
ŽAL Virgil Šček: Lokavske starine I [Antiquities of Lokev I], Lokev, s. a. (manuscript).

Other Sources

Oral traditions were collected by students of the Lokev branch school of the Srečko Kosovel Sežana Primary School as part of a school project in 1995.
Interviewees:
Bernetič M, 1 Dec 2017
Čok F, 22 Aug 2018
Frankovič J, 31 Aug 2017
Jordan Guštin, 5 Mar 2019
Lavrenčič S, 19 Nov 2019

Mljač F, 4 Mar 2018
Rože S, 19 Nov 2019; 29 Nov 2019
Sila C, 3 Sept 2017
Skok P, 3 Sept 2017
Stopar I, 28 Aug 2017, 3 Sept 2017; 14 Oct 2017; 15 Jan 2023; 29 Jan 2023
Stopar J, 12 Nov 2017; 22 Nov 2017
Stopar M, 1 Nov 2017
Stopar M, 2 Sept 2017; 1 June 2019
Ščuka M, 29 Nov 2019
Škabar L, 7 Jan 2019
Štok D, 29 Nov 2019
Tavčar D, 22 Oct 2019
Neda Tavčar Lah, 12 Dec 2019
Vojska T, 29 Dec 2017
Žvab R, 19 Nov 2019
Written or audio recordings are kept by the author of the article

Literature

Babij V (2005) Naravne danosti Krasa: Uvod = Natural environment of the Karst: Introduction. In: Mihevc A (ed) Kras: voda in življenje v kamniti pokrajini = Water and life in a rocky landscape. Založba ZRC, Ljubljana, pp 17–19

Babij V et al (2005) Flora in vegetacija kalov in lokev na Krasu = Flora and vegetation of Karstic ponds 'kali and lokve' (Kras, Slovenia). In: Mihevc A (ed) Kras: voda in življenje v kamniti pokrajini = Water and life in a rocky landscape. Založba ZRC, Ljubljana, pp 83–99

Belingar E (2007a) O kalih kot o kulturni dediščini [About kals as cultural heritage]. In: Maher I (ed) Okrogla voda: priročnik o kalih. Institute of the Republic of Slovenia for nature conservation, Ljubljana, pp 116–128

Belingar E (2007b) Voda v življenju Kobjeglavcev in Tupelčanov [L' aqua nella vita degli abitanti di Kobjeglava e Tupelče]. Kronika 55(2):433–452

Belingar E (2011a) Uporaba kamna v življenju Kraševcev [The use of stone in the life of the Karst people]. AR: Arhitektura, raziskave 11(3):71–78

Belingar E (2011b) Vodovod v Danah [Water distribution system in Dane]. Kronika 59(1):75–90

Belingar E (2012) Vodnjaki na Krasu: Pojav, način gradnje, oblike [Wells in the Karst: occurrence, methods of construction, shapes]. Kronika 60(2):285–298

Bogić M (1998) Pregled razvoja železniškega omrežja v Sloveniji in okolici [Overview of the development of the railway network in Slovenia and its surroundings]. Slovenske železnice, Železniški muzej, Ljubljana

Borisov P (2009) Zgodovina medicine: poskus sinteze medicinske misli [History of medicine: an attempt at a synthesis of medical thought]. Pivec, Maribor

Car P (2017) Slovenci—junaki avstrijskih armad in njihova odlikovanja (od 1757 do 1918) [Slovenes—heroes of the Austrian armies and their decorations]. Signum Laudis, Ljubljana

Čelik T et al (2005) Inventarizacija kalov in lokev na Krasu ter njihov pomen za biotsko raznovrstnost = Inventory of karstic ponds (kal and lokev) and their importance for biotic diversity. In: Mihevc A (ed) Kras: voda in življenje v kamniti pokrajini = water and life in a rocky landscape. Založba ZRC, Ljubljana, pp 72–82

Foucault M (1978) The history of sexuality: volume 1: an introduction. Pantheon Books, New York

Furiesi A (1999) L'acqua a Volterra. Storia dell'approvvigionamento idrico della città. Nuova imagine editrice, Siena

Gabrič A (2009) Sledi šolskega razvoja na Slovenskem [Traces of school development in Slovenia]. Pedagoški inštitut, Ljubljana. http://www.pei.si/Sifranti/StaticPage.aspx?id=17

Gams I (2003) Kras v Sloveniji v prostoru in času [Karst in Slovenia in space and time]. Založba ZRC, Ljubljana

Giulia P (2009) L'acquedotto Teresiano di Trieste. Società Adriatica di speleologia, Trieste

Gligić D (2011) Zgodovinski pregled proizvodnje cementa in uporabe betona v Sloveniji: diplomska naloga [Historical overview of cement production and use of concrete in Slovenia: thesis]. University of Ljubljana, Faculty of Civil and Geodetic Engineering, Ljubljana

Hren M (1942) Bezuljak: Lokev, ki je ni več [1942 Bezuljak—Lokev that no longer exists]. Accessed 29 Jan 2021. https://stareslike.cerknica.org/2015/09/05/bezuljak-lokev-ki-je-ni-vec/.

Holz E (2015) Cestne povezave na Krasu od konca 17. do sredine 19. stoletja [Road connections in the Karst from the end of the 17th to the mid-19th century]. Kronika 63(3):495–512

Jurkovšek B et al (1996) Formacijska geološka karta južnega dela Tržaško-komenske planote [Kartografsko gradivo] = Geological map of the southern part of the Trieste-Komen plateau (Slovenia). Institute of Geology, Geotechnics and Geophysics, Ljubljana

Kalc A (1999) Tržaško podeželje in policijski red iz leta 1777: kratek sprehod med črko in stvarnostjo [The Trieste Countryside and The Police Rule of 1777: a Short Walk between letter and reality]. Annales: Series historia et sociologia 9(18):271–288

Kavrečič P (2015) Turistični razvoj jame Vilenica in Škocjanskih jam po poročanju planinskih društev in vodnikov v drugi polovici 19. Stoletja [Tourism development of Vilenica and Škocjan caves according to reports by alpine societies and guides in the second half of the 19th century]. Kronika 63(3):547–560

Kosi M (1998) Potujoči srednji vek: cesta, popotnik in promet na Slovenskem med antiko in 16. stoletjem [The traveling Middle Ages: road, traveler, and transport in Slovenia between antiquity and the 16th century]. Založba ZRC, Ljubljana

Kostelac M (2009) Povijest vodovoda i vodoopskrbe Liburnije 1884–2009. (povodom 125. obljetnice gradnje prvog Liburnijskog vodovoda)—1. dio [History of plumbing and water supply in Liburnia 1884–2009. (On the occasion of the 125th anniversary of the construction of the first Liburnian plumbing)—part 1]. Hrvatske vode: Časopis za vodno gospodarstvo 17(69–70):269–280

Kranjc A (1994) O imenu in zgodovini pokrajine Kras [About the name and history of the Karst province]. Ann Ser Historia Naturalis 4(4):131–134

Kranjc A (1997) Introduction. In: Kranjc A (ed) KRAS: slovene classical karst. Založba ZRC, Ljubljana, pp 7–17

Kristan S, Skrinjar P (1998) Kraški vodovod Sežana: 50 let: 1948–1998 [Karst plumbing: 50 years: 1948–1998]. Kraški vodovod, Sežana

Lisec A, Miran F (2017) 200 let od začetka parcelno orientiranega katastra na Slovenskem [200 years from the origins of the parceloriented land cadastre in Slovenian lands]. Geodetski vestnik 61(1):76–90

Maher I (2007) Kali na Primorskem: kaj, kakšni in zakaj so? [Kals in Primorska: what they are, what do they look like and why?]. In: Maher I (ed) Okrogla voda: priročnik o kalih. Institute of the Republic of Slovenia for nature conservation, Ljubljana, pp 10–44

Muznik A (2000) Goriško podnebje = Clima Goritiense. Institute for History of Medicine of Medical Faculty, Scientific Society of Slovenian Health Culture, Ljubljana

Opredelitev krajinske pestrosti in krajinskih značilnosti, pomembnih za ohranjanje biotske raznovrstnosti: Ciljni raziskovalni program (CRP): Zagotovimo si hrano za jutri (2011–2020): Končno poročilo projekta [Definition of landscape diversity and landscape characteristics important for the preservation of biodiversity: target research program: "Let's ensure food for tomorrow" (2011–2020): Final project report] (2015). University of Ljubljana, Biotehnical Faculty, Ljubljana

Panjek A (2006) Človek, zemlja, kamen in burja: zgodovina kulturne krajine Krasa (oris od 16. do 20. stoletja) [Man, earth, stone and burja: the history of the cultural landscape of the Karst (outline from the 16th to the 20th century)]. Univerza na Primorskem, Znanstveno-raziskovalno središče, Koper

Panjek A (2015) Paesaggio culturale e ambiente del Carso: L'uso delle risorse naturali in età moderna. University of Primorska Press, Koper
Pavlovec R (1997) A.C. Morlot, a geologist and a less known researcher of karst phenomena. Acta Carsologica 26(2):167–173
Perko V (2015) Muzej Krasa: med željami, potrebami sodobne družbe, teorijo in prakso [Museum of Kras region and contemporary society: between desires and needs, theory and practice]. Kronika 63(3):717–724
Petrič M (2005) Hidrogeološke značilnosti Krasa = Hydrogeological characteristics of Kras. In: Mihevc A (ed) Voda in življenje v kamniti pokrajini = Water and life in a rocky landscape. Založba ZRC, Ljubljana, pp 21–28
Remec M (2015) Podrgni, očedi, živali otrebi: higiena in snaga v dobi meščanstva [Hygiene and power in the age of the bourgeoisie]. Institute of Contemporary History, Ljubljana
Renčelj S, Lah L (2008) Kraška hiša in arhitektura Krasa med očarljivostjo in vsakdanom [Karst house and Karst architecture between the charming and the everyday life]. Libris, Koper
Ribnikar P (1982) Zemljiški kataster kot vir za zgovino [Das Grundsteuerkataster als Geschichtsquelle]. Zgodovinski časopis 36(4):321–337
Rože A (2018a) Najstarejši ohranjeni lokavski oklep vodnjaka v ilirskobistriškem parku [The Oldest preserved Lokev's shell of the well in the park of Ilirska Bistrica]. Kraški obzornik: glasilo občine Sežana 20(2):9–10
Rože A (2018b) Preskrba z vodo v vaseh Lokev in Prelože na Krasu v prvi polovici 19. stoletja [Water supply in villages of Lokev and Prelože in Kras Plateau in first half of the nineteenth century]. Goriški letnik 42:223–240
Rože A (2019) Ko je voda po vodovodnih ceveh prvič pritekla v domove Lokavcev in Preložcev [When water first flowed into the homes of Lokavci and Preložci through water pipes]. Kraški obzornik: glasilo občine Sežana 21(3):16–17
Rože A (2020a) Franciscejski kataster kot vir za preučevanje zgodovine vodne infrastrukture in preskrbovanja z vodo [Franciscean cadastre as a source for studying the history of water infrastructure and water supply: tax district Sežana]. Kronika 68(1):99–116
Rože A (2020b) Občinski vodnjak v vaškem središču Na placu v Dutovljah [Municipal well in the Village Centre 'Na placu'in Dutovlje]. Goriški letnik 44:115–140
Rože A (2020c) Izgradnja zadnjega občinskega vodnjaka v nekdanji občini Lokev [Construction of the last municipal well in the former municipality of Lokev]. SLO: časi, kraji, ljudje: slovenski zgodovinski magazin 28:38–43
Rože A (2020d) Prispevek o kalu 'Kaluža' v Lokvi [Article on the 'Kaluža' kal in Lokev]. Accessed 1 Feb 2021. http://www.lokev.si/2020/06/14/prispevek-o-kalu-kaluza-v-lokvi/
Serše S (1994) Pooblaščeni civilni inženirji na Kranjskem: do leta 1918 [Certified civil engineers in Carniola: until 1918]. Arhivi 17(1-2):74–76
Shaw TR (1998) Early tourists at Škocjanske jame—18th century to 1914. Acta Carsologica 27(2):235–264
Slatnar J (2009) Tehnološki list—Poraba vode pri domačih živalih [Technology sheet—Water consumption by domestic animals]. Accessed 11 Feb 2023. https://lj.kgzs.si/novice/ArtMID/915/ArticleID/251
Slovenian Environment Agency, "Atlas okolja" [Environmental atlas of Slovenia], Climate, Precipitation, Mean annual corrected precipitation 1971–2000. Accessed 22 Jan 2023 http://gis.arso.gov.si/atlasokolja/profile.aspx?id=Atlas_Okolja_AXL@Arso&culture=en-US
Štih P (2009/2010) Kmečko prebivalstvo v grofijah Goriška in Gradiščanska je sicer delavno, vendar zelo revno ...: Poročilo dvornega vojnega sveta o socialnem in gospodarskem stanju na Goriškem in Gradiščanskem iz leta 1770 [La popolazione contadina delle Contee di Gorizia e diGradisca è comunque laboriosa, ma molto povera ... : Le relazioni del Consiglio di guerra di Corte sulle condizioni socialied economiche nel Goriziano e nel territorio di Gradisca nel 1770]. Marušičev zbornik: zbornik prispevkov v počastitev 70-letnice prof. dr. Branka Marušiča. Goriški letnik 33/34(1):319–341

Valvasor JV (2009) Čast in slava vojvodine Kranjske [Die Ehre Dess Hertzogthums Crain], first book. Zavod Dežela Kranjska, Ljubljana

Vigarello G (1999) Čisto in umazano: telesna higiena od srednjega veka naprej [Le Propre et le Sale: L'hygiène du corps depuis le Moyen Âge]. Založba *cf, Ljubljana

Vilfan S (1957) Podobe iz nekdanje živinoreje med Trstom in Slavnikom [Images from the former livestock farming between Trieste and Slavnik]. Kronika 5(2):69–87

Vodeb M (2018) Ukrivljene konstrukcije [Curved structures]. University of Maribor, Faculty of Civil Engineering, Transportation Engineering and Architecture, Maribor

Zobec M (2007) Pogled v preteklost kalov na Krasu [A look into the past of the kals in the Karst]. In: Maher I (ed) Okrogla voda: priročnik o kalih. Institute of the Republic of Slovenia for nature conservation, Ljubljana, pp 193–207

Zobec M (2008) Človek, živina in kali: pogled v preteklost vodnih zajetij na Krasu [Uomo, bestiame e stagni: uno sguardo al passato dei depositi d'aqua sul Carso]. Kronika 56(3):495–503

Zorn M, Erhartič B, Komac B (2009) La Slovénie, berceau du géotourism karstique. Karstologia 54:1–10

Zorn M, Erhartič B, Komac B (2012) The beginnings of karst geotourism—karst geotourism in Slovenia. In: Giusti C (ed) Géomorphosites 2009: imagerie, inventaire, mise en valeur et vulgarisation du patrimoine géomorphologique: volume des actes = Geomorphosites 2009: raising the profile of geomorphological heritage through iconography, inventory and promotion: proceedings volume. Sorbonne Université, Paris, pp 348–355

Zwitter Ž (2014) Okolje na Kranjskem v 17. stoletju po Slavi vojvodine Kranjske [Die Umwelt in Krain im 17. Jahrhundert in der Ehre Deß Hertzogthums Crain]. In: Weiss J (ed) Studia Valvasoriana: zbornik spremnih študij ob prvem integralnem prevodu Die Ehre Deß Hertzogthums Crain v slovenski jezik. Zavod Dežela Kranjska, Ljubljana, pp 611–702

Zwitter Ž (2015) Izbrani vidiki okoljske zgodovine kobilarne Lipica od ustanovitve do začetka tridesetih let 17. stoletja [Selected environmental history aspects of the Lipica stud farm from its foundation to the early 1630s]. Kronika 63(3):457–474

Žarnić R (2005) Lastnosti gradiv [Properties of materials]. University of Ljubljana, Faculty of Civil and Geodetic Engineering, Ljubljana

Žontar J (1980) Nastanek in razvoj upravnih okrajev na Slovenskem do leta 1849 [Entstehung und Entwicklung der Verwaltungsbezirke in Slowenien bis zum Jahre 1849]. Zgodovinski časopis 34(1–2):119–155

Aneja Rože Kravanja She graduated in history and art history. After completing her master's studies in history, she passed the exam as a curator. During her internship, she started researching the history of water supply on Kras Plateau. She presented her findings in Slovenian scientific journals and raised awareness among the local population. Recently she started researching the history of spatial landmarks. It is worth highlighting her exhibition *In the embrace of poisonous fumes: the gas attack in the 12th Battle of Soča/Isonzo front*.

Chapter 12
Climate and Discharge Trends, and Flood Hazard in Slovenia's Dinaric Karst Region Since the Mid-Twentieth Century

Mauro Hrvatin and Matija Zorn

Abstract The temperature, precipitation, and snow-cover data collected at the meteorological stations in the Slovenia's Dinaric Karst region between 1961 and 2020 indicate rapid and clear climate change. During this period, temperatures increased by 2.5 °C on average, precipitation decreased by 50–400 mm on average, and the number of days with snow cover declined by 30–60%. Several studies have confirmed that these changes have a strong impact on water conditions. What is alarming is especially a decrease in the average minimum and mean discharges, and a downward trend can also be observed for many average maximum discharges.

Over the past sixty years, the climate in Slovenia's Dinaric Karst region has become warmer and drier, and there has been a significant reduction in the volume of water at most gauging stations in the area studied. Based on the trends described, a significantly smaller flood hazard would be expected, but data from the past decades suggest the opposite. The following are highlighted as possible reasons for the current flood hazard: upward trends in absolute maximum discharges in individual river basins, a slightly downward trend in absolute maximum discharges and their simultaneously greater variability, various forms of land development that used to be avoided due to the risk of flooding, and more frequent short-term extreme precipitation events causing flash floods on smaller streams.

Keywords Climate change · Hydrology · Discharge · Floods · Dinaric Slovenia

M. Hrvatin · M. Zorn (✉)
ZRC SAZU Anton Melik Geographical Institute, Ljubljana, Slovenia
e-mail: matija.zorn@zrc-sazu.si

M. Hrvatin
e-mail: mauro.hrvatin@zrc-sazu.si

B. Fuerst-Bjeliš et al. (eds.), *Environmental Histories of the Dinaric Karst*, Environmental History 17, https://doi.org/10.1007/978-3-031-56089-7_12

12.1 Introduction

The Dinaric Karst region lies in southern Slovenia and covers over a fourth of Slovenia's territory. It is mainly composed of karstified plateaus and hills, and lowlands and plains strewn between them. In terms of elevation, it can be divided into the high and low Dinaric Karst region (Fig. 12.1). The high karst region covers nearly three-fifths of the entire Dinaric region, or 3140 km^2, and the low karst region extends over more than two-fifths of the Dinaric region, covering 2302 km^2. Because of the extensive wooded and almost completely unpopulated high Dinaric plateaus, settlement density is nearly half the Slovenian average (22 people per km^2 in the high Dinaric Karst region and 93 people per km^2 in the low Dinaric Karst region, compared to the 2017 Slovenian average of 109 people per km^2).[1]

Slovenia's Dinaric Karst region is part of the northwestern Dinaric Alps, which predominantly extend in a typical northwest–southeast Dinaric direction. Its main characteristics include the predominance of karstified landscape and subterranean water flow, karst poljes, intermittent lakes, high precipitation and forest density, livestock- and forestry-oriented private farming, scant settlement, sparse population concentrated in the fertile parts of poljes and plains, a traditional woodworking industry, and exceptional environmental vulnerability.[2]

This chapter examines the period from 1961 to 2020 for trends in average annual temperature, annual precipitation, days with precipitation over 0.1 mm, days with snow cover, minimum annual discharges, mean annual discharges, maximum annual discharges, and absolute maximum annual discharges, as well as discharge regimes and flood hazard, using the Dinaric Karst of Slovenia as an example.

12.2 Methods

To determine the trend of change in selected climate and hydrological variables (Table 12.1) from 1961 to 2020, the Mann–Kendall test and Theil–Sen estimator (also known as Sen's slope estimator) were used at selected temperature, precipitation, and gauging stations (Tables 12.2 and 12.3; Fig. 12.1). The Mann–Kendall test is a nonparametric test used to detect monotonic trends. It is insensitive to data outliers and is based on the test statistics. A positive test statistic implies an upward trend, and a negative test statistic indicates a downward trend. Sen's slope estimator is the most frequent nonparametric test used for detecting linear time trends.[3] Compared to linear regression, Sen's slope is more accurate for asymmetric data distribution

[1] Perko and Ciglič (2020); Perko et al. (2021).

[2] Kladnik (1998).

[3] Kraner Šumenjak and Šuštar (2011); Hrvatin and Zorn (2020), p. 16; Hrvatin and Zorn (2022), p. 88.

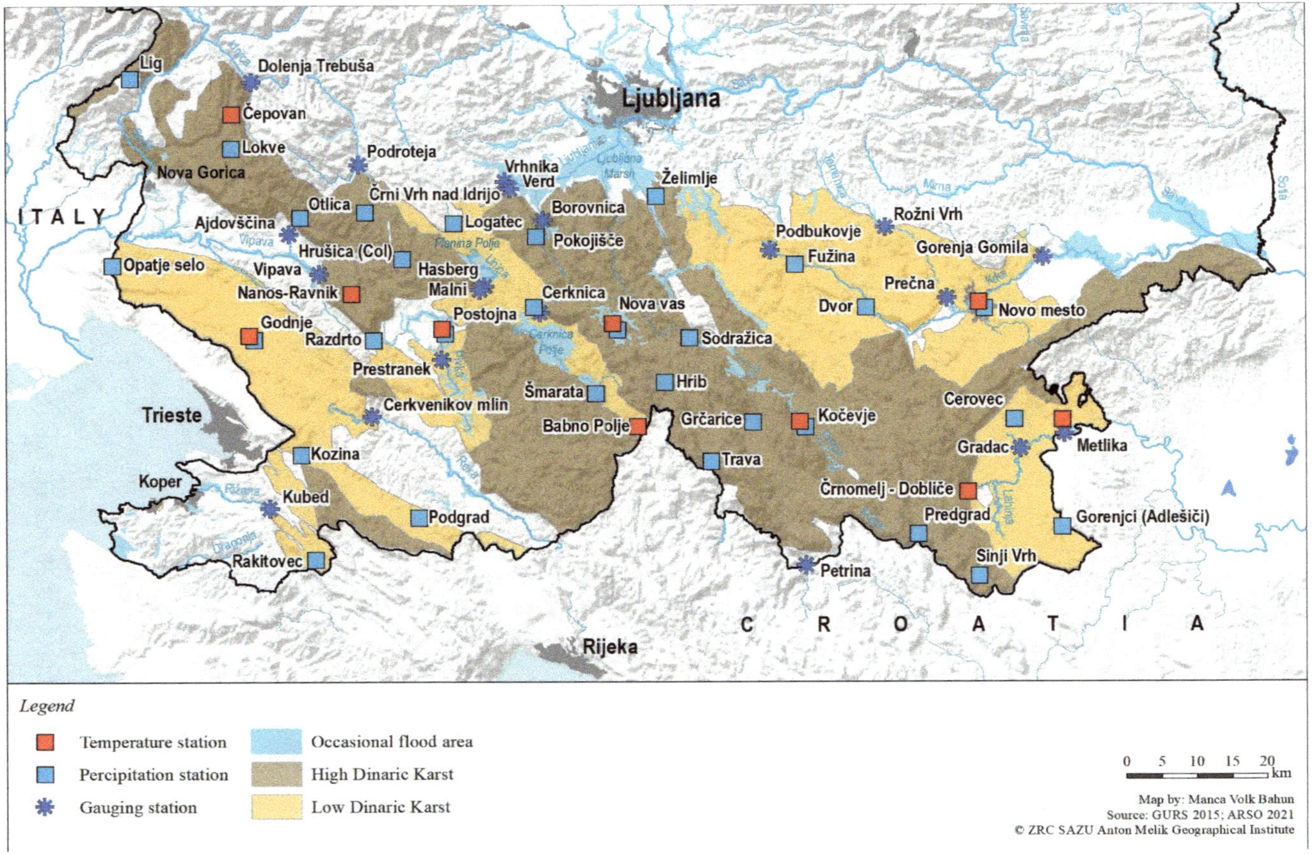

Fig. 12.1 Locations of temperature, precipitation, and gauging stations included in the analysis. *Source* authors

and, with normal data distribution, it yields results that are completely comparable to those obtained by applying the least square method.[4]

The free Excel template MAKESENS (Mann–Kendall test for trend and Sen's slope estimates) was used to calculate the Mann–Kendall test and Sen's slope estimator values.[5]

In addition to the Mann–Kendall test and Sen's slope values, the tables with climate and hydrological variables also show the confidence levels, the (initial) 1961 trend value, the (final) 2020 trend value, and absolute and relative trend differences.[6]

In statistics, the confidence level indicates the probability that the confidence interval calculated includes the value of the parameter estimated. In this specific case, the higher the confidence level, the higher the probability that the upward or downward trend in the selected variable detected actually exists.[7]

The initial 1961 trend value represents the value of a selected variable in 1961 read from the trend line, and the final 2020 trend value is the value of a selected variable in 2020 read from the trend line. The absolute trend difference is the difference between the final and initial trend values, and the relative trend difference is the difference between the final and initial trend values expressed as a percentage.[8]

The trend value per year can be calculated using the following equation[9]:

trend value per year x = Sen's slope * (trend year x − initial trend year) + initial trend value.

Table 12.1 Climate and hydrological variables examined

Variable type	Variable
Climate	Average annual air temperature
	Annual precipitation
	Annual days with precipitation over 0.1 mm
	Annual days with snow cover
Hydrological	Minimum annual discharge
	Mean annual discharge
	Maximum annual discharge
	Absolute maximum annual discharge

[4] Kovačič (2016); Kovačič et al. (2016); Hrvatin and Zorn (2020), p. 163; Hrvatin and Zorn (2022), p. 88.

[5] Salmi et al. (2002).

[6] Hrvatin and Zorn (2020), p. 163; Hrvatin and Zorn (2022), p. 88.

[7] Ibid.

[8] Ibid.

[9] Ibid.

Table 12.2 Weather stations with the time series analyzed

Station type	Weather station	Municipality	Elevation, m	Time series	No. of annual measurements
Temperature	Babno polje	Loška dolina	755	1961–2020	60
	Čepovan	Nova Gorica	593	1961–2020	60
	Črnomelj-Dobliče	Črnomelj	210	1961–2020	60
	Godnje	Sežana	295	1961–2020	60
	Kočevje	Kočevje	467	1961–2020	60
	Metlika	Metlika	210	1961–2020	60
	Nanos	Vipava	915	1961–2020	60
	Nova vas na Blokah	Bloke	722	1961–2020	60
	Novo mesto	Novo mesto	220	1961–2020	60
	Postojna	Postojna	533	1961–2020	60
Precipitation	Cerknica	Cerknica	569	1961–2020	60
	Cerovec	Semič	190	1961–2020	60
	Črni Vrh	Idrija	688	1961–2020	58
	Dvor	Žužemberk	203	1963–2020	57
	Fužina	Ivančna Gorica	243	1961–2020	60
	Godnje	Sežana	295	1961–2020	57
	Gorenjci pri Adlešičih	Črnomelj	232	1961–2020	59
	Grčarice	Ribnica	520	1961–2020	59
	Hrib	Loški Potok	827	1961–2020	59
	Hrušica pri Colu	Ajdovščina	872	1961–2020	58
	Kočevje	Kočevje	467	1961–2020	58
	Kozina	Hrpelje-Kozina	484	1961–2020	55
	Lig	Kanal ob Soči	620	1961–2017	55
	Logatec	Logatec	485	1961–2020	60
	Lokve	Nova Gorica	955	1961–2020	57
	Nova vas na Blokah	Bloke	720	1961–2020	60
	Novo mesto	Novo mesto	220	1962–2020	58
	Opatje selo	Miren-Kostanjevica	165	1961–2020	60
	Otlica	Ajdovščina	826	1961–2020	56
	Podgrad	Ilirska Bistrica	596	1961–2015	55
	Pokojišče	Vrhnika	716	1961–2016	55
	Postojna	Postojna	533	1962–2020	59
	Predgrad	Kočevje	375	1961–2020	58
	Rakitovec	Koper	504	1961–2020	59

(continued)

Table 12.2 (continued)

Station type	Weather station	Municipality	Elevation, m	Time series	No. of annual measurements
	Razdrto	Postojna	577	1961–2020	60
	Sinji Vrh	Črnomelj	365	1961–2020	59
	Sodražica	Sodražica	542	1961–2020	57
	Šmarata	Loška dolina	579	1962–2020	58
	Trava	Loški Potok	765	1962–2020	55
	Želimlje	Škofljica	309	1961–2020	60

12.3 Data

12.3.1 Climate Data

Climate data were obtained from the Slovenian Environment Agency.[10] The analysis included ten temperature stations and thirty precipitation stations in the Dinaric part of Slovenia (Table 12.2, Fig. 12.1) that had measurement data spanning several decades.

12.3.2 Hydrological Data

Hydrological data were obtained from the Slovenian Environment Agency.[11] The analysis included twenty gauging stations in the Dinaric part of Slovenia (Table 12.3, Fig. 12.1) that had measurement data spanning several decades.

12.4 Results

12.4.1 Climate Variables

The following were analyzed in terms of climate variables (Table 12.1): (1) average annual air temperature trends, (2) annual precipitation trends, (3) trends in annual days with precipitation over 0.1 mm, and (4) trends in annual days with snow cover.

---

[10] Agencija Republike Slovenije za okolje, "Arhiv meteoroloških podatkov".

[11] Agencija Republike Slovenije za okolje, "Arhiv hidroloških podatkov".

Table 12.3 Rivers with the time series analyzed

River	Gauging station	Municipality	Elevation, m	Time series	No. of annual measurements	Drainage area, km^2	Mean annual discharge 1961–2020, m^3/s
Borovniščica	Borovnica	Borovnica	295	1961–2020	53	34.89	1.09
Cerkniščica	Cerknica	Cerknica	560	1961–2020	57	47.29	1.03
Hubelj	Ajdovščina	Ajdovščina	107	1961–2020	58	93.15	2.92
Idrijca	Podroteja	Idrija	328	1961–2020	59	112.84	9.01
Kolpa	Metlika	Metlika	127	1961–2020	60	2002.00	70.27
Kolpa	Petrina	Kostel	220	1961–2020	58	460.00	25.63
Krka	Gorenja Gomila	Šentjernej	149	1961–2020	58	1865.71	43.84
Krka	Podbukovje	Ivančna Gorica	259	1961–2020	59	321.44	7.88
Lahinja	Gradac	Metlika	129	1961–2020	60	221.32	5.65
Ljubija	Verd	Vrhnika	286	1961–2020	59	karst*	6.78
Ljubljanica	Vrhnika	Vrhnika	285	1961–2020	59	karst*	23.60
Malenščica	Malni	Postojna	442	1961–2020	56	karst*	6.48
Pivka	Prestranek	Postojna	520	1961–2020	60	karst*	2.73
Prečna	Prečna	Novo mesto	164	1961–2020	59	294.17	4.31
Reka	Cerkvenikov mlin	Divača	342	1961–2020	60	377.89	8.13
Rižana	Kubed	Koper	58	1961–2020	57	204.50	3.93
Temenica	Rožni Vrh	Trebnje	268	1961–2020	59	80.87	0.80
Trebuščica	Dolenja Trebuša	Tolmin	186	1961–2020	58	54.70	2.70
Unica	Hasberg	Postojna	445	1961–2020	60	karst*	21.99
Vipava	Vipava	Vipava	96	1961–2020	60	131.90	6.48

karst*: the drainage area was not determined or assessed due to karst conditions

Table 12.4 Average annual air temperature trends, 1961–2020

Temperature station	Mann–Kendall test, Z	Confidence level, %	Sen's slope, Q	1961 trend value, °C	2020 trend value, °C	1961–2020 trend difference, °C	1961–2020 trend difference, %
Babno Polje	5.89	99.9	0.034	5.57	7.58	2.01	0.72
Čepovan	6.20	99.9	0.039	7.88	10.16	2.28	0.81
Črnomelj-Dobliče	6.04	99.9	0.048	9.21	12.07	2.86	1.01
Godnje	5.92	99.9	0.043	9.96	12.50	2.54	0.90
Kočevje	6.10	99.9	0.032	7.87	9.74	1.87	0.67
Metlika	6.45	99.9	0.041	9.44	11.85	2.41	0.85
Nanos	6.51	99.9	0.044	6.43	9.00	2.57	0.92
Nova vas na Blokah (Fig. 12.2a)	6.47	99.9	0.046	6.23	8.94	2.71	0.97
Novo mesto	6.56	99.9	0.050	8.80	11.75	2.95	1.05
Postojna	5.88	99.9	0.046	7.78	10.52	2.74	0.98

The trend difference as a percentage is calculated based on the absolute (Kelvin) temperature scale

12.4.1.1 Average Annual Air Temperature

The average annual air temperature trends from 1961 to 2020 are similar at all ten temperature stations analyzed, and they indicate a clear upward trend (Table 12.4, Fig. 12.2a). An exceptionally high confidence level is typical, amounting to 99.9% at all temperature stations.

During the period studied, from 1961 to 2020, the temperature at the selected temperature stations rose by 0.032 to 0.050 °C a year on average, which means that over the past six decades the temperatures at these stations increased by 1.87 to 2.95 °C. Taking all the selected temperature stations into account, the temperature rose by 2.49 °C on average. The temperature difference was the smallest at Kočevje, where the temperature rose by 1.87 °C, and the greatest at Novo mesto, where it rose by 2.95 °C.

12.4.1.2 Annual Precipitation

In contrast to the upward temperature trends, the annual precipitation trends from 1961 to 2020 indicate a downward trend at twenty-four out of the thirty precipitation stations studied (Table 12.5, Fig. 12.2b). The confidence level is fairly modest because at as many as twenty-two stations it does not even reach 90%, which is far from statistical significance (95%). All eight stations with an at least 95% confidence level show a downward annual precipitation trend.

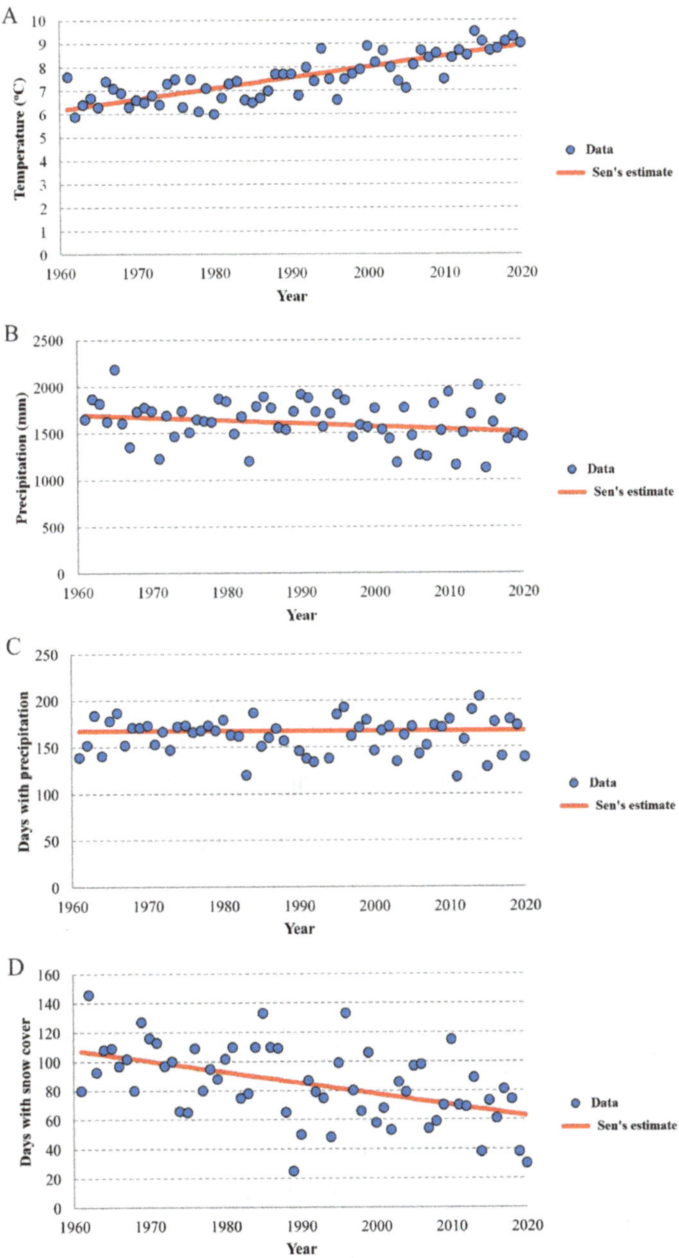

Fig. 12.2 Climate trends at selected temperature and precipitation stations: **a** average annual temperature trend at the Nova vas na Blokah temperature station, 1961–2020, **b** annual precipitation trend at the Cerknica precipitation station, 1961–2020, **c** trends in annual days with precipitation at the Kočevje precipitation station, 1961–2020, and **d** trend in annual days with snow cover at the Predgrad precipitation station, 1961–2020. *Source* Authors

Table 12.5 Annual precipitation trends, 1961–2020

Precipitation station	Mann–Kendall test, *Z*	Confidence level, %	Sen's slope, *Q*	1961 trend value, mm	2020 trend value, mm	1961–2020 trend difference, mm	1961–2020 trend difference, %
Cerknica (Fig. 12.2b)	−1.86	90.0	−3.068	1694.61	1513.61	−181.00	−10.68
Cerovec	0.41	<90.0	0.671	1159.05	1198.64	39.59	3.42
Črni Vrh	−1.34	<90.0	−3.332	2641.67	2445.06	−196.61	−7.44
Dvor	−0.56	<90.0	−0.938	1243.15	1187.78	−55.37	−4.45
Fužina	−0.98	<90.0	−1.104	1296.63	1231.51	−65.12	−5.02
Godnje	−1.34	<90.0	−2.951	1485.21	1311.12	−174.09	−11.72
Gorenjci pri Adlešičih	−1.18	<90.0	−1.942	1327.05	1212.45	−114.60	−8.64
Grčarice	−0.31	<90.0	−0.503	1582.08	1552.43	−29.65	−1.87
Hrib	−2.35	95.0	−3.367	1597.20	1398.57	−198.63	−12.44
Hrušica pri Colu	−2.65	99.0	−7.555	2222.23	1776.49	−445.74	−20.06
Kočevje	−1.65	90.0	−2.616	1561.47	1407.11	−154.36	−9.89
Kozina	−2.83	99.0	−5.064	1586.06	1287.27	−298.79	−18.84
Lig	−0.58	<90.0	−1.717	2349.20	2247.92	−101.28	−4.31
Logatec	−2.01	95.0	−3.793	1973.78	1750.02	−223.76	−11.34
Lokve	−1.27	<90.0	−4.587	2546.21	2275.60	−270.61	−10.63
Nova vas na Blokah	0.72	<90.0	0.898	1482.03	1535.02	52.99	3.58
Novo mesto	0.08	<90.0	0.167	1165.48	1175.32	9.84	0.84
Opatje selo	0.00	<90.0	−0.034	1401.58	1399.55	−2.03	−0.14
Otlica	−2.10	95.0	−6.748	2572.94	2174.80	−398.14	−15.47
Podgrad	−2.45	95.0	−6.161	1677.11	1313.62	−363.49	−21.67
Pokojišče	−2.35	95.0	−4.559	1656.06	1387.09	−268.97	−16.24
Postojna	−0.69	<90.0	−0.939	1595.61	1540.20	−55.41	−3.47
Predgrad	1.02	<90.0	1.650	1499.93	1597.28	97.35	6.49
Rakitovec	−1.92	90.0	−3.877	1618.54	1389.80	−228.74	−14.13
Razdrto	−2.56	95.0	−5.435	1766.81	1446.15	−320.66	−18.15
Sinji Vrh	0.44	<90.0	0.780	1361.52	1407.54	46.02	3.38
Sodražica	−0.75	<90.0	−1.260	1478.90	1404.57	−74.33	−5.03
Šmarata	−1.19	<90.0	−1.813	1525.91	1418.92	−106.99	−7.01
Trava	0.61	<90.0	1.105	1731.75	1796.95	65.20	3.76
Želimlje	−1.22	<90.0	−1.233	1412.93	1340.21	−72.72	−5.15

Most differences in the annual precipitation are moderate. At eighteen precipitation stations out of thirty, these differences do not exceed 10%, and at twenty-eight, they do not exceed 20%. A downward annual precipitation trend is the most distinct at Podgrad (−363.5 mm or −21.7%), Hrušica pri Colu (−445.7 mm or −20.1%), and Kozina (−298.8 mm or −18.8%), and an upward trend is the most evident at Predgrad (+97.4 mm or +6.5%), Trava (+65.2 mm or +3.8%), and Nova vas na Blokah (+53.0 mm or +3.6%).

12.4.1.3 Annual Days with Precipitation Over 0.1 mm

From 1961 to 2020, the annual number of days with precipitation over 0.1 mm increased at twelve precipitation stations, decreased at fourteen, and remained the same at the stations in Godnje, Gorenjci pri Adlešičih, Kočevje, and Želimlje (Table 12.6, Fig. 12.2c). The confidence level is mostly very modest because at twenty-two stations, it does not even exceed 90%. The confidence level is at least 95.0% at the remaining eight precipitation stations, of which three have an upward trend and five have a downward trend in the annual days with precipitation.

Twenty-one stations show smaller negative or positive trend deviations, not reaching 10%. The decrease in the annual days with precipitation over 0.1 mm is the most distinct at Predgrad (−46.7 days or −28.9%), Trava (−36.6 days or −21.7%), and Dvor (−31.8 days or −19.9%), and the increase is the greatest at Podgrad (24.6 days or 22.1%), Otlica (19.7 days or 14.8%), and Cerknica (19.7 days or 13.8%).

12.4.1.4 Annual Days with Snow Cover

Between 1961 and 2020, the number of days with snow cover decreased significantly at all precipitation stations where it snows every year (Table 12.7, Fig. 12.2d). The confidence level at most of these stations is high and statistically significant.

At the Godnje and Opatje selo stations, the snow cover is more of an exception due to their low elevation and the proximity of the sea, and therefore their trends cannot be compared with other stations. At twenty-two of the remaining twenty-eight precipitation stations, the number of days with snow cover decreased by 30 to 60%, or on average by thirty-five days a year. The only station where it decreased by less than 30% was Otlica (−20.4%), whereas a more than 60% decrease was recorded at Podgrad (−71.9%), Dvor (−63.1%), Novo mesto (−61.3%), Fužina (−61.1%), and Cerovec (−60.9%).

The duration of snow cover expressed in the number of days decreased the most at Lokve (−62.9 days), Črni Vrh (−55.6 days), and Logatec (−52.3 days).

Table 12.6 Trends in annual days with precipitation, 1961–2020

Precipitation station	Mann–Kendall test, *Z*	Confidence level, %	Sen's slope, *Q*	1961 trend value, days	2020 trend value, days	1961–2020 trend difference, days	1961–2020 trend difference, %
Cerknica	2.37	95.0	0.333	142.33	162.00	19.67	13.82
Cerovec	−1.03	<90.0	−0.131	129.66	121.93	−7.73	−5.96
Črni Vrh	0.67	<90.0	0.116	147.36	154.22	6.86	4.65
Dvor	−3.61	99.9	−0.538	159.46	127.69	−31.77	−19.92
Fužina	−0.61	<90.0	−0.074	153.48	149.11	−4.37	−2.85
Godnje	0.08	<90.0	0.000	131.00	131.00	0.00	0.00
Gorenjci pri Adlešičih	−0.08	<90.0	0.000	132.00	132.00	0.00	0.00
Grčarice	0.89	<90.0	0.118	145.65	152.59	6.94	4.77
Hrib	−0.13	<90.0	−0.021	154.27	153.04	−1.23	−0.80
Hrušica pri Colu	−0.65	<90.0	−0.111	154.44	147.89	−6.55	−4.24
Kočevje (Fig. 12.2c)	−0.10	<90.0	0.000	167.50	167.50	0.00	0.00
Kozina	1.05	<90.0	0.152	116.12	125.06	8.94	7.70
Lig	−0.82	<90.0	−0.118	138.06	131.12	−6.94	−5.03
Logatec	−0.46	<90.0	−0.053	151.18	148.08	−3.10	−2.05
Lokve	1.14	<90.0	0.200	141.40	153.20	11.80	8.35
Nova vas na Blokah	−2.07	95.0	−0.273	166.14	150.05	−16.09	−9.68
Novo mesto	−1.42	<90.0	−0.188	161.06	150.00	−11.06	−6.87
Opatje selo	1.36	<90.0	0.143	107.64	116.07	8.43	7.83
Otlica	1.93	90.0	0.333	132.83	152.50	19.67	14.81
Podgrad	2.59	99.0	0.417	111.33	135.92	24.59	22.08
Pokojišče	1.40	<90.0	0.190	142.95	154.19	11.24	7.86
Postojna	−0.93	<90.0	−0.143	151.14	142.71	−8.43	−5.58
Predgrad	−4.42	99.9	−0.792	161.90	115.19	−46.71	−28.85
Rakitovec	1.19	<90.0	0.196	106.36	117.35	10.99	10.34
Razdrto	−0.95	<90.0	−0.167	146.42	136.58	−9.84	−6.72
Sinji Vrh	−2.39	95.0	−0.304	150.13	132.17	−17.96	−11.96
Sodražica	2.12	95.0	0.307	133.22	151.31	18.09	13.58
Šmarata	0.73	<90.0	0.087	148.22	153.35	5.13	3.46
Trava	−3.34	99.9	−0.621	168.97	132.34	−36.63	−21.68
Želimlje	−0.09	<90.0	0.000	142.00	142.00	0.00	0.00

Table 12.7 Trends in the annual days with snow cover, 1961–2020

Precipitation station	Mann–Kendall test, Z	Confidence level, %	Sen's slope, Q	1961 trend value, days	2020 trend value, days	1961–2020 trend difference, days	1961–2020 trend difference, %
Cerknica	−2.57	95.0	−0.459	72.04	44.93	−27.11	−37.63
Cerovec	−2.90	99.0	−0.458	44.40	17.35	−27.05	−60.92
Črni Vrh	−4.21	99.9	−0.943	106.04	50.41	−55.63	−52.46
Dvor	−3.75	99.9	−0.845	79.04	29.19	−49.85	−63.07
Fužina	−3.72	99.9	−0.776	74.85	29.09	−45.76	−61.14
Godnje[a]	−2.96	99.0	−0.167	12.17	2.33	−9.84	−80.85
Gorenjci pri Adlešičih	−2.64	99.0	−0.515	66.92	36.53	−30.39	−45.41
Grčarice	−3.24	99.0	−0.650	91.05	52.70	−38.35	−42.12
Hrib	−3.86	99.9	−0.818	110.73	62.45	−48.28	−43.60
Hrušica pri Colu	−2.91	99.0	−0.600	95.60	60.20	−35.40	−37.03
Kočevje	−2.81	99.0	−0.588	83.97	49.26	−34.71	−41.34
Kozina	−1.72	90.0	−0.130	17.48	9.78	−7.70	−44.04
Lig	−2.13	95.0	−0.229	25.97	12.49	−13.48	−51.91
Logatec	−3.99	99.9	−0.886	91.81	39.56	−52.25	−56.91
Lokve	−3.84	99.9	−1.067	111.80	48.87	−62.93	−56.29
Nova vas na Blokah	−3.84	99.9	−0.753	107.25	62.83	−44.42	−41.41
Novo mesto	−4.01	99.9	−0.872	84.01	32.54	−51.47	−61.27
Opatje selo[a]	−0.52	<90.0	0.000	1.00	1.00	0.00	0.00
Otlica	−0.91	<90.0	−0.250	72.25	57.50	−14.75	−20.42
Podgrad	−2.58	99.0	−0.342	28.08	7.89	−20.19	−71.90
Pokojišče	−3.17	99.0	−0.700	71.50	30.20	−41.30	−57.76
Postojna	−3.01	99.0	−0.500	52.00	22.50	−29.50	−56.73
Predgrad (Fig. 12.2d)	−3.19	99.0	−0.596	76.28	41.13	−35.15	−46.08
Rakitovec	−1.31	<90.0	−0.059	10.35	6.88	−3.47	−33.55
Razdrto	−1.73	90.0	−0.294	41.88	24.53	−17.35	−41.43
Sinji Vrh	−3.23	99.0	−0.636	76.55	39.00	−37.55	−49.05
Sodražica	−3.11	99.0	−0.652	94.19	55.74	−38.45	−40.82
Šmarata	−3.57	99.9	−0.800	80.60	33.40	−47.20	−58.56
Trava	−2.94	99.0	−0.656	111.53	72.81	−38.72	−34.72
Želimlje	−2.69	99.0	−0.586	67.00	32.41	−34.59	−51.63

[a] due to low elevation and the proximity of the sea, the trends are not comparable with others

12.4.2 Hydrological Variables

The following were examined in terms of hydrological variables (Table 12.1): (1) minimum annual discharge trends, (2) mean annual discharge trends, and (3) maximum annual discharge trends.

12.4.2.1 Minimum Annual Discharge

The minimum annual discharge trends from 1961 to 2020 were downward for all the watercourses analyzed except the Pivka and Reka (Table 12.8, Fig. 12.3a). The confidence level varies greatly: it does not exceed 90% at seven gauging stations, it reaches 95% for the Unica, and it reaches at least 99% at twelve gauging stations.

The average minimum annual discharge in the period observed decreased by 0.05 to 7.48 m^3/s, with an average decrease of 1.37 m^3/s or 26.7%. One gauging station recorded a relative decrease of up to 10%, four stations recorded a relative decrease of 10 to 20%, five stations of 20 to 30%, and eight stations over 30%. The greatest absolute decrease was recorded for the Kolpa at Metlika, where the discharge decreased by 7.48 m^3/s, and the relative difference was the greatest for the Temenica at Rožni Vrh, where the average minimum discharge decreased by 62.2%, and for the Borovniščica at Borovnica, where the decrease was 58.6%. For the Pivka and Reka, which are the only Slovenian Dinaric rivers with a positive trend, the average minimum annual discharge increased by a negligible 0.02 m^3/s and 0.01 m^3/s, respectively.

12.4.2.2 Mean Annual Discharge

The mean annual discharge trends from 1961 to 2020 were downward for all rivers and at all the gauging stations observed (Table 12.9, Fig. 12.3b). The confidence level varies from only 90% or lower at seven stations to 95% at five stations and to at least 99% for the Borovniščica, Kolpa at Petrina, Ljubljanica, Malenščica, Rižana, Temenica, Trebuščica, and Vipava.

During the period studied, the mean annual discharge decreased by 0.30 to 14.61 m^3/s. Most rivers recorded a relative decrease of 10 to 30%, with a major decrease only observed for the Borovniščica (−51.0%), Temenica (−36.5%), and Rižana (−35.5%). The absolute trend difference in the mean annual discharge between 1961 and 2020 was the greatest for the Kolpa at Metlika, where the discharge declined by 14.61 m^3/s, and the greatest relative trend difference can be observed for the Borovniščica at Borovnica, where the discharge decreased by 51.0%.

Table 12.8 Minimum annual discharge trends, 1961–2020

River	Gauging station	Mann–Kendall test, Z	Confidence level, %	Sen's slope, Q	1961 trend value, m^3/s	2020 trend value, m^3/s	1961–2020 trend difference, m^3/s	1961–2020 trend difference, %
Borovniščica	Borovnica	–4.77	99.9	–0.005	0.50	0.21	–0.29	–58.55
Cerkniščica	Cerknica	–1.03	<90.0	–0.001	0.35	0.30	–0.05	–14.61
Hubelj (Fig. 12.4a)	Ajdovščina	–1.52	<90.0	–0.003	0.99	0.83	–0.17	–16.77
Idrijca	Podroteja	–1.57	<90.0	–0.006	2.55	2.19	–0.37	–14.31
Kolpa	Metlika	–3.68	99.9	–0.127	26.08	18.60	–7.48	–28.67
Kolpa	Petrina	–1.80	90.0	–0.021	7.71	6.44	–1.27	–16.42
Krka	Gorenja Gomila	–3.42	99.9	–0.095	18.43	12.84	–5.59	–30.32
Krka	Podbukovje	–3.77	99.9	–0.022	3.71	2.43	–1.27	–34.39
Lahinja	Gradac	–3.17	99.0	–0.008	1.65	1.19	–0.47	–28.22
Ljubija	Verd	–0.75	<90.0	–0.005	3.52	3.22	–0.29	–8.39
Ljubljanica (Fig. 12.4b)	Vrhnika	–3.10	99.0	–0.061	9.79	6.18	–3.61	–36.85
Malenščica	Malni (Fig. 12.3a)	–3.26	99.0	–0.024	5.41	3.99	–1.42	–26.27
Pivka	Prestranek	0.15	<90.0	0.000	0.49	0.51	0.02	4.13
Prečna	Prečna	–3.96	99.9	–0.018	2.83	1.78	–1.05	–37.10
Reka (Fig. 12.4c)	Cerkvenikov mlin	0.06	<90.0	0.000	1.88	1.89	0.01	0.74
Rižana (Fig. 12.4d)	Kubed	–3.82	99.9	–0.008	0.99	0.52	–0.47	–47.22

(continued)

Table 12.8 (continued)

River	Gauging station	Mann–Kendall test, Z	Confidence level, %	Sen's slope, Q	1961 trend value, m^3/s	2020 trend value, m^3/s	1961–2020 trend difference, m^3/s	1961–2020 trend difference, %
Temenica	Rožni Vrh	–4.11	99.9	–0.005	0.44	0.17	–0.28	–62.15
Trebuščica	Dolenja Trebuša	–3.01	99.0	–0.007	1.30	0.89	–0.40	–31.07
Unica (Fig. 12.4e, f)	Hasberg	–2.43	95.0	–0.041	9.83	7.39	–2.45	–24.87
Vipava	Vipava	–2.98	99.0	–0.009	2.27	1.74	–0.53	–23.27

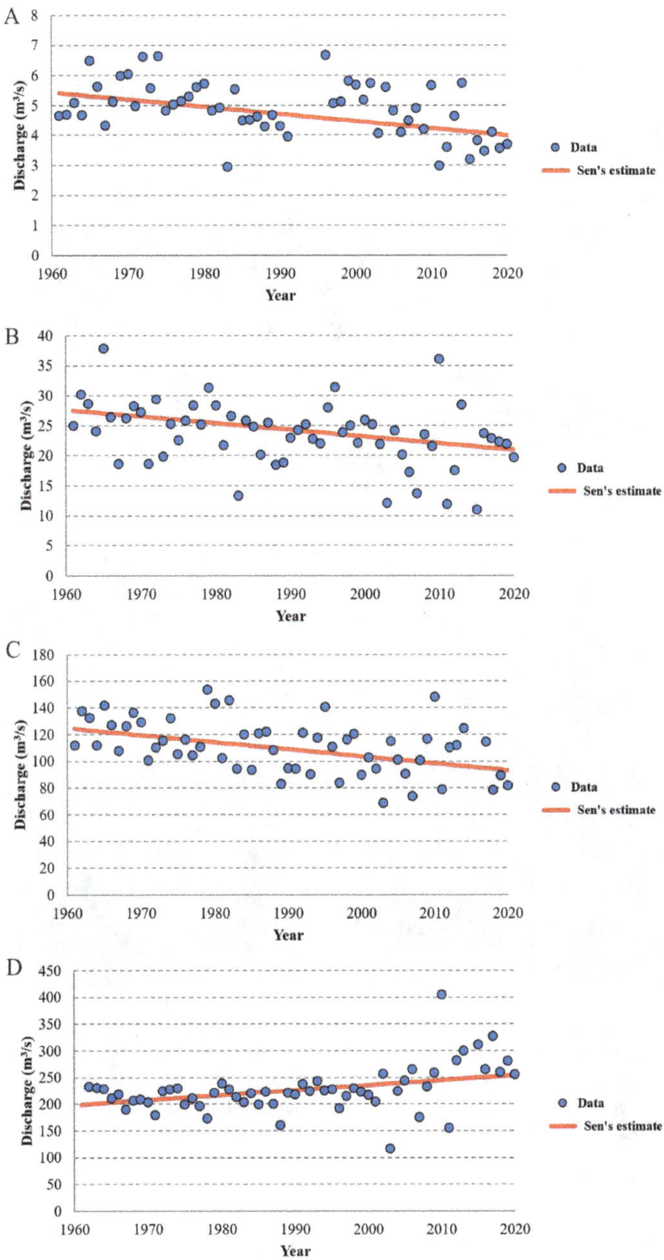

Fig. 12.3 Discharge trends at selected gauging stations: **a** minimum annual discharge trend of the Malenščica River at the Malni gauging station, 1961–2020, **b** mean annual discharge trend of the Ljubljanica River at the Vrhnika gauging station, 1961–2020, **c** maximum annual discharge trend of the Kolpa River at the Petrina gauging station, 1961–2020, **d** absolute maximum annual discharge trend of the Krka River at the Gorenja Gomila gauging station, 1961–2020. *Source* Authors

Fig. 12.4 Some of the watercourses in Slovenia's Dinaric Karst region: **a** the Hubelj karst spring at Ajdovščina, **b** the Ljubljanica karst spring at Vrhnika, **c** the Reka, when it disappears underground at Škocjan Caves; **d** the Rižana karst spring has been developed to supply drinking water, and **e** the source of the Unica at Planina Cave. All photos were taken when the discharge was below average (*Photo* Matija Zorn, GIAM ZRC SAZU Archive). **f** The Unica flooding the Planina Polje (*Photo* Miha Pavšek, GIAM ZRC SAZU Archive. Reproduced with permission.)

12.4.2.3 Maximum Annual Discharge

The maximum annual discharge trends from 1961 to 2020 were downward for all the watercourses analyzed except for the Lahinja, Prečna, and Temenica (Table 12.10, Fig. 12.3c). The confidence level varies from only 90% or even lower at six gauging stations to 95% on the Kolpa at Melika, Krka at Podbukovje, Reka, and Unica,

Table 12.9 Mean annual discharge trends, 1961–2020

River	Gauging station	Mann–Kendall test, *Z*	Confidence level, %	Sen's slope, *Q*	1961 trend value, m^3/s	2020 trend value, m^3/s	1961–2020 trend difference, m^3/s	1961–2020 trend difference, %
Borovniščica	Borovnica	–4.67	99.9	–0.013	1.45	0.71	–0.74	–51.04
Cerkniščica	Cerknica	–2.43	95.0	–0.005	1.15	0.85	–0.30	–26.11
Hubelj (Fig. 12.4a)	Ajdovščina	–2.33	95.0	–0.010	3.24	2.64	–0.61	–18.66
Idrijca	Podroteja	–1.81	90.0	–0.034	9.66	7.66	–1.99	–20.63
Kolpa	Metlika	–2.21	95.0	–0.248	77.48	62.87	–14.61	–18.86
Kolpa	Petrina	–2.59	99.0	–0.089	28.37	23.11	–5.26	–18.53
Krka	Gorenja Gomila	–2.24	95.0	–0.157	48.02	38.74	–9.28	–19.32
Krka	Podbukovje	–2.55	95.0	–0.037	8.56	6.36	–2.20	–25.73
Lahinja	Gradac	–1.08	<90.0	–0.014	5.93	5.12	–0.81	–13.71
Ljubija	Verd	–1.95	90.0	–0.017	7.19	6.19	–1.00	–13.88
Ljubljanica (Fig. 12.4b)	Vrhnika (Fig. 12.3b)	–3.38	99.9	–0.112	27.49	20.89	–6.59	–23.99
Malenščica	Malni	–3.44	99.9	–0.023	7.06	5.72	–1.34	–19.00
Pivka	Prestranek	–1.82	90.0	–0.012	3.01	2.27	–0.73	–24.36
Prečna	Prečna	–1.86	90.0	–0.014	4.49	3.64	–0.85	–18.91
Reka (Fig. 12.4c)	Cerkvenikov mlin	–1.31	<90.0	–0.021	8.89	7.65	–1.24	–13.90
Rižana (Fig. 12.4d)	Kubed	–3.08	99.0	–0.029	4.76	3.07	–1.69	–35.52

(continued)

Table 12.9 (continued)

River	Gauging station	Mann–Kendall test, Z	Confidence level, %	Sen's slope, Q	1961 trend value, m^3/s	2020 trend value, m^3/s	1961–2020 trend difference, m^3/s	1961–2020 trend difference, %
Temenica	Rožni Vrh	–2.73	99.0	–0.006	0.97	0.62	–0.35	–36.49
Trebuščica	Dolenja Trebuša	–3.60	99.9	–0.015	3.15	2.28	–0.87	–27.76
Unica (Fig. 12.4e, f)	Hasberg	–1.55	<90.0	–0.067	23.66	19.71	–3.95	–16.68
Vipava	Vipava	–2.58	99.0	–0.021	7.13	5.92	–1.21	–17.03

and at least 99% at ten stations. During the period observed, the maximum annual discharge largely decreased by 10 to 40%. A decrease over 40% was recorded for the Borovniščica, Rižana, and Trebuščica. The greatest absolute decrease was recorded for the Kolpa at Metlika, where the discharge declined by 61.49 m^3/s, and the relative difference was the greatest for the Trebuščica at Dolenja Trebuša, where the average maximum discharge decreased by 45.6%. The greatest increase in the maximum annual discharge was observed for the Prečna at Prečna—that is, 2.13 m^3/s or 24.5%.

12.5 Discharge Regimes

Long-term changes in temperature and precipitation not only affect the volume of the minimum, mean, and maximum discharge but also have a significant impact on changes in the discharge regime.[12] Among the climate indicators examined, changes in temperature seem especially important because they affect evapotranspiration, as do changes in the number of days with snow cover because they strongly affect all discharge regimes with a distinct snow component. In Dinaric Slovenia, the snow share is important for rivers with headwaters on the high Dinaric plateaus (Fig. 12.1), such as the Trnovo Forest Plateau (*Trnovski gozd*), the Javorniki Hills, and the Snežnik, Bloke, Menišija, Velika Gora, and Kočevje Rog plateaus.

In classifying the discharge regimes based on the 1961–1990 dataset,[13] rivers in Dinaric Slovenia were categorized into four groups. The Borovniščica, whose headwaters extend high up to the shady slopes of the Menišija high plateau, was categorized among rivers with an Alpine rain–snow regime typical of areas in central Slovenia with a temperate continental climate.[14] Among all the discharge regimes in Slovenia, this regime stood out in terms of modest discharge variation and unpronounced discharge maximums and minimums. The Borovniščica's main discharge maximum in November and a secondary one in April only slightly exceeded the average annual discharge. Similarly, its secondary discharge minimums in January and February were also only slightly below the annual average. Only its main discharge minimum in August deviated more significantly from the annual average.[15]

The Alpine rain–snow regime was very similar to the Dinaric–Alpine rain–snow regime, characteristic of the Cerkniščica, Krka, Ljubija, Prečna, and Temenica. This regime, too, was typical of areas in central Slovenia with a temperate continental climate,[16] but it had slightly more pronounced discharge maximums and minimums.

[12] Hrvatin (1998); Frantar (2005); Frantar and Hrvatin (2005); Hrvatin and Zorn (2017); Hrvatin and Zorn (2022).

[13] Hrvatin (1998).

[14] Ogrin (1996).

[15] Hrvatin (1998).

[16] Ogrin (1996).

Table 12.10 Maximum annual discharge trends, 1961–2020

River	Gauging station	Mann–Kendall test, Z	Confidence level, %	Sen's slope, Q	1961 trend value, m^3/s	2020 trend value, m^3/s	1961–2020 trend difference, m^3/s	1961–2020 trend difference, %
Borovniščica	Borovnica	–3.71	99.9	–0.044	5.87	3.25	–2.62	–44.60
Cerkniščica	Cerknica	–1.80	90.0	–0.020	4.96	3.78	–1.18	–23.82
Hubelj (Fig. 12.4a)	Ajdovščina	–3.29	99.0	–0.071	13.72	9.52	–4.20	–30.64
Idrijca	Podroteja	–3.24	99.0	–0.387	59.75	36.93	–22.82	–38.19
Kolpa	Metlika	–2.23	95.0	–1.042	303.08	241.59	–61.49	–20.29
Kolpa	Petrina (Fig. 12.3c)	–3.42	99.9	–0.528	124.20	93.07	–31.13	–25.07
Krka	Gorenja Gomila	–0.62	<90.0	–0.100	119.46	113.56	–5.90	–4.94
Krka	Podbukovje	–1.96	95.0	–0.079	24.87	20.21	–4.67	–18.76
Lahinja	Gradac	0.15	<90.0	0.008	22.02	22.49	0.47	2.13
Ljubija	Verd	–3.00	99.0	–0.043	13.64	11.09	–2.55	–18.69
Ljubljanica (Fig. 12.4b)	Vrhnika	–3.10	99.0	–0.202	60.10	48.18	–11.92	–19.84
Malenščica	Malni	–2.84	99.0	–0.017	8.55	7.53	–1.02	–11.96
Pivka	Prestranek	–1.74	90.0	–0.027	6.70	5.08	–1.62	–24.15
Prečna	Prečna	1.70	90.0	0.036	8.66	10.79	2.13	24.54
Reka (Fig. 12.4c)	Cerkvenikov mlin	–2.09	95.0	–0.175	41.21	30.91	–10.30	–25.00
Rižana (Fig. 12.4d)	Kubed	–3.17	99.0	–0.134	19.42	11.54	–7.88	–40.57

(continued)

Table 12.10 (continued)

River	Gauging station	Mann–Kendall test, Z	Confidence level, %	Sen's slope, Q	1961 trend value, m^3/s	2020 trend value, m^3/s	1961–2020 trend difference, m^3/s	1961–2020 trend difference, %
Temenica	Rožni Vrh	0.23	<90.0	0.002	2.15	2.26	0.11	5.16
Trebuščica	Dolenja Trebuša	–3.69	99.9	–0.123	15.92	8.66	–7.26	–45.62
Unica (Fig. 12.4e, f)	Hasberg	–2.11	95.0	–0.143	46.41	37.98	–8.42	–18.15
Vipava	Vipava	–2.89	99.0	–0.107	28.93	22.60	–6.33	–21.88

The main discharge maximum occurred in March or April, and a secondary one in November or December.[17]

The largest group of rivers, composed of the Hubelj, Idrijca, Kolpa, Lahinja, Ljubljanica, Trebuščica, Unica, and Vipava, had a Dinaric rain–snow regime typical of the temperate continental climate of western and southern Slovenia.[18] Their spring and fall discharge maximums were fairly the same, which is why some of them had their main discharge maximum in April and a secondary one in November, and the others exactly the opposite. Their main discharge minimum was in the summer, and their secondary minimums in January or February were close to the average annual discharge.

The Pivka, Reka, and Rižana, whose basins are close to the Adriatic, had a simple rain regime. Due to the impact of the sub-Mediterranean climate,[19] they had their discharge maximum in the fall, after which the volume of water did not decrease significantly up until April. They had a highly pronounced discharge minimum in the middle of the summer.[20]

After three decades, the discharge regime of Dinaric rivers changed significantly, but notably less than that of Alpine rivers. The greatest changes in the discharge regime in Slovenia can be observed in rivers formerly characterized by highly pronounced snow retention in late fall and winter, and a notable increase in the discharge in the second half of spring due to the melting of the snow cover, which at that time was still ample.[21] The discharge conditions of Dinaric rivers, which only a few decades ago had been classified under three rain–snow regime versions, are now significantly more uniform. For all of them, the November or December discharge maximums significantly exceed the spring maximums, and the August minimums significantly exceed the January or February secondary minimums. Moreover, the secondary winter discharge minimums of most Dinaric rivers are already above the annual average.

Relatively small changes have been observed in the discharge conditions of rivers with a simple rain regime (i.e., the Pivka, Reka, and Rižana). Their discharge maximums have remained in November and December, and their discharge minimums have remained in July and August. Because of thinner snow cover on the Snežnik high karst plateau and its faster melting, the secondary discharge maximum for the Reka has moved from April to February. Similar conditions can be observed for the Rižana, which obtains a portion of its water from the Brkini Hills, which used to receive significantly more snow.

A comparison of the discharge regimes based on the 1961–1990 dataset with the discharge regimes based on the 1991–2020 dataset shows the following differences (Table 12.11, Fig. 12.5):

---

[17] Hrvatin (1998).

[18] Ogrin (1996).

[19] Ibid.

[20] Hrvatin (1998).

[21] Cf. Hrvatin and Zorn (2017); Hrvatin and Zorn (2018); Hrvatin and Zorn (2020); Hrvatin and Zorn (2022).

Table 12.11 1961–1990 and 1991–2020 monthly discharge coefficients (blue shading shows the main and secondary minimums, and red shading shows the main and secondary maximums); the correlation between the two datasets is based on the Pearson correlation coefficient

River	Gauging station	Data set	Jan	Feb	Mar	Apr	May	Jun	Jul	Aug	Sep	Oct	Nov	Dec	Pearson's r
Borovniščica	Borovnica	1961–1990	0.88	0.88	1.22	1.25	1.06	1.17	0.74	0.64	0.80	1.05	1.31	1.01	0.64
		1991–2020	0.92	1.23	1.28	1.01	0.89	0.68	0.38	0.35	0.85	1.31	1.58	1.52	
Cerkniščica	Cerknica	1961–1990	1.09	1.04	1.30	1.40	0.92	0.96	0.68	0.55	0.73	0.87	1.23	1.23	0.83
		1991–2020	1.05	1.20	1.26	1.13	0.85	0.75	0.39	0.35	0.81	1.12	1.57	1.54	
Hubelj	Ajdovščina	1961–1990	0.91	0.95	1.21	1.62	1.06	0.89	0.50	0.43	0.72	1.18	1.38	1.16	0.85
		1991–2020	1.07	1.01	1.15	1.27	0.93	0.62	0.41	0.31	0.79	1.18	1.67	1.58	
Idrijca	Podroteja	1961–1990	1.12	0.98	1.28	1.37	0.81	0.79	0.58	0.54	0.78	1.10	1.50	1.16	0.87
		1991–2020	1.02	1.18	1.27	1.05	0.75	0.58	0.41	0.37	0.87	1.14	1.71	1.64	
Kolpa	Metlika	1961–1990	1.03	1.08	1.35	1.50	0.95	0.70	0.43	0.45	0.70	1.05	1.42	1.34	0.93
		1991–2020	1.09	1.25	1.29	1.25	0.81	0.51	0.33	0.30	0.79	1.20	1.57	1.60	
Kolpa	Petrina	1961–1990	1.04	0.98	1.20	1.51	0.96	0.67	0.40	0.41	0.74	1.16	1.54	1.38	0.94
		1991–2020	1.08	1.13	1.20	1.25	0.76	0.48	0.32	0.26	0.84	1.27	1.69	1.72	
Krka	Gorenja Gomila	1961–1990	0.87	1.02	1.40	1.50	0.91	0.87	0.68	0.56	0.77	0.99	1.25	1.19	0.86
		1991–2020	0.99	1.07	1.30	1.21	0.86	0.73	0.51	0.48	0.85	1.20	1.40	1.40	
Krka	Podbukovje	1961–1990	0.91	1.04	1.40	1.38	0.91	0.90	0.75	0.60	0.73	0.98	1.26	1.13	0.84
		1991–2020	0.95	1.12	1.29	1.19	0.83	0.73	0.49	0.44	0.89	1.17	1.47	1.43	
Lahinja	Gradac	1961–1990	1.02	1.19	1.59	1.42	0.91	0.62	0.44	0.41	0.63	1.00	1.40	1.37	0.96
		1991–2020	1.08	1.35	1.42	1.31	0.82	0.50	0.29	0.31	0.76	1.11	1.45	1.60	
Ljubija	Verd	1961–1990	1.10	1.03	1.15	1.37	1.02	0.90	0.66	0.50	0.71	0.98	1.30	1.29	0.90
		1991–2020	1.20	1.12	1.19	1.12	0.86	0.69	0.48	0.38	0.71	1.15	1.58	1.51	
Ljubljanica	Vrhnika	1961–1990	1.09	1.03	1.19	1.46	0.96	0.84	0.54	0.45	0.69	1.01	1.40	1.34	0.89
		1991–2020	1.20	1.02	1.17	1.16	0.74	0.57	0.35	0.28	0.76	1.31	1.73	1.70	
Malenščica	Malni	1961–1990	1.04	0.97	1.04	1.18	1.10	1.05	0.93	0.76	0.82	0.91	1.07	1.13	0.84
		1991–2020	1.17	1.09	1.12	1.13	1.05	0.91	0.73	0.61	0.73	1.00	1.21	1.25	
Pivka	Prestranek	1961–1990	1.28	1.14	1.20	1.58	1.00	0.58	0.16	0.09	0.47	1.02	1.80	1.67	0.92
		1991–2020	1.54	1.24	1.20	1.08	0.59	0.25	0.03	0.02	0.42	1.09	2.22	2.32	
Prečna	Prečna	1961–1990	1.01	1.09	1.36	1.26	0.92	0.87	0.78	0.68	0.77	0.91	1.16	1.18	0.86
		1991–2020	1.06	1.15	1.21	1.12	0.87	0.78	0.62	0.57	0.89	1.07	1.27	1.38	
Reka	Cerkvenikov mlin	1961–1990	1.34	1.33	1.29	1.34	0.82	0.66	0.26	0.27	0.53	1.08	1.64	1.44	0.94
		1991–2020	1.33	1.43	1.15	0.98	0.74	0.45	0.21	0.23	0.53	1.03	2.01	1.90	
Rižana	Kubed	1961–1990	1.36	1.35	1.33	1.36	0.83	0.75	0.26	0.32	0.59	0.94	1.52	1.40	0.95
		1991–2020	1.24	1.51	1.29	1.10	0.86	0.52	0.20	0.24	0.59	0.95	1.74	1.76	
Temenica	Rožni Vrh	1961–1990	1.02	1.00	1.22	1.19	0.96	0.96	0.86	0.70	0.85	0.90	1.20	1.12	0.82
		1991–2020	0.99	1.18	1.34	1.02	0.82	0.79	0.61	0.48	0.85	1.13	1.40	1.40	
Trebuščica	Dolenja Trebuša	1961–1990	1.08	0.97	1.18	1.38	1.02	0.88	0.60	0.55	0.73	1.09	1.35	1.16	0.91
		1991–2020	1.08	1.11	1.22	1.19	0.94	0.74	0.53	0.41	0.75	1.05	1.54	1.46	
Unica	Hasberg	1961–1990	1.17	1.07	1.14	1.50	1.05	0.85	0.52	0.38	0.61	0.95	1.36	1.39	0.90
		1991–2020	1.35	1.16	1.24	1.20	0.80	0.55	0.31	0.24	0.60	1.10	1.69	1.76	
Vipava	Vipava (Fig. 12.5)	1961–1990	1.01	1.02	1.27	1.49	0.96	0.91	0.55	0.53	0.71	1.01	1.35	1.20	0.85
		1991–2020	1.05	1.11	1.19	1.14	0.87	0.70	0.45	0.41	0.81	1.17	1.62	1.48	

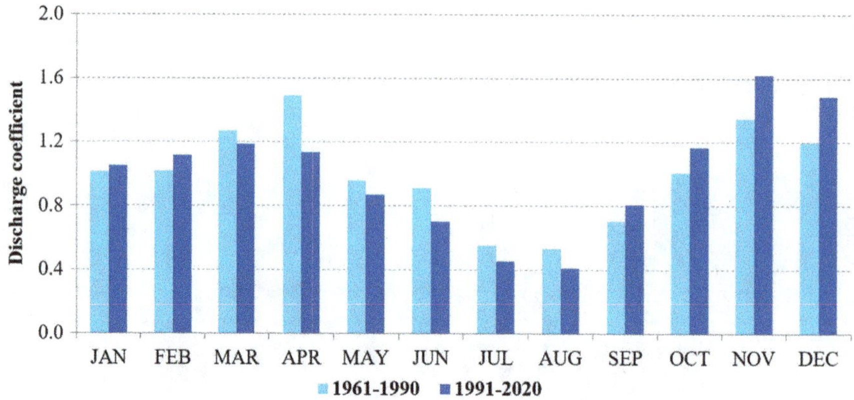

Fig. 12.5 Changes in the discharge regime of the Vipava River at the Vipava gauging station between the 1961–1990 and 1991–2020 periods. *Source* Authors

- For rivers that used to have a Dinaric–Alpine and Dinaric snow–rain regime (i.e., the Cerkniščica, Hubelj, Kolpa, Krka, Lahinja, Ljubija, Ljubljanica, Malenščica, Prečna, Temenica, Trebuščica, and Vipava), the main discharge maximum has moved from the spring to the fall, which shows a significantly reduced impact of snow cover;
- The impact of winter snow retention has significantly decreased and can now only be observed in rivers with their catchment area on high karst plateaus;
- Summer minimums are becoming more pronounced for all rivers;
- The November and December water levels have increased significantly, implying that winter is "running late."

The intensity of changes in the monthly discharge coefficients for individual rivers between the 1961–1990 and 1991–2020 periods was also ascertained with the Pearson correlation coefficient (Table 12.11). By far the lowest values were calculated for the Borovniščica (0.64), where the monthly discharge coefficients between June and August decreased significantly. At the same time, the November, December, and February coefficients notably increased as a result of gradual transition in precipitation from snow to rain. At the Pokojišče precipitation station close to the upper reaches of the Borovniščica, the number of days with snow cover decreased from 70 to 30, or to less than half, between 1961 and 2020.

Pearson correlation coefficients ranging from 0.83 to 0.90 were ascertained for the Cerkniščica, Hubelj, Idrijca, Krka (at Podbukovje and Gorenja Gomila), Ljubija, Ljubljanica, Malenščica, Prečna, Temenica, Unica, and Vipava. These are mostly rivers whose basins extend onto high karst plateaus, where both the volume and duration of the snow cover have been declining significantly over the past decades.

The highest Pearson correlation coefficients, ranging between 0.91 and 0.96, are typical of the Kolpa, Lahinja, Pivka, Reka, Rižana, and Trebuščica. This small group includes very diverse rivers. For example, the Kolpa and Lahinja are karst streams into which water flows through long and branched underground karst channels, which predominantly mitigates major discharge fluctuations. In turn, the Reka and Rižana have a major portion of their basins in the Mediterranean low hills, where snow cover is relatively rare or even an exception, and therefore a smaller volume of snow has not had a significant impact on discharge changes in recent decades.

Among other things, the fact that changes in the discharge regimes of Dinaric rivers between the 1961–1990 and 1991–2020 periods has primarily been the result of stronger evapotranspiration and changes in the duration and volume of snow cover can be observed at the gauging station at the source of the Vipava River (Fig. 12.5). The November rain maximum has already strongly exceeded the spring high waters, which are primarily the consequence of snow melting on the Nanos and Hrušica high karst plateaus. Due to high temperatures and strong evapotranspiration, its main summer minimum has become even more pronounced and, due to a smaller volume of snow retained, its secondary winter minimum already somewhat exceeds the average annual discharge.

Differences in the monthly discharge coefficients between the 1961–1990 and 1991–2020 periods (Table 12.12) are largely consistent for all the rivers analyzed in Slovenia's Dinaric Karst region. They indicate a gradual decrease in the spring and summer discharges and an increase in the fall and, to a slightly smaller extent, winter discharges. The greatest decrease in the discharge occurs in April and from June to August, and the greatest increase occurs in November, December, and February. In addition to increased late spring and summer evapotranspiration, an important reason for these developments is the gradual temperature increase and the resulting thinner and briefer snow cover. In the past, a significant share of fall precipitation was in the form of snow, which caused snow retention on the rivers, and the thicker snow cover on higher karst plateaus continued melting until late spring.

12.6 Floods

Because of the modest river network and the predominance of subterranean water flow, floods only pose a threat to a small portion of Dinaric Slovenia. They are the most frequent in the numerous karst poljes, where they usually build up gradually and the water then remains there for several days, weeks, or even months. They develop because the karst groundwater level rises above the surface or because the volume of water flowing into the underground drainage channels exceeds their capacity.[22] For example, flood-prone areas in the Cerknica Polje cover 2370 ha of land, in the Planina Polje (Fig. 12.4f) they cover 950 ha, and in the Lož Polje they cover 300 ha.[23] Because floods are annual and long-term in some karst poljes and in some parts of karst plateaus, the term "intermittent lakes" has been established in such cases.

[22] Natek (2005).

[23] Komac et al. (2008).

Table 12.12 Changes in the monthly discharge coefficients between the 1961–1990 and 1991–2020 periods (blue shading indicates decreasing ratios, and red shading shows increasing ratios)

River	Gauging station	Jan	Feb	Mar	Apr	May	Jun	Jul	Aug	Sep	Oct	Nov	Dec
Borovniščica	Borovnica	0.04	0.35	0.05	−0.23	−0.17	−0.49	−0.36	−0.29	0.05	0.26	0.27	0.51
Cerkniščica	Cerknica	−0.04	0.16	−0.05	−0.28	−0.07	−0.21	−0.29	−0.20	0.08	0.25	0.33	0.31
Hubelj	Ajdovščina	0.16	0.06	−0.05	−0.35	−0.13	−0.27	−0.09	−0.12	0.07	0.00	0.29	0.42
Idrijca	Podroteja	−0.10	0.20	0.00	−0.32	−0.05	−0.21	−0.16	−0.17	0.09	0.04	0.21	0.48
Kolpa	Metlika	0.06	0.17	−0.06	−0.25	−0.14	−0.19	−0.10	−0.14	0.08	0.15	0.15	0.26
Kolpa	Petrina	0.04	0.15	0.00	−0.26	−0.20	−0.19	−0.07	−0.15	0.10	0.11	0.14	0.35
Krka	Gorenja Gomila	0.11	0.05	−0.10	−0.29	−0.05	−0.14	−0.17	−0.08	0.08	0.22	0.16	0.21
Krka	Podbukovje	0.04	0.08	−0.11	−0.19	−0.08	−0.17	−0.26	−0.16	0.15	0.19	0.21	0.29
Lahinja	Gradac	0.06	0.15	−0.17	−0.11	−0.09	−0.12	−0.15	−0.10	0.13	0.12	0.04	0.24
Ljubija	Verd	0.10	0.09	0.04	−0.25	−0.16	−0.21	−0.17	−0.11	0.00	0.17	0.28	0.22
Ljubljanica	Vrhnika	0.11	−0.01	−0.02	−0.30	−0.21	−0.27	−0.20	−0.16	0.07	0.30	0.33	0.36
Malenščica	Malni	0.12	0.12	0.09	−0.05	−0.05	−0.14	−0.20	−0.15	−0.08	0.09	0.14	0.12
Pivka	Prestranek	0.26	0.10	0.00	−0.51	−0.41	−0.33	−0.13	−0.07	−0.05	0.07	0.41	0.64
Prečna	Prečna	0.05	0.06	−0.15	−0.14	−0.04	−0.09	−0.15	−0.11	0.11	0.16	0.11	0.20
Reka	Cerkvenikov mlin	−0.01	0.10	−0.14	−0.36	−0.08	−0.20	−0.05	−0.04	0.00	−0.04	0.37	0.45
Rižana	Kubed	−0.12	0.16	−0.05	−0.25	0.03	−0.23	−0.06	−0.08	−0.01	0.01	0.22	0.37
Temenica	Rožni Vrh	−0.03	0.18	0.12	−0.17	−0.15	−0.17	−0.26	−0.22	0.00	0.23	0.20	0.28
Trebuščica	Dolenja Trebuša	−0.01	0.14	0.04	−0.19	−0.08	−0.14	−0.06	−0.15	0.02	−0.05	0.19	0.30
Unica	Hasberg	0.18	0.09	0.10	−0.31	−0.25	−0.30	−0.21	−0.14	0.00	0.14	0.33	0.37
Vipava	Vipava	0.04	0.10	−0.08	−0.36	−0.09	−0.20	−0.10	−0.12	0.10	0.16	0.27	0.29

The best-known Dinaric intermittent lakes include Lake Cerknica in the Cerknica Polje[24] and the Pivka Lakes along the upper reaches of the Pivka River.[25]

In addition to floods in karst poljes, flashfloods also occur occasionally along individual streams. They are the most common on the Iška, Borovniščica, Cerkniščica, (Sodražica) Bistrica, and (Podbočje) Sušica.[26] Along the lower courses of major Dinaric rivers, lowland floods (Fig. 12.1) also occur, mostly posing a threat to the neighboring areas. They develop due to the difference in the speed of incoming karst high waters and the discharge capacities of the riverbeds.[27] For example, along its lower course, the Krka River threatens the Pannonian Krka Basin, and the Vipava and Hubelj rivers pose a threat to the Mediterranean Vipava Valley.[28]

Floods are closely connected with the absolute maximum river discharge. In this regard, the upward trends in certain watercourses are the main cause for concern.[29] Between 1961 and 2020, the absolute annual maximum discharge trends were downward for the Borovniščica, Cerkniščica, Hubelj, Idrijca, Kolpa, Lahinja, Ljubija, Pivka, Rižana, and Trebuščica, and upward for the Krka, Ljubljanica, Malenščica, Prečna, Reka, Temenica, Unica, and Vipava (Table 12.13, Fig. 12.3d). The confidence level is low, not even reaching 90% at fifteen out of twenty gauging stations. The only exceptions are the stations at Petrina on the Kolpa (99.9%), Gorenja Gomila on the Krka (99.9%), Vrhnika on the Ljubljanica (95%), Malni on the Malenščica

[24] Kovačič (2010).

[25] Petrič and Kogovšek (2005); Stepišnik (2017).

[26] Trobec (2016).

[27] Natek (2005).

[28] Komac et al. (2008).

[29] Oblak et al. (2021).

(99%), and Prečna on the Prečna (99.9%). At twelve gauging stations, the absolute maximum annual discharge in the period observed decreased or increased by less than 10%.

In terms of water share, the absolute maximum annual discharge decreased the most at Petrina on the Kolpa (i.e., by 176.34 m^3/s or 34.3%), at Kubed on the Rižana (i.e., by 9.94 m^3/s or 18.7%), and at Borovnica on the Borovniščica (i.e., by 4.93 m^3/s or 18.3%). On the other hand, it increased the most at Prečna on the Prečna (i.e., by 6.74 m^3/s or 44.4%), at Rožni Vrh on the Temenica (i.e., by 1.78 m^3/s or 35.9%), and at Gornja Gomila on the Krka (i.e., by 55.49 m^3/s or 27.9%).

The coefficients of variation in the maximum annual discharge between the 1961–1980, 1981–2000, and 2001–2020 periods are important for understanding flood occurrence. Specifically, the coefficients of variation show the dispersion of data: the higher the ratio, the greater the dispersion.[30] In terms of river discharge, the higher the variation ratio, the higher the probability of the occurrence of both minimum and maximum discharges.

The calculations showed that the lowest coefficients of variation were approximately evenly distributed between the 1961–1980 and 1981–2000 periods. For the Kolpa (at the Petrina gauging station), Lahinja, and Temenica, they were the highest between 1961 and 1980, for the Idrijca and Reka, they were the highest between 1981 and 2000, and for the Borovniščica, Cerkniščica, Hubelj, Kolpa (at the Metlika gauging station), Krka, Ljubija, Ljubljanica, Malenščica, Pivka, Prečna, Rižana, Trebuščica, Unica, and Vipava, they were the highest in the most recent period (i.e., from 2001 to 2020). Based on this, greater deviations in the maximum annual discharge, including occasional "catastrophic" events, can be expected for these rivers (Table 12.14).

For example, in Fig. 12.3d, which shows the absolute maximum annual discharge trend of the Krka at the Gorenja Gomila gauging station, the trend line varies between 198.9 m^3/s and 254.4 m^3/s, with certain annual values standing out significantly in both the downward and upward directions. For example, the 2003 absolute maximum annual discharge was only 116.2 m^3/s, whereas the 2010 one reached 404.7 m^3/s.

With regard to the potential occurrence of floods, the increasing fall maximum discharge should also be highlighted; it already exceeds the spring maximum for all rivers because it is primarily the November and December water levels that are increasing as a result of winter "running late." At the same time, the summer (August) minimum is decreasing increasingly (Table 12.11). The former indicates an increased probability of high water and floods in the fall months and a smaller probability in the spring. The latter indicates the occurrence of summer droughts. Other researchers also recognize the increased probability of floods in the fall and winter months.[31]

---

30 Sagadin (2003).

31 Ministrstvo za okolje in prostor, "Ocena tveganja za poplave.".

Table 12.13 Absolute maximum annual discharge trends, 1961–2020

River	Gauging station	Mann–Kendall test, Z	Confidence level, %	Sen's slope, Q	1961 trend value, m^3/s	2020 trend value, m^3/s	1961–2020 trend difference, m^3/s	1961–2020 trend difference, %
Borovniščica	Borovnica	–1.32	<90.0	–0.084	26.88	21.95	–4.93	–18.33
Cerkniščica	Cerknica	–0.52	<90.0	–0.043	25.88	23.32	–2.56	–9.90
Hubelj	Ajdovščina	–0.40	<90.0	–0.025	33.52	32.02	–1.49	–4.46
Idrijca	Podroteja	–0.22	<90.0	–0.128	201.22	193.66	–7.56	–3.76
Kolpa	Metlika	–1.65	90.0	–1.760	852.86	749.04	–103.82	–12.17
Kolpa	Petrina	–3.43	99.9	–2.989	514.71	338.37	–176.34	–34.26
Krka	Gorenja Gomila (Fig. 12.3d)	3.53	99.9	0.941	198.93	254.42	55.49	27.90
Krka	Podbukovje	0.79	<90.0	0.076	54.38	58.87	4.49	8.26
Lahinja	Gradac	–0.89	<90.0	–0.099	67.19	61.33	–5.86	–8.73
Ljubija	Verd	–0.89	<90.0	–0.012	20.97	20.27	–0.70	–3.33
Ljubljanica	Vrhnika	2.21	95.0	0.135	91.43	99.38	7.96	8.70
Malenščica	Malni	2.73	99.0	0.013	9.35	10.12	0.77	8.27
Pivka	Prestranek	–0.08	<90.0	–0.002	14.75	14.61	–0.14	–0.98
Prečna	Prečna	3.60	99.9	0.114	15.19	21.93	6.74	44.40
Reka	Cerkvenikov mlin	0.42	<90.0	0.201	182.09	193.96	11.87	6.52
Rižana	Kubed	–1.33	<90.0	–0.169	53.32	43.38	–9.94	–18.65
Temenica	Rožni Vrh	1.16	<90.0	0.030	4.97	6.75	1.78	35.87
Trebuščica	Dolenja Trebuša	–0.40	<90.0	–0.041	41.15	38.73	–2.43	–5.89
Unica	Hasberg	0.34	<90.0	0.036	73.39	75.49	2.10	2.86
Vipava	Vipava	1.49	<90.0	0.117	49.77	56.68	6.90	13.87

Table 12.14 Absolute maximum annual discharge coefficients of variation (DCV), 1961–2020

River	Gauging station	DCV 1961–1980	DCV 1981–2000	DCV 2001–2020
Borovniščica	Borovnica	19.69	42.97	48.99
Cerkniščica	Cerknica	37.79	37.03	67.24
Hubelj	Ajdovščina	26.80	21.08	29.59
Idrijca	Podroteja	26.43	35.99	35.70
Kolpa	Metlika	21.76	13.28	22.79
Kolpa	Petrina	25.34	16.70	24.22
Krka	Gorenja Gomila	8.80	8.54	25.28
Krka	Podbukovje	17.96	21.49	36.45
Lahinja	Gradac	28.72	13.21	26.45
Ljubija	Verd	6.14	5.11	7.53
Ljubljanica	Vrhnika	3.39	6.11	12.96
Malenščica	Malni	3.21	3.66	6.76
Pivka	Prestranek	23.03	25.58	63.09
Prečna	Prečna	13.93	11.08	40.89
Reka	Cerkvenikov mlin	27.83	31.74	30.35
Rižana	Kubed	19.94	30.54	51.17
Temenica	Rožni Vrh	50.98	27.55	48.75
Trebuščica	Dolenja Trebuša	41.55	30.66	48.64
Unica	Hasberg	10.21	8.26	16.08
Vipava	Vipava	16.18	18.39	24.53

However, it is not only climate change that leads to changes in the hydrological regime and the consequent occurrence of floods. One must not neglect the anthropogenic impacts, such as land-use changes, river training, dams, and urbanization.[32]

12.7 Conclusion

The key findings about changes in the selected climate and hydrological variables between 1961 and 2020 largely agree with the trends established by other authors.[33] They can be summarized as follows:

[32] Ulaga (2002); Bormann (2010); Zampieri et al. (2015); Šraj et al. (2016); Kovačič et al. (2020).

[33] Cf. Ulaga (2002); Uhan (2007); Frantar et al. (2008); Ulaga et al. (2008); Kobold et al. (2012); Kovačič et al. (2016); Makor (2016); Hrvatin and Zorn (2017); Hrvatin and Zorn (2018); Hrvatin and Zorn (2020); Hrvatin and Zorn (2022); Ravbar et al. (2017); Dolinar (2018); Jelen et al. (2020); Oblak et al. (2021); *Podnebne spremembe* (2021).

- The average annual air temperature from 1961 to 2020 increased by 0.032 to 0.050 °C a year at all temperature stations analyzed, which means that during that period, the average annual air temperature warmed up by approximately 2.5 °C. The temperature difference was the smallest at the Kočevje station, where the temperature increased by 1.87 °C over six decades, and the largest at the Novo mesto station, where the temperature increased by 2.95 °C.
- From 1961 to 2020, annual precipitation decreased at twenty-four out of thirty precipitation stations. Most differences in annual precipitation are moderate, not exceeding 10% at eighteen stations and 20% at twenty-eight. The downward trend was the most pronounced at Podgrad and Hrušica pri Colu (the differences slightly exceeding −20%), and the upward trend was the most pronounced at Predgrad (a nearly +7% difference).
- From 1961 to 2020, the annual number of days with precipitation over 0.1 mm increased at twelve precipitation stations, decreased at fourteen, and remained the same at four. At twenty-one stations, the negative or positive trend deviations were smaller, not reaching 10%. The most pronounced downward trend was recorded at the Predgrad, Trava, and Dvor stations (differences below −20%), and the upward trend was the strongest at Podgrad (differences over +20%), Otlica, and Cerknica (differences over +10%).
- During the 1961–2020 period, the number of days with snow cover decreased significantly at all precipitation stations, where it typically snows every year. At twenty-two out of twenty-eight stations, it decreased by 30 to 60%, or by an average of thirty-five days a year. It only decreased by less than 30% at Otlica, whereas a decrease of over 60% was recorded at Podgrad, Dvor, Novo mesto, Fužina, and Cerovec. The duration of snow cover expressed in days declined the most at Lokve (−63 days), Črni Vrh (−56 days), and Logatec (−52 days).
- During the 1961–2020 period, the average minimum annual discharge trends were downward for all the rivers analyzed, except for the Pivka and Reka. One gauging station recorded a decrease lower than 10%, four between 10 and 20%, five between 20 and 30%, and eight over 30%. On the Temenica, the average minimum annual discharge decreased by a full 62% and on the Borovniščica by 59%.
- From 1961 to 2020, the average mean annual discharge trends were downward for all the rivers observed. The mean annual discharge of Slovenian Dinaric rivers decreased by 0.30 to 14.61 m^3/s. A relative decrease of 10 to 30% was recorded for most rivers, with a more pronounced decrease only being ascertained on the Borovniščica (−51%), Temenica (−37%), and Rižana (−36%).
- From 1961 to 2020, the average maximum annual discharge trends were downward for all the rivers analyzed, except for the Lahinja, Prečna, and Temenica, largely showing a 10 to 40% decrease. The average maximum annual discharge decreased by over 40% on the Borovniščica, Rižana, and Trebuščica. Among the rivers with a downward trend, the discharge declined the most on the Trebuščica (−46%), and among the rivers with an upward trend it increased the most on the Prečna (+25%).

- Between 1961 and 2020, the absolute maximum annual discharge trends were downward at eleven gauging stations and upward at nine. At half of the stations, the absolute maximum annual discharge decreased or increased by less than 10%. Among the rivers with a downward trend, it decreased the most on the Kolpa at Petrina (−34%) and on the Rižana (−19%), and among the rivers with a positive trend, it increased the most on the Prečna (+44%), Temenica (+36%), and Krka at Gorenja Gomila (+28%).
- A comparison between the 1991–2020 discharge regime dataset and the 1961–1990 dataset shows that the impact of snow cover is decreasing rapidly. At all gauging stations, the main discharge maximum due to fall rain has already strongly exceeded the spring secondary maximum, in which snow melting plays an important role. The summer minimum is growing more distinct at all stations and, with regard to the secondary winter minimum, the discharge is increasing and in many places it already exceeds the annual average, implying that winter is "running late."
- The increasing absolute maximum annual discharge on some rivers, the increasing discharge variability, and the increasing fall discharge, despite a decrease in the total volume of water, may indicate a greater flood hazard.

Acknowledgements This study was conducted as part of the research program The Geography of Slovenia (P6-0101), financed by the Slovenian Research and Innovation Agency.

References

Sources

Agencija Republike Slovenije za okolje. Arhiv hidroloških podatkov [Archive of hydrological data]. Accessed 17 Aug 2021. http://www.arso.gov.si/vode/podatki/arhiv/hidroloski_arhiv.html

Agencija Republike Slovenije za okolje. Arhiv meteoroloških podatkov [Archive of meteorological data]. Accessed 11 Aug 2021. http://meteo.arso.gov.si/

Ministrstvo za okolje in prostor. Ocena tveganja za poplave [Flood risk assessment]. Accessed 19 Nov 2021. https://www.gov.si/assets/ministrstva/MOP/Dokumenti/Voda/NZPO/ocena_tveganj_poplave_2016.pdf

Literature

Bormann H (2010) Runoff regime changes in German rivers due to climate change. Erdkunde 64(3):257–279. https://doi.org/10.3112/erdkunde.2010.03.04

Dolinar M (ed) (2018) Ocena podnebnih sprememb v Sloveniji do konca 21. stoletja: sintezno poročilo—prvi del [Assessment of climate change in Slovenia until the end of the 21st century: synthesis report—part one]. Agencija Republike Slovenije za okolje, Ljubljana. https://meteo.arso.gov.si/uploads/probase/www/climate/text/sl/publications/OPS21_Porocilo.pdf

Frantar P (2005) Pretočni režimi slovenskih rek in njihova spremenljivost [River flow regimes of Slovene rivers and their fluctuation]. Ujma 19:145–153

Frantar P, Hrvatin M (2005) Pretočni režimi v Sloveniji med letoma 1971 in 2000 [Discharge regimes in Slovenia from 1971 to 2000]. Geografski Vestnik 77(2):115–127

Frantar P, Kobold M, Ulaga F (2008) Trend pretokov [Discharge trends]. In: Frantar P (ed) Vodna bilanca Slovenije 1971–2000. Agencija Republike Slovenije za okolje, Ljubljana, pp 50–60

Hrvatin M (1998) Discharge regimes in Slovenia. Geografski Zbornik 38:59–87

Hrvatin M, Zorn M (2017) Trendi pretokov rek v slovenskih Alpah med letoma 1961 in 2010 [River discharge trends in Slovenian Alps between 1961 and 2010]. Geografski Vestnik 89(2):9–35. https://doi.org/10.3986/GV89201

Hrvatin M, Zorn M (2018) Recentne spremembe rečnih pretokov in rečnih režimov v Julijskih Alpah [Recent changes in river discharges and discharge regimes in the Julian Alps]. In: Zorn M et al (ed) Triglav 240. Založba ZRC, Ljubljana, pp 107–129. https://doi.org/10.3986/9789610500841

Hrvatin M, Zorn M (2020) Climate and hydrological changes in Slovenia's Podravje region between 1961 and 2018. Podravina 19(38):161–180

Hrvatin M, Zorn M (2022) Climate change impacts on hydrology in the Mediterranean part of Slovenia. In: Filho WL, Manolas E (eds) Climate change in the Mediterranean and Middle Eastern Region. Springer, Cham, pp 85–118. https://doi.org/10.1007/978-3-030-78566-6_4

Jelen M, Mikoš M, Bezak N (2020) Kraški izviri v Sloveniji: analiza trendov [Karst springs in Slovenia: trend analysis]. Acta Hydrotechnica 33(58):1–12. https://doi.org/10.15292/acta.hydro.2020.01

Kladnik D (1998) Dinarski svet [Dinaric landscapes]. In: Perko D, Orožen Adamič M (eds) Slovenija: pokrajine in ljudje. Mladinska knjiga, Ljubljana, pp 296–309

Kobold M, Dolinar M, Frantar P (2012) Spremembe vodnega režima zaradi podnebnih sprememb in drugih antropogenih vplivov [Water regime changes due to climate change and other anthropogenic impacts]. In: Brilly M (ed) 1. kongres o vodah Slovenije. Fakulteta za gradbeništvo in geodezijo, Ljubljana, pp 7–22

Komac B, Natek K, Zorn M (2008) Geografski vidiki poplav v Sloveniji [Geographical aspects of floods in Slovenia]. Založba ZRC, Ljubljana. https://doi.org/10.3986/9789612545451

Kovačič G (2010) An attempt towards an assessment of the Cerknica Polje water balance. Acta Carsologica 39(1):39–49. https://doi.org/10.3986/ac.v39i1.111

Kovačič G (2016) Trendi pretokov rek jadranskega povodja v Sloveniji brez Posočja [Discharge trends of the Adriatic Sea basin rivers in Slovenia, excluding the Soča river basin]. Geografski Vestnik 88(2):9–29. https://doi.org/10.3986/GV88201

Kovačič G, Kolega N, Brečko Grubar V (2016) Vpliv podnebnih sprememb na količine vode in poplave morja v slovenski Istri [Climate change impacts on water quantities and sea flooding in Slovene Istria]. Geografski Vestnik 88(1):21–36. https://doi.org/10.3986/GV88102

Kovačič G, Petrič M, Ravbar N (2020) Evaluation and quantification of the effects of climate and vegetation cover change on karst water sources: case studies of two springs in South-Western Slovenia. Water 12(11):1–20. https://doi.org/10.3390/w12113087

Kraner Šumenjak T, Šuštar V. (2011) Parametrični in neparametrični pristopi za odkrivanje trenda v časovnih vrstah [Parametric and nonparametric approach for trend detection in time series]. Acta Agriculturae Slovenica 97(3):305–312

Makor S (2016) Trendi spreminjanja pretokov rek v Sloveniji [Trends of river discharge changes in Slovenia]. BSc Thesis, University of Ljubljana

Natek K (2005) Poplavna območja v Sloveniji [The flood areas of Slovenia]. Geografski Obzornik 52(1):13–18

Oblak J, Kobold M, Šraj M (2021) The influence of climate change on discharge fluctuations in Slovenian rivers. Acta Geogr Slov 61(2):155–169. https://doi.org/10.3986/AGS.9942

Ogrin D (1996) Podnebni tipi v Sloveniji [The climate types in Slovenia]. Geografski Vestnik 68:39–56

Perko D, Ciglič R (2020) Slovenia's landscapes. In: Perko D, Ciglič R, Zorn M (eds) The geography of Slovenia: small but diverse. Springer, Cham, pp 211–225. https://doi.org/10.1007/978-3-030-14066-3_14

Petrič M, Kogovšek J (2005) Hidrogeološke značilnosti območja presihajočih Pivških jezer [Hydrogeological characteristics of the area of intermittent karst lakes of Pivka]. Acta Carsologica 34(3):599–618. https://doi.org/10.3986/ac.v34i3.255

Perko D, Ciglič R, Hrvatin M (2021) Landscape macrotypologies and microtypologies of Slovenia. Acta Geogr Slov 61(3):7–89. https://doi.org/10.3986/AGS.10384

Podnebne spremembe 2021: Fizikalne osnove in stanje v Sloveniji. Poročilo IPCC 2021, Povzetek za odločevalce z dodanim opisom stanja v Sloveniji [Climate change 2021: Physical basis and situation in Slovenia. IPCC Report 2021, Summary for decision-makers with an added description of the situation in Slovenia]. Agencija Republike Slovenije za okolje, Ljubljana, 2021. https://meteo.arso.gov.si/uploads/probase/www/climate/text/sl/publications/2021_11-Poro%C4%8Dilo%20IPPC%20Podnebje%202021.pdf.

Ravbar N, Kovačič G, Petrič M, Kogovšek J, Brun C, Koželj A (2017) Climatological trends and anticipated karst spring quantity and quality: case study of the Slovene Istria. Geol Soc London Spec Publ 466:295–305. https://doi.org/10.1144/SP466.19

Sagadin J (2003) Statistične metode za pedagoge [Statistical methods for educators]. Obzorja, Maribor

Salmi T, Määttä A, Anttila P, Ruoho-Airola T, Amnell T (2002) Detecting trends of annual values of atmospheric pollutants by the Mann-Kendall test and Sen's slope estimates—the excel template application MAKESENS. Finnish Meteorological Institute, Helsinki

Šraj M, Menih M, Bezjak N (2016) Climate variability impact assessment on the flood risk in Slovenia. Phys Geogr 37(1):73–87. https://doi.org/10.1080/02723646.2016.1155389

Stepišnik U (2017) Dinarski kras: plitvi kras Zgornje Pivke [The Dinaric Karst: Shallow karst of Upper Pivka]. Znanstvena založba Filozofske fakultete, Ljubljana. https://doi.org/10.4312/978 9612379223

Trobec T (2016) Spatio-temporal distribution of flash floods in Slovenia. Dela 46:21–39. https://doi.org/10.4312/dela.46.1.5-39

Ulaga F (2002) Trendi spreminjanja pretokov slovenskih rek [Trends in discharge changes of Slovenian rivers]. Dela 18:93–114. https://doi.org/10.4312/dela.18.93-114

Uhan J (2007) Trendi velikih in malih pretokov rek v Sloveniji [Trends in high and low river flows in Slovenia]. Ujma 21:164–166

Ulaga F, Kobold M, Frantar P (2008) Trends of river discharges in Slovenia. IOP Conf Ser Earth Environ Sci 4(1). https://doi.org/10.1088/1755-1307/4/1/012030

Zampieri M, Scoccimarro E, Gualdi S, Navarra A (2015) Observed shift towards earlier spring discharge in the main Alpine rivers. Sci Total Environ 503–504:222–232. https://doi.org/10.1016/j.scitotenv.2014.06.036

Mauro Hrvatin The former Head of the Department of Physical Geography of the Anton Melik Geographical Institute of the Research Centre of the Slovenian Academy of Sciences and Arts (Ljubljana, Slovenia). He is a geographer and a sociologist, Ph.D. His main research interests include physical geography (especially geomorphology, hydrogeography and climate geography) and geographic information systems.

Matija Zorn Head of the Anton Melik Geographical Institute of the Research Centre of the Slovenian Academy of Sciences and Arts (Ljubljana, Slovenia), and a lecturer at the Faculty of Humanities of the University of Primorska (Koper, Slovenia). He is a geographer and a historian, Ph.D. His main research interest includes physical geography (especially geomorphology), natural hazards, land degradation, environmental history, historical geography and geographic information systems. Most recently he co-edited a book titled *The Geography of Slovenia: Small But Diverse* (Springer Nature). He is Chairman of the *Commission on Land Degradation and Desertification of the International Geographical Union.*

Part IV
Environmental Degradation

Chapter 13
Empire, Extraction, and Externalization. Wood Impregnation in Early Twentieth Century Bosnia and Herzegovina and Its Precarious Legacy

Martin Meiske

Abstract At the turn of the century, the development of high-pressure impregnation techniques and the use of the coal tar oil creosote substantially prolonged the life of railway ties and other wooden infrastructure materials. The rising wood impregnation industry played a vital role in Austro-Hungarian imperial endeavors, such as in Bosnia and Herzegovina. The construction and maintenance of infrastructure helped to establish and maintain a quasi-colonial rule and exploit the country's natural resources. These practices had long-term impacts on the environment and society. Considering that large parts of the region consist of karst ecologies that are especially vulnerable to pollution, close environmental monitoring and sustainable management are needed. This paper examines questions of (post-)colonialism, occupational health, environmental justice, and the long-term environmental impact of toxic impregnation substances, following the activities of the Austrian impregnation company Guido Rütgers KG. It reflects on how environmental history can contribute to reconstructing changing human-nature relations in the Dinaric Karst, trace back sources of pollution, and trigger debates for a sustainable management of the region's ecologies.

Keywords Infrastructures · (Post-)colonialism · Environmental justice · Contaminated sites · Occupational health · Environmental history · Creosote

13.1 Introduction

On a fall day in the year 1916, two iron tank wagons slowly started to move south, leaving behind the smokestacks of Vítkovice—one of the largest steelworks in the Austro-Hungarian Empire. In the shadow of this landmark of industrialization in the

M. Meiske (✉)
Research Institute for the History of Science and Technology at the Deutsches Museum, Munich, Germany
e-mail: m.meiske@deutsches-museum.de

B. Fuerst-Bjeliš et al. (eds.), *Environmental Histories of the Dinaric Karst*, Environmental History 17, https://doi.org/10.1007/978-3-031-56089-7_13

Ostrava region, the company Julius Rütgers ran a coal tar refinery. In the second half of the nineteenth century, the coal tar creosote—a side product of gas and coal distillation which had long been a pollutant for rivers in European industrial zones—had become a new standard for the conservation of wooden railway ties, as well as for telegraph and electricity poles. Thanks to the development of new high-pressure impregnation techniques and the use of creosote, untreated oak ties, which had been exchanged on average after 17.5 years in track, now resisted environmental and mechanical stress for 28 years. Untreated Beech ties, which often had been rotten away after 2.5 years, even were in service for 35 years after being impregnated.[1] Two brothers coming from the vicinity of Cologne, Julius and Guido Rütgers, ran dozens of wood impregnation facilities across Europe, reaching from Berlin to Bilbao, from Bohemia to Budapest.[2] They capitalized on the update of an allegedly "old" infrastructure material in the "age of coal and steel."[3] The two wagons with the numbers 503,537 and 503,587, which contained 24,000 kg of creosote oil, were meant for the facility in Han Compagnie-Vitez, owned by the Guido Rütgers KG. They were only two of dozens of wagons that were heading toward Bosnia and Herzegovina between the end of the 1880s and the early 1920s.[4]

Researching the history of creosote and its precarious legacy in the Dinaric Karst poses questions of empire, extraction, and externalization, that I will touch upon in this paper. The substance is connected today with dozens of contaminated sites, thousands of tons of hazardous waste, and burning environmental justice issues. To reconstruct a history of creosote and its use as a wood preservative for railway sleepers, I take up recent impulses from the history of technology and science and technology studies considering topics of repair and maintenance that illuminate the lifecycles and "afterlifes" of technology beyond the established innovation focus.[5] The history of creosote's ambivalent material links with plants, animals, and humans brings them together with issues of environmental history and (new/neo) material history.[6]

Methodologically, reconstructing environmental impacts of wood impregnation in Bosnia and Herzegovina posed various challenges. As the country's Federal Ministry of Tourism and Environment declared in a report in 2010, "there are no relevant and

---

[1] See Bach (1950), p. 424.

[2] The long, but potentially not complete, lists of the companies' impregnation facilities can be found in Giese (1968), pp. 43–44.

[3] See Meiske and Zumbrägel (2021), pp. 251–285.

[4] StLA, Rütgers, Box 167, letter from Rosenstiel to the k.u.k. Zentraltransportleitung, October 11, 1916.

[5] Some of the key publications in the growing body of literature on maintenance and repair are Edgerton (2006); Jackson (2014), pp. 221–240; Denis et al. (2015), pp. 5–15; Russell and Vinsel (2018), pp. 1–25; Henke and Sims (2020); Krebs and Weber (2021).

[6] See e.g. Coole and Frost (2010); LeCain (2017); Bennett (2004); Bennett (2010). Here I also refer to "Stoffgeschichte," a branch of material history that gained popularity mainly in German-speaking research communities in the last years, see Böschen et al. (2004), pp. 19–25; Espahangizi and Orland (2014), pp. 11–35; Haumann (2018), pp. 97–122; Soentgen (2019); Haumann et al. (2023).

reliable information on land contamination (heavy metals, persistent organic pollution/polluters [POPs], etc.), or maps of the most important land damages, regardless of their cause." The ministry explained that the "problem of land contamination is not systematically presented or tracked," there are "no basic preconditions for planning of elimination / minimizing of negative effects on human health and biodiversity."[7] While according to the UNEP country report from 2012, the surface water monitoring had made "a significant progress over the last ten years," the information gap for groundwater seemed to be still significant.[8] In the same year, the DIKTAS report on the Dinaric Karst concluded that "monitoring of groundwater quality has never been very well established in Bosnia and Herzegovina (B&H), so there is no clear understanding on impact to groundwater environment." The researchers further underlined that "monitoring equipment in B&H is almost completely destroyed during the war activities, so the current activities on water quality monitoring are brought to a minimum."[9] The lack of precise measurement data necessarily adds a speculative dimension to this research. However, I hope to show that under these conditions, which many countries still suffer from, environmental history could contribute to identifying potential risk areas and chemicals that could be part of the contamination. This, of course, would not substitute any structured in situ analyses of soil samples or groundwater measurements but could provide first hints for further investigation.

This paper is making use of rich archival material from the business papers of the Guido Rütgers KG that are stored at the Styrian regional archives in Graz, printed historical documents, and hydrogeological maps from the DIKTAS project on the Dinaric Karst. Finally, contamination cases of former impregnation facilities, published by the *Umweltbundesamt* in Austria, are also analyzed as an indicator of general socioenvironmental impact dimensions of pollution.

The central protagonist of this paper is the impregnation company *Guido Rütgers KG*. The company with headquarters in Vienna operated several facilities in Bosnia, among which some were mobile impregnation plants, that moved between locations and operated for several years in Dobrljin (ca. 16 km north of Novi Grad), Prijedor, Omarska (ca. 22 km south-east of Prijedor), Banja Luka, and Podgrab/Prača (ca. 33 km south-east of Sarajevo). In Han Compagnie-Vitez (ca. 16 km south-west of Zenica), however, a permanent impregnation facility was opened in 1892. Although this study focuses on creosote oil, as it became the standard for impregnation at the turn of the century, it is important to note that Rütgers often also applied a mixture

---

[7] Federal Ministry of Tourism and Environment (2010).

[8] UNEP (2012).

[9] DIKTAS (2012). The DIKTAS project (Protection and Sustainable Use of the Dinaric Karst Transboundary Aquifer System) aimed at the sustainable management of karst groundwaters contained in the Dinaric Karst. While the project focussed on activities in Albania, Bosnia and Herzegovina, Croatia and Montenegro, also other aquifier bearing countries joined the project for an international exchange. The DIKTAS project run between 2010 and 2014, implemented by UNDP and executed by UNESCO. See: DIKTAS (2014). The DIKTAS Information Management System (IMS) created during this project provides a map of the Dinaric Karst with several helpful visual layers, such as a hydrogeological map: ggis.un-igrac.org/view/diktas, last accessed June 30, 2023.

of fluids during the impregnation process.[10] Variations in the combination were not only the result of assumed higher efficiency due to new research coming from the company's laboratories. They were also determined by the properties of specific kinds of timber and the limited supply of certain chemicals, especially during war times. In the case of Bosnia, the main chemical used by Rütgers besides creosote, though in lower quantities than the coal tar oil, was Zinc chloride–another wood preservative categorized by the EPA as "harmful if swallowed," that could cause "severe skin burns and eye damage" to humans and animals, and that is proven to be "very toxic to aquatic life with long-lasting effects."[11]

Some of the impregnation facilities in Bosnia were located inside, others outside karst geologies. This offers the opportunity to discuss the specific vulnerability to pollution that characterizes karst aquifer systems due to their low self-cleansing capacity.[12] Taking into account that karst aquifer systems still supply more than half of the Bosnian population with drinking water today and considering the long time persistence of polycyclic aromatic hydrocarbons (PAH) components of creosote in groundwater, points to questions of environmental justice and the legacy of Austro-Hungarian reign in the region that require further examination.[13]

The first part of this paper traces back the political and socioeconomic foundations of empire, extraction, and externalization in the context of infrastructure construction and maintenance in Bosnia and Herzegovina at the turn of the century. The second part reconstructs labor and occupational health history aspects of the wood impregnation industry, which predominantly employed people living in the Dinaric Karst. The third section finally engages with the environmental impacts of creosote impregnation. It reflects on various ways of pollution and discusses their potential threat in the specifically vulnerable karst environments. The paper concludes with a discussion on the opportunities and limits of environmental history as a tool for reconstructing pollution histories of the Dinaric Karst.

[10] Details of the double impregnation procedure with both chemical that the Guido Rütgers KG are for example described in the manuscript of an offer made to the Bosnisch-Hercegovinischen Staatsbahnen from 1909. See StLA, Rütgers, Box 356, "Beschreibung des sogenannten Doppelverfahrens d.i. die Imprägnierung mit Chlorzinklauge und Steinkohlenteeröl in aufeinanderfolgender Behandlung," n.p.

[11] Environmental Protection Agency—GHS Data: "Zink chloride." Like creosote oil, zink chlorid was first used for wood impregnation in the late 1830s. At the turn of the century, creosote oil gained dominance in the wood impregnation sector for railway ties. A historical sketch of its use until the 1950s can be found in a manuscript of Rütger's head chemists: StLA, Rütgers, Box 161, Sigurd Prokopp: Chlorzink als Holzkonservierungsmittel, n.p. [ca. 1948].

[12] Petrić (2011), pp. 18–22; Calò and Parise (2009), pp. 63–72; Drew and Höltzl (1999).

[13] Stefanović (2018), p. 225.

13.2 Imperial (Infra-)Structures in Bosnia and Herzegovina

In the nineteenth century, the long leading power in the Balkans, the Ottoman Empire, experienced a severe loss of territorial control and influence. By the 1840s, some states, such as Serbia, had already gained profound autonomy from Constantinople, uprisings in the wider region such as in Bosnia and Herzegovina (1875) continued and flared up in the Russo-Turkish War (1877/8).[14] In 1878, European Great Powers (including Germany, Austro-Hungary, Russia, the Ottoman Empire, Italy, France, and Great Britain) met in Berlin to reorder the territories of southeastern Europe. Although some of the states of the region were allowed to send representatives (Serbia, Romania, Greece, and Montenegro), they had no voting rights at the Berlin Congress. This colonial modus operandi was repeated ten years later in the Berlin Conference (1884–1885), which regulated European colonial rule in West Africa—a remarkable parallel aside from the somewhat different geopolitical context. This political attitude of New Imperialism also left its trace in foreign rule in Bosnia and Herzegovina, which according to the agreement in 1878, formally stayed Ottoman territory, but practically fell under Austro-Hungarian rule. Soon after the treaty was signed, Vienna started an occupation campaign, which—after several battles and violent encounters with parts of the local population—was completed in October the same year.

The motives for the forced integration into the Habsburg Monarchy were manifold. A key factor was the fear of the further spread of national movements in the region that could lead to the formation of a sovereign Yugoslavian state. Consequently, the unification of Serbs and Croats would block Austro-Hungarian's direct access to ports of the Adria, which were of significant geostrategic and economic value for the empire. Furthermore, Vienna was interested in modernizing regulations, and initiating economic development after a "European model," as these were essential preconditions for the intensification of the forest economy, as well as for coal production and mining. By 1907, transportation infrastructures were expanded by around 2000 km of streets and 1000 km of railways.[15]

A report of the former head of the Mariabrunn Forestry Research Institute, Ludwig Dimitz, illustrates the Austro-Hungarian aspiration for its impact on forest management as well as the economic development in Bosnia and Herzegovina:

> The emergence of large mining, smelters, and other industrial operations required provisions for meeting the growing demand for wood and coal. The economic upswing of the cities and markets, the revival of building activity by the state and the private sector everywhere, also took up forest production. [...] The Bosnian land became an emporium of the most advanced timber industry and a factor in the world timber trade. The state forests gradually entered into a regular annuity, the dead land of the mountains came to life and where formerly the

[14] See Calic (2019), pp. 25–27.

[15] See Calic (2019), pp. 31–33; Schmid (1914)—especially the chapter on the mobility sector, pp. 567–617.

> shepherd only drove his flocks from time to time, rails meandered, locomotives steamed, and the electric spark flashed.[16]

These descriptions of allegedly "dead land" that was either called "empty" or inhabited by "nomads" in contrast to the "vitalizing power" of "European" modernization and culture are typical examples of widely described "civilizing mission"[17] narratives that Western actors have mobilized to legitimize colonization. In the same attitude, Dimitz claimed that "the Turkish reign" had "lulled the people into the half-activity of the Orientals." According to him, also the "Karst question" was "a cultural question in the widest sense of the word."[18] It is "the lack of need of the people, their nomadic tendencies, the constant unrest" which he claimed to have hindered the development of the "Kulturland."[19] Part of this program was the revitalization and development of karst wells and the melioration of poljes.[20] Accordingly, Dimitz called modernizing forestry in Bosnia and Herzegovina "similar" to tasks in Taiwan, which Japan had conquered and colonized during the wave of New Imperialism in 1895.[21] As Iva Lučić has recently shown in a well-researched paper focusing on the legal regulation of forest ownership and usage rights, the formation of Habsburg imperial governance in Bosnia and Herzegovina was in no way straightforward. Instead, it was shaped by complex "Ottoman legacies, conflicting economic interests, and the imperial need to form alliances with the locals."[22]

In this context of empire, exploitation, forest management, and timber trade, wood impregnation played a central role. The engagement of Rütgers in Bosnia and Herzegovina touches upon a key field of imperial rule, namely the construction and maintenance of infrastructures. Infrastructures and technological systems of communication, such as railroads or telegraph lines, have been described as "tools of empire" which drove the European exploitation of natural resources and human labor in the colonies in the nineteenth century.[23] More recent research has, however, also called to complicate "this straightforward, 'diffusionist' perspective" and suggested that there is much to gain from "unpacking the changing motives and ideologies behind technology transfer" and "the manifold ways in which infrastructures reflected and (re)produced colonial spaces and identities."[24]

In recent years, various researchers have picked up impulses from postcolonial studies to reevaluate the role of Austria and the Habsburg Empire in a globalized world in the past and the present. In contrast to imperial powers like Great Britain, France, or Germany, Austria had no comparable colonial land outside of Europe under

[16] Dimitz (1905), p. 101.

[17] For the Austrian case see Feichtinger (2016), pp. 147–181; Telesko (2015), pp. 35–48. More general on the concept, see Barth and Osterhammel (2005); Fischer-Tiné and Mann (2004).

[18] Dimitz (1905), p. 321.

[19] Dimitz (1905), p. 87.

[20] Dimitz (1905), p. 321.

[21] Dimitz (1905), p. IV.

[22] Lučić (2022), p. 607.

[23] Headrick (1981).

[24] Van der Straeten and Hasenöhrl (2016), pp. 355–356.

its control. Nevertheless, as we know from the Swiss case, there exist various forms of a "colonialism without colonies."[25] Recent postcolonial studies have focused, among others, on a reevaluation of the "quasi-colonial power relations" in the case of Bosnia and Herzegovina under Habsburg rule.[26] In this context, Clemens Ruthner has argued with ten points why Austro-Hungarian colonialism should be interpreted "as a critical term beyond the contemporary rhetoric": (1) A military conquest followed after a mandate provided by an international summit (Berlin Congress 1878)—a circumstance that repeated in the Berlin Congo Conference (1884/85). (2) The legal status of the territory stayed behind other regions of the empire and had no legal representation in the two parliaments. (3) Habsburg established an indirect rule through the participation and gradual "reformation" of existing elites—similar to British reign in India. (4) A patronizing civil administration, mainly in the hands of foreigners, was put in place. (5) Like many other colonial powers, Austria tried to establish an epistemic regime, especially through the new subject of Volkskunde, a new Landesmuseum, and the attempt to (re)write a Bosnian regional history. (6a) In connection with that, the administration tried to establish a top-down unifying Bosnian identity. (6b) At the same time, Bosnians were "othered" through popular Orientalist discourses to justify structures of governance that were less democratic than in the motherland. (7) The occupied territories experienced paternalism in the field of economics, limitations in free trade, and the creation of modern infrastructure. (8) The region functioned as a "lab of modernity," and was at the same time administered by a conservative administration. (9) Similar to the Indian Gurkhas of the Royal Army, the Bosnian population was exploited for the war efforts of the empire. And finally, (10) settlers (e.g., from Germany) were encouraged to move to Bosnia and form "role model-villages."[27]

A recently edited volume that further theorizes and illustrates the case of Bosnia and Herzegovina as a colony also discusses the long-term effects of the occupation period.[28] However, none of the authors engage with questions of environmental impact on regional ecologies or questions of environmental justice. "It is a pervasive condition of empires," writes Rob Nixon in *Slow Violence and the Environmentalism of the Poor*, "that they affect great swathes of the planet without the empire's populace being aware of the impact—indeed, without being aware that many of the affected places even exist."[29] The German sociologist Stephan Lessenich has described this broad ignorance toward the extraction of natural resources and the export of toxic industries and waste from the Global North to the Global South as "externalization."[30] Taking into account these impulses, I will try to show that environmental history can

---

[25] Purtschert et al. (2012).

[26] Burton (2021), p. 327; Nikolić (2021), pp. 47–58; Müller-Funk et al. (2002); Feichtinger et al. (2003); among the older publications discussing colonial patterns are Williamson (1991), see the chapter "Aehrenthal's Legacy: Bosnian Colonial Success and the Italo-Turkish War," pp. 58–81, and Kann, "Trends towards Colonialism in the Habsburg Empire," pp. 164–180.

[27] Ruthner (2014), pp. 156–169; Ruthner (2018), pp. 7–9.

[28] Ruthner and Scheer (2018), for the long-term impacts see the contributions in the chapter "Nachwirkungen," pp. 441–530.

[29] Nixon (2011), p. 35.

[30] Lessenich (2019).

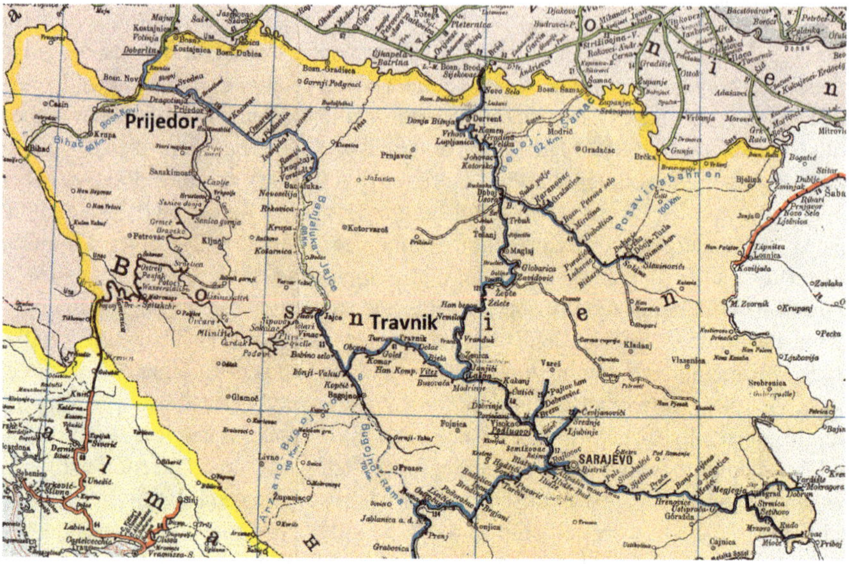

Fig. 13.1 Map section of Bosnia depicting the main railway lines and stations that have been part of an expanding infrastructure network at the eve of World War, with Prijedor and Travnik enlarged in lettering. *Source* Karl Peucker, *Artaria's Eisenbahnkarte von Österreich-Ungarn und den Balkanländern*, Wien 1918.—Wikimedia commons: https://commons.wikimedia.org/wiki/File:1918_map_-_Artaria%27s_Eisenbahnkarte_von_%C3%96sterreich-Ungarn_und_den_Balkanl%C3%A4ndern.jpg, last accessed on March 30, 2023

add to the long-term perspectives on this matter and unpack a history of the Dinaric Karst that engages with the entanglements of the material world of this fascinating environment with the socioeconomic development of the region.

The first project that Rütgers took up in the region was the impregnation of railway ties for the *k.u.k. Militärbahn Banjaluka-Doberlin*. The company started in 1889 with an annual output of 19,000 beechwood ties, impregnated with Zinc chloride alone or mixed with creosote. After a short conflict about wood quality and the end of the collaboration at the beginning of the century, it regained the contract in 1903, now mainly with the coal tar oil creosote as wood preservative and an annual output between 10,000 and 16,000 ties, that the company impregnated until the early 1920s.[31] To fulfill the contractual obligations, the company often used mobile impregnation facilities. One of such temporary plants was set up at the central station in Banja Luka in 1889, before it was sent to Omarska in 1902 and later stationed in Prijedor and Podgrab. In 1892/3, a permanent facility was opened in Han Compagnie-Vitez, which was also able to impregnate telegraph poles, wooden pavements blocks and other forms of timber (Fig. 13.1).

The list of clients in Bosnia and Herzegovina was long and diverse. Infrastructure projects such as railway lines planned and run by the (Austro-Hungarian) military,

[31] StLA, Rütgers, Box 167, letter from Rütgers to the B.H. Landesbahnen, September 12, 1903.

the (occupation) government and its Bosnische Staatsbahnen, as well as several logging railroads (initiated by private companies or the state), took the largest part of the production capacity. The demand was so high that different institutions, such as the Bosnische Staatsbahnen and the k.u.k. Militärpost- und Telegraphendirektion at times clashed over the use of Rütgers' impregnation plants.[32] Impregnated poles for communication technology and electricity formed the second-largest business. However, among the list of clients, one also finds other actors, such as a tobacco production site or a mining plant.[33]

The key to the commodification of Bosnian natural resources was the construction of transport infrastructures. Following earlier judgments of Austrian forest experts, Dimitz underlined that more than in other regions, in the occupied territories of Bosnia, railways are the "mediators of economic development," and only the "extensive use" of logging railways would help to exploit the forest.[34] Rütgers' plants in the country indeed not only helped to construct and maintain general railway lines but also several logging railways. In the year 1900 alone, different regional forest administrations asked the company for 10,000 impregnated railway ties to erect a logging railway in the vicinity of Han Compagnie, and another 7000 for a line planned in the Kruščica valley.[35] Another partner of Rütgers was the Steinbeisbahn (Fig. 13.2)—a project initiated by the German entrepreneur Otto Steinbeis, who had gained a concession to exploit Bosnian forest land for his company, the Bosnische Forstindustrie Aktiengesellschaft (BIFAG).[36]

A report to the k.u.k. Heeresverwaltung also uncovers a long list of clients of Austro-Hungarian military authorities in the Balkans that Rütgers supplied with infrastructure material, especially since the start of WWI. In the first two years of the conflict alone, the company sent more than 3200 impregnated telegraph poles to the military engineering and telegraph departments of the fortresses in Pula, Sarajevo and Mostar, and impregnated wood blocks to Belgrade, which was temporarily under Austro-Hungarian control during the war.[37] In a letter to the ministry of war, the company itself underlined its critical role for the military, as it fulfilled several tasks that were "of value" for the "security and readiness of the army." To prevent material getting confiscated and workforce from getting mobilized, Rütgers emphasized its vital part for the "maintenance and expansion of railway lines and telegraph lines" as well as "for the construction of prison camps, and the like."[38]

---

[32] StLA, Rütgers, Box 316, letter from the B.H. Landesbahnen to Rütgers, September 23, 1910.

[33] StLA, Rütgers, Box 119, letters from Oscar Thomann, July 5, 1916 & May 3, 1917.

[34] Dimitz (1905), p. 220.

[35] StLA, Rütgers, Box 354, letter to the Forstverwaltung in Busovac, Bosnia, February 22, 1900; Box 354, letter from Landesregierung für Bosnien und die Hercegovina to Rütgers, August 15, 1900.

[36] StLA, Rütgers, Box 356, letter to the Staatsbahnen, August 13, 1909; Box 337, letter to the k.u.k. Bezirksamt, Prijedor, June 26, 1915.

[37] StLA, Rütgers, Box 167, letter from Rütgers to the k.u.k. Heeresverwaltung, August 16, 1916.

[38] StLA, Rütgers, Box 167, letter from Rütgers to the k.u.k. Kriegsministerium, July 10, 1915.

Fig. 13.2 At the turn of the century, the exploitation of natural resources in Bosnia and Herzegovina increased significantly. A key actor in this region was Otto Steinbeis, an entrepreneur from Germany and client of the Guido Rütgers KG. His company, the BIFAG, not only constructed a railway line (the so-called Steinbeisbahn) and exported timber but also ran a cellulose plant in Drvar. The postcard shows a logging railway at the station Potoci, which was part of the Steinbeisbahn (undated, around 1905). *Source* Wikimedia commons—https://commons.wikimedia.org/wiki/File:Station_Potoci_der_Steinbeisbahn.jpg, last accessed March 30, 2023

In so far, the territories of the Dinaric Karst belonging to Bosnia and Herzegovina have experienced specific socioeconomic and geopolitical dynamics at the turn of the century. The impregnation industry played a central role in building and maintaining imperial infrastructures that allowed an intensified exploitation of natural resources. Like most colonial endeavors, no transformation of scale takes place without incorporation of at least parts of the local population. For the construction and maintenance of (imperial) infrastructures, thousands of workers were needed over a longer period of time. The majority of Rütgers' employees in the Balkans called the limestone terrains of the Dinaric Karst their home. This industry not only left its traces in the environment by enforcing the exploitation of resources but also in the bodies of their inhabitants. However, the workers and the local population in Bosnia did not passively endure the "tools of empire." Again and again, they also stood up and resisted when exploitation and externalization of toxic industrial practices endangered the health of humans and the environment in the Dinaric Karst.

13.3 Precarious Labor in a Hazardous Work Environment

As Gregg Mitman, Michelle Murphy, and Christoph Sellers have underlined in their Osiris special issue on "landscapes of exposure," uncertainty is "perhaps the single most pervasive characteristic of the history of exposure."[39] Invisibility, delayed physical reactions and difficult-to-detect causal chains and interdependencies of chemical substances, which often occur mixed, pose great challenges for historical analyses. Agnotology—the cultural production of ignorance, as described by Proctor and Schiebinger—might have contributed to black-boxing the effects of creosote as well.[40] In the meantime, however, vast medical and occupational health literature provided valuable insights into creosote's effects on humans. Rich sources on labor conditions dating back to the turn of the century are rare, but in the case of the Guido Rütgers KG, surprisingly dense material could be found. When gaps occurred in the Bosnian case studies I discuss here, I carefully tried to bridge them with reports from other Rütgers impregnation facilities of the same period.

Skin, eye, and respiratory diseases as a result of direct contact with tar oil have been proven in numerous medical studies and were first mentioned even before the turn of the century.[41] "Phototoxic reactions are common in outdoor workers exposed to coal tar or its crude extracts, such as railroad workers who must handle creosote-impregnated ties," write Zenz et al. in their handbook on occupational medicine.[42] Even if protective clothing has been shown to reduce the risk of skin injury, people and animals that have intensive contact with creosote have an increased risk of skin cancer.[43] Impregnation plant workers carried dripping sleepers, which weighed up to 90 kg, on their shoulders during the loading process, handled tensioning straps soaked with creosote or walked through puddles of tar oil that had formed in the shunting area of the ironing trolleys and in the storage area. Especially auxiliary workers who loaded and unloaded the wagons with heavy timber appear disproportionately often in the accident statistics of Rütgers' impregnation plants.[44]

How did the company leadership perceive the risk posed to the workers in Rütgers' facilities across Europe at the turn of the century? Karl Schmook, who followed Guido Rütgers as the director of the company, stated in the files of a workers' accident insurance for his impregnation facility in St. Marein that "high degrees of heat, toxins, and harmful gases" do not occur.[45] In 1914, however, an operations manager of another Rütgers plant wrote to Schmook that "strong" people "insensitive to the effects of tar oil" were needed and not the "inferior material" that the k.u.k. employment agency

[39] Mitman et al. (2004), p. 13.

[40] Proctor and Schiebinger (2008).

[41] Mackenzie (1898), p. 417; Cookson (1924), p. 368; Lenson (1956), pp. 520–522.

[42] Zenz et al. (1994), p. 114.

[43] Carlsten et al. (2005), pp. 96–97.

[44] StLA, Rütgers, Box 101, see the cases in the file "Arbeiter-Unfallversicherung, 1914–1939".

[45] StLA, Rütgers, Box 101, letter from Rütgers to the Arbeiterunfallversicherungsanstalt für Steiermark und Kärnten, March 10, 1922.

offered—a first hint that the fluid was indeed a known occupational health hazard.[46] Four years later, another plant manager asked for "sole repair material" for the shoes of "cadre soldiers and prisoners of war," in order to be able to protect their feet "from the penetration of the caustic impregnation liquid."[47] However, the boundaries between (conscious) ignorance and uncertainty can probably not always be resolved entirely in the sources of these early historical episodes.

What do we know about Rütgers' employees in Bosnia? While the personnel in Vitez is relatively well documented, the archival files say relatively little about workers of the mobile impregnation plants. Files on Prijedor show that around 15 workers were employed in mobile plants and that the campaign only lasted for several months. The works were suspended during the Winter months. Files on Omarska uncover that soon after establishing the plant, the impregnation had to be stopped. Workers from Omarska and other regions of Bosnia refused to work with creosote oil, and Rütgers had to bring in experienced workers from far away facilities. This was a clear hint that they were aware of the hazardous effects of the chemical, and at the same time, it showed that they were not passive "human resources" but able to interrupt the construction and maintenance of infrastructures.[48]

In the permanent facility in Vitez, the total number of machinists, stockers, general workers, and auxiliary workers fluctuated between 25 and 50 over the years. Most of the men employed were between 16 and 60 years old. Early employment lists (1893–1898) that include information on the religious affiliation of workers mention 60% Catholics and 40% Muslims, but no Orthodox.[49] During the impregnation campaign, they often worked 10 to 11 h per day.[50] A similar list from 1916 illustrates the exploitation of the local population for the Austro-Hungarian war effort. 15 out of 39 employees are marked as mobilized for military service.[51] In 1912, a group of workers went on strike, and when they demanded answers to their pay raise request from the headquarters in Vienna, the manager chased them off the land with his revolver. However, eventually the workers, who were beforehand described in the letter to Schmook as "too stupid" to achieve anything without the supervision of the foreman, received a raise, even though it was rather moderate.[52]

During the war, the food situation significantly worsened, and living costs increased. In 1917, a report to the Rütgers headquarters stated that workers would receive 9 kg and their families additional 6 kg of cornmeal.[53] But already the

---

[46] StLA, Rütgers, Box 128, handwritten remark on a double letter from an employee in Kötsch to Rütgers, September 4/5, 1914.

[47] StLA, Rütgers, Box 167, letter from Rütgers to the k.u.k. Kriegsministerium, July 18, 1918.

[48] StLA, Rütgers, Box 337, letter to the k.u.k. Bezirksamt, Prijedor, June 26, 1915, and letter to the Direktion der k.u.k. Militärbahn Banjaluka-Doberlin, September 26, 1907.

[49] StLA, Rütgers, Box 159, see the file "Arbeiterunterstützungsfonds 1889-1910".

[50] StLA, Rütgers, Box 129, letter from Reschke to Schmook, March 18, 1918.

[51] StLA, Rütgers, Box 160, "Verzeichnis der Maschinisten, Heizer und Arbeiter," November 26, 1916.

[52] StLA, Rütgers, Box 128, letter from Reschke to Rütgers, July 21, 1912.

[53] StLA, Rütgers, Box 119, letter from Thomann to Schmook, May 3, 1917.

following year when a military supervisor complained to Vienna about inadequate provisions, deficient clothing, and a "life-endangering work situation" in Vitez, plant manager Reschke showed little understanding. Instead, he claimed in his letter to Schmook that even as he tried to organize food from all over the region, the local population was very selective in its diet. They "did not know how to make food with milk, meat and vegetables" and only ate polenta and rarely potatoes. Furthermore, providing food at the impregnation facility would have been very complicated because the "Turks" supposedly would not eat with others and rejected food prepared in the same bins. While some of these preferences might have existed, it seems very unlikely that, facing weeks of famine, workers at large turned down staple food offered by Rütgers. Furthermore, Reschke feared that "all 20 workers would bring their wives and kids" and that "would go too far." According to him, also working shoes were not appropriately maintained and sometimes given to relatives instead of being used at the plant.[54] As the picture below indicates, and as we know from other impregnation facilities, workers had very close contact with the impregnation oil (Fig. 13.3). Considering what we know today about its effects on human bodies, it seems safe to assume that some of them paid a high price for this exposure.[55]

The facilities in Vitez continued to experience a changeful history. When the Austro-Hungarian empire fell apart after WWI, the newly formed Kingdom of Yugoslavia started sequestration processes in Bosnia and Herzegovina. While Rütgers was able to keep its plant in Kötsch/Koce, Slovenia (and through this was able to keep on doing business in the Balkans), the mobile plant in Prijedor and the impregnation facility in Vitez were confiscated by the newly formed SHS state.[56] After WWII, it continued to be in state ownership throughout the existence of the Socialist Federal Republic of Yugoslavia. During the Balkan wars of the 1990s, large parts of the site became the headquarters of a logistical division of UN troops. When the wave of privatization hit the Balkans in the early 2000s, the Austrian Alpe Adria Group and two private investors from Austria bought the site—a remarkable episode from a long-term postcolonial perspective. Soon after the "Impregnacija-HOLZ d.o.o. Vitez" closed forever. Today, only ruins of the buildings are left, and the grounds have become an unofficial waste dump.[57] This, however, is likely not the last chapter of the history of the impregnation industry in Vitez. As other cases of impregnation plants show, the afterlife of this technology raises critical questions regarding the long-term impact for society as much as for the vulnerable karst ecologies.

[54] StLA, Rütgers, Box 129, letter from Reschke to Schmook, March 18, 1918.

[55] See e.g. Carlsten et al. (2005).

[56] StLA, Rütgers, Box 358, see files "Südslavische Sequestration, 1919–23" and "Beschlagnahme Prijedor, 1918–1922".

[57] Čilić (2015).

Fig. 13.3 This picture in front of the Han Compagnie-Vitez impregnation facility shows a group of workers in clothing soaked in impregnation fluid. The photo was taken in 1904 during a visit of Karl Schmook, the director of the Guido Rütgers KG, who is standing slightly elevated in the center of the scene. *Source* Giese (1968), p. 109. Reproduced with permission of G. Rütgers and Kunstverlag Wolfrum

13.4 Environmental Impacts for Dinaric Karst Environments

According to the DIKTAS IMS map, all of the locations of Rütgers' impregnation facilities run until the End of WWI that could be identified in historical documents, lay on the territory of the Dinaric Karst region. None of them was located on karst aquifers, which are specifically vulnerable due to their high permeability and low self-cleansing capacity. Most of them were positioned fully or partially on intergranular porous aquifers with a "good permeability," and one was positioned on a locality with no aquifer which is classified as "nearly impermeable" in DIKTAS (Fig. 13.4). An *aquifer* is a water-bearing formation that contains groundwater—one of the primary sources of human water exploitation, from drinking water to irrigation. Consequently, in any case, the protection of these essential areas for communal life is of high importance.

As was already discussed above regarding aspects of occupational health, contact with creosote can directly affect humans or have long-term consequences for their physical condition. This also counts for its impact on the environment and its flora

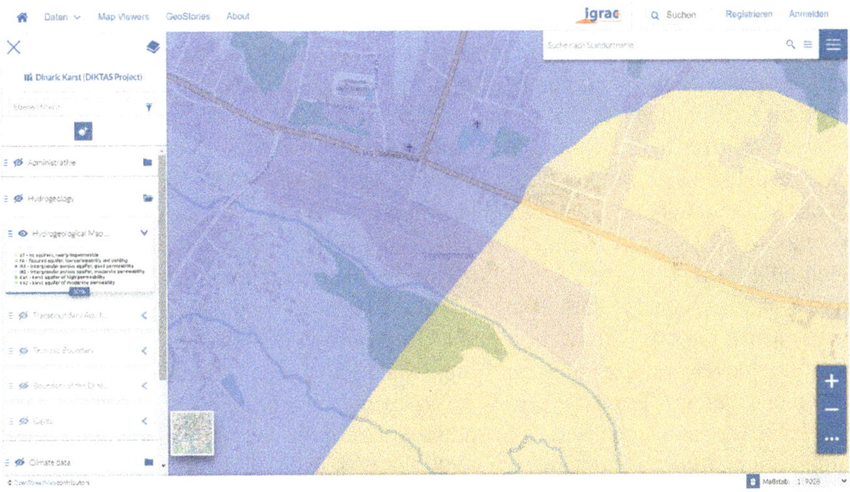

Fig. 13.4 According to the DIKTAS IMS map, Rütgers' impregnation plant in Han Compagnie-Vitez is located on the boundary of an intergranular porous aquifer with good permeability (blue) and an area with no aquifers that are nearly impermeable (beige). Due to the close vicinity to the river Lašva, also aquatic life might have also been affected by the impregnation industry in this part of the karst region. *Source* Screenshot from https://ggis.un-igrac.org/view/diktas, taken on September 29, 2021. Reproduced with permission

and fauna. Creosote is characterized by high persistence—it can stay in the soil or in the sediment of water bodies for very long time periods as will be discussed below. This means that former impregnation facilities can still pose a threat to surrounding ecological systems and the local population for decades and longer after they have been closed down.

Three kinds of cases can be discussed here. Thanks to the rich archival material on Rütgers, a specific historical case of river pollution from the early twentieth century could be traced back. However, questions of contaminated sites will keep a more speculative character, as already mentioned above. Again, as there is no register of contaminated sites for Bosnia and Herzegovina, the only chance is to look at the long-term impact of comparable Rütgers facilities in other countries that established such registers. As all five permanent Rütgers facilities in Austria have been declared contaminated sites in the publicly available *Altlastenatlas*, Vitez, and other impregnation facilities in the Dinaric Karst should also be treated as suspected cases. Finally, the impact of impregnated railway ties on track for the surrounding ecology will be discussed.

In June 1907, a report alarmed the State Government for Bosnia and Herzegovina. The local police in Podgrab and the fishing association in Sarajevo had observed suspicious fish mortality in the river Prača. Soon the mobile impregnation plant that Rütgers had established in the region in 1904 was identified as a suspected source of contamination. As the correspondence between the company and the authorities show, Rütgers recognized the loss of around 200 kg of tar oil that had entered the water body. These files also contain remarkable insight into the company's damage control strategies—not regarding the pollution, but its responsibility. Firstly, an earlier draft of the response to the government contained a more detailed calculation, according to which not 200 but 700 kg of tar oil were lost. This more significant number, however, seems to have never been communicated. Rütgers described the accusations in its response as "erroneous" and even the number of 200 kg as "exaggerated." Secondly, the company significantly downplayed the effect of tar oil—knowingly or unknowingly—when stating that only a tiny soluble part of the oil could have potentially harmed the fish stock, and the heavier parts would have sunk to the ground and thoroughly washed away in the next rainstorm. Thirdly, according to the company, the oil likely went into the river during the night, which would mean that only thieves or other individuals that would like to damage the firm could have caused this incident. For these events, no company could be held responsible, argued Rütgers.[58] We can only speculate about the extent of the damage and the real reason behind the incident because for now, only the files from the Rütgers archives have been accessed. However, as we know from other cases of river pollution with creosote oil, it is very persistent and continues to be a threat to the environment and aquatic life for decades, and sometimes even more than a hundred years.[59] A case of river pollution at the Elizabeth River in southeastern Virginia showed that toxins such as creosote oil resulted in "severe skin, fin, and gill erosion, reduced hematocrits, and pancreatic and liver lesions" in local fish populations, and that the coal tar oil even co-shaped evolutionary pathways of fish species.[60]

The second case of potential environmental pollution in the Dinaric Karst regards the permanent impregnation facility in Han Compagnie-Vitez. Although the business files remain silent with respect to environmental incidents in this facility, the high rate of impregnation plants turning to contaminated sites in other countries urges us to also consider this location as potentially contaminated. As owner structures of impregnation facilities change over time, it should be kept in mind, however, that questions of responsibility for environmental impacts and potential remediation cases should be addressed to all of them. Every single of the five permanent

[58] StLA, Rütgers, Box 356, draft of a letter from Rütgers to the Eisenbahn-Baudirektion d. bosn. herzog. Staatsbahnen, August 3, 1907 & letter from Rütgers to the Landesregierung für Bosnien und die Herzegovina, August 9, 1907.

[59] "Chemicals in creosote break down in water very slowly. They tend to cling to particles of matter, and as such, sediments are considered the primary location for these contaminants to collect in aquatic environments. Many PAHs [polycyclic aromatic hydrocarbons—the primary chemicals of concern in creosote, MM] do not migrate far from the point of contamination and accumulate at that primary location. […] For some organisms, low PAHs are acutely toxic but may be considered non-cancer causing. High PAHs however, are not as toxic, but to many organisms—such as fish, birds, amphibians, mammals—can cause cancer, mutation or malformation of embryo/fetus."—Washington State Department of Natural Resources, Aquatic Resources—Science of Creosote, October 14, 2019, 1. The case of the Island End River in Everett, Massachusetts shows, that the impacts of a coal tar processing plant continue to be a threat to the watershed into the twenty-first century, even though its operations ended in the 1950s. See National Oceanic and Atmospheric Administration, Island End River.

[60] See Di Giulio and Clark (2015), pp. 259–298.

impregnation facilities that the Guido Rütgers KG (at least temporarily) operated in Austria suspected to be hazardous, have by now been officially identified as contaminated sites (*Altlast*) by the Umweltbundesamt. One facility in Angern, which also contained a tar distillery, was classified as priority class 1 and needed a rapid remediation response. Due to the high amount of tar oil in the ground on which a residential area had been erected, the upper 50 cm of the soil had to be exchanged. The cost of 5 million Euros for the first remediation was covered by money generated through the *Altlastenbeitrag*—a tax collected since 1989 in Austria for the use of landfills, waste reprocessing, and the export of waste. The final remediation will occur only once the final question regarding the culpability in the pollution case has been resolved.[61]

The other four facilities (Enns, St. Marein, Amstetten, and Gerasdorf)—which are maybe more comparable to the one in Vitez in terms of scale—are also described as highly hazardous for humans and the local ecology.[62] This means that remediation is planned but not undertaken immediately. However, the classification and the report generally come with detailed advice regarding limits of land use until the soil is cleaned up. Such information is essential for the local community living in the close vicinity, for potential land buyers, and the organization of environmental monitoring programs.

A closer look at the case of the facility in Enns (Upper Austria) illustrates the impact structures. At this location, north of a railway station, railway ties and utility poles have been impregnated between 1869 and 1972—until the end of the nineteenth century with a mobile plant, later in a permanent facility. The report of the Umweltbundesamt is the result of in situ studies conducted between 1996 and 1998 and 2005 and 2010. One hundred twenty-two dry core drillings, more than 240 sediment samples, and dozens of groundwater samples were collected and analyzed (Fig. 13.5). Like in Vitez, both coal tar oil and zinc chloride were used as impregnating agents in Enns. According to the Umweltbundesamt, "the heavily contaminated areas pose a significant threat to human health and the environment."[63] The remediation, however, is still pending.

The third case is geographically much more challenging to locate. Rütgers delivered impregnated railway ties, utility poles and other timber all around the Dinaric Karst region. Even today, after stricter regulations have been established in Germany, an average oak sleeper is still impregnated with 6 kg and a beech sleeper with 16 kg of creosote oil. During their use, both variants lose no less than 5 kg of tar oil, most

[61] See Umweltbundesamt, Altlastenatlas, Altlast N53: Teerfabrik Rütgers—Angern, https://www.altlasten.gv.at/atlas/verzeichnis/Niederoesterreich/Niederoesterreich-N53.html, last accessed June 30, 2023.

[62] See Umweltbundesamt, Altlastenatlas, Altlast O45: Imprägnieranstalt Enns, https://www.altlasten.gv.at/atlas/verzeichnis/Oberoesterreich/Oberoesterreich-O45.html; Altlast ST31: Holzimprägnierwerk Rütgers St. Marein, https://www.altlasten.gv.at/atlas/verzeichnis/Steiermark/Steiermark-ST31.html; Altlast N93: Holzimprägnierung Rütgers Amstetten Werk I, https://www.altlasten.gv.at/atlas/verzeichnis/Niederoesterreich/Niederoesterreich-N93.html; and Altlast N73: Holzimprägnierung Rütgers Gerasdorf, https://www.altlasten.gv.at/atlas/verzeichnis/Niederoesterreich/Niederoesterreich-N73.html; all last accessed June 30, 2023.

[63] See Umweltbundesamt, Altlastenatlas, Altlast O45.

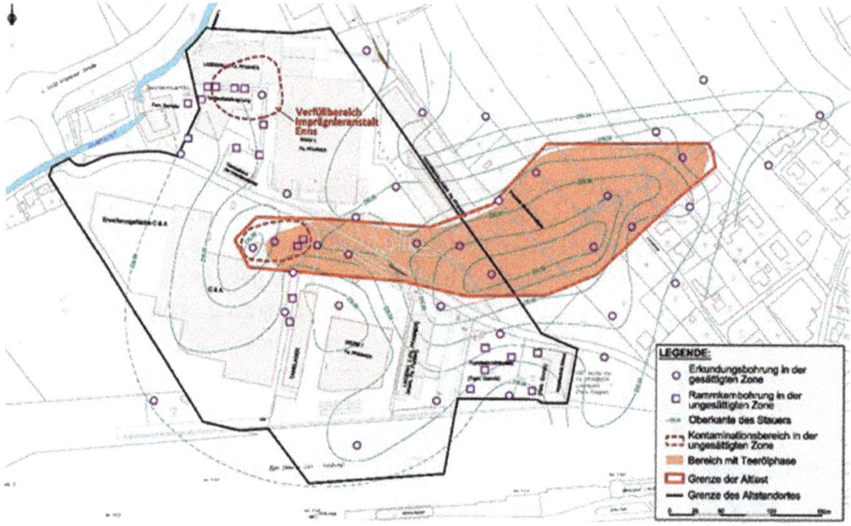

Fig. 13.5 The image shows the former boundaries of Rütgers impregnation facility (framed in black) and the contaminated land (framed in red)–a phase containing around 2 million liters of tar oil had spread in an area of 30,000 m^2. In this area, the groundwater was "massively contaminated with polycyclic aromatic hydrocarbons (PAHs) and heterocyclic aromatic hydrocarbons and subordinately with aromatic hydrocarbons (BTEX)." According to the Umweltbundesamt, "the heavily contaminated areas pose a significant threat to human health and the environment." *Source* See Umweltbundesamt, Altlastenatlas, Altlast O45: Imprägnieranstalt Enns, https://www.altlasten.gv.at/atlas/verzeichnis/Oberoesterreich/Oberoesterreich-O45.html, last accessed June 30, 2023. Reproduced with permission

of which ends up in the soil surrounding the thresholds.[64] The amount of creosote Rütgers used per railway tie for its impregnation in Bosnia varied greatly, but is very comparable.[65] In the first year of operation and especially when exposed to the sun, the tar oil comes to the surface of the railway sleepers and thus becomes an open source of danger for the maintenance staff and the local ecosystem. Older medical studies from the USA report numerous cases in which the access of cows and sheep to freshly impregnated creosote wood on railway sleepers, telephone poles or fences is given as the likely cause of death.[66] A recent study was able to demonstrate the long-term changing effects on complex food chains and ecological processes through soil contamination by creosote.[67]

Finally, it is essential to remark that old impregnated railway ties also have a long history of reuse after their service in track. For example, at least since the

---

[64] Bayerisches Landesamt für Umwelt, "Bahnschwellen" (2018), p. 2.

[65] Rütgers' offer to the k.u.k. Miltärbahn from 1903 mentioned the use of 10 kg of creosote per beech wood railway tie. See StLA, Rütgers, Box 167, letter from Rütgers to the Direktion der k.u.k. Militärbahn Banjaluka-Doberlin, September 12, 1903.

[66] Harrison (1959), pp. 89–98; Olafson and Leutritz (1959), pp. 54–57.

[67] Blakely et al. (2002), pp. 71–88.

end of the 1920s, old railway ties were sold to construction firms, farmers, and private individuals.[68] They used the long-lasting timber for fences, or to build small huts at their farms or in their gardens. Even as the leakage from these impregnated railway ties is reduced after decades and locally limited, it can still pose a threat to the environment and society. Therefore, impregnated railway ties are declared hazardous waste, and in the European Union, their trade is reduced to industrial clients.[69] When tracing back the environmental impact of creosote impregnation, this potential afterlife of technological infrastructures should also be further investigated, and—if necessary—regulated to protect humans and ecologies in the Dinaric Karst.

13.5 Conclusion

Following the history of wood impregnation with creosote in Bosnia and Herzegovina and its precarious legacy has uncovered complex entanglements between the material world and sociopolitical spheres. Through the lens of the coal tar oil, questions of power and empire, the exploitation of natural resources and labor, occupational health, as much as river, soil, and groundwater pollution have been examined. Such material research perspectives offer the opportunity to analyze short-term conflicts and negotiations, as much as they allow studying long-term transformations and environmental impacts. Environmental history in following its topics often crosses national borders and urges us to question well-established periodizations and uncover diverse temporalities. Furthermore, it invites interdisciplinary conversations about how the past and the present are connected and how sustainable futures could be shaped.

The case of Rütgers has illustrated how Bosnia and Herzegovina has been integrated into the imperial territory as a "quasi-colony." Infrastructures played a central role in Austro-Hungary's approaches to "modernize" the Southeastern European state and exploit its natural resources. Constructing and maintaining them was vital for the economic and military interests, and the wood impregnation industry provided essential services for these endeavors. Leading managers of the business showed clear colonial mindsets and repeatedly mobilized anti-Bosnian and especially anti-Muslim stereotypes in their correspondence. Contrary to impregnation plants in other countries, Rütgers did not seem to have used forced labor during the occupation. However, workers were mobilized for the empire's army during WWI. A closer look at the working conditions for local labor in impregnation plants uncovered significant occupational health hazards. In some cases, locals refused to work with creosote oil, while others—not being aware of its side effects or simply driven by the economic

---

[68] See Schulz (1930), pp. 201–204.

[69] Regulation (EC) No. 1907/2006 of the European Parliament and of the Council of 18 December 2006 concerning the Registration, Evaluation, Authorisation and Restriction of Chemicals (REACH), paragraph 31.

hardship of the region at the beginning of the twentieth century—worked for Rütgers for several years.

The environmental impacts of the wood impregnation business became visible in various aspects. During the occupation, it helped to establish and maintain an exploitative logging regime that substantially transformed the ecologies of the Dinaric Karst. Furthermore, the archival material gave insights into a contemporary river pollution case caused by one of Rütgers' impregnation plants. Combining the information gained from archival material and modern scientific data from environmental monitoring, karst aquifer assessments, and contaminated site analyses promises to give essential hints to fill information gaps in the study of environmental pollution in Bosnia and Herzegovina. A better understanding of the historical standards of the impregnation industry and one of its leading actors in the region might help anticipate potential pollutants. A comparison with plants of the same company that have been declared contaminated sites in Austria allowed for discussing potential contamination scenarios for cases in Bosnia and Herzegovina. Finally, the question of the afterlife of old impregnated railway ties that still carry toxic impregnation fluids and their circulation on the second-hand market has been raised.

While Bosnia and Herzegovina has made significant steps forward in monitoring surface water, the information gap regarding land and groundwater contamination still poses a substantial danger for the environment and society. Projects like DIKTAS helped significantly to evaluate the current stage of the Dinaric Karst, exchange information with other karst regions in the world, and develop sustainable management goals. In this endeavor, environmental history could help to understand the changing relations of humans with the region's karst ecologies in the past and raise awareness of pollution and its origins. Such histories pose burning questions of responsibility and environmental justice—aspects that could also help to trigger broader debates on how a sustainable future of the Dinaric Karst could look like.

Acknowledgement This work was funded by the Deutsche Forschungsgemeinschaft (DFG) under the project title, 'Cultures and costs of maintenance. The rise of Creosote and its precarious legacy', grant number 457488187.

References

Sources

DIKTAS Information Management System (IMS). https://ggis.un-igrac.org/view/diktas/. Last accessed June 30, 2023

StLA (Steiermärkisches Landesarchiv), Firmenarchiv Guido Rütgers KG

Umweltbundesamt, Altlastenatlas. https://www.altlasten.gv.at/atlas.html. Last accessed June 30, 2023

Literature

Bach K (1950) Eisenbahnoberbau. In: Mahlke F, Troschel E, Liese J (eds) Handbuch der Holzkonservierung. Springer, Berlin/Heidelberg, pp 418–441

Barth B, Osterhammel J (eds) (2005) Zivilisierungsmission. Imperiale Weltverbesserung seit dem 18. Jahrhundert. UVK, Konstanz

Bayerisches Landesamt für Umwelt, "Bahnschwellen" (2018). infoBlätter Kreislaufwirtschaft 5:1–7

Bennett J (2004) The force of things: steps toward an ecology of matter. Polit Theor 32(3):347–372

Bennett J (2010) Vibrant matter: a political ecology of things. Duke University Press, Durham, NC

Blakely JK et al (2002) Soil invertebrate and microbial communities and decomposition as indicators of polycyclic aromatic hydrocarbon contamination. Appl Soil Ecol 21(1):71–88

Böschen S, Reller A, Soentgen J (2004) Stoffgeschichten—eine neue Perspektive für transdisziplinäre Umweltforschung. Gaia 13(1):19–25

Burton E (2021) Postkolonialismus. In: Gräser M, Rupnow M (eds) Österreichische Zeitgeschichte—Zeitgeschichte in Österreich. Eine Standortbestimmung in Zeiten des Umbruchs. Böhlau, Wien/Köln, pp 321–347

Calic M-J (2019) A history of Yugoslavia. Purdue University Press, West Lafayette, IN

Calò F, Parise M (2009) Waste management and problems of groundwater pollution in karst environments in the context of a post-conflict scenario: the case of Mostar (Bosnia Herzegovina). Habitat Int 33:63–72

Carlsten C et al (2005) Squamous cell carcinoma of the skin and coal tar creosote exposure in a railroad worker. Environ Health Persp 113(1):96–97

Čilić Z (2015) Impregnacija: Austrijanci izgradili pa i uništili, June 25, 2015. https://www.vitez.info/impregnacija-austrijanci-izgradili-pa-i-unistili/. Last accessed June 30, 2023

Cookson HA (1924) Epithelioma of skin after prolonged exposure to creosote. BMJ 1:368

Coole D, Frost S (eds) (2010) New materialisms: ontology, agency, and politics. Duke University Press, Durham, NC

Denis J, Mongili A, Pontille D (2015) Maintenance & repair in science and technology studies. Tecnoscienza 6(2):5–15

Di Giulio RT, Clark BW (2015) The Elizabeth river story: a case study in evolutionary toxicology. J Toxicol Environ Health Part B 18:259–298

Dimitz L (1905) Die forstlichen Verhältnisse und Einrichtungen Bosniens und der Hercegovina. Wilhelm Frick, Wien

DIKTAS (2012) Protection and sustainable use of the dinaric karst transboundary aquifer system, national report: Bosnia and Herzegovina. http://diktas.iwlearn.org/resources/projectdocuments/diktas-country-report-bosnia-and-herzegovina/view. Last accessed June 30, 2023

DIKTAS (2014) TDA—Transboundary Diagnostic Analysis (TDA), Summary. http://diktas.iwlearn.org/resources/projectdocuments/transboundary-diagnostic-analysis-summary. Last accessed June 30, 2023

Drew D, Höltzl H (eds) (1999) Karst hydrogeology and human activities: impacts, consequences and implications. A.A. Balkema, Rotterdam

Edgerton D (2006) The shock of the old: technology and global history since 1900. Profile Books, London

Environmental Protection Agency–GHS Data: "Zink chloride." https://comptox.epa.gov/dashboard/dsstoxdb/results?search=DTXSID2035013#safety. Last accessed June 30, 2023

Espahangizi K, Orland B (2014) Pseudo-Smaragde, Flussmittel und bewegte Stoffe. Überlegungen zu einer Wissensgeschichte der materiellen Welt. In: Espahangizi K, Orland B (eds) Stoffe in Bewegung. Beiträge zu einer Wissensgeschichte der materiellen Welt. Diaphanes, Zürich, pp 11–35

Federal Ministry of Tourism and Environment (2010) State of the environment in the federation of Bosnia and Herzegovina—2010 Report. https://www.fmoit.gov.ba/upload/file/okolis/State%20of%20The%20Environment%20in%20Federation%20of%20Bosnia%20and%20Herzegovina.pdf. Last accessed June 30, 2023

Feichtinger J, Prutsch U, Csáky M (eds) (2003) Habsburg postcolonial: Machtstrukturen und kollektives Gedächtnis. StudienVerlag, Innsbruck

Feichtinger J (2016) Modernisierung, Zivilisierung, Kolonisierung als Argument. Konkurrierende Selbstermächtigungsdiskurse in der späten Habsburgermonarchie. In: Dejung C, Lengwiler M (eds) Ränder der Moderne. Neue Perspektiven auf die europäische Geschichte (1800–1930). Böhlau, Cologne/Weimar/Vienna, pp 147–181

Fischer-Tiné H, Mann M (eds) (2004) Colonialism as civilizing mission. Cultural ideology in British India. Anthem, London/New York/New Delhi

Giese U (1968) Hundert Jahre Holzkonservierung in Österreich. Festschrift zum hundertjährigen Bestehen der Fa. G. Rütgers. Wolfrum, Wien

Harrison DL (1959) The toxicity of wood preservatives to stock. N Z Vet J 7:89–98

Haumann S (2018) Kritische Rohstoffe. In: Engels JI (ed) Was heißt Kritikalität? Zu einem Schlüsselbegriff der Debatte um Kritische Infrastrukturen. Transcript, Bielefeld, pp 97–122

Haumann S, Roelevink E-M, Thorade N, Zumbrägel C (eds) (2023) Perspektiven auf Stoffgeschichte. Materialität, Praktiken, Wissen. Transcript, Bielefeld

Headrick DR (1981) The tools of empire. Technology and European imperialism in the nineteenth century. Oxford University Press, New York

Henke CR, Sims B (eds) (2020) Repairing infrastructures. Maintenance of materiality and power. MIT Press, Cambridge, MA

Jackson SJ (2014) Rethinking repair. In: Gillespie T, Boczkowski PJ, Foot KA (eds) Media technologies: essays on communication, materiality, and society. MIT Press, Cambridge, pp 221–240

Kann RA (1977) Trends towards Colonialism in the Habsburg Empire, 1878–1918: The Case of Bosnia-Herzegovina, 1878–1914. In: Rowney DK, Orchard GE (eds) Russian and Slavonic History. Slavica Publishers, Columbus, OH, pp 164–180

Krebs S, Weber H (eds) (2021) The persistence of technology. Histories of repair, reuse and disposal. Transcript, Bielefeld

LeCain TJ (2017) The matter of history. How things create the past. Cambridge University Press, New York

Lenson N (1956) Multiple cutaneous carcinoma after creosote exposure. N Engl J Med 254:520–522

Lessenich S (2019) Living well at others' expense. The hidden costs of western prosperity (Nick Somers with the assistance of Stephan Lessenich, trans.). Polity Press, Cambridge

Lučić I (2022) Law of the Forest: early legal governance in Bosnia-Herzegovina during the inter-imperial transition between Ottoman and Austro-Hungarian Rule, 1878–1901. Slav Rev 81(3):585–608

Mackenzie S (1898) A case of tar eruption. Br J Dermatol 10:417

Meiske M, Zumbrägel C (2021) Holz im Zeitalter von Kohle und Stahl. Zur Persistenz und Wandelbarkeit eines Werkstoffes in der Hochindustrialisierung. Technikgeschichte 88(3):251–285

Mitman G, Murphy M, Sellers C (eds) (2004) Osiris 19. Landscapes of exposure: knowledge and illness in modern environments

Müller-Funk W, Plener P, Ruthner C (eds) (2002) Kakanien revisited. Das Eigene und das Fremde (in) der österreichisch-ungarischen Monarchie. A. Francke, Tübingen

National Oceanic and Atmospheric Administration, Island End River. https://darrp.noaa.gov/hazardous-waste/island-end-river. Last accessed June 30, 2023

Nikolić A (2021) Austro-Hungarian cultural imperialism in Bosnia and Herzegovina: the case of Kosta Hörmann and the provincial museum. In: Pavlović VG (eds) The Balkans in the age of new imperialism and beyond. Proceedings of the session held at the 12th international congress of South-East European studies (Bucharest, 2–6 Sept 2019). Editura Istros a Muzeului Brăilei "Carol I", Brăila, pp 47–58

Nixon R (2011) Slow violence and the environmentalism of the poor. Harvard University Press, Cambridge, MA

Olafson P, Leutritz J (1959) The toxicity of creosote and creosote–pentachlorophenol mixtures to cattle. In: Proceedings of the annual meeting of the American wood-preservers' association 55:54–57

Petrić M (2011) The importance and protection of Karst waters. In: Prelovšek M, Zupan Hajna N (eds) Pressures and protection of the underground Karst: Cases from Slovenia and Croatia. Karst Research Institute ZRC SAZU, Postojna, pp 18–22

Proctor RN, Schiebinger L (eds) Agnotology. The making and unmaking of ignorance. Stanford University Press, Stanford, CA

Purtschert P, Lüth B, Falk F (eds) (2012) Postkoloniale Schweiz. Formen und Folgen eines Kolonialismus ohne Kolonien. Transcript, Bielefeld

Regulation (EC) No 1907/2006 of the European Parliament and of the Council of 18 December 2006 concerning the Registration, Evaluation, Authorisation and Restriction of Chemicals (REACH)

Russell AL, Vinsel L (2018) After innovation, turn to maintenance. Technol Cult 59(1):1–25

Ruthner C (2014) Bosnia-Herzegovina, 1878–1918: a colony of a multinational empire. In: Healy R, Dal Lago E (eds) The shadow of colonialism on Europe's modern past. Palgrave, London, pp 156–169

Ruthner C (2018) Habsburg's only colony? Bosnia-Herzegovina and Austria-Hungary, 1878–1918. SEEU Review 13(1):1–14

Ruthner C, Scheer T (eds) (2018) Bosnien-Herzegowina und Österreich-Ungarn, 1878–1918. Annährungen an eine Kolonie. Narr Francke Attempto, Tübingen

Schmid F (1914) Bosnien und die Herzegovina unter der Verwaltung Österreich-Ungarns. Veit & Comp, Leipzig

Schulz A (1930) Was geschieht mit den alten hölzernen Bahnschwellen? Die Holzschwelle 11:201–204

Soentgen J (2019) Konfliktstoffe. Über Kohlendioxid, Heroin und andere strittige Substanzen. Oekom, München

Stefanović Z (2018) Global distribution and use of water from karst aquifers. In: Parise M, Kaufmann G, Ravbar N (eds) Advances in Karst research: theory, fieldwork and applications. The Geological Society Publishing House, London, pp 217–243

Telesko W (2015) Colonialism without colonies. The civilizing missions in the Habsburg Empire. In: Falser M (ed) Cultural heritage as civilizing mission. From decay to recovers. Springer, New York/Vienna, pp 35–48

UNEP (2012) State of the environment report of Bosnia and Herzegovina. https://www.unep.org/resources/report/state-environment-report-bosnia-and-herzegovina-2012. Last accessed June 30, 2023

Van der Straeten J, Hasenöhrl U (2016) Connecting the empire. New research perspectives on infrastructures and the environment in the (Post)Colonial world. NTM 24:355–356

Washington State Department of Natural Resources, Aquatic Resources—Science of Creosote, Oct 14, 2019. https://www.dnr.wa.gov/publications/aqr_rest_creosote_science_brief.pdf. Last accessed June 30, 2023

Wikimedia commons: https://commons.wikimedia.org/wiki/File:1918_map_-_Artaria%27s_Eisenbahnkarte_von_%C3%96sterreich-Ungarn_und_den_Balkanl%C3%A4ndern.jpg. Last accessed on March 30, 2023

Wikimedia commons: https://commons.wikimedia.org/wiki/File:Station_Potoci_der_Steinbeisbahn.jpg. Last accessed March 30, 2023

Williamson SR Jr (1991) Austria-Hungary and the origins of the first world war. St. Martin's Press, New York

Zenz C et al (1994) Occupational medicine. Principles and practical application. Mosby, St. Louis, MO

Chapter 14
Examples of Human Disturbances in the Dinaric Karst Landscape Since the Early Twentieth Century: A Case Study of Bela Krajina

Daniela Ribeiro and Matej Simčič

Abstract Human activities have been impacting karst environments for thousands of years, but as human pressure increased, so has the human disturbance of the karst landscapes. Karst landscapes are highly vulnerable due to the nature of their hydrological system, and once "damaged" can be extremely difficult to "repair." The chapter gives an overview of the disturbance history in Bela Krajina. The management of litter-raking forests, the illegal deposition of waste condensers, the agricultural land improvement, and the use of fertilizers are among the greatest environmental pressures to which the Bela Krajina's landscapes were subjected in the twentieth century. The greatest disturbances in this century were identified as the spreading of liquid manure on farmland and the leveling of dolines.

Keywords Karst vulnerability · Disturbance to karst landscapes · Human impacts · Slovenia

14.1 Introduction

Karst landscapes provide valuable ecosystem services, particularly storing pristine water, providing local building materials, and mitigating climate change. However, the sustainability of karst systems is threatened by increasing human activities in these vulnerable regions.[1] Like no other landscape, the karst environment is uniquely

[1] Brinkmann and Garren (2011), p. 362.

D. Ribeiro (✉)
ZRC SAZU Anton Melik Geographical Institute, Ljubljana, Slovenia
e-mail: daniela.ribeiro@zrc-sazu.si

M. Simčič
Institute of the Republic of Slovenia for Nature Conservation, Novo Mesto Regional Unit, Novo Mesto, Slovenia
e-mail: matej.simcic@zrsvn.si

B. Fuerst-Bjeliš et al. (eds.), *Environmental Histories of the Dinaric Karst*, Environmental History 17, https://doi.org/10.1007/978-3-031-56089-7_14

vulnerable to human impact due to the connectivity between the underground and surface.[2] Human activities have been impacting on karst environments for thousands of years,[3] but as human pressure has increased, human disturbance upon the karst landscapes has also increased. Activities such as quarrying, pollution, groundwater extraction, construction, and agriculture are major motivators for disturbing surface and subsurface karst.[4] Due to the nature of the karst hydrology, karst landscapes are particularly vulnerable to overuse and misuse, and once "damaged" can be extremely difficult to "repair."[5] Recognizing this problem, the European Union has made efforts to protect the quality and quantity of groundwater resources[6]; the IUCN has also formulated guidelines for the protection of caves and karst,[7] and for sustainable tourism in protected areas.[8]

In spite of this, present policies in Slovenia (e.g., Environmental Protection Act,[9] Nature Conservation Act[10]) do not take into account karst specificities and its vulnerability. At a national level, the only legal act that states karst protection is the Underground Cave Protection Act,[11] which regulates cave activities and cave protection; however, there is a lack of implementation of this act.[12]

Karst landscapes in Slovenia cover over 44% of the country's area and are characterized by rocky surfaces with dolines, collapse dolines, solution valleys, poljes, corrosion plains, and dry and blind valleys.[13]

The aim of the chapter is to present human impacts in the karst landscape of Bela Krajina (SE Slovenia) in the twentieth and twenty-first centuries.

14.1.1 Study Area

According to the geological, hydrological, and morphological conditions, the Slovenian karst can be divided into: the Alpine Karst, the Dinaric Karst, and isolated karst.[14] The Dinaric Karst is the largest karst area in Slovenia, representing around two thirds of the whole karst area in the country,[15] with Bela Krajina as our study

---

[2] Ibid., p. 366.

[3] See also Chap. 4, in this volume: Mlekuž Vrhovnik and Fabec (2024); Zorn et al. (2015), p. 562.

[4] Van Beynen and Townsend (2005), p. 104.

[5] Ford and Williams (2007), p. 472.

[6] Zwahlen (2004).

[7] Watson et al. (1997), p. 1.

[8] Eagles et al. (2002), p. 1.

[9] "Zakon o varstvu okolja".

[10] "Zakon o ohranjanju narave".

[11] "Zakon o varstvu podzemnih jam".

[12] Ribeiro (2020), p. 26.

[13] Gams (2003), p. 461.

[14] Ibid., p. 488; Habič (1969), p. 79.

[15] Gams (2003), p. 489.

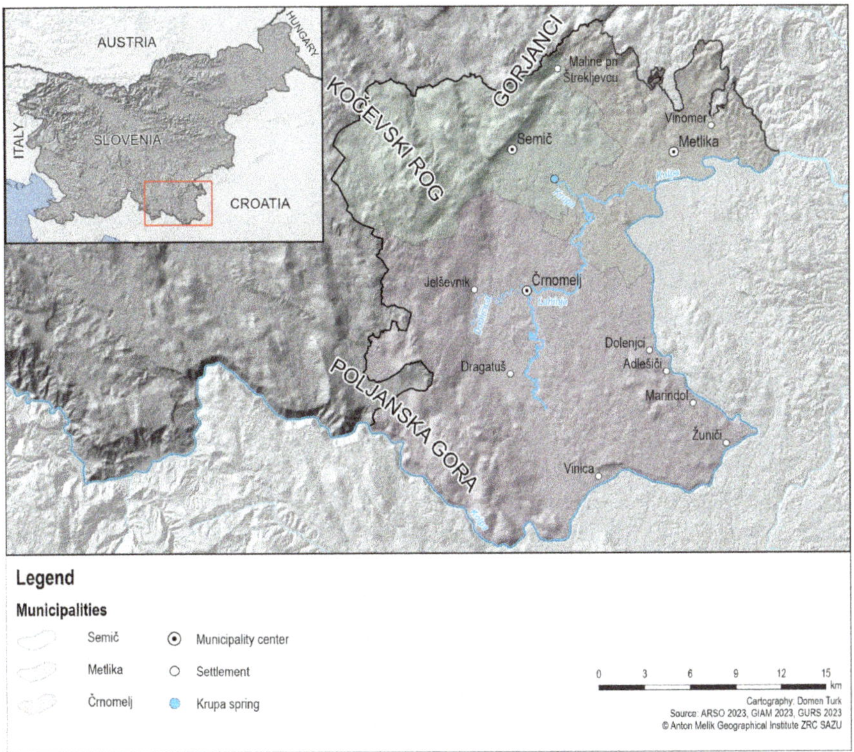

Fig. 14.1 Map of the study area

region being part of it. Bela Krajina is located in the southeastern part of Slovenia, along the Slovenian-Croatian border, covering 595 km^2 (including the municipalities of Črnomelj, Metlika and Semič; Fig. 14.1). It is characterized by karst features, predominantly forested landscape and rich biodiversity.[16]

One of the major geomorphological features of Bela Krajina is its pronounced leveled surface, the Črnomelj Plain.[17] The Črnomelj Plain is surrounded by the high karst plateaus of the Gorjanci Hills to the north, Poljanska Gora to the west and Kočevski Rog to the northwest.[18] Because Bela Krajina is distinctly karstified, a lot of caves have developed, and most of the largest ones and pits have developed on high karst plateaus.[19]

---

[16] Ribeiro (2020), p. 13.

[17] Gams (2003), p. 408; Stepišnik and Natek (2014), p. 37.

[18] Plut (1984), pp. 26–27; Stepišnik and Natek (2014), p. 37.

[19] Natek and Stepišnik (2008), p. 61.

14.2 Contemporary Human Impacts in Bela Krajina's Karst Landscapes

14.2.1 Disturbances in Twentieth Century

The management of litter-raking forests, the deposit of waste condensers in the wider karst hinterland of the Krupa spring, the agricultural land improvement, and the utilization of fertilizers are among the major environmental pressures that Bela Krajina' landscapes faced in twentieth century.

14.2.1.1 Management of Litter-Raking Forests

The limited possibilities for arable farming in the karst terrain of Bela Krajina[20] contributed to extensive, largely grazing-based livestock farming playing a prominent role within the polyculture farming in the long run. The natural forest in the lowland of Bela Krajina is an oak-hornbeam forest (including the typical forms of the *Asperulo odoratae–Carpinetum betuli* and *Abio albae–Carpinetum betuli*). Near the settlements, the farmers used the forest litter (removal of fallen leaves, brushwood, and undergrowth from the forest) for agricultural needs (use of litter as livestock bedding in stables and enriched with excrements, as manure for crops). In this way, the forest was constantly deprived of organic material, which is important for the formation of the humus layer in the soil. As a result, the soils in the litter-raking forests became poor and poorly supplied with nutrients. In addition to nutrient input, tree litter plays an important role in forest ecosystems by protecting the soil from erosion and compaction and buffering soil microclimatic fluctuations.[21] The long historical period in which litter raking was common has further acidified the soils. In these areas, a degraded forest has developed, dominated by birch, and bracken (*Pteridio–Betuletum pedulae*; Fig. 14.2); the latter was used as litter. Thus, the litter-raking forests are a secondary formation created by the impact of agriculture on the forest.[22] Nevertheless, the litter-raking forest has long been a typical landscape element of Bela Krajina. Intensive littering began at the end of the eighteenth century and spread in the middle of the nineteenth century when the prevailing free grazing was replaced by seasonal livestock grazing.[23] By 1960, there were slightly less than 10,000 ha of litter-raking forests in the region.[24] Just under half of this area (4865 ha) was designated for conversion to forests and slightly more (5114 ha) for further development of agricultural land. Štublar[25] estimated that around 1980, there

[20] Ciglič et al. (2012), p. 79.
[21] Glatzel (1991), pp. 2–3.
[22] Štublar (1981), p. 41.
[23] Šilc et al. (2008), p. 71.
[24] Miklavžič (1965), p. 3.
[25] Štublar (1981), p. 50.

Fig. 14.2 Litter-raking forests in Marindol. *Photo* Matej Simčič

were between 5200 and 6600 ha of litter-raking forest areas. Today, the litter-raking forests are being overgrown by shrubs and trees as ecological succession continues on the abandoned agricultural land. At the beginning of the twenty-first century, there are only a good 100 ha of litter-raking forest areas left in Bela Krajina in three larger areas (Vinomer, Marindol, and Žuniči). Recently, some litter-raking forest areas have become protected and the traditional management stimulated.[26] And so, to preserve this characteristic element of the cultural landscape, the overgrowth of former litter-raking forest areas is prevented, especially in these areas. Their management is the responsibility of the public institution of the Kolpa Landscape Park (Marindol and Žuniči) and the local community from Drašiči (Vinomer).

14.2.1.2 Deposition of Waste Condensers in the Wider Karst Hinterland of the Krupa Spring

While the example of the litter-raking forests presented in Sect. 14.2.1.1 was a pre-industrial human intervention in the landscape, this section is about industrial pollution.

In the wider hinterland of the spring of the Krupa River, condensers containing polychlorinated biphenyls (PCBs), used for impregnation at the Iskra Kondenzatorji factory in Semič, were dumped in dolines in 1962–1985. They were improperly stored and unprotected. The approximate locations of these landfills mapped based on fieldwork[27] can be seen in Fig. 14.3. The total amount of emissions and dumped

[26] Šilc et al. (2008), p. 77.

[27] Flajšman (1987).

waste was about 70 tons.[28] It is estimated that 13 tons of PCBs were washed into the karst underground in the hinterland of Krupa.[29] In 1985, the remediation of the landfills began. The negative effects of PCB contamination have led to health problems such as increased morbidity of certain diseases, ecological and infrastructural problems such as loss of a potential source of water supply, and further economic problems such as the contamination of locally produced food with PCBs. Last but not least, conservation problems such as the groundwater and surface water pollution due to continuous leaching of PCB sediments in the hinterland of the Krupa River during heavy rains, affecting subsurface fauna (e.g., *Proteus anguinus*) and river fauna.[30] Long-term monitoring measurements show that the pollution with PCBs has gradually declined. For example, the PCB concentration in the soil in 2010–2011 (<0.005 mg/kg) decreased significantly compared to the period 1978–1991 (0.045 mg/kg).[31] Nowadays, local agricultural products such as meat, milk, and fodder crops are safe for humans and livestock unless regular monitoring is abandoned. Nonetheless, the PCB concentrations in the area of the Krupa River and in the immediate vicinity of the Iskra factory in Semič are still high (the limit value of 125 ng/g for fish is exceeded, in some cases by more than 20 times).[32] The Krupa River is still the main source of PCB emissions into the environment.[33] Even several decades after the end of the improper disposal of condensers and remediation in the 2014–2019 period, the water quality status assessment was rated as poor or "moderate" due to the presence of the specific PCB pollutant.[34]

14.2.1.3 Agricultural Land Improvement

In 1946, the Črnomelj Agricultural Cooperative was founded, which included farmers from the municipalities of Črnomelj and Semič. The Agricultural Cooperative was an intermediary between farmers (producers) and buyers (domestic food production/processing companies—Žito Ljubljana, Belsad Črnomelj). It was easier for farmers because they had a well-organized central purchasing system with good prices. The Agricultural Cooperative acted as a kind of extension of the agricultural advisory service, because it was in constant and good contact with the farmers, knew the situation on the ground, and was aware of the needs and trends. The representative of the Agricultural Cooperative was also a member of the Municipal Committee for Agriculture, which prepared the development program for agriculture in the Municipality of Črnomelj. Part of this development program was also agricultural land improvement.

[28] Polič (2005), p. 33.

[29] Harlander and Miljavac (2011), p. 5.

[30] Ibid., p. 5; Pezdirc et al. (2011), p. 990; Lapajne et al. (2012), p. 7.

[31] Lapajne et al. (2012), p. 22.

[32] Ibid., p. 7.

[33] Pezdirc et al. (2011), p. 990.

[34] ARSO (2021).

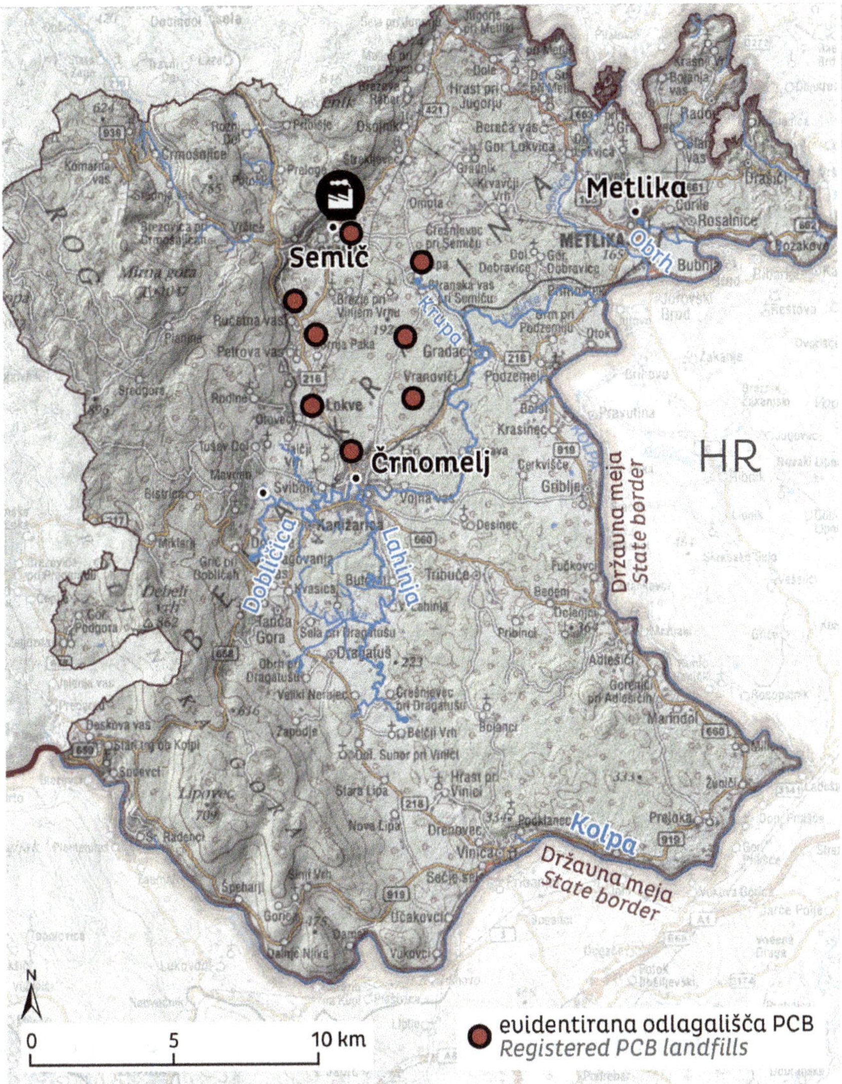

Fig. 14.3 Map of Bela Krajina with approximate locations of landfills where waste condensers containing PCBs were deposited. Map by Nina Jankovič, ZRSVN Archive[35]

In order to improve the quality of agricultural land, land improvement works began in 1980 on agricultural land in the municipalities of Črnomelj and Semič. The work included excavation and removal of stones, construction of dirt roads, and liming and fertilization of agricultural land due to the high acidity of the soil. The works were carried out on the basis of prepared plans, which, in addition to agricultural land

[35] See also Flajšman (1987), p. 12.

improvement works, often included land consolidation works and, in the case of the stream Podturnščica on the section near Dragatuš (1986), hydromelioration works. The works were systematically carried out with state funds. At the local level, the process was managed by the Črnomelj Agricultural Cooperative. In practice, the work was carried out in such a way that the Agricultural Cooperative selected a contractor to realize the plan. The farmers (landowners) bore 25% of the work, and the rest were government grants. For this purpose, farmers received special vouchers with which they paid the contractor for the machine hours worked. Based on the vouchers received, the contractor received the remaining funds. Farmers used these vouchers to pay for clearing stones in fields and meadows, as well as for the construction of paths.[36]

Most of the excavated stones were buried in larger holes dug at the bottom of larger dolines, from which soil had been previously excavated. When they filled the bottom of the doline with excavated stones, they covered it with soil.[37] Eventually, the terrain later settled in some places due to inadequate ongoing consolidation.

As a result of agricultural land improvement works (Fig. 14.4a), most of which were carried out in the period 1980–1997, the structure of the karst landscape changed, allowing farmers to practice more intensive agriculture than they would have done in the previous landscape. As part of the agricultural land improvement work, huge amounts of protruding stones typical of karst plains with a thick soil layer were removed from agricultural land (Fig. 14.4b).[38] Indeed, if farmers wanted to have as much grass as possible for their cattle, they had to remove the stones protruding from the soil, but if they wanted to plow the land, they had to remove the stones to the depth of the plow. The plow size and the depth of plowing increased significantly during the twentieth century. In Bela Krajina, the stones were removed mainly in flat areas, while in hilly areas the stones were not removed; an example of such an area is the so-called stone forest near the village of Vrtača pri Semič (Fig. 14.4c).

14.2.1.4 Utilization of Fertilizers

Bela Krajina is the most outstanding among the karst regions in Slovenia in terms of the share of agricultural land. In the Slovenian karst regions, the share of agricultural land is about 30%, and in Bela Krajina, it is 55%. Among the agricultural land (Fig. 14.4d), the predominant land use categories are meadows and pastures. Even though the most intensive agricultural production takes place in fields, vineyards, and orchards, and therefore these areas are the most affected by agricultural pollution, fertilization in areas occupied by meadows is necessary to achieve an adequate level of production.[39] Another reason for environmental pollution by nutrients, in addition to

---

[36] Mušič, interview.

[37] Ibid.

[38] Gams (2003), p. 217.

[39] Rejec Brancelj (1999), p. 103.

Fig. 14.4 **a** Agricultural land improvement works. *Photo* Matej Simčič; **b** protruding stones were removed from agricultural land. *Photo* Petra Gostinčar; **c** stone forest near the village of Vrtača pri Semič. *Photo* Andrej Hudoklin, ZRSVN Archive; **d** intensive fields with a biogas plant north of Črnomelj. *Photo* Matej Simčič

the excessive amount of fertilizers added, is the timing of their application in relation to the needs of the crops grown.[40]

Liming and fertilization (especially, NPK fertilizers; nitrogen, phosphorus, potassium) were carried out in Bela Krajina in large quantities and over a long period of time, especially since mid-twentieth century. There were cases (e.g., Dragovanja vas) when farmers in good faith overdid it with fertilization and their cattle were poisoned.[41] In 1995, Bela Krajina, with 151 kg of nitrogen per hectare on average, was among the regions that came quite close to the limit value for annual nitrogen input of 210 kg/ha according to the Regulation on the Limitation of Hazardous Substances and Fertilizers in Soil (1996).[42]

[40] Ibid., p. 113.

[41] Mušič, interview.

[42] Rejec Brancelj (1999), p. 116.

Fig. 14.5 Application of liquid manure on farmland. *Photo* Matej Simčič

14.2.2 Disturbances in the Twenty-First Century

Main disturbances to karst landscapes of the region in this century have been the application of liquid manure on farmland and leveling of dolines.

14.2.2.1 Application of Liquid Manure on Farmland

In order to fertilize, the farmers need a fertilizer plan that takes into account a range of factors in addition to soil analysis. However, in practice the decision on what amount of liquid manure to be applied on agricultural land in Bela Krajina is currently left to farmers. The application of liquid manure in Bela Krajina (Fig. 14.5) is problematic because of the shallow karst, and karst groundwater with a low self-purification capacity can flow as close to the surface as up to one meter below the ground. Therefore, reducing the quality of groundwater is one of the main negative effects of farming. Besides this, the polluted water can also have a fatal effect on the black *Proteus anguinus*, an endemic species to Bela Krajina (and other underground animals). This issue has been widely covered by the media.[43]

[43] Božič Kranjec and Žist (2010); Bezek-Jakše (2011); Pihlar (2011); STA (2013); Rajšek (2016); Virant (2020).

Fig. 14.6 Cultivated dolines of Vrhovci–Marindol. *Photo* Matej Simčič

14.2.2.2 Leveling of Dolines

The dolines are the predominant surface karst form in the lowland of Bela Krajina. According to the latest data, there are slightly more than 69,000 dolines[44] on the territory of the Bela Krajina, which corresponds to an average density of 116 dolines/ km^2. Locally, the density of dolines can be much higher.

Since there is usually a thicker layer of soil at the bottom of the dolines, in many dolines man has planted fields on the leveled ground and used the sloping hillsides for grass (so-called "cultivated doline"). In order to reach the bottom of the dolines, farmers often created an access path to the more or less steep slopes. By creating cultivated dolines, farmers have reshaped dolines to a lesser extent in the past, influencing the course of natural processes (especially, the leaching of soil toward the bottom of the doline and into the subsoil).[45] Such examples are known in practically all settlements of the region where there is a lack of land for crops, but the most picturesque was the cultivated dolines west of Marindol and Vrhovci, the so-called *Vrhovske vrtače* (Fig. 14.6). Nowadays, most of the dolines in the area of Vrhovci–Marindol have no fields left at their bottoms, because agriculture is being abandoned as a consequence of the aging of farming population.

[44] Mihevc and Mihevc (2021).

[45] Gams et al. (1971), p. 227.

With easier access to construction machinery, dolines have been filled in certain areas since 2000 to level the surface for easier agriculture or to create space for other purposes (e.g., manipulation areas for industry (warehouses, parking lots), tourist complexes, landfills). The recorded backfillings of dolines are known mainly from the wider surroundings of the settlements Črnomelj (foundry sand near Jelševnik), Adlešiči (Adlešiči, Dolenjci), initiatives for larger backfillings of dolines took place in the upland (e.g., Maline pri Štrekljevcu) and in the lowlands of the Semič municipality (hinterland of the Krupa spring). In addition to the need for modifying the landscape for economic purposes, leveling of dolines usually occurs during large infrastructure projects where there is a large surplus of material. In the area of Bela Krajina, there have been no such projects in the last decades; however, with the planned construction of expressway on the route Novo Mesto–Metlika and further in the direction of Črnomelj and Vinica, this may change.

14.3 Conclusions

Bela Krajina is characterized by rich biodiversity and a diverse cultural landscape, which is mainly due to the balance between nature protection and human activities. This mosaic-like landscape, expressed by its environmental sensitivity and vulnerability, as well as karst features that limit human activities, has been affected in this century mainly by the negative impacts of the spreading of liquid manure on farmland and leveling of dolines. The major twentieth-century disturbances presented here are the deposition of waste condensers in the wider karst hinterland of the Krupa spring and land use practices that improve the productivity and usability of agricultural lands (management of litter-raking forests, use of fertilizers, and agricultural land improvement). Most of the reported human impacts on the Bela Krajina's landscape are related to agricultural practices. Although on the one hand these human activities have an impact on the environment and especially on vulnerable karst ecosystems, on the other hand some of them contribute to the preservation of the cultural landscapes of Bela Krajina.[46]

Acknowledgements This study was conducted as part of the postdoctoral research project Sustainable Landscape Management: From Science to Action (Z7-1885), as well as research program The Geography of Slovenia (P6-0101), both financed by the Slovenian Research Agency.

[46] Ribeiro and Zorn (2021), pp. 13–14.

References

Sources

ARSO (2021) Ocena ekološkega stanja vodotokov za obdobje 2014–2019 [Assessment of the ecological state of watercourses for the period 2014–2019]. Agencija Republike Slovenije za okolje, Ljubljana. https://www.gov.si/assets/organi-v-sestavi/ARSO/Vode/Stanje-voda/Ekolosko_stanje_reke_2014-19.pdf

Harlander D, Miljavac B (2011) Posledice vpliva PCB na okolje v Beli krajini z oceno tveganja za zdravje ljudi zaradi uživanja doma pridelanih živil (jajca, mleko, perutnina) in rib iz reke Krupe, glede na vsebnost PCB [Consequences of the impact of PCBs on the environment in Bela Krajina with an assessment of the risk to human health due to the consumption of home-grown foods (eggs, milk, poultry) and fish from the Krupa River, according to the PCB content]. Zavod za zdravstveno varstvo Novo mesto, Novo mesto

Lapajne V, Hrženjak V, Vončina E, Ivanoš J, Žerjal E, Arnuš L, Harlander D (2012) Pregledna ocena stanja obremenitev okolja s PCB v Beli krajini, z njimi povezanih tveganj za zdravje ljudi, predlog priporočil in ukrepov za prebivalce Bele krajine in za druge deležnike, povezane s prehrano prebivalcev [A comprehensive assessment of the state of environmental contamination with PCBs in Bela Krajina, the associated risks to human health, proposals for recommendations and measures for the residents of Bela Krajina and for other stakeholders related to the residents' diet]. Zavod za zdravstveno varstvo Maribor, Maribor

Mušič M (2022) Agromelioration in former Črnomelj Municipality, 1980–1997. Interview by Matej Simčič. Dragovanja vas, 22 Feb 2022

Zakon o ohranjanju narave [Nature conservation act]. Uradni list Republike Slovenije, no 96 (2004)

Zakon o varstvu okolja [Environmental protection act]. Uradni list Republike Slovenije, no 41 (2004)

Zakon o varstvu podzemnih jam [Underground cave protection act]. Uradni list Republike Slovenije, no 2 (2004)

Literature

Bezek-Jakše M (2011) Črnomaljski svetniki predvsem o bioplinski gnojevki [Črnomelj municipal councilors mainly about biogas slurry]. Doljenski list, 11 Mar 2011. https://www.dolenjskilist.si/2011/03/11/39505/novice/bela_krajina/56/

Božič Kranjec J, Žist V (2010) Vse bolj zastrupljena [More and more poisoned]. Žurnal24.si, 7 Oct 2010. https://www.zurnal24.si/slovenija/vse-bolj-zastrupljena-97466

Brinkmann R, Garren SJ (2011) Karst and sustainability. In: van Beynen PE (ed) Karst management. Springer, Dordrecht, pp 361–378. https://doi.org/10.1007/978-94-007-1207-2_16

Ciglič R, Hrvatin M, Komac B, Perko D (2012) Karst as a criterion for defining areas less suitable for agriculture. Acta Geogr Slov 52(1):61–98. https://doi.org/10.3986/AGS52103

Eagles PFJ, McCool SF, Haynes CD (2002) Sustainable tourism in protected areas: guidelines for planning and management. IUCN, Gland, Cambridge

Flajšman B (1987) Je v Beli krajini ogroženih nekaj tisoč ljudi? [Are several thousand people at risk in Bela Krajina?]. Dolenjski list, 29 Jan 1987

Ford D, Williams P (2007) Karst hydrogeology and geomorphology. Wiley, Chichester. https://doi.org/10.1002/9781118684986

Gams I (2003) Kras v Sloveniji v prostoru in času [Karst in Slovenia in Space and Time]. Založba ZRC, Ljubljana

Gams I, Lovrenčak F, Ingolič B (1971) Krajna vas: študija o prirodnih pogojih in agrarnem izkoriščanju Krasa [Krajna vas: a study of the natural conditions and of agrarian land utilization on the Kras]. Geografski zbornik 12:223–263

Glatzel G (1991) The impact of historic land use and modern forestry on nutrient relations of Central European forest ecosystems. Fertilizer Res 27:1–8. https://doi.org/10.1007/BF01048603

Habič P (1969) Hidrografska rajonizacija krasa v Sloveniji [Hydrographic regionalization of karst in Slovenia]. Krš Jugoslavije 6:79–91

Miklavžič J (1965) Premena belokranjskih steljnikov v gozdove. Zbornik Inštituta za gozdno in lesno gospodarstvo Slovenije 4:1–87

Mihevc A, Mihevc R (2021) Morphological characteristics and distribution of dolines in Slovenia, a study of a lidar-based doline map of Slovenia. Acta Carsologica 50(1):11–36. https://doi.org/10.3986/ac.v50i1.9462

Mlekuž Vrhovnik D, Fabec T (2024) The formation of the Kras landscape from prehistory to the early modern period. In: Fuerst-Bjeliš B, Mrgić J, Petrić H, Zorn M, Zwitter Ž (eds) Environmental histories of the Dinaric Karst. Springer, Cham, pp 81–107

Natek K, Stepišnik U (2008) Geomorfološke značilnosti Bele krajine in Krajinskega parka Lahinja [Geomorphological characteristics of Bela Krajina and Lahinja Landscape Park]. In: Plut D (ed) Bela krajina in Krajinski park Lahinja. Znanstvena založba Filozofske fakultete, Ljubljana, pp 53–70

Pezdirc M, Heath E, Bizjak Mali L, Bulog B (2011) PCB accumulation and tissue distribution in cave salamander (*Proteus anguinus anguinus*, Amphibia, Urodela) in the polluted karstic hinterland of the Krupa River, Slovenia. Chemosphere 84(7):987–993. https://doi.org/10.1016/j.chemosphere.2011.05.026

Pihlar T (2011) Gnojevke iz bioplinarne Lokve kmetje ne smejo več uporabljati [Farmers may no longer use manure from the Lokve biogas plant]. Dnevnik, 6 Apr 2011, https://www.dnevnik.si/1042436171

Plut D (1984) Zemljepisni oris Bele krajine [Geographical outline of Bela Krajina]. In: Bezenšek A, Plut D (eds) Vodnik ekskurzije po Beli krajini in osrednji Dolenjski. Geografsko društvo Slovenije, Ljubljana, pp 26–33

Polič S (2005) Onesnaženost in ekološke obremenitve okolja kraškega območja reke Krupe s polikloriranimi bifenili—PCB [PCB pollution dynamic and environmental impact in the Krupa river karstic region]. Ph.D. dissertation, University of Ljubljana

Rajšek B (2016) Stariha za smrad krivi prebavne težave prašičev [Stariha blames the stench on the pigs' digestive problems]. Delo, 19 Sept 2016. http://www.delo.si/novice/slovenija/stariha-za-smrad-kriviprebavne-tezave-prasicev.html

Rejec Brancelj I (1999) Environmental protection aspects of agriculture in landscape regions of Slovenia. Geografski zbornik 39:93–132

Ribeiro D (2020) Bela Krajina—Sustainability in a Karst landscape. Založba ZRC, Ljubljana. https://doi.org/10.3986/9789610504627

Ribeiro D, Zorn M (2021) Sustainability and Slovenian karst landscapes: evaluation of a low karst plain. Sustainability 13(4):1655. https://doi.org/10.3390/su13041655

STA (2013) V Beli krajini bodo določili, kje bo prepovedano uporabljati gnojevko iz črnomaljske bioplinarne [In Bela Krajina, they will determine where it will be prohibited to use slurry from the Črnomelj biogas plant]. STA, 16 June 2013. https://www.sta.si/1898529/v-beli-krajini-bodo-dolocili-kje-bo-prepovedano-uporabljati-gnojevko-iz-crnomaljske-bioplinarne

Stepišnik U, Natek K (2014) Geomorphological characteristics of the Bela Krajina Plain. In: Štangelj M, Ivanovič M (ed) Natural heritage of Bela Krajina, Slovenia. Belokranjski muzej, Metlika, pp 37–39

Šilc U, Čarni A, Košir P, Marinšek A, Zelnik I (2008) Litter-raking forests in SE Slovenia and in Croatia. Hacquetia 7(1):71–88. https://doi.org/10.2478/v10028-007-0008-4

Štublar B (1981) Steljarjenje v zasebnih belokrajnskih gozdovih kot družbeno ekonomski problem [Littering in private forests in Bela Krajina as a socio-economic problem]. BSc Thesis, University of Ljubljana

Van Beynen P, Townsend KA (2005) Disturbance index for karst environments. Environ Manag 36:101–116. https://doi.org/10.1007/s00267-004-0265-9

Virant J (2020) Prdec Zelenega Jurija [Green George's fart]. Radio Odeon, 22 Aug 2020. https://www.radio-odeon.com/novice/prdec-zelenega-jurija/

Watson J, Hamilton-Smith E, Gillieson D, Kiernan K (1997) Guidelines for Cave and Karst protection. IUCN, Glang, Cambridge

Zorn M, Kumer P, Ferk M (2015) Od gozda do gozda ali kje je goli, kamniti Kras? [From forest to forest or where is the bare rocky Karst?]. Kronika 63(3):561–574

Zwahlen F (ed) (2004) Vulnerability and risk mapping for the protection of carbonate (Karst) aquifers. Office for Official Publications of the European Communities, Luxembourg

Daniela Ribeiro Research fellow, Ph.D. She graduated with a university degree in Environment and Natural Resources Engineering at the University of Trás-os-Montes and Alto Douro in Portugal; obtained a M.Sc. degree in Environmental Sciences at the Graduate School of the University of Nova Gorica, Slovenia and a Ph.D. in Environmental Protection at the University of Ljubljana, Slovenia. She mostly focuses on biogeography, landscape sustainability, nature conservation and human disturbances on karst environments.

Matej Simčič Senior nature conservation consultant. He graduated with a university degree in Geography at the Faculty of Arts, University of Ljubljana, Slovenia. He has much expertise on the communication with stakeholders involved in river and ponds restoration, nature conservation and flora and fauna monitoring.

Chapter 15
From Karstology to War Crimes: Third Reich's Military and Scientific Exploitation of the Dinaric Karst

Klemen Kocjančič

Abstract This study focuses on the Dinaric Karst under the occupation of the Third Reich from 1943 to 1945. Even before Italy's capitulation in September 1943, Waffen-SS formed a specialized mountain unit, *Karstjäger*, for warfare on rugged karst terrain. The unit was proposed and initially led by Hans Brand, a speleologist, and a geologist, who explored this area for military purposes during the First World War as part of the German military-geological unit, sent here to help Austro-Hungarian troops in a fight with Italians. *Karstjäger* were, and are still, the only military unit created for the sole purpose of warfare in karst area(s). After his dismissal as a commander, Brand established a research center that continued the scientific exploration of the Dinaric Karst for military purposes. Using wartime documents and post-war literature, the author presents this unique science-military nexus, which started with the geographical exploration in Austria-Hungary, peaked during the Isonzo front of the First World War, and ended with numerous war crimes committed by the *Karstjäger* upon civilians and partisans alike. Current literature on this topic is scarce, focusing only on the military aspect or the military-geological aspect, while at the same time ignoring the archival documents dealing with this subject.

Keywords Dinaric Karst · Waffen-SS · Second world war · *Karstjäger* · Military geography · War crimes · Mountain warfare

15.1 Introduction

During the recorded history, the area of Western Slovenia and North-eastern Italy was and still is important for military strategy due to its specific location, enabling the easiest passage from/to Italy, South-eastern Europe, and Central Europe south

K. Kocjančič (✉)
Defence Research Centre, University of Ljubljana's Faculty of Social Sciences, Ljubljana, Slovenia
e-mail: klemen.kocjancic@fdv.uni-lj.si

B. Fuerst-Bjeliš et al. (eds.), *Environmental Histories of the Dinaric Karst*, Environmental History 17, https://doi.org/10.1007/978-3-031-56089-7_15

Fig. 15.1 Map of southern and central Primorska and neighboring regions. *Source* Own elaboration. With permission by Lenart Štaut

of the Alps. The Alps, stretching from France in the west to Austria and Slovenia in the east, form a natural barrier that affects raids, mass migration, and military operations.[1] The same applies to the Dinaric Alps, linking the North-eastern Italy and the Western Kosovo (Figs. 1.1 and 15.1).

This contribution focuses on the period of the Second World War and the presence of the Third Reich's (para)military forces in the karst border areas of today's Italy, Slovenia, and Croatia. In April 1941, Germany attacked the Kingdom of Yugoslavia, followed by additional attacks by Italy, Hungary, and Bulgaria. After a short fight, Yugoslavia capitulated and was divided among the victors. The northern territories of present-day Slovenia were occupied by Germany, southern by Italy, while Hungary received the north-eastern part of present-day Slovenia and the newly formed Independent State of Croatia several villages in the south-eastern part of Slovenia.[2]

With the progress of war and Allied advances in Africa and the invasion of Sicily, the Kingdom of Italy officially capitulated on 8 September 1943, which resulted in the German occupation of Italian and former Italian-occupied territories. Further, in the area of Northern Italy, Western and Southern Slovenia, and the northwestern

[1] For instance, Dzino (2010), p. 71; Štekar (2013); Gračanin (2006), pp. 84–85; Grafenauer (1970–1971); Guštin (2004).

[2] Kocjančič (2021), p. 77.

Fig. 15.2 Operational zone of the Adriatic Littoral. *Source* Own elaboration, the backgrounds are cited in the figure. With permission by Lenart Štaut

part of Croatia, Germans created on 10 September 1943 a special administrative district, the Operational Zone of the Adriatic Littoral (*Operationszone Adriatisches Küstenland*; OZAK, Fig. 15.2). The OZAK formed the former Italian provinces of Udine, Gorizia, Trieste (all in Italy), Pula, Rijeka (both in present-day Croatia), and Ljubljana (Slovenia).[3]

Due to the rough terrain[4] and presence of local partisans, Germans sent a mountain warfare unit, specializing in karst topography, formed by the *Waffen-SS*, a military wing of the German Nazi party (*Nationalsozialistische Deutsche Arbeiterpartei*; NSDAP). This was followed by the deployment of several institutions specializing in applied military-geography research, providing the necessary information for the formation of military tactics and strategy. However, this specialization ultimately resulted in war crimes committed by German forces against the local population.

Based on archival documents and post-war literature, this study analyzes the creation and employment of these specialized units and institutions connected with

[3] Ferenc (1967), pp. 352–374.

[4] Jurkovšek et al. (2016).

the Dinaric Karst in the OZAK and the consequences of their actions. While separate research has already been conducted on German military units and military-geological activities during the Second World War,[5] this contribution focuses on the OZAK area.

15.2 Environmental Histories of Fascism and Nazism

According to Marco Armiero, fascism was interested in nature, not only in nature conservation, but also in nature reclamation (turning marshes into agricultural land), deforestation, and building critical infrastructure and new settlements, while at the same time "blending ideas of race, landscape, history, modernity, and ruralism."[6] This fascist concept of nature includes not only rhetorical but also geographical space. For example, their investment in land works (improving degraded farmlands, better management of forests and waterways, etc.) resulted not only in the improvement of living and working spaces, and economy, but fascism was also considered a progressive political movement by its supporters. The nature, in spite of its unspoiled beauty, "had to be" adapted to the will of humans, even though in the course of this "progress" some natural landscape was destroyed. This is because, "the land was stained by some kind of sin and needed to be redeemed—that is, to be saved from its impure condition."[7] Furthermore, treatment of nature by fascists was also a blueprint for treating people, as it "transformed people and places into something to be domesticated and shaped to reassemble the racial and ecological matrix."[8] Regarding the mountains, the Fascist regime envisioned them as a harsh type of environment, which can produce a "true and better stock of Italians; that environment made them strong, healthy, pure, and disciplined."[9]

Similarly, Nazism also approached nature as a space that had to be remodeled and subjugated to the needs of humankind (in the form of mass landscape projects) while at the same time as an object of (secular) worship and admiration. For example, as the urban areas were considered decadent and impure, Nazism turned its focus to nature, especially the forested areas, which were considered the birthplaces of German culture. Based on the history of the ancient Germans, naturalism was combined with the *völkisch* ideals, where "the forest becomes a metaphor for the German people (…), the German *Volk* is symbolically identified with the nation—a nation centered around a Maypole capped with a mantle of swastika-bearing flags."[10] While German conservationists supported Nazism as per their worship of nature, the notion

[5] See below in this chapter.

[6] Armiero (2014a), p. 242.

[7] Armiero and von Hardenberg (2013), pp. 286–287.

[8] Armiero et al. (2021), p. 13.

[9] Armiero (2014b), p. 261.

[10] Lee and Wilke (2005), p. 25.

also arose that nature "could and had to be" remodeled according to the wishes and needs of the people, especially when it came to military matters and war needs.[11]

15.3 Twentieth-Century Military (Geography) and the Dinaric Karst

There is a distinct lack of scientific research on the general military geography of the northwestern Dinaric Karst, while there is more literature dealing with specific periods of military history, especially the First World War.

15.3.1 The First World War's Isonzo Front

With the entry of Italy into the First World War in May 1915, the new (Italian) front opened, which ran from the Swiss border to the Adriatic Sea, passing through or by "Cima d'Asta, Colbricon, Marmolada, the Falzarego pass, the mountain ridge of the Carnic Alps, fortifications at Predilpass, the Isonzo river to Montfalcone." The front was present in the Alps and the Dinaric Karst, which presented several challenges to (mountain) warfare: fighting at high altitudes, exposure to inclement climate (snow, rain, low temperature, bora wind, etc.), limited access to drinking water, hygiene, and rough terrain. To combat this, the Austro-Hungarian army employed war geologists (and geographers) "to solve daily problems, such as water supply for the troops, cavern construction, building of paved roads, and construction of cableways for supply."[12]

Rough terrain also hindered the movement of troops, facing the enemy in entrenched positions on the top of the mountains, ridges, and plateaus, while at the same time constructing fortified positions for the state's own troops, artillery, and supply system.[13] Another major challenge was sourcing drinking water (most likely with drilling walls or running pipelines from streams and rivers), while also considering potential contamination of water through war material (as in explosives, ammunition) or casualties (especially near the front lines).[14] A special tactic employed in this area was the underground mining of enemy positions.[15]

While the first six Italian offensives were relatively successful (conquering the Gorizia/Görz in August 1916), the next five offensives resulted in stalemates.[16]

---

[11] Uekötter (2007), pp. 280–281.

[12] Angetter and Hubmann (2013), p. 7; Fasching (1995), p. 83.

[13] Zečević and Jungwirth (2007), pp. 58–61.

[14] Angetter and Hubmann (2014), pp. 119–121; Bernik (2014), pp. 39–40; Meerschman et al. (2011), pp. 417–424.

[15] Angetter (2009), pp. 297–299.

[16] Häusler (2013), p. 31.

In 1918, it was clear that Austria–Hungary on itself would not be able to defeat Italy; therefore, the German Empire sent troops and its own specialists to prepare for a decisive offensive.[17] This was the first formal German military research mission in the Dinaric Karst. Among the German specialists was Hans Brand (1879–1959). He studied mining, engineering, geology, and drawing and worked as a professor at different schools. When World War I began in 1914, he became an infantry officer. Until 1916, he was a company commander, highly decorated for frontline bravery, and then transferred to artillery as an intelligence officer. In 1918, he was once again reassigned to the military surveying detachment, and with it, he was also sent to the hinterland of the Isonzo Front to provide a survey of terrain in the Upper Carniola region (today mostly in northwestern Slovenia).[18]

After the war, he settled in the Franconian Switzerland (*Fränkische Schweiz*); he lived in Pottenstein and Bayreuth, and in Munich, all in Bavaria. He continued his geological research. In the interwar years, he led several research expeditions to Yugoslavia, Iceland, Norway, and Denmark. He was also a member of the paramilitary antirepublican *Stahlhelm*. In July 1933, he joined the National-Socialist Teachers' Union (*Nationalsozialistischer Lehrerbund*), and in May 1935, the NSDAP.[19]

15.3.2 Cave Research in the Third Reich Before the Second World War

Heinrich Himmler (1900–1945), *Reichsführer-SS* from 1929, in his pursuit of scientific evidence for Nazi Aryan theory, founded in July 1935 a research think tank *Ahnenerbe*.[20] In April 1937, Himmler ordered Dr. Walther Steinhäuser (1908–1946), then the SS officer in the SS Race and Settlement Main Office (*Rasse- und Siedlungshauptamt der SS*), to start "systematic register and research of caves" in Germany. Steinhäuser created a small *Sondergruppe der SS für kulturhistorische Höhlenforschung* located in Berlin, which was responsible for the ideological direction of speleology in the Third Reich. Only two months after Austria's *Anschluß*, in May 1938, the new *Ahnenerbe*'s Research station for karst and caves studies (*Forschungsstäte für Karst- und Höhlenkunde*) was formed in Salzburg, initially under the leadership of another geology researcher and SS officer Walter Steinhäuser (1908–1946). In July 1938, the research station was moved to Munich, a central location for research in Germany and Austria. In August of the same year, Brand assumed the leadership of the station.[21]

---

[17] Schaffler (1991), p. 39.

[18] Engelbrecht (1998), pp. 80–81; Ferenc (1979), pp. 191–192.

[19] Engelbrecht (1998), pp. 81–82; INZ, Nemška okupacija 139, Hans Brand's file, "Lebenslauf".

[20] Kater (2006), p. 11.

[21] BA, NS 21/Vorl. 331, "Neuordnung der Großdeutschen Karst- und Höhlenforschung," 3 December 1938; BA, NS 21/Vorl. 332, "Voläufige Vollzugsmeldung über die Einrichtung der

At that time, the research station had the following responsibilities: promotion of cooperation between specialist and amateur scientists, elaboration of a speleological research methodology, creation of the "Greater German Central Cave Cadaster," creation of a specialist library, establishment of a central speleological museum (with traveling exhibitions and providing training courses), work on speleological photography, creation of laboratories for investigation of found artifacts, establishing its archive, work on technical aids and material, nature conservation and monument preservation, role of caves in medicine and agriculture, development of caves for tourism and educational purposes, publication of the research/scientific journal *Mitteilungen über Höhlen- und Karstforschung* and maintaining ideological, National-Socialist overview in the field of speleology.[22] While Brand was cooperating with the SS for years, he only officially joined the SS on the first day of the Second World War (1 September 1939) with the rank of captain (*SS-Hauptsturmführer*), assigned to Himmler's personal staff (*Persönlicher Stab Reichsführer-SS*).[23]

In 1941, the research station comprised six departments: (1) prehistoric speleology, (2) karst geology, (3) general karst and cave studies (tourism and survey), (4) military-geological speleology, (5) medicine, and (6) mythology and local customs. Brand also started to ask for practical deployment of the research station, to perform "specialist military duties in karst areas of Slovakia or on the West Front."[24]

The research station was again relocated in September 1943 to Pottenstein, where it remained until the end of the war.[25]

15.4 Karstjäger

As the war was progressing and Brand wanting more practical duties, he sent—starting in early 1942—Himmler several proposals for the creation of specialist military unit for mountain warfare in karst regions, using his expertise regarding this terrain type. He envisioned the use of such a unit in the Balkans and the Soviet Union, especially in Crimea (the Black Sea region).[26]

Forschungsstätte für Karst- und Höhlenkunde, München, Güllstr. 3/I," 12 April 1939; Danner (2017), pp. 17–20.

[22] Danner (2017), pp. 20–21.

[23] Engelbrecht (1998), p. 85; INZ, Nemška okupacija 139, Hans Brand's file, "Lebenslauf".

[24] Schaffler (1991), pp. 42–43.

[25] Danner (2017), p. 28.

[26] BA, NS 21/Vorl. 785, "Aktenverkmerk über Ausbildungs-, Einsatz- und Organisationsfragen in der SS-Karstwehrtruppe," 15 June 1943; BA, NS 21/Vorl. 799, "Karst- und Höhlenkunde usw.," 5 June 1943; NARA, T-175 R-70, "Dinarische Höhlenkartierung und SS-Karstwehrtruppe," 15 February 1943; Danner (2017), p. 30; Engelbrecht (1998), p. 92.

15.4.1 The Formation of Karstjäger

Brand's wish was finally granted on 3 July 1942, when Himmler issued an order for the creation of one *SS-Karstwehr-Kompanie*, which was directly subordinated to him and Brand was its commander.[27] Brand created a company composed of three distinct groups: specialized shock troops (*SS-Karstjäger*), defense construction specialists (*SS-Karstfestungspioniere*), and military scientists (*SS-Karstwehrwissenschaftler*).[28] At first, the unit was located near Dachau (Bavaria), but was then relocated to Pottenstein, the seat of the research station's dislocated unit; the station personnel became involved in military training. The training was conducted in Pottenstein and Pegnitz (Bavaria). Training (Fig. 15.3) was organized by the *Karstschulungsstätte* (also *Karstwehrschulungsstätte*) on newly constructed shooting ranges, in climbing centers, practice/training caves, shafts, sinkholes (Fig. 15.4), and training water treatment facilities. Later, part of the training (mid- and high-mountain warfare courses) was conducted in Mühlbach, Hochkönig, and Werfenweng (all in the province of Salzburg, Austria). Brand suggested and granted the expansion of the unit to battalion strength in November 1942.[29]

To better manage these facilities, a dislocated concentration camp (as part of the Flossenbürg camp) was created in Pottenstein, with internees providing manual labor. In Werfenweng, a similar camp for Italian prisoners of war was created later. *The Konzentrationslager-Außenlager Pottenstein* was formed in October 1942 and was operational until 16 April 1945, when the area was liberated by American troops. At first, the camp had around 40 inmates; until March 1945, the number rose to 369 inmates, mainly Russians, Poles, and Czechs, as well as Germans, Belgians, Dutch, Italians, Yugoslavs, Slovaks, Hungarians, and some Jews. Inmates built training, living, research, and supply facilities for SS troops, the concentration camp itself, roads, swimming pools, etc.[30]

In early 1943, Brand suggested that the unit be ready for combat deployment, and he advised that it should be sent to Upper Carniola to be utilized against Slovenian partisans. However, this did not happen, and *SS-Karstwehr-Bataillon* was meant to be sent to Bosnia. Italy capitulated on September 1943, followed by the German occupation of Northern Italy.[31]

---

[27] BA, NS 21/Vorl. 837, "Rechenschaftsbericht über die Schaffung der Ausbildungs- und Unterkuftsanlangen für das SS-Karstwehrbataillon in Pottenstein," 30 June 1944.

[28] BA, NS 21/Vorl. 837, "Die SS-Karstwehrschulungsstätte Pottenstein," 20 March 1944; NARA, T-175 R-110, "Aufstellung einer SS-Karstwehrkompanie," 3 July 1942; NARA, T-175 R-110, "Aufstellung der SS-Karstwehrkompanie," 24 July 1942.

[29] BA, NS 21/Vorl. 837, "Rechenschaftsbericht über die Schaffung der Ausbildungs- und Unterkuftsanlangen für das SS-Karstwehrbataillon in Pottenstein," 30 June 1944; NARA, T-175 R-110, "Aufstellung eines SS-Karstwehr-Btls. Für das SS-Gen.Kdo.(Pz.)," 19 November 1942; Danner (2017), p. 29.

[30] Engelbrecht (1998), pp. 13–30.

[31] Kocjančič (2021), p. 99.

Fig. 15.3 *Karstjäger* during climbing training. *Source* BA, Bildarchiv, Bild 146-1991-068-11/ CC-BY-SA 3.0. https://commons.wikimedia.org/wiki/File:Bundesarchiv_Bild_146-1991-068-11,_SS-Karstwehr-Bataillon,_Ausbildung.jpg. With permission by Bundesarchiv

15.4.2 Combat Operations

The first combat action of the *Karstjäger* occurred on 9 September 1943, while securing the Tarvisio-Udine railway (today in North-easternmost Italy, Fig. 15.2). The next day, one combined company was sent to occupy the Slovenian town of Bovec (in the Upper Soča/Isonzo catchment, Fig. 15.2) and then participated in the elimination of the Kobarid Republic, a Partisan-held area around Kobarid, famous for the Austro-German breakthrough during the First World War's 12th Isonzo Offensive. Later, *Karstjäger* was stationed between Tarvisio and Bovec, conducting counterinsurgency operations before moving to Grahovo ob Bači (today in Prealps of Western Slovenia, Fig. 15.2) and in December 1943 to Gradisca d'Isonzo in Italy (Fig. 15.2). In late November 1943, Himmler also decided that *SS-Karstwehr-Bataillon* would be stationed in the OZAK, where it would be used to combat partisans, while also establishing "two strong garrisons" (*zwei feste Garnisonen*).[32]

For almost 2 years, *Karstjäger* has participated in numerous counterinsurgency battles in the border areas of Italy and Slovenia, the Dinaric Karst. In June 1944, they

[32] NARA, T-175 R-81, Himmler's order, 26 November 1943; Kocjančič (2021), pp. 99–102.

Fig. 15.4 *Karstpioniere* while drilling. *Source* BA, Bildarchiv, Bild 146-1994-024-20/CC-BY-SA 3.0. https://commons.wikimedia.org/wiki/File:Bundesarchiv_Bild_146-1994-024-20,_SS-Karstwehr-Bataillon,_Karst-Festungspioniere.jpg. With permission by Bundesarchiv

were relocated to the Italian town of Tolmezzo, where the battalion was reorganized into divisions (*24. Waffen-Gebirgs[Karstjäger]-Division der SS*). While the division was expected to reach around 6600 men, in late September 1944, it could not even muster 2000 men. This division was stationed in Moggio Udinese, Pontebba, Cividale del Friuli, Malborgetto, Palmanova, and Tolmezzo (today all in north-eastern Italy, Fig. 15.2). In December 1944, the division was downgraded to the brigade level, but in February 1945, it was upgraded (on paper) to the divisional level. In May 1945, *Karstjäger* managed to reach Austria, where they were captured by the Allied forces.[33]

15.4.3 War Crimes

According to Slovenian historian Tone Ferenc, *the Karstjäger* formation committed the largest number of war crimes in the Slovenian territory of all the German units that operated against the Slovenian Partisans.[34]

[33] Kocjančič (2021), pp. 102–114.

[34] Ferenc (1977); Ferenc (1979). For more on this topic, see BA, NS 21/Vorl. 785, different letters/correspondence; Engelbrecht (1998), p. 108.

On 11 October 1943, *Karstjäger* committed their first war crime in Slovenian territory; they shot 16 men, evicted other villagers, and burned down the village of Strmec (south of Idrija, today in Western Slovenia, Fig. 15.2). This was done as retaliation for partisan ambush, conducted the previous day, resulting in several dead and wounded *Karstjägers*. Later, the Brand's unit participated in the destruction of several more Slovenian villages (Lokve, Komen, Rihemberk, Peternel, Šebrelje, etc.—today all in Western Slovenia and north-easternmost Italy, Fig. 15.2), while at the same time they also murdered, robbed, and deported local inhabitants.[35] The last recorded war crime, committed by *Karstjäger*, occurred on May 2, 1945, in the Italian village of Avasinis (near the Tagliamento River, North-eastern Italy), where they killed 51 villagers, including five children (aged between 2 and 12 years) and two women, 75 and 81 years old.[36] On 9 June 1944, *Karstjäger* committed one of the most notorious war crimes that happened in Slovenian territory: they tortured (burned with cigarettes, gouging out eyes) and then beheaded two captured Slovenian partisans.[37]

15.4.4 Back to Research

Brand became involved in the conflict with Odilo Globocnik (1904–1945),[38] who held the position of Higher SS and Police Leader in the OZAK. Their disagreement was due to different visions on the proper usage and future of *Karstjäger*; at the same time, Brand was sending scolding letters on Globocnik to Himmler and SS leadership in Berlin. However, this backfired on him, and in August 1944, he was formally dismissed as commander of the *Karstjäger* formation. Brand wanted to create a special inspection, which would supervise all *SS-Karstwehrformationen*, including training and creation of a specialized military-scientific institute.[39]

The SS leadership decided to completely remove Brand from any supervision/command role; thus, he was ordered to return to research. As part of the Institute for Applied Military Research (*Institut für wehrwissenschaftliche Zweckforschung*), already created in 1942, a new Karst Military Research Institute (*Karstwehrwissenschafliches Institut*) was founded in Pottenstein. Brand was given five researchers from Ahnenerbe's research station, who were already part of the *Karstjäger*, and 52 soldiers, for administrative work and guarding the inmates. During the war different designations for this institute were also used: "*SS-Karstwehrtruppe. Wehrwissenschaftliches Institut für Karst- und Höhlenforschung. Höhlennachweis-Abteilung*," "*Wehrwissenschaftliches Institut der Waffen-SS. Abteilung Karst-*

[35] Kocjančič (2021), pp. 101, 230.

[36] Engelbrecht (2021).

[37] Ferenc (1977), pp. 520–538.

[38] For more on Globocnik, see: Poprczeczny (2004); Pucher (1997); Rieger (2007); Sachslehner (2016).

[39] Danner (2017), p. 30; Ferenc (1979), p. 207.

und Höhlenforschung," and "*Institut für wehrwissenschaftliche Zweckforschung. Abteilung Karstwesen*". The primary mission of this institute was to identify all suitable caves for use as bomb shelters and underground (factory) facilities, while simultaneously providing *Karsttruppe* with relevant scientific information (especially in the fields of geography, physics, chemistry, and surveying). Additionally, they produced maps of the Dinaric caves and found solutions to provide water in karst areas. Brand considered this institute as the "stem cell of all karst and cave research in the world" (*die Stammzelle der gesamten Karst- und Höhlenforschung der Welt*).[40]

In February 1943, another research center was created in Pottenstein, dedicated to fortification research (*SS-Fortifikations-Forschungstelle*). Based on this research center, Himmler in June 1944 ordered a creation of SS Fortification Research Centre for Exploring Italian Border Fortifications (*SS-Fortifikationsforschungsstelle zur Erkundung der italienischen Grenzwehranlagen*), which was to be subordinated to Higher SS and Police Leader for Italy (*Höchster SS- und Polizeiführer Italien*). The center was initially stationed in the town of Fusine (near Tarvisio, Fig. 15.2), with the mission of exploring and critically evaluating already-gathered material on existing fortifications in Northern Italy. The primary haul of material came from the Italian Speleological Institute (*Istituto Italiano di Speleologia*), which was stationed in Postojna (today in the Classical Karst of Southwestern Central Slovenia, Figs. 15.1 and 15.2). To accomplish this, Himmler, in August of the same year, ordered that all relevant maps, blueprints, drawings, and other documents on Italian defense matters (originals or copies) should be handed over to the research center. The center, composed of around 40–50 personnel (including Italian researchers and former prisoners of war), was led by Dr. Kurt Scharlau (1906–1964). In October 1944, Brand announced that the research center would produce detailed maps regarding the defense of North Italy and the Dinaric Karst. Owing to the progress of the war, all personnel of the *Karstwehrwissenschafliches Institut* were in 1945 transferred to the research center to create a registry of all underground water sources in the Monfalcone-Gorizia-Ljubljana-Kočevje-Novi (in the eastern Gulf of Kvarner)-Adriatic Sea area (today in North-easternmost Italy, central and southern Slovenia, and Croatia—Fig. 15.2).[41]

The work of these two Brand-led institutions was primarily used to create fortification systems in the Po Valley and wider areas. The building of this German last defense line in Northern Italy started in June 1944, and on this project, at one point worked more than 120,000 workers. During construction, they also utilized Italian and Austrian fortifications (or their remnants) from the First World War and Italian fortifications from the interwar period.[42]

[40] BA, NS 21/Vorl. 785, "Karst- und Höhlen-Forschungsstätte," 13 June 1944; BA, NS 21/Vorl. 799, "Errichtung einer Karst- und Höhlen-Forschungsstätte und SS-Fortifikationsforschungsstelle," 13 June 1944; BA, NS 21/Vorl. 799, "Karst- und Höhlenforschungsstätte," 5 July 1944; Danner (2017), p. 31. For more on the underground facilities, see: Knolle et al. (2013), pp. 66–68.

[41] BA, NS 21/Vorl. 838, "Bestätigung," 18 August 1944; BA, NS 21/Vorl. 838, Brand's letters, 26 October, 16 November, 6 December 1944; Danner (2017), pp. 31–32; Ferenc (1979), p. 209.

[42] BA, NS 21/Vorl. 799, Brand's letter, 7 March 1945; Ferenc (1979), p. 212.

Another Brand's institution was formed in Slovenian territory. As part of the *Forschungsstätte für Karst- und Höhlenkunde*, in March 1944 Himmler ordered a creation of yet another research center, this time in Ljubljana, dedicated to the research of caves and karst in the Carniola and Adriatic Littoral (*Forschungsstätte für Karst- und Höhlenkunde, Abteilung Krain und adriatisches Küstenland*), while at the same time carrying out defense research in the assigned area. Richard Spöcker (1897–1975) was named as the deputy leader and supervisor of karst show caves (*Karstschauhöhlen*), while Brand was named the leader of the center, with Franz Mühlhofer (1881–1955) as technical and research advisor. Italian speleologist Franco Anelli (1899–1977), the director of the *Istituto Italiano di Speleologia*, was an advisor for all Italian karst research. The research center dealt primarily with the issue of water supply/regulation (regarding building a dam on the Pivka River inside the Postojna Cave to flood nearby flatlands) and the registry of caves in Čičarija, the Northernmost Istrian Peninsula.[43]

In both Brand's institutions (the research center and the institute), there were only a small number of people (17 in autumn 1944). This also resulted in limited work success, which led Brand to request transfer to the *Wunderwaffen* project, specifically to the supervision of building underground facilities for V-1 and V-2 rockets. His request was denied, and instead, he was offered supervision of building defense fortifications in the Alpine region, but he turned it down to his age (65 at that time).[44]

During the two years of research in the wider area of the OZAK, Brand-led institutions produced[45]:

1. 452 pages of maps and blueprints regarding fortifications between the Croatian port city of Rijeka (Fig. 15.2) and Slovenian mountain of Mojstrovka (Julian Alps).
2. 143 pages of maps and blueprints regarding underground and ground-level water sources in the Monfalcone-Gorizia-Ljubljana-Kočevje-Novi-Adriatic Sea area.
3. 96 pages of maps on the most important caves (for military use) with a 1:25,000 scale.

In summer 1945, Brand was captured around Pottenstein and spent two years in the internment camp. Later, in July 1947, Brand was declared a lesser offender (*Minderbelastete*), despite his involvement with concentration camps and war crimes in today's Slovenia and Italy. He was released, and continued his research until his death in January 1959.[46] Despite his wartime service, he was still prominent and respected in Pottenstein. In 2001, a road in Pottenstein, named after him (*Der Dr.-Hans-Brand-Ring*), was renamed.[47]

[43] BA, NS 21/Vorl. 799, correspondence between Brand and Spöcker; BA, NS 21/Vorl. 799, telegraph message, 15 December 1944; Danner (2017), pp. 32–33; Ferenc (1979), pp. 213–214.

[44] NARA, T-175 R-70, letter to Brand, 21 November 1944; Ferenc (1979), pp. 211–212.

[45] Ferenc (1979), pp. 212–213; NARA, T-175 R-70, "Leistungsbescheinigung," 20 November 1944.

[46] Engelbrecht (1998), pp. 123–133.

[47] Fritz (2015), p. 100.

15.5 Other German (Military-)Geographical Presence During the Second World War

In addition to the Brand-related military units and institutions that were present in the OZAK or Northern Italy, there were also some other military-geographical units and institutions present in that area.

15.5.1 *Forschungsstaffel Z.B.V.*

Brand's research station was not the only German military-geographical research institution present in Slovenia during the Second World War. The second institution was *Forschungsstaffel zur besonderen Verwendung des Oberkommandos der Wehrmacht* (Research Command for Special Deployment of the Armed Forces High Command), a unique unit that combined military and civilian personnel.

The creator of this unit was Dr. Otto Schulz-Kampfhenkel (1910–1989), a German researcher in the field of zoogeography, filmmaker, and pilot. In July 1940, Schulz-Kampfhenkel, an officer in the German Air Force (*Luftwaffe*), formed a society (*Forschungsgruppe Schulz-Kampfhenkel e.V.*), which was created as a research society in the field of research expeditions, natural sciences, and ethnology. In the autumn of the same year, he started to work for *Amt Ausland/Abwehr*, the military intelligence service of the *Oberkommando der Wehrmacht* (OKW), planning an expedition to North Africa. He continued his research, and in April 1941, defended his dissertation at the University of Würzburg. As part of the advance party, he was, in November/December 1941, sent to Italy, Libya, and Greece, where he worked until September 1942, when he was named the leader of the research group in *Sonderkommando Dora*, a military-geographical research unit of the OKW.[48]

Sonderkommando Dora was formed to provide scientific information for the German *Afrikakorps*, especially in the form of maps of central and south Libyan Sahara Desert and the Northern French Equatorial Africa. Schulz-Kampfhenkel was the leader of 16 researchers working in the fields of meteorology, climatology, zoology, biology, cartography, medicine, geography, geology, road engineering, topography, ethnology, astronomy, navigation, and land surveys. They were tasked with finding the best routes across the desert and the water supply points.[49] During their work in Africa, they created 23 specialized maps of different scales: "Morphologic-geologic maps were drawn based on the geologic mapping and reconnaissance flights, highlighting hamada (high rocky plateau, often with scarps), serir (gravel desert), ergs (areas of shifting sand dunes), dunes and oases."[50]

---

[48] Häusler (2018a), pp. 13–14.

[49] Häusler (2018a), pp. 18–19.

[50] Häusler (2018b), p. 38.

As Italo-German forces were driven out of Africa, *Sonderkommando Dora* pulled back into Europe and disbanded. The research personnel were regrouped into *Forschungsstaffel z.b.V.*, with Schulz-Kampfhenkel naming its commander in May 1943. The focus of their research changed from Africa to Europe, with an analysis of photographic material and scientific literature regarding the terrain and making different maps for strategic, tactical, or specialized needs (for example, maps showing terrain suitability for the usage of tank troops). In the last year of the war, *Forschungsstaffel z.b.V.* was composed of four research commands (West, South, East, and Camouflage) and three research groups (Lapland, North Russia, and Ukraine). They also produced a map of the North Italian lowlands between the Isonzo and Po rivers and maps of the karst regions of Istria and Dalmatia.[51]

The Research Command South (*Forschungskommando Süd*) was in Ljubljana (then *Laibach*), covering the South, Southeast and Central Europe (specifically Yugoslavia, Albania, Greece, Western Hungary and south side of Eastern Alps in Austria and Italy); until March 1944, the command was located in Austrian town of Graz. In April 1944, Spöcker became the commander of this command, which was composed of more than 20 experts from different fields (cartography, aerial photography, phytosociology, geology, hydrology, land survey, and road engineering), including some translators for the Slovenian language. They were stationed in the villa, confiscated from the Slovenian Jew, located in Šentvid near Ljubljana. Their primary mission was to create specialized maps regarding the usage of tank troops (for Dalmatia, Istria, and a wider area of Ljubljana), the presence of water sources in the Dinaric Karst, and possible seaborne landing sites along the coast of the Adriatic Sea. This was done using the existing scientific literature and contemporary aerial photography of these areas.[52]

Spöcker, himself an expert in karstology and a member of the *Organization Todt* (a paramilitary organization specializing in the construction of major projects), spent the majority of his research in the area of the Postojna Cave (Classical Karst) and Čičarija, and thus became connected with Brand's research center, ultimately becoming a deputy leader (before the end of the year, he left the center because of the conflict with Brand). For Čičarija, he made report regarding the cave system in the area, and its possible use for military purposes (water supply and storage space).[53]

15.5.2 SS-Wehrgeologen

According to some sources, *SS-Wehrgeologen-Bataillon 500* was also present in the OZAK. This military-geological battalion of Waffen-SS was formed in 1941 to perform specialized tasks such as building fortifications, underground bunkers, and drilling water wells. In the Italian port city of Trieste, this battalion supposedly

---

[51] Häusler (2018a), pp. 19–23.

[52] BA, NS 21/Vorl. 799, "Gutachten," 22 November 1944; Plewnia (2011), pp. 306–309.

[53] Plewnia (2011), pp. 313–316.

participated in the creation of the underground system of shelters and tunnels known as Little Berlin (*Kleines Berlin*).[54]

15.6 Conclusions

While the activities of *Karstjäger* are known, and we also know the usage of scientific findings of military-geography institutions in building fortifications in Northern Italy, there is little information on such usage in the OZAK. One known example is the world-famous Postojna Cave, which was used by Germans as a storage for aviation fuel until Slovenian partisans managed to infiltrate the cave system and then blew up the fuel in April 1944. The explosion and subsequent fire, which lasted for seven days, also damaged the cave itself.[55] From German wartime correspondence, it is clear that they were considering more extensive use of caves for military purposes, but most likely due to the progress of war and insufficient building materials, this did not happen. Some of the karst phenomena (such as caves and shafts) were used during the war (and even more immediately after the war) for the disposal of murdered people (as in the *foibe* and Kočevski Rog shafts in the Northern Dinaric Karst).[56]

In relation with the First World War, when there was a greater need for the utilization of military-geographical research, during the Second World War, we can clearly see a lesser level of such activities for military use. This is clearly a direct consequence of more mobile warfare during the Second World War, in contrast to the static warfare of the First World War. Long-term deployment and fighting of military units in the mountain and karst terrain made it necessary to employ military geographers and geologists. These military scientists were also present during the Second World War, but in a much diminished role.

The presented case of Hans Brand, *Karstjäger*, and related karst research for military purposes shows an ethical dilemma and misuses of science for war crimes. While starting with "pure" scientific research into geographical/geological subjects, this has turned, under the indoctrination of political ideology, into the usage of concentration camp inmates to provide labor force for supporting military organizations. The worst abuse of scientific information for the creation of specialized military units became infamous for war crimes against the local population. Thus, instead of participating in the development of human society, scientific knowledge ultimately resulted in mass killing and destruction. Brand, which was directly involved in several war crimes and contributed his knowledge of karst to military usage, went unpunished after the war and continued his research without any hindrance. Further research in the lives of these people is necessary, and in their wartime activities, which is now partly hindered by the lack of relevant wartime documentation, now lost, or destroyed long time ago.

---

[54] Kocjančič (2021), p. 221.

[55] Petelin (1985), p. 88; Zupančič et al. (2011), pp. 307–317.

[56] Ferenc (2008), pp. 155–160; Korb (2014), p. 122.

References

Published sources

BA (Bundesarchiv, The German Federal Archives), Bildarchiv, Bilder 146-1991-068-11 (https://commons.wikimedia.org/wiki/File:Bundesarchiv_Bild_146-1991-068-11,_SS-Karstwehr-Bataillon,_Ausbildung.jpg), 146-1994-024-20 (https://commons.wikimedia.org/wiki/File:Bundesarchiv_Bild_146-1994-024-20,_SS-Karstwehr-Bataillon,_Karst-Festungspioniere.jpg)

Copernicus (2023) European digital elevation model (EU DEM) v 1.1. European Environment Agency, Copenhagen

GURS [Surveying and Mapping Authority, Slovenia] (2023) Državna meja [State border]. Geodetska uprava Republike Slovenije, Ljubljana

Archival sources

BA (Bundesarchiv, The German Federal Archives)

BA, NS 21/Vorlage 331, 332, 785, 799, 837, 838

INZ (Inštitut za novejšo zgodovino, Institute of Contemporary History, Slovenia)

INZ, box Nemška okupacija 139

NARA (National Archives and Records Administration, USA)

NARA, T-175 rolls R-70, R-81, R-110

Literature

Ajlec K, Repe B (2023) Dismembered Slovenia. University of Ljubljana Press, Ljubljana

Angetter D (2009) Geologische Aspekte in der Kriegsführung des Ersten Weltkriegs. Jahrb Geol Bundesanst 149:291–300

Angetter D, Hubmann B (2013) 'K & K war geologists' and combat mission at the Southern Front during World War I. Berichte der Geologischen Bundesanstalt 101:7–9

Angetter D, Hubmann B (2014) Bedeutende österreichische Kriegsgeologen im Einsatz an der Südfront des Ersten Weltkrieges. Geo.Alp 11:117–134

Armiero M (2014a) Introduction: fascism and nature. Mod Italy 19(3):241–245

Armiero M (2014b) Making Italians out of rocks: Mussolini's shadows on Italian mountains. Modern Italy 19(3):261–274

Armiero M, Biasillo R, Guimarâes P (2021) Editorial: environmental histories of Mediterranean fascisms. Perspectivas, Special issue, Environmental histories of Mediterranean fascisms:9–16

Armiero M, von Hardenberg WG (2013) Green rhetoric in blackshirts: Italian fascism and the environment. Environ Hist 19:283–311

Bernik V (2014) Kemično orožje: posebnost prve svetovne vojne ali še danes preteča nevarnost [Chemical weapons: peculiarity of World War I or a still imminent danger]. Sodobni vojaški izzivi 16(4):31–48

Danner P (2017) Die Neuordnung der Großdeutschen Höhlenforschung und die Höhlenforschung in Salzburg von 1938 bis 1945. Geologische Bundesanstalt, Wien

Dzino D (2010) Illyricum in Roman politics 229BC–AD68. Cambridge University Press, Cambridge

Engelbrecht P (2021) SS-Karstwehr ermordete in Avasinis 51 Einwohnerinnen. Resistenza.de. https://resistenza.de/ss-karstwehr-ermordete-in-avasinis-51-einwohnerinnen/. Accessed 20 Sept 2021

Engelbrecht P (1998) Touristenidylle und KZ-Grauen. Vergangenheitsbewältigung in Pottenstein. Dritte unveränderte Auflage. Verlag C. u. C. Rabenstein, Bayreuth

Fasching GL (1995) Wehrgeologie in Österreich 1915 bis 1995. Arbeitstagung der Geologischen Bundesanstalt 1995:83–91

Ferenc M (2008) Secret world war two mass graves in Slovenia. In: Jambrek P (ed) Crimes committed by totalitarian regimes. Slovenian Presidency of the Council of the European Union, Ljubljana, pp 155–160

Ferenc T (1967) Kapitulacija Italije in narodnoosvobodilna borba v Sloveniji jeseni 1943 [Capitulation of Italy and national-liberation struggle in Slovenia in autumn of 1943]. Obzorja, Maribor

Ferenc T (1977) 'Dva od njih so obglavili s sekiro': Okupatorjev zločin v Idrijskih Krnicah ['Two of them were beheaded with an axe': the occupator's crime in Idrijske Krnice]. Borec 10:517–550

Ferenc T (1979) Krasoslovec in jamar, polkovnik in zločinec [Karstologist and speleologist, colonel and criminal]. Goriški letnik 6:191–216

Fritz U (2015) Die 'deutsche Stammzelle des Karstwehrwesens': Hans Brand, die SS-Karstschulungsstätte und das KZ-Außenlager in Pottenstein. In: Hofmann R, Kraus J (eds) Fürchten, Bangen, Hoffen. Leben um 1945 auf dem Land am Beispiel der Fränkischen Schweiz: Aufsätze. Fränkische-Schweiz-Museum, Pottenstein, pp 87–100

Gračanin H (2006) Goti i južna Panonija [Goths and south Pannonia]. Scrinia Slavonica 6:83–126

Grafenauer B (1970–1971) Naselitev Slovanov v Vzhodnih Alpah in vprašanje kontinuitete [The settlement of Slavs in the Eastern Alps and the question of continuity]. Arheološki vestnik 21–22:17–32

Guštin M (2004) Začetki slovanskega naseljevanja na Slovenskem [Beginnings of Slavic settlement in Slovenian territory]. Časopis za zgodovino in narodopisje 75:253–265

Häusler H (2013) Oberleutnant in der Reserve Dr. Artur Winkler von Hermaden: Leiter der Geologengruppe 'Isonzo' im Jahr 1918". Berichte der Geologischen Bundesanstalt 103:24–51

Häusler H (2018a) Geographen im Zweiten Weltkrieg: Die 'Forschungsstaffel z.b.V'. Nutzung der Potenziale des Georaumes für militärische Zwecke. Mitteilungen der Österreichischen Geographischen Gesellschaft 160:9–56. https://doi.org/10.1553/moegg160s9

Häusler H (2018b) Sonderkommando Dora: special military geoscientific unit of the German counter-intelligence service in north Africa 1942. Sci Militaria 46(1):37–57. https://doi.org/10.5787/46-1-1224

Jurkovšek B, Biolchi S, Furlani S, Kolar-Jurkovšek T, Zini L, Jež J, Tunis G, Bavec M, Cucchi F (2016) Geology of the classical karst region (SW Slovenia–NE Italy). J Maps 12(S1):352–362

Kater MH (2006) Das 'Ahnenerbe' der SS 1935–1945: Ein Beitrag zur Kulturpolitik des Dritten Reiches, 4th edn. R. Oldenbourg Verlag, München

Knolle F, Kempe S, Travassos LEP (2013) Nazi military use of German caves, Dr. Benno Wolf and the World cave registry project. In: Filippi M, Bosák P (eds) Proceedings of the 16th international congress of speleology, Czech Republic, Brno. July 21–28, 2013: vol 1. Czech Speleological Society, Praha, pp 65–70

Kocjančič K (2021) Red mrtvaške glave pod Alpami: enote in ustanove Waffen-SS na Slovenskem med drugo svetovno vojno [The order of the death's head beneath the Alps: Waffen-SS units and institutions in Slovenia during World War II]. Inštitut za novejšo zgodovino, Ljubljana

Korb A (2014) The disposal of corpses in an ethnicized civil war: Croatia, 1941–1945. In: Dreyfus J-M, Anstett E (eds) Human remains and mass violence: methodological approaches. Manchester University Press, Manchester, pp 106–128

Lee RG, Wilke S (2005) Forest as *Volk*: *Ewiger Wald* and the religion of nature in the Third Reich. J Soc Ecol Bound 1(1):21–46

Meerschman E, Cockx L, Islam MM, Meeuws F, Van Meirvenne M (2011) Geostatistical assessment of the impact of World War I on the spatial occurrence of soil heavy metals. Ambio 40:417–424

Petelin S (1985) Enaintrideseta divizija [The thirty-first division]. Borec and Partizanska knjiga, Ljubljana

Plewnia K (2011) Das Forschungskommando 'Süd' der Forschungsstaffel z.b.V. in Laibach: Feldforschungen im Karstgebiet Jugoslawiens 1944/45. Eine Fallstudie. In: Flachowsky S, Stoecker H (eds) Vom Amazonas an die Ostfront: Der Expeditionsreisende und Geograph Otto Schulz-Kampfhenkel (1910–1989). Böhlau Verlag, Köln, pp 303–320

Poprczeczny J (2004) Odilo Globocnik, Hitler's man in the east. McFarland & Comp, London

Pucher S (1997) In der Bewegung führend tätig: Odilo Globočnik, Kämpfer für den "Anschluß", Vollstrecker des Holocaust. Drava, Celovec

Rieger B (2007) Creator of nazi death camps: the life of Odilo Globocnik. V. Mitchell, London

Sachslehner J (2016) Dva milijona smo jih pospravili: Odilo Globočnik: Hitlerjev menedžer smrti [We liquidated two million of them: Odilo Globočnik: Hitler's manager of death]. Cankarjeva založba, Ljubljana

Schaffler H (1991) Die 'Höhlenforschung' im Dritten Reich. Karst und Höhle 1989(90):33–98

Štekar A (2013) Poskus lociranja bitke pri Frigidu leta 394 na območju med Sanaborjem in Colom [Attempt to locate the Battle of the Frigidus of 394 in the area between Sanabor and Col]. Annales: Series Historia et Sociologia 23(1):1–14

Uekötter F (2007) Green Nazis? Reassessing the environmental history of nazi Germany. Germ Stud Rev 30(2):257–287

Zečević M, Jungwirth E (2007) The influence of geology on battlefield terrain and [its] [e]ffects on military operations in mountains and karst regions: examples from WWI and Afghanistan. Rudarsko-geološko-naftni zbornik 19:57–66

Zupančič N, Šebela S, Miler M (2011) Mineralogical and chemical characteristics of black coatings in Postojna Cave system. Acta Carsol 40(2):307–317

Klemen Kocjančič He holds a university degree in theology, a master's degree in defence studies, and a PhD in history. His research interests include the military history of Slovenia (especially the Second World War), extremism, radicalization, terrorism, and military sociology (military families, military chaplains, etc.). Between 2019 and 2021 he was a research assistant at the Defense Research Centre of the University of Ljubljana's Faculty of Social Sciences. His research monograph *Red mrtvaške glave pod Alpami: Enote in ustanove Waffen-SS na Slovenskem med drugo svetovno vojno* [*The Order of the Death's Head beneath the Alps: Waffen-SS Units and Institutions in Slovenia During World War II*] was published in 2021.

Part V
Conclusion

Chapter 16
The Dinaric Karst in Environmental History Perspective: Conclusions, Contextualisations, and Outlook

Žiga Zwitter, Borna Fuerst-Bjeliš, Jelena Mrgić, Hrvoje Petrić, and Matija Zorn

Abstract This chapter concludes the volume *Environmental Histories of the Dinaric Karst*, which examines the historical interactions between humans and the rest of nature from prehistory to the twenty-first century. The monograph at hand focuses on the most prominent features of karstic cultural landscapes, environmental conditions, and environmental change, as well as on the dynamics of the hazards and risks through time. This section provides further long-term and socio-economic history contextualisations to the environmental history data that is analysed, synthesised, interpreted, and explained in the previous chapters. This represents a starting point for a critical reflection on the environmental degradation and improvement narratives in the history of the Dinaric Karst. In the end, a selection of fundamental research desiderata is listed.

Keywords Environmental history · Society-nature relations · Karst · Environmental degradation narratives · Environmental improvement narratives · Long-term research · Dinaric Karst · Kras · Eastern Adriatic · Southeast Europe

---

Ž. Zwitter (✉)
Department of History, University of Ljubljana's Faculty of Arts, Ljubljana, Slovenia
e-mail: ziga.zwitter@ff.uni-lj.si

B. Fuerst-Bjeliš (✉)
Department of Geography, University of Zagreb Faculty of Science, Zagreb, Croatia
e-mail: bornafb@geog.pmf.unizg.hr

J. Mrgić
Department of History, University of Belgrade Faculty of Philosophy, Belgrade, Serbia
e-mail: jmrgic@f.bg.ac.rs

H. Petrić
Department of History, University of Zagreb Faculty of Humanities and Social Sciences, Zagreb, Croatia
e-mail: hrvoje.petric@ffzg.unizg.hr

M. Zorn
ZRC SAZU Anton Melik Geographical Institute, Ljubljana, Slovenia
e-mail: matija.zorn@zrc-sazu.si

B. Fuerst-Bjeliš et al. (eds.), *Environmental Histories of the Dinaric Karst*, Environmental History 17, https://doi.org/10.1007/978-3-031-56089-7_16

16.1 Conclusions

The Dinaric Karst is a shared yet fragmented and diversified region, as highlighted in its historical geographies and environmental histories.[1] This is not only due to geomorphology, hydrology, or vegetation history but also because of some common and diverse historical conditions and cultural heritage. Therefore, the Dinaric Karst is not just a geomorphological and hydrological phenomenon or a complex physical environmental concept but also a sociocultural one.[2] Its broad and complex nature and meaning inevitably require a multifaceted approach.

This volume focuses on three crucial areas of environmental history: the material and, to a lesser extent, policy-related and intellectual environmental history. As observed by John Robert McNeill, "*The subject matter of environmental history is often just the same as the subject matter in historical geography or historical ecology*".[3] This is also the case here. While some chapters rely primarily on traditional historical sources, others approach the study questions with different methodologies, in line with the research subject and the authors' expertise.

The contents of this volume convey a variety of Dinaric Karst "stories" of the specific and often unique relations between human society and nature within the broad time span of the Anthropocene, the Age of Humans in its broadest sense.[4] The present monograph discusses the impacts of several major turning points in human history, designated as the potential beginnings of the Anthropocene: the transition from hunting and gathering to agriculture, the environmental impacts of industrialisation and use of persistent industrial chemicals, and the Great Acceleration starting in the 1950s.[5]

The chapters—i.e. the case studies—deal with some of the crucial issues of environmental history and its variety of discourses: cultural landscapes holistically, i.e. the landscapes physically transformed (or altered) by human activity over time, which involves the materialisations of social or political ideologies[6] (Chaps. 2–6); woodland and forests (Chaps. 7–9); water and climate (Chaps. 10–12); and environmental degradation (Chaps. 13–15). The monograph shows that the environment and people as an integral part of it are mutually dependent and co-evolve in time and space. Each human society experiences and reshapes the natural environment through culture, composed of mental and physical elements: cognitive systems, ideas, beliefs, preconceptions, values, and—more directly, through labour—tools and technology. Hence, the notion of "natural resources" is socioculturally constructed. Due to their short lifespans, people try to accumulate and often share the dynamically evolving transgenerational traditional ecological knowledge, which significantly contributes to the system's sustainability. Though this buffer strategy worked relatively well in

[1] See all other chapters of this monograph.

[2] Rogić (1976).

[3] McNeill (2011), pp. 160–161 (quote from page 161); see also Fuerst-Bjeliš and Petrić (2024).

[4] Ellis (2018); Lewis and Maslin (2015); Tripathy-Lang (2021); Zalasiewicz et al. (2021).

[5] Head et al. (2022).

[6] Mitchell (2000); Fuerst-Bjeliš and Vukosav (2019).

several ways in the preindustrial sociometabolic regime, some periods were, more markedly than others, subject to various hazards and risks to societies whose actions and interventions in their environment had uncertain outcomes.[7] In the industrial sociometabolic regime, the order of magnitude of various types of human interventions in the environment increased rapidly. This was combined with the abandonment of species-rich agroecosystems of traditional landscapes dependent on labour-intensive agriculture. These processes brought about fundamental environmental transformations, including serious hazards and increased risks in some cases. The story of creosote and the hazardous environmental legacy of the historical industrial sociometabolic regime in Bosnia-Herzegovina, as presented in Chap. 13, the industrial karst water pollution in Bela Krajina (Slovenia) in Chap. 14, the massive military fuel explosion and fire in a sensitive cave environment (Chap. 15), and the magnified fire risk (Chaps. 5 and 6) are among striking examples.

Information on historical cultural landscapes is often traceable in situ and in archival sources; the former becomes increasingly apparent with the advance of the relevant technologies. The benefits thereof are also used for discovering "hidden landscapes". This is an analytic term for many archaeological sites, i.e. remnants of cultural landscapes, which used to be inaccessible or non-evident. In the Dinaric Karst, this was usually due to the ecological succession or fairly low relief contrast between the preserved remnants and the surrounding landscape, but can now be explored using new, non-invasive terrain survey methods, GIS, and geosciences,[8] as presented in Chap. 4 and in the following subsection 16.2.

16.2 Additional Contextualisations of the Book Contents

As stated in Chap. 1, most contributions in this volume focus on the northern Dinaric Karst. Half of them—Chaps. 1, 2, 4, 8, 10, 11, and 12—deal with the Kras Plateau as the main focus or one of the areas under consideration. As the Kras Plateau is simultaneously the region that, for intellectual history reasons, gave the name to the processes of rock dissolution and the resulting landforms and groundwater systems worldwide (Chap. 1), the following selection of fresh insights focuses on the Plateau and the Dinaric Karst in its surroundings.

[7] Here is a mere selection of works from the ever-expanding collection of relevant literature: Winiwarter and Knoll (2007), p. 129, *et passim*; Jaritz and Winiwarter (1997), pp. 91–111; Mrgić (2013), pp. 22–31, with extensive references to the subject; Babai et al. (2014); Babai and Molnár (2014); Mrgić (2014), pp. 52–72; Fuerst-Bjeliš (2003); Fuerst-Bjeliš and Glamuzina (2021); Zwitter (2020).

[8] Leusen et al. (2011). Compare Ivanišević et al. (2015); Krevs (2021).

16.2.1 *Three Strongest Phases of Human Landscape Reorganisation in the Northern Dinaric Karst*

Chapters 2, 3, 4, 5, 6, 7, and 8 discuss the history of land use and land cover in the Dinaric Karst. Studies in the northern Dinaric Karst identified late prehistory as **the first among the most fundamental phases of human landscape reorganisation** (Chaps. 3 and 4). On the Kras Plateau, the Bronze Age of roughly the second half of the third and the second millennium BCE produced the second most open cultural landscape ever (Chap. 4). This general finding should not obscure the fact that, due to the transformations in settlement and land-use distribution patterns through time, the Bronze-Age or Iron-Age agricultural land use in the vicinity of certain hillforts was the most intensive in history (see also Chap. 4). In the north-westernmost Dinaric Karst, (fortified) hilltop settlements were founded during the Bronze Age and Early Iron Age; the latter started in this area around 800 BCE.[9] At that time, hillforts were a widespread settlement type in the Dinaric Karst.[10] In general, the Early Iron Age in the region of today's south-western Slovenia was characterised by "*economic development, rise of hillforts, population growth*[,] *and marked social stratification*". The region's Iron-Age hillfort communities participated in trade with Pannonia, the Dinaric Alps, the Eastern Alps, and the Apennine Peninsula—the central and northern Italian centres, for instance. The hillfort communities reached a significant demographic peak in the fifth-to-fourth century BCE.[11] The population density in these hillforts appears to have declined later but was considerable again in the waning Late Iron Age, mid-second to first-century BCE. Some hillfort regions in the northern Dinaric Karst, for example, the one along a ten-kilometre ridge between Šilentabor (Fig. 16.1) in the north and the eastern hinterland of Ilirska Bistrica in the south, were characterised by settlement continuity until the Late Antiquity.[12] The formation of hillfort communities in the Late Bronze and especially the Early Iron Age "*probably went hand in hand with the development of… agricultural and pastoral* [agrokarst] *landscape*".[13] At several places in the Dinaric Karst, agricultural land belonging to late-prehistoric hilltop settlements was located at places that are currently no longer used for agriculture.[14] A hidden late-prehistoric cultural landscape has been well preserved around several hillforts between Šilentabor and the eastern hinterland of Ilirska Bistrica, as established by a detailed LiDAR-based digital elevation model and archaeological fieldwork. Figure 16.1 shows structures such as linear earthworks, parallel banks, and enclosures as integral parts of the multi-period, largely late-prehistoric (mainly Iron Age, partly Late-Bronze-Age) agrokarst landscape palimpsest, which evolved gradually. They are the legacy of

[9] Laharnar (2022), pp. 350, 352; Laharnar et al. (2019), pp. 263, 265. See also Gams (1991), pp. 16–18; and Chap. 4 in this volume: Mlekuž Vrhovnik and Fabec (2024).

[10] E.g. Fuerst-Bjeliš (2018), p. 139; Kranjc (2009), p. 18.

[11] Laharnar (2022), p. 350.

[12] Laharnar (2022), pp. 347–348, 354, 360, 362, 364; Laharnar et al. (2019), pp. 266, 269.

[13] Laharnar (2022), pp. 144–145, 352 (quote); compare Panjek (2015), pp. 39, 102–103.

[14] Kranjc (2009), p. 18.

the labour-intensive agriculture of the agrarian sociometabolic regime's controlled solar-energy system. The age of the sampled structures was confirmed by prehistoric pottery shards. The land surface structures include arrangements associated with field cultivation (hoe farming) and livestock farming, which involved pasture enclosures and, possibly, hay meadows cleared of stone, potentially also arboriculture. An excavated bank revealed that it was not a simple clearance cairn but rather a mixture of soil and stones, so some banks and linear earthworks might have offered deeper soil layers for cultivars, sites less prone to summer drought.[15] In the High and Late Middle Ages, villages that still exist nowadays were not located on hilltops in this area. Farm settlements were founded at lower elevations, while the traces of the late-prehistory-to-antiquity agrokarst landscape on shallow soils on the slopes below the hillfort ruins often belonged to outfield grasslands.[16] At some locations, for example, at the hillforts of Šilentabor and Grmada south of it or Stražica north-east of Ilirska Bistrica, the areas of the preserved legacies of late-prehistoric agrokarst landscape were used for pastures in the early nineteenth century.[17] Thus, land use in these marginal areas was more labour-intensive at the peak of the demographic and economic developments in the Early Iron Age than in the early nineteenth century, an era characterised by a very high agricultural density (see below in this subsection). In this case, transitions in human activities represented one of the crucial drivers of agroecosystem change in the long term. It remains to be seen whether the intensity of land use on the slopes below Šilentabor reached the late-prehistoric magnitude during the late medieval and Early Modern temporary uses of the large hillfort located in the same place as the late-prehistoric one, a shelter for the inhabitants from thirteen villages in the broader surroundings.[18] However, the scarce evidence studied so far and the merely temporary character of the late medieval to Early Modern fort, excluding the few inhabitants of the castle that was erected in the hillfort site, do not support this hypothesis—fields and meadows at lower elevations were largely cultivated also when people temporarily retreated to the hillfort.[19] This suggests that the late-prehistoric land use there was indeed most intensive.

Similar examples can be found at some other hillforts in the broader region[20]: a comparison between the LiDAR-based insights into the hidden prehistoric landscape in the left central section of Fig. 4.4 and the early-nineteenth-century land use in the same area[21] reveals many structures on the slopes of this marginal area—not

[15] Laharnar (2022), pp. 26, 28, 88–89, 92–93, 108–109, 117, 138–139, 144–145, 148, 150, 352; Laharnar et al. (2019), pp. 263–270; Laharnar, interview. A general presentation of historical sociometabolic regimes: Krausmann et al. (2016).

[16] Compare Laharnar et al. (2019), p. 269 and precise early-nineteenth-century cadastre in "Arcanum Maps", Cadastral maps, Habsburg Empire—Cadastral maps (XIX. [!] century).

[17] "Arcanum Maps", Cadastral maps, Habsburg Empire—Cadastral maps (XIX. [!] century).

[18] About the Early Modern hillfort there, see Fister (1975), pp. 18, 33, 81, 96–97; ÖStA AVAFHKA, Hoffinanz und Niederösterreichische Kammer, Niederösterreichische Kammerakten, bundle 15: July 1547.

[19] Compare Šumrada (1986), pp. 21–23.

[20] E.g. Kokalj and Hesse (2017), pp. 56–57.

[21] "Arcanum Maps", Cadastral maps, Habsburg Empire—Cadastral maps (XIX. [!] century).

Contemporary Landscape
Šilentabor
Map by: Lenart Štaut
Sources: ARS 1823; GURS 2023;
Laharnar, *From Orca to Albion*, 88
© ZRC SAZU Anton Melik Geographical Institute
Early 19th Century Landscape
Built-up area
Cultivated field
Hay meadow
Pasture
Water
Cart track
Prehistory Landscape
0
125
250 m

◂**Fig. 16.1** (1) Left: the late-prehistoric hillfort of Šilentabor in the north-western Dinaric Karst (red arrows point to its outer ramparts) and the largely prehistoric hidden landscape palimpsest, revealed by a detailed LiDAR-based digital elevation model (with Boštjan Laharnar's permission). Note that the figure label is simplified: the left figure section does not only show the prehistoric legacies. Instead, it presents a palimpsest landscape with elements from prehistory until today, though prehistoric structures are widely present in this figure section (2) Centre: digitised georeferenced early nineteenth century land use in the same area based on precise cadastral data, made partly transparent to allow for easier comparison with the map on the left, which also serves as a background here. (3) Right: a digital orthophoto of the same landscape nowadays, showing modern landscape structures, such as the road network, easily recognisable on the digital elevation model in the left figure section. Many of the structures, which were located in pastures in the early nineteenth century and cannot be identified as fences around meadows and cultivated fields existing in that period or as present structures, are remnants of this landscape's most intensive prehistoric agricultural use. *Sources* as noted in the Figure; the map in the middle: own elaboration

only enclosures but also field boundaries as legacies of agricultural practices, more intensive than simple grazing in the area used as a pasture in the early nineteenth century (Chap. 4). Since the High Middle Ages, examples of land use more intensive than in the early nineteenth century have been only exceptional and limited to a few permanently deserted villages and places where farmers could not cope with water and wind erosion in the cultivated fields, which they therefore abandoned.[22]

The second, Early Modern phase of landscape reorganisation as a consequence of a substantial increase in agricultural density is presented in Chap. 4 based on the soil accumulation in *dolines*. These results are fundamental and relevant in the broad Dinaric Karst context, so they need to be embedded in the economic and social history processes proven by written sources to integrate environmental history into the total history frame.[23]

In the Late Middle Ages, until around the fourteenth century, the Kras Plateau was most likely covered by relatively extensive forests next to some agricultural land. Archival sources match palynological results indicating a medieval expansion of oak forests.[24] Namely, during the Late Middle Ages, the Kras Plateau was severely depopulated for many reasons, such as the broader economic conditions in Europe, epidemics, harvest failures, military incursions from the expanding Ottoman Empire, the War of the League of Combrai (1508–1516), or emigration to urban settlements,[25] for example, the Venetian coastal town of Koper (It. *Capodistria*), by far the largest urban settlement in the vicinity and one of the largest towns along the eastern Adriatic coast with some 7000–10,000 inhabitants from the Late Middle Ages to the mid-sixteenth century,[26] or the Habsburg port of Trieste, a town of considerable size, though much smaller than Koper at the time.[27] In the seigneury of Senožeče (the

[22] Cf. Vilfan (1957a), pp. 63, 69. See examples of permanently abandoned medieval villages in Kos (1976), p. 7, and Šumrada (1986), pp. 19–21.

[23] Compare Fuerst-Bjeliš and Glamuzina (2021).

[24] Panjek (2015), pp. 66, 101–104.

[25] Kos (1976), pp. 9–10; Šumrada (1986); Panjek (2015), pp. 104–105; Bajt and Vidic (2011), p. 99; Lane (1973), pp. 242–245; Voje (1978), but see also Schmitt (2016).

[26] Bertoša (2007), p. 107; Mlacović (2014), p. 99.

[27] Kos (1976), p. 10; Kalc (2018), p. 160; Luzzatto-Fegiz (1929), p. 15.

central settlement is located on the eastern Kras Plateau), roughly 40% of farms were uninhabited as recorded in 1531—a percentage representative of the broader area and a result of the processes lasting for more than a century, including the severe demographic losses between the late 1460s and 1531. Most empty farms were used as extensively managed grasslands, while a minority were completely abandoned.[28] In the era of late medieval and early-sixteenth-century depopulation, quite a lot of forests were located on the Kras Plateau.[29] By the second half of the sixteenth century, the demographic conditions recovered in a period of strengthened political stability, and, therefore, a significant increase in the number of farm households compared to the time before the late-medieval crisis began. Simultaneously, on the increasingly populated Kras Plateau, another relevant farming institution was founded in 1580: the princely stud farm in Lipica, a new local and regional economic and environmental actor, which significantly raised the demand for grassland resources. In the second half of the 1610s, the Lipica stud farm kept about 150–200 mares, foals, and stallions, and significantly more at the peak of the early seventeenth century. The number then decreased due to the 1625 stud farm reform. In diverse stages, the general trend of the increasing agricultural density on the Kras Plateau and the growing number of smallholdings continued until the nineteenth century. This led to land-use intensification, for example, the conversion of pastures and forests to meadows, vineyards, and cultivated fields as of the sixteenth century, which were often situated in marginal environmental conditions. Thus, even though villages on the Kras Plateau originate from the Middle Ages, the magnified Early Modern agricultural density, in interaction with natural processes and living beings, created an incomparably more open cultural landscape than medieval colonisation. These processes belonged to the widespread deforestation in the (Sub)Mediterranean environments in the early modern Europe, including the (Sub)Mediterranean areas of the Dinaric Karst. However, it is important to emphasise that the rising population numbers were not directly linked to the increase in farmland: not only were the farming orientations and the economic standard not constant through time, but the integrated peasant economy also involved several kinds of agricultural and non-agrarian income. In the case of the Kras Plateau, the latter included the mesoregional trade with Habsburg and Venetian port towns in northern Istria, crafts, temporary jobs in urban settlements, etc. Thus, the pressure of labour-intensive agriculture on farmland rose particularly heavily in times of economic depression in the non-agricultural economy. Nonetheless, the broader trade conditions, like the demand in the rapidly growing Trieste, the Habsburg Monarchy's main port in the mercantilist eighteenth and nineteenth centuries, also resulted in strengthened market-oriented agriculture on the Kras Plateau.[30]

[28] Šumrada (1986).

[29] Panjek (2015), pp. 66–68, 81–84, 94–102, 104–105.

[30] Fuerst-Bjeliš (2018), pp. 139–141; Moritsch (1969), pp. 73–74, 123, 139, 146; Panjek (2015), pp. 25, 26–28, 33, 57–58, 61–66, 68–70, 76, 105–108; Kalc (2017) (see also Kalc (2018) and Bertoša (1986); Bajt and Vidic (2011), p. 79; Šumrada (1986), p. 20; Vilfan (1957a), pp. 61–64, 67–69; Vilfan (1957b), pp. 72–73, 76, 78; Voje (1996) (but see also Schmitt (2016)); Zwitter (2015).

An approximate comparison of livestock recorded in eight villages between the Kras Plateau east of Trieste (Vrhpolje) and Črni Kal east of Koper (Capodistria) in 1627 and in 1955 (or 1948), most of which are located on limestone karst, shows a considerable increase in the number of total livestock held at farms in these villages. The approximate number of livestock units in the mid-twentieth century (about 967) amounted to approximately 235% of the one officially inventoried in 1627 (around 411), with huge differences between villages. In the same interval, the number of holdings more than doubled (from 135 to 302). In addition, within this chronological interval, the livestock size expanded significantly, so the environmental impacts of the rising livestock numbers were even greater. Nonetheless, it is clear that several farms did not keep as many of their own livestock as they could in 1627 because several domestic animals perished due to an infectious disease, a severe bora storm that killed tens of domestic animals in a pasture, etc. In 1627, these farmers thus also reared some foreign livestock owned by burghers or other farmers, for example.[31] It is also likely that some livestock was hidden in 1627. Therefore, it seems that the inventoried livestock numbers did not fully reach the carrying capacity of the 1627 cultural landscape. All in all, the livestock present in the mid-twentieth century most likely demanded significantly more fodder than the 1627 cultural landscape could deliver, but not as much more as the figures presented would suggest. We should add that in the late nineteenth and early twentieth century, cattle replaced sheep in most places on the Kras Plateau due to several economic factors. The development of the global wool and cotton market and raised milk demand in Trieste were important elements in this process.[32] One of the main landscape impacts of the transition from sheep to cattle on the Kras Plateau was an increased stone clearing in pastures where sheep and some old cattle breeds could graze, while the pastures were unsuitable for modern cattle breeds. Stone clearing was also a consequence of the frequent land-use change from pastures to hay meadows, namely more hay was required because cattle was not involved in the transhumance and the role of barn-feeding in cattle husbandry strengthened.[33] Looking deeper into the relevant history, Early Modern sheep farming on the Kras Plateau, like in most of the Dinaric Karst, was market oriented to a significant degree. Despite the meagre appearance of the species-rich dry calcareous grasslands and low yield quantities, they provided quality, aromatic fodder, which resulted in tasty meat. Thus, the lambs from the Kras Plateau were massively exported to Venice, where they reached exceptionally high prices, as observed in 1780[34]—an evidence of how the considerable demand co-shaped the local economy. Cereals from this plateau also had their merits—granted, the grains were usually small, but the seed hull was thin and the quantity of flour per

[31] Vilfan (1957b), pp. 78–80, 84–87; Novak (1970), pp. 387–389; see also Chap. 11 in this volume: Rože Kravanja (2024).

[32] About the switch from sheep to cattle: Moritsch (1969), especially page 138; Panjek (2015), pp. 73, 79; Vilfan (1957b), pp. 78, 85–86; Schöller (2010).

[33] Gams (1993), p. 112; Gams (1991), pp. 20, 22–23, 81, 90.

[34] Compare Panjek (2015), pp. 28–29, 63–64, 92–94; Vilfan (1957b), pp. 78, 84, 86; Cvijić (1925), p. 3; Breg Valjavec et al. (2018), p. 3830; Zwitter and Rasran, forthcoming.

unit of grains was high, ensuring a considerable wheat market price. Namely, despite the limited surfaces suitable for field cultivation on the Kras Plateau, resulting in regular cereal purchases to fulfil the needs of many households, significant wheat quantities from the plateau were regularly sold, as the landlords or tithe lords sold much of the rent they had received in wheat dues. Later, farmers would sell their wheat surplus directly, while their daily diet was based on cheaper cereals.[35]

The accelerated Early Modern soil erosion, demonstrated in Chap. 4 based on the *doline* infills near Divača, reflects, on the one hand, the soil erosion in the grasslands, especially overgrazed ones, leading, for instance, to a gradual increase in barren karst (see below). On the other hand, soil erosion in cultivated fields was also intense. In addition to the expanding grasslands and agricultural fields, the high soil erosion rates in the Early Modern period resulted from the increasingly marginal site conditions in the areas where farming was introduced or land use intensified in the context of the Early Modern farming expansion (Chap. 4).[36] The Little Ice Age climate might have also contributed to higher soil erosion rates. The regular appearance of stones in agricultural land, caused by pluvial and wind erosion and mitigated by the occasionally repeated stone removal from the fields and the transport of the eroded soil back to the plots of its origin, was usually observed in many Dinaric Karst regions with agriculture on shallow or rocky soils—farmers in various places of the macroregion, including the Kras Plateau, used to say that "*stones grow*".[37] The removed rocks were used for dry stone walls or deposited in clearance cairns (Chap. 4). Soil erosion in the cultivated fields on shallow soils that covered gentle slopes was so strong that the soil had to be collected in the depressions and transported back to the fields after heavy rain as well as regularly, every second or third year, to keep the land productive for field crops in times of labour-intensive agriculture of the agrarian sociometabolic regime's controlled solar-energy system. This process was mitigated by terraces and walls stabilising the soil that farmers had brought to some locations to create fields artificially.[38]

In the global context of karst macroregions, the geological composition of the Dinaric Karst is exceptionally complex.[39] This volume therefore deals with several different karst environments, as outlined in Chap. 1. Even on the Kras Plateau, a connection between environmental and social conditions resulted in two agricultural regions on the plateau in the last centuries (Fig. 16.2). The "*Teran* Kras" is named after the local red wine called *teran*. In the Early Modern Period, it was intended for local consumption and the mesoregional market, as it was not as highly appreciated as quality white wines. During the nineteenth century, the attitude towards *teran* changed: it became highly valued in the city of Trieste and has remained a prized wine variety ever since. In the early nineteenth century, the open or wooded, often rocky pastures, primarily used for sheep grazing, covered more than half of the *Teran* Kras,

[35] Panjek (2015), pp. 21, 29, 42, 63, 68; Moritsch (1969), 75.

[36] Panjek (2015), pp. 21, 27, 29–30, 41–42, 44–45, 49, 55, 65, 72.

[37] Cvijić (1925), p. 3; Gams (1991), p. 29.

[38] Panjek (2015), pp. 21, 27, 29–30, 41–45, 49, 55, 72; Gams (1991), pp. 11–13 *et passim*.

[39] Ford (2007), p. 683.

with significant differences in the percentage of pastures in the total agricultural land at the village level.[40] Nonetheless, the *Teran* Kras was characterised by a significantly higher percentage of cultivated fields and vineyards than the so-called Grassland Kras, i.e. the other agricultural region on the plateau. The considerable share of wood meadows in the *Teran* Kras delivered higher grassland yields than the open meadows in the local environment. Wood meadows also produced wood, acorn for pigs, ash-leaf hay (mainly for sheep), and leaf litter for bedding. The tree shade alleviated the summer drought, resulting in denser grassland, while fallen leaves contributed to nutrient cycling in the meadow agroecosystem.[41] Wood meadows have a long tradition there: in the 1630s, the care for the high-quality grassland and the good condition of the oak stand in the local wood meadow was recorded in the sources regarding the Lipica stud farm.[42] In some areas of the *Teran* Kras, meadows were so extensive and haymaking so labour-intensive that the locals had to rely on an external haymaking workforce, at least in the nineteenth and early twentieth centuries—similarly to some sections of the Grassland Kras, as presented below. One such example was the cadastral municipality of Tomaj (see its central settlement in Fig. 16.2), where hay meadows, most of them wooded, covered the majority of the territory in the early nineteenth century; the microregion of Tomaj is highly fertile for agriculture in the broader Kras context. Only 7% of the 346 ha of hay meadows in the cadastral municipality of Tomaj, however, allowed for mowing by scythe only. Elsewhere, the rocks were not entirely cleared, so haymakers would use scythes as well as sickles (47% of the meadows) or had to cut the meadow plants using sickles only (46% of the meadows).[43]

The Grassland Kras, characterised by a meagre percentage of cultivated fields, covered the plateau's south-eastern, north-western, and western parts (Fig. 16.2). The Divača Kras, where soil samples referred to in Chap. 4 were taken, belongs to this type of Kras. The Grassland Kras was characterised by a high percentage of pastures and hay meadows. Pastures, including some barren land, covered about 70% of the area. Barren rocks were less common in the Grassland Kras at higher elevations than at lower altitudes. In the cadastral municipality of Divača (see its central settlement in Fig. 16.2), most hay meadows needed to be cut by sickle. As the hay meadows were so extensive, seasonal workers from the Upper Soča valley (today north-western Slovenia), located in the south-eastern Alps more than forty kilometres away, as well as those from some not so distant parts were usually hired to help with haymaking, as reported for the south-east of the Grassland Kras in the nineteenth and early twentieth centuries. Hay from several parts of the Grassland Kras (see the surroundings of Senožeče in Chap. 10) would be sold in the coastal urban settlements and villages, an activity which had started centuries before the

[40] Moritsch (1969), pp. 75, 78, 82, 131–132, 138, 144; Panjek (2015), pp. 23, 25–26, 37, 40–42, 63, 69, 85–88, 90, 92–94, 97, 109–110.

[41] Moritsch (1969), pp. 71, 76, 100, 131–132, Karte 5; Panjek (2015), pp. 45–47, 82–87, 88, 90–96, 100–101; Zwitter and Rasran, forthcoming.

[42] Zwitter (2015), pp. 467–468.

[43] Moritsch (1969), pp. 70, 73, 75–76, 79, 103, 132–133, Karte 5.

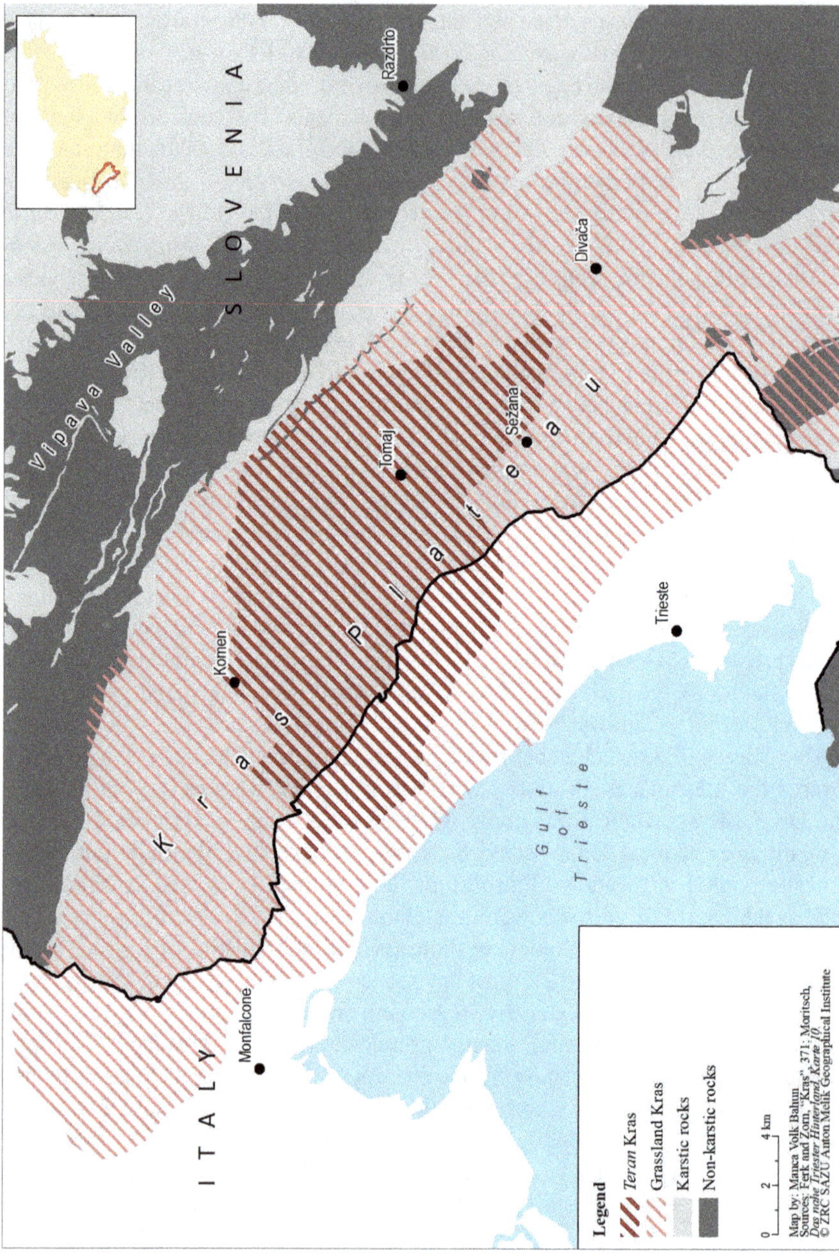

Fig. 16.2 Two main agricultural regions on the Kras Plateau: the "*Teran* Kras" and the "Grassland Kras". *Source* Own elaboration, adapted from Moritsch (1969), Karte 10

Habsburg Monarchy built the railway connection between Vienna and Trieste in 1857 (see Chap. 1).[44]

In several parts of both the Grassland and *Teran* Kras, the local demand for hay was reduced by transhumance. Chap. 1 emphasises that transhumance used to be a common practice in the agricultural history of the Dinaric Karst in the *longue durée* perspective. A recent cultural landscape history synthesis on the Kras Plateau underestimated its local significance in the Early Modern Period but rightly emphasised the potential changes in the presence and trends of transhumance involving Kras through time.[45] In the Early Modern Period and later, transhumance linked several diverse parts of the Kras Plateau (elevations between 200 and 500 m ASL) and the summer pastures on the high Dinaric Karst plateaus or their slopes—e.g. Nanos (about 1200 m ASL) and probably Javorniki (similarly, with the highest peaks at about 1200 m ASL) and Snežnik (even higher)—the Vremščica Hill (about 1000 m ASL), as well as some parts of the Kras Plateau and the lowland winter pastures in Istria (nowadays located in south-western Slovenia and north-western Croatia) and Friuli (today in north-eastern Italy). It is, however, clear that a lot (or most) of livestock overwintered on the Kras Plateau.[46] Historical transhumance involving sheep and goats from the plateau deserves further research attention.

The third fundamental phase of human-induced landscape reorganisation started in the late nineteenth century, with extensive afforestation projects: initially, the black pine plantations on the Kras Plateau, followed by further examples in several other places of the Dinaric Karst. Focusing on the forest transition on the Kras Plateau and in many other Dinaric Karst regions, the afforestation was coupled with the ecological succession due to deagrarianisation. These processes are presented in Chaps. 8, 5, and 6, which focus also on the socio-economic triggers and environmental consequences.[47] Another crucial aspect of this landscape transformation phase, mainly since the second half of the twentieth century, is the industrialised farming intensification in times of the Great Acceleration, contributing its share to the losses of species-rich agroecosystems and causing serious environmental degradation. However, the spatial scale of this process differs significantly from one Dinaric Karst mesoregion to the other. While the intensive industrialised farming has, to a certain degree, affected the agroecosystems in all the regions where the farming has not ceased, its impacts in several areas with considerable karstic environmental constraints on farming, such as the Kras Plateau[48] or most of the Dalmatian karst (Chaps. 5 and 6), have been much less profound than in some other

---

[44] Moritsch (1969), pp. 79, 98, 100, 103, 114, 132–133; Panjek (2015), pp. 29, 85–87, 88, 90–95, 109–110; Gams (1991), pp. 20, 80; Vilfan (1957b), p. 76.

[45] Panjek (2015), p. 109.

[46] Klemenčič (1959), pp. 104–114; Leban (1950); Matičetov (1948), p. 21; Moritsch (1969), pp. 77, 114, 123, 133; Smerdel (1989), p. 13; Umek (1957), pp. 71–76; Vilfan (1957a), p. 68; Vilfan (1957b), p. 82.

[47] About the north-westernmost Dinaric Karst, see Moritsch (1969).

[48] Kaligarič and Ivajnšič (2014), pp. 152–157; but see Gams (1991), p. 81 *et passim*, and Moritsch (1969), pp. 108, 115, 145.

Dinaric Karst regions that are more favourable for or more easily adapted to modern intensive farming (Chap. 14).[49]

The processes presented here have had a diverse and extensive impact on the contents of this volume. For instance, the network of aquatic and wetland ecosystems of artificial waterbodies on the Kras Plateau changed thoroughly exactly when it was heavily populated by livestock and people and the agricultural density was enormous[50] (Chap. 11). The prevalent local smallholders struggled to cope with the subsistence crisis of the mid-1860s in a heavily used environment. Hence, many farmers were forced to look elsewhere for work—often in the port city of Trieste and some in the distant forests of what is now northern Croatia and elsewhere (Chap. 10). Due to climate change with increasing aridity, altered river beds, and higher evapotranspiration because of the expansion of forests, the persistence of the northern Dinaric Karst floods of 1864 is difficult to imagine today (compare Chaps. 10 and 12).

Modern karst explorations began in the Dinaric Karst (Chap. 1) when it was more or less at its barest and when the surface karst phenomena were therefore most obvious.[51] At that time, these phenomena were thus incomparably more noticeable in the landscape than today, more than a century after the massive afforestation with black pine and the relevant ecological succession. This is yet another example of the complex interactions between the natural processes and human society (Chaps. 5–8), which brings our attention to the environmental degradation and improvement narratives in the Dinaric Karst's history or the topic of the subsection 16.2.2.

16.2.2 The Reconsidered Environmental Degradation and Improvement Narratives

On the one hand, instead of an unreflected "*lament over loss and damage as people did awful things to flora, fauna and the land* [...], *a more mature examination of a wider, more complex range of relationships with the rest of the nature*" became characteristic for the global mainstream environmental history research.[52] On the other hand, historiographical narratives of humans-induced environmental "improvement" that followed the historical scientific and policy discourse need to be reconsidered (Chap. 8).[53] Both approaches were applied in the recent environmental historiography of the Kras Plateau.

Considering the degradation narratives in scientific writing, this text focuses on the history of barren karst. Barren rocky ground was characteristic of some sections

49 See also Moritsch 1969, pp. 143–144.

50 Compare Panjek (2015), p. 107 *et passim*.

51 The connection between landscape barrenness and karst investigations: Ferk and Zorn (2015), p. 370; Gams (1993); Kranjc (2009), pp. 15, 19; and Panjek (2015), p. 68.

52 Coates (2004), p. 410, see also pages 415, 418–420, 424; see also McNeill (2011), pp. 164–165.

53 Panjek (2015), pp. 110–111.

of the Dinaric Karst, mainly on Adriatic islands, the Adriatic-facing mountain slopes, and the various lowlands in the Dalmatian hinterland (Zagora), as observed in Early Modern administrative records and scientific writing.[54] There, grassland plants were often limited to depressions in karren. Barrenness resulted from the interactions between unsustainable forest uses (forest fires, sometimes excessive felling) and overgrazing in a sensitive agrokarst grassland or forest environment. In Dalmatia, with its Mediterranean climate, regular summer droughts and fierce, dry bora wind gusts that easily blow away the topsoil (especially on the islands facing the main bora corridors; see Chaps. 1 and 7), the barren area was larger, and the degree of the landscape's barrenness at the sites most void of vegetation was somewhat higher than on the Kras Plateau.[55] However, apart from the natural environmental conditions and the general course of agricultural and forest history, the centuries-long Early Modern border position (between the Ottoman, Venetian, and Habsburg territories) significantly influenced the Dalmatian karst in the environmental-historical sense, as it contributed to the lack of possibilities for continuous resource management and the unsustainable use of the scarce resources (Chap. 5). In the Early Modern Period, the borderland position in Dalmatia resulted in wars, frequent border changes until the early eighteenth century, insecurity for persistent sedentary settlement, migrations, and favoured migratory pastoralism on the mainland. Unsurprisingly, the barren area in the Dalmatian hinterland (Zagora) was most likely larger in the early eighteenth century than in the times of Cvijić's fieldwork almost two centuries after the end of the Ottoman–Venetian wars. For the islands, the political instability until the early eighteenth century meant the influx of settlers from the war-torn mainland and thus heavier environmental pressure on arable land, hoed parcels, pastures, vineyards, olive groves, and woodlands in terms of more labour-intensive uses and, in several places, intensification of the land-use categories. These processes contributed to the expansion of barren areas, especially in the pasture commons.[56]

However, at least in some parts of the Dinaric Karst, such as the Kras Plateau, barren land became a topos, and its extent was presented in an exaggerated manner.[57] This topos still appears in recent karstology publications.[58] In addition to the rocky

[54] For instance, Cvijić (1925), pp. 1–3, 9–11 and Figs. 9 and 28, Fuerst-Bjeliš (2003, 2018); and examples in Chap. 7 in this volume: Pavelić et al. (2024).

[55] Cvijić (1925), pp. 1–3, 9–10, and Fig. 28; see also Kranjc (2009), p. 21, and Moritsch (1969), pp. 76, 110.

[56] See also McNeill (1992), p. 275; Fuerst-Bjeliš (2020), pp. 82–85; Fuerst-Bjeliš and Glamuzina (2021), pp. 100–109, 133–136; Carter (1994), pp. 7–47; Kulušić (1965), pp. 223–230; Kulušić (2000), pp. 43–65; Kulušić (2000), pp. 99, 101–108; compare Chap. 5 in this volume: Tekić et al. (2024) with Cvijić (1925), pp. 9–10.

[57] Panjek (2015), pp. 22–24, 26–27, 29–30, 33, 79–80, 101–103, 105–106; see also Chap. 8 in this volume: Remec (2024).

[58] Chapter 8 in this volume: Remec (2024). For a conspicuous example, see Kranjc (2009), pp. 19–20, describing a photo of an agrokarst landscape around Postojna (an elevated temperate continental Dinaric Karst site in the south-west of today's central Slovenia) as a "barren limestone desert"—but there are extensive agriculturally used grasslands in that photo providing huge amounts of fodder and containing only fairly limited rocky sites; the area was located far from the barren rocky karst—compare the eighteenth-century and late-nineteenth or early-twentieth century observations from

topos, coppice—which produced firewood, vineyard stakes, tree-leaf hay, and grassland plants—and (pollarded) solitary trees were interpreted as a massively degraded agroenvironment in accordance with the historical forestry science standards that opposed traditional agroforestry uses, including those that maintained the species-rich wooded ecosystems in the long term and provided (most of) the wood needed in the local economy. In fact, the Early Modern records attest to extensive traditional ecological knowledge that ensured functional and often relatively sustainable uses of such agroecosystems on the Kras Plateau. These were thoroughly different from natural forests but were nevertheless species-rich. Therefore, such a transformed species-rich environment that enabled economic uses in the context of labour-intensive agriculture of the agrarian sociometabolic regime's controlled solar-energy system cannot be deemed as degradation. Around 1830, when the deforestation of the Kras Plateau was at its worst, wooded grasslands and forests—i.e. land with at least some woody vegetation (excluding vineyards and orchards)—covered one-third of Kras (forests grew on 13% of the plateau). Moreover, open pastures covered another 38% of the surface, and open meadows represented 7%. Altogether, only these land-use categories covered 76% of the landscape. Granted, many of the pastures and some of the meadows were exceedingly rocky: in the cadastral municipality of Divača—as an example of a microregion with an exceptionally high percentage of barren karst in the very south of its territory in the broader Kras context—pastures delivering only a quarter of the usual amount of grazing resources due to the vast rocky area covered almost three quarters of the territory in early nineteenth century (see the central settlement in Fig. 16.2).[59] The soil erosion leading to such conditions is also reflected in the analysed *doline* infills—several sampling sites yielding the results presented in Chap. 4 are located exactly there.[60] This was one of the barren karst microregions on the Kras Plateau often observed by visitors, as it was crossed by the old commercial road from Vienna to Trieste that was improved in the mercantilist eighteenth century. Since 1857, the area under consideration was located near the Divača station of the Vienna–Trieste railway; since 1876, this barren karst was crossed by the railroad connection from Vienna to Pula (southern Istria, today in Croatia).[61] The impressions from this region must have contributed to the notion of Kras as a barren, stony landscape. Nonetheless, the Kras Plateau as a whole has never been a rocky desert where agricultural land would be largely confined to the "oases" in *dolines*.[62] Even the cadastral municipality of Divača sold some hay to the coastal

this area in Cvijić (1925), pp. 9–10, and in Kaligarič and Ivajnšič (2014), p. 149; see also Moritsch (1969), p. 133; Zorn et al. (2015), p. 566.

[59] Panjek (2015), especially pages 79–87, 92, 94–101, 105–112; Moritsch (1969), pp. 110, 114; Zorn et al. (2015), pp. 567–569; see also Zwitter (2020).

[60] Compare Fig. 4.7 in this volume with ARSO, "Atlas okolja" [Environmental atlas of Slovenia]: Backgrounds, Lidar hillshade & Layers, Spatial units, Cadastral areas (and Names of the cadastral areas).

[61] Compare "Arcanum Maps", Maps of Europe, Europe in the XVIII. [!] century; "Arcanum Maps", Maps of Europe, Europe in the XIX. [!] century; Bajt and Vidic (2011), pp. 92, 114, 120, 132, 152–153; Moritsch (1969), pp. 110–114; Zwitter (2014a), pp. 146–150.

[62] Compare Panjek (2015), the pages cited earlier in this paragraph.

settlements, as recorded in the early nineteenth century.[63] During the hot and dry summer weather, resulting in the desolate appearance of grasslands and the absence of surface water, visitors felt as if they had been travelling around a desert-like, rocky environment.[64] Meanwhile, the cultivated fields located in *dolines* were mostly out of sight of casual travellers,[65] who thus often had an impression of a landscape that was very different from reality. While hay scarcity called for hay purchases in several cadastral municipalities on the Kras Plateau, the following further evidence proves that the historical agroecosystems on several parts of the plateau were comparatively healthy: vast quantities of hay were sold from some parts of the region of Grassland Kras (Fig. 16.2); the agroecosystems supported large numbers of sheep; and from a fairly limited part of the plateau, even firewood was sold to Trieste, as recorded in the nineteenth century, at the peak of the Kras Plateau's deforestation.[66]

Another example of agroecosystems in the northern Dinaric Karst, which are usually considered degraded, are the leaf litter forests of Bela Krajina, a region now located in south-eastern Slovenia. In the *longue durée* perspective, these socio-natural sites[67] were altered by the labour-intensive agriculture. The resulting ecosystems are characterised by distinct species compositions (Chap. 14). Litter raking in these environments started way back in the controlled solar-energy system of the agrarian sociometabolic regime and intensified in the eighteenth and the nineteenth centuries, when the role of barn-keeping in livestock husbandry increased. Forest litter raking mostly ceased in the second half of the twentieth century with the spread of other livestock bedding types, the introduction of slatted barn floor, or the abandonment of livestock farming. In the long run, litter raking changed the forest ecosystems, mainly due to the regular removal of nutrients. Nonetheless, with several protected species occurring there—for example, the *Helleborus dumetorum* hellebore species, stag's-horn clubmoss (*Lycopodium clavatum*), sweet william (*Dianthus barbatus*), and European cyclamen (*Cyclamen purpurascens*)—the existent litter forests are nowadays considered ecosystem types worth preserving for biodiversity reasons, at least in limited landscape sections. Soil acidification as a consequence of litter-raking created more favorable site conditions for stag's-horn clubmoss, an organism occuring only at acidic, nutrient poor sites (Aeschimann et al. 2004).[68] Thus, following the traditional-forestry-science-inspired narrative of a degraded environment of litter forests is not the only option (Chap. 14). The plurality of possible interpretations results from more than one sensible ecological baseline set as the reference point.[69] On the one hand, the increased soil acidity and decreased nutrient availability reduced the species diversity and created conditions less fertile for many plants, which had grown there

[63] Moritsch (1969), p. 114.

[64] Ferk and Zorn (2015), p. 373.

[65] Gams (1993), p. 112.

[66] Panjek (2015), pp. 90–91, 94–95, 107, 109–111.

[67] On this concept, see Winiwarter et al. (2013), pp. 105–108.

[68] Šilc et al. (2008); Aeschimann et al. (2004), pp. 48–49.

[69] Compare Ureta et al. (2020).

prior to the litter raking.[70] However, on the other hand, this allowed for an increased abundance of other organisms occurring in the "traditional" cultural landscape of the nineteenth century and earlier, including the listed rare species.

Focusing on the narratives of improvement, the classic karstology literature claims that the project of the afforestation of the Kras Plateau with black pine, taking place in the Austro-Hungarian Empire, the Kingdom of Italy, and the socialist Yugoslavia (see the contexts in Chap. 8) between the late-nineteenth and the mid-twentieth century, "*was long-lasting and successful, resulting in the selection of appropriate species, the development of new techniques for planting, and for the protection of plantations. It was influential in forestry practice elsewhere in Europe*".[71] Such discourse was inspired by historical forestry science.[72] In fact, the afforestation of the Kras Plateau with black pine in Austria-Hungary was the first extensive afforestation project in the (Sub-)Mediterranean, as established by Alfred Thomas Grove and Oliver Rackham. It inspired several subsequent afforestation projects (see also Chaps. 5 and 7).[73] It did establish forests in grasslands and even in previously barren land, where soil was brought in the framework of that project. Nonetheless, the afforestation with a non-native species that became naturalised in Kras opened up new and unforeseen long-term challenges in the interaction between humans and the rest of nature. These included the following: insecticide application in the sensitive karst environments in the framework of the afforestation; pine processionary moth infestation; the environmental impacts of the chemical suppression of organisms perceived as pests in pine stands; consequences for human health resulting from pine processionary moth suppression; unfulfilled expectations regarding the black pine's impact on the soil; the increased fire risk in pine forests; the influences of the altered land use on the local farm economy; and decreased biodiversity in the monoculture black pine plantations (Chap. 8). "*From today's point of view, the planting of alien tree species, despite preventing further soil erosion, is considered an inappropriate measure by most ecologists and many foresters.*"[74] However, the holistic picture of the afforestation projects' environmental impacts at the various Dinaric Karst sites differed significantly, even where the same tree species was involved. In the case of the southern slopes of the Svilaja Mountain in the Dalmatian hinterland (Zagora) during the 1960s in the Socialist Federal Republic of Yugoslavia, the black pine that was planted was an autochthonous species there, and the goal was to stop and prevent excessive flooding in one of the most important fertile areas of Mućko Polje after the unsustainable use of resources, deforestation, and subsequent massive erosion.[75]

[70] Šilc et al. (2008).

[71] Gams (1993), p. 112.

[72] Panjek (2015), pp. 97, 110–112.

[73] Panjek (2015), p. 111, also citing Grove and Rackham's *The nature of Mediterranean Europe: an ecological history*.

[74] Kaligarič and Ivajnšič (2014), p. 156.

[75] Durbešić and Milković (2005), pp. 133–143; see also the case of the reforestation with the non-native Aleppo pine in the Dalmatian littoral and the islands since the nineteenth century, which became invasive, in Fuerst-Bjeliš (2018), pp. 141–142.

The examples given here focus on the barren land and forest. However, several other environmental histories of the Dinaric Karst deserve a similar holistic perspective. This topic brings us to the last subsection.

16.3 Paving the Way for Future Research: From the Local to the Global Perspective

As explained in Chap. 1, this volume is the first multiauthored inter- and multidisciplinary monograph on the history of the Dinaric Karst environments and the local interactions between society and nature, spanning several historical periods. Thus, several gaps remain that will need to be bridged in the future.

Since this book focuses on material, some policy-related, and some intellectual environmental history, the fourth main environmental history study field—cultural environmental history[76]—largely remains a desideratum, along with further material, policy, and intellectual environmental history studies. A few Early Modern examples of cultural history data on the northern Dinaric Karst caves are briefly presented. In some parts of the Dinaric Karst, people would regularly get water from caves, for example, at Bločice east of Cerknica Polje (today in Slovenia), a place that will be mentioned again below. At many other locations, they would descend to the water in caves in times of drought.[77] Farmers also visited the caves below the bottoms of intermittent lakes in some karst *poljes* to fish or merely collect the fish that got stuck there when the water was sinking—full cartloads in some cases. Regular cave visits, however, did not mean that the fear of the karstic underground world vanished.[78] We can all imagine the monsters appearing on the cave walls and speleothems in the flickering light and shadows as the visitors found their way to the water carrying torches.[79] Visits to caves resulted in many stories of cave ghosts, such as the one allegedly reported by a man fishing in a cave below the Cerknica Polje in the seventeen century: as a thunderstorm approached and the thunder echoed in the cave, the fisherman got scared and ascended to the surface. When he returned, he found his fishing net broken and claimed a ghost had destroyed it. The endemic cave fauna—the olm (*Proteus anguinus*), a cave-dwelling salamander that occasionally appears in some karst springs—was sometimes interpreted as dragon younglings from the karst underground. Caves that were said to get annoyed if provoked and send out storms are another cultural environmental history topic, and annual blessings were held at some such caves in the early modern period and later to prevent the storms.[80] The cultural history of fears of the karst underground could result in exceedingly tragic episodes such as the witch trial killing many innocent inhabitants of the village of

[76] McNeill (2011), pp. 160–161.

[77] Cvijić (1925), p. 4; Gams (1981–1982), p. 38; Zwitter (2014b), pp. 650, 655, 674.

[78] Zwitter (2014b), pp. 655, 660.

[79] Palladino and Bidovec (2008), p. 91.

[80] Gams (1981–1982), p. 36; Zwitter (2014b), pp. 639, 655, 659–661.

Bločice in the seventeenth century, who reportedly claimed they had met the Devil in the caves where they supplied themselves with water.[81] Witch hunts have recently received much environmental historiography attention in Central Europe, focused on the climate history perspective[82]; a systematic karst-specific approach to the topic is a desideratum. A cultural history of caves can range from perceptions of fear, and their broader impacts to positive attitudes towards the karst underground, such as healing cave water springs and uses of speleothems in folk medicine. The history of some caves as holy mass sites[83] could address the question of whether holding holy masses in some caves was primarily a continuation of pre-Christian cult sites,[84] a sign of taming the eerie underground world, or simply a means of appreciating the beauty of the speleothem-filled caves.

Focusing on general environmental history desiderata, some authors in this volume showed remarkable talent for interdisciplinary cooperation. However, optimally, "[e]*nvironmental history is about as interdisciplinary as intellectual pursuits can get.*"[85] The desire to ensure even greater interdisciplinarity is not specific to the Dinaric Karst: it has recently been emphasised by global mainstream environmental historians.[86] To quote Verena Winiwarter's response at a roundtable debate *Practicing interdisciplinarity in environmental history: potentials and pitfalls*, held at the *Third world congress of environmental history* in Florianópolis, Brazil: "*If you want to go fast – go alone. But, if you want to go far – go together!*".[87]

Another global observation that John Robert McNeill's environmental historiography emphasises is that more inspiration by the publications written by scholars working on similar or related subjects, concepts, and methodological approaches elsewhere in the world "*will also propel environmental history forward*".[88] This is particularly pressing in several texts on the Dinaric Karst. Namely, in the times when environmental historiography production is so extensive that following U.S. publications alone has become a full-time job; when environmental history is published in many languages; and when certain academic environments maintain their tradition as centres of knowledge while the achievements of diligent authors, published in states perceived as peripheral (like the Dinaric Karst countries, for example), receive much less worldwide attention, and the inadequate knowledge of some studies from other continents in the Dinaric Karst countries and of some Dinaric Karst literature abroad is sometimes a challenge.[89]

---

[81] Zwitter (2014b), pp. 655–656.

[82] Pfister and Wanner (2021), pp. 267–270, 274–276; see also Tratnik-Volasko and Košir (1995).

[83] Gams (1981–1982), pp. 35, 37; Malez (1984), p. 93, see also pages 94–95; Zwitter (2014b), p. 655.

[84] Compare Frelih (1997); Malez (1984), p. 93.

[85] McNeill (2003), p. 9.

[86] E.g. McNeill (2011), p. 174.

[87] Conference discussion following the roundtable Carruthers et al. (2023); see Zwitter (2019), p. 161.

[88] McNeill (2011), pp. 174–175 (quote from page 175).

[89] Compare McNeill (2011), pp. 163, 174–175.

This challenge also results in a lack of a theoretical coherence within this interdisciplinary monograph. Workshops focusing on the issue at hand, participated in by all the authors involved, could have led to a higher level of theoretical consistency "*while maintaining the richness of various scientific approaches and flavo*[u]*rs*". Nonetheless, considerable theoretical uniformity in an interdisciplinary volume tackling diverse research topics, written by several authors used to diverse epistemological frameworks, is a highly time-consuming goal even under ordinary conditions,[90] let alone during the times of the COVID-19 pandemic when such an effort would have resulted in excessive delays. Hence, the decision was made to skip this stage in the first book of this kind focusing on the Dinaric Karst.

In conclusion, let us quote John Robert McNeill: "[E]*nvironmental history as a field needs more integration, both more integration with other disciplines, and more integration within itself, among scholars at work on different regions and different problems.*"[91] This is one of the key desiderata regarding the environmental history of the Dinaric Karst.

Acknowledgements The work on this chapter was funded by (1) the Ministry of Science, technological development, and innovations of the Republic of Serbia, University of Belgrade—Faculty of Philosophy (No. 451-03-47/2023-01/200163), (2) the Slovenian Research and Innovation Agency, research programmes No. P6-0101 (Geography of Slovenia) and No. P6-0235 (Slovene History), and (3) the University of Zagreb's Grant for research on the influence of paleo and recent climate changes on processes in marine and continental environments in 2020–2023 (Nos. 20285705; 20286205; 20286474; 20286585). The authors wish to thank Borut Praper, who copy-edited the text, and Lenart Štaut and Manca Volk Bahun, who drew the maps.

References

Sources

Published Sources

"Arcanum Maps", Cadastral maps, Habsburg Empire—Cadastral maps (XIX. [!] century). Accessed on 21 July 2023. https://maps.arcanum.com/en/map/cadastral/?layers=3%2C4&bbox=1794877.5643639148%2C6125569.198257975%2C1867760.27083383%2C6149035.3659415245

"Arcanum Maps", Maps of Europe, Europe in the XVIII. [!] century. Accessed on 21 July 2023. https://maps.arcanum.com/en/map/europe-18century-firstsurvey/?layers=163%2C165&bbox=1545481.2389516747%2C5726941.901501992%2C1564141.4284817935%2C5734146.0914178705

"Arcanum Maps", Maps of Europe, Europe in the XIX. [!] century. Accessed on 21 July 2023. https://maps.arcanum.com/en/map/europe-19century-secondsurvey/?layers=here-aerial%2C158%2C164&bbox=1545258.343866624%2C5724786.826674126%2C1562412.971807913%2C5731409.760349359

[90] Compare Haberl et al. (2016), p. xliv (including the quote).

[91] McNeill (2011), p. 175.

ARS [Archives of the Republic of Slovenia]: SI AS 176, Franciscejski kataster za Kranjsko [Franziszeischer Kataster für Krain]: A119 (cadastral municipality of Parje) and A151 (cadastral municipality of Zagorje), both from 1823

ARSO [Slovenian Environment Agency], Atlas okolja [Environmental atlas of Slovenia]. Accessed on 14 August 2023. http://gis.arso.gov.si/atlasokolja/profile.aspx?id=Atlas_Okolja_AXL@Arso&culture=en-US.

GURS [Surveying and Mapping Authority of the Republic of Slovenia] (2023) Digitalni ortofoto [Digital ortophoto]. Ljubljana

Archival Source

ÖStA AVAFHKA (Österreichisches Staatsarchiv: Allgemeines Verwaltungsarchiv – Finanz- und Hofkammerarchiv): Hoffinanz und Niederösterreichische Kammer, Niederösterreichische Kammerakten, boundle 15

Literature

Aeschimann D, Lauber K, Moser DM, Theurillat J-P (2004) Flora alpina: Ein Atlas sämtlicher 4500 Gefäßpflanzen der Alpen: Mit Farbfotografien von Konrad Lauber und Skizzen von André Michel. Band 1, Lycopodiaceae–Apiaceae. Haupt Verlag, Bern, Stuttgart, and Wien

Babai D, Molnár Z (2014) Small-scale traditional management of highly species-rich grasslands in the Carpathians. Agricult, Ecosyst Environ 182:123–130. https://doi.org/10.1016/j.agee.2013.08.018

Babai D, Molnár A, Molnár Z (2014) Ahogy gondozza, úgy veszi hasznát": Hagyományos ökológiai tudás és gazdálkodás Gyimesben [Traditional ecological knowledge and land use in Gyimes (Eastern Carpathians)]. Vácrátót, Budapest

Bajt D, Vidic M (eds) (2011) Slovenski zgodovinski atlas [Historical atlas of the Slovenian territory]. Nova revija, Ljubljana

Bertoša M (1986) Jedna zemlja, jedan rat: Istra 1615/1618 [A land in a war: Istria 1615–1618]. Istarska naklada, Pula

Bertoša M (2007) Istarsko rano novovjekovlje. Razvojne smjernice od 16. do 18. stoljeća [Early Modern Istria: development trajectories from the 16th to the 18th century]. In: Budak N (ed) Dalmacija, Dubrovnik i Istra u ranome novom vijeku, Vol 3 of Hrvatska povijest u ranome novom vijeku. Barbat and Leykam International, Zagreb, pp 80–120

Breg Valjavec M, Zorn M, Čarni A (2018) Human-induced land degradation and biodiversity of Classical Karst landscape: on the example of enclosed karst depressions (dolines). Land Degradat Develop 29(10):3823–3835. https://doi.org/10.1002/ldr.3116

Carruthers E, Hou S, Ribeiro de Oliveira R, Winiwarter V, Slavin P (2023) Practicing interdisciplinarity in environmental history: potentials and pitfalls. In: 3rd world congress of environmental history. Accessed on 3 August 2023. https://convention2.allacademic.com/one/iceho/iceho19/index.php?cmd=Online+Program+View+Session&selected_session_id=1453130&PHPSESSID=k7g0qnqmvfstt3bss3m2e9i6tr

Carter FW (1994) Settlement and population during Venetian rule (1420–1797): Hvar Island, Croatia. J European Econ Hist 23(1):7–47

Coates P (2004) Emerging from the wilderness (or, from redwoods to bananas): recent environmental history in the United States and the rest of America. Environ Hist 10(4):407–438. https://doi.org/10.3197/0967340042772676

Cvijić J (1925) Karst i čovek [Karst and man]. Glasnik geografskog društva 11:1–11

Durbešić A, Milković I (2005) Pošumljavanje neobraslog šumskog zemljišta na južnim padinama Svilaje—Muć s ciljem protuerozijskog djelovanja [Afforesting unstocked forest land on the southern slopes of Svilaja—Muć mountain for anti-erosive purposes]. Šumarski list 129:133–143

Ellis E (2018) Anthropocene: a very short introduction. Oxford University Press, Oxford. https://doi.org/10.1093/actrade/9780198792987.001.0001

Ferk M, Zorn M (2015) Kras—večpomenski izraz z mednarodno veljavo [Kras—a polysemous term with international recognition]. Kronika: časopis za slovensko krajevno zgodovino 63(3):367–378

Fister P (1975) Arhitektura slovenskih protiturških taborov [The architecture of anti-Ottoman peasant forts (tabors) in Slovenia]. Slovenska matica, Ljubljana

Ford D (2007) Jovan Cvijić and the founding of karst geomorphology. Environmental Geol 51:675–684. https://doi.org/10.1007/s00254-006-0379-x

Frelih M (1997) The prehistoric cave sanctuary Mušja jama in Slovenia: an entrance to the reign of Hades? Slovenian Science Foundation, Ljubljana

Fuerst-Bjeliš B (2003) Reading the Venetian cadastral record: an evidence for the environment, population and cultural landscape of the 18th century Dalmatia. Hrvatski geografski glasnik 65(1): 47–62. https://doi.org/10.21861/HGG.2003.65.01.03

Fuerst-Bjeliš B (2018) Deforestacija i reforestacija u hrvatskome Dinarskom kršu: odnosi i implikacije [Deforestation and reforestation in Croatian Dinaric Karst: human-environmental relations and implications]. Ekonomska i Ekohistorija 14:136–144

Fuerst-Bjeliš B (2020) Environmental response to marginality: between borderlands and littoralization in the eastern Adriatic. In: Fuerst-Bjeliš B, Leimgruber W (eds) Globalization, marginalization and conflict. Political, economic and social processes. Springer, Cham, pp. 75–87. https://doi.org/10.1007/978-3-030-53218-5_6

Fuerst-Bjeliš B, Vukosav B (2019) Landscapes of identities in shared spaces of the borderlands. Europa Reg 26(2):5–13. https://nbn-resolving.org/urn:nbn:de:0168-ssoar-66829-3

Fuerst-Bjeliš B, Glamuzina N (2021) The historical geography of Croatia: territorial change and cultural landscapes. Springer, Cham. https://doi.org/10.1007/978-3-030-68433-4

Fuerst-Bjeliš B, Petrić H (2024) Environmental history in Croatia – the origins, state and perspectives: views from geography and history. Econ and Ecohist: J Econ Hist Environ Hist 20: forthcoming

Gams I (1981–1982) Izraba jam v Sloveniji skozi stoletja [Use of the caves in Slovenia through the centuries]. Naše jame 23–24:35–41

Gams I (1991) Sistemi prilagoditve primorskega dinarskega krasa na kmetijsko rabo tal [Systems of adapting the littoral Dinaric Karst to agrarian land use]. Geografski zbornik 31:5–106

Gams I (1993) Origin of the term 'karst,' and the transformation of the Classical Karst (kras). Environ Geol 21:110–114. https://doi.org/10.1007/BF00775293

Haberl H, Erb K-H, Fischer-Kowalski M, Groß R, Krausmann F, Plutzar C, Schmid M, Winiwarter V (2016) Introduction. In: Haberl H, Fischer-Kowalski M, Krausmann F, Winiwarter V (eds) Social ecology: society-nature relations across time and space. Springer, Cham, xli–lxii

Head MJ, Steffen W, Fagerlind D, Waters CN, Poirier C, Syvitski J, Zalasiewicz JA, Barnosky AD, Cearreta A, Jeandel C et al (2022) The Great Acceleration is real and provides a quantitative basis for the proposed Anthropocene Series/Epoch. Episodes 45(4):359–376. https://doi.org/10.18814/epiiugs/2021/021031

International Union of Geological Sciences: International Commission on Stratigraphy (2015) Subcommission on Quaternary Stratigraphy, Working group on the 'Anthropocene'. http://quaternary.stratigraphy.org/working-groups/anthropocene/

Ivanišević V, Veljanovski T, Cowley D, Kiarszys G, Bugarski I (eds) (2015) Recovering lost landscapes. Institute of Archaeology and Aerial Archaeology Research Group, Belgrade

Jaritz G, Winiwarter V (1997) On the perception of nature in a renaissance society. In: Teich M, Porter R, Gustafsson B (eds) Nature and society in historical context. Cambridge University Press, Cambridge, pp 91–111

Kalc A (2017) Peasant population and income integration: the case of the Trieste port-town hinterland (Karst, 16th-20th centuries). In: Panjek A, Larsson J, Mocarelli L (eds) Integrated peasant economy in a comparative perspective: Alps, Scandinavia and beyond. University of Primorska Press, Koper, pp 205–227

Kalc A (2018) Prebivalstvo in gospodarski viri v tržaškem kraškem zaledju (16.–20. stoletje) [Population and economic resources in the Trieste hinterland of Kras (16th–20th century)] In: Panjek A, Lazarević Z (eds) Preživetje in podjetnost: integrirana kmečka ekonomija na Slovenskem od srednjega veka do danes. Založba Univerze na Primorskem, Koper, pp 155–188

Kaligarič M, Ivajnšič D (2014) Vanishing landscape of the 'classic' Karst: changed landscape identity and projections for the future. Landsc Urban Plan 132:148–158. https://doi.org/10.1016/j.landurbplan.2014.09.004

Klemenčič V (1959) Pokrajina med Snežnikom in Slavnikom: gospodarska geografija [La région entre le Snežnik et le Slavnik: géographie économique]. Slovenska akademija znanosti in umetnosti, Ljubljana

Kokalj Ž, Hesse R (2017) Airborne laser scanning raster data visualization. A guide to good practice. Založba ZRC, Ljubljana. https://doi.org/10.3986/9789612549848

Kos M (1976) Pivka v srednjem veku [Pivka in the Middle Ages]. Kronika: časopis za slovensko krajevno zgodovino 24(1):5–10

Kranjc A (2009) History of deforestation and reforestation in the Dinaric Karst. Geogr Res 47(1):15–23. https://doi.org/10.1111/j.1745-5871.2008.00552.x

Krausmann F, Weisz H, Eisenmenger N (2016) Transitions in sociometabolic regimes throughout human history. In: Haberl H, Fischer-Kowalski M, Krausmann F, Winiwarter V (eds) Social ecology: society-nature relations across time and space. Springer, Cham, pp 63–92. https://doi.org/10.1007/978-3-319-33326-7_3

Krevs M (ed) (2021) Hidden geographies. Springer Nature, Cham. https://doi.org/10.1007/978-3-030-74590-5

Kulušić S (1965) Kornatska otočna skupina [The Kornati archipelago]. Geografski glasnik 27:215–245. https://hrcak.srce.hr/56136

Kulušić S (2000) Naseljenost i vrednovanje Kornatskih otoka od drevnih vremena do danas [Settlement and evaluation of the Kornati islands since antiquity to the present day]. Hrvatski geografski glasnik 62(1):43–65. https://hrcak.srce.hr/36924

Kulušić S (2001) Naseljenost i vrednovanje Kornatskih otoka od drevnih vremena do danas (Nastavak) [Settlement and evaluation of the Kornati islands since Antiquity to the present day (Continuation)]. Hrvatski geografski glasnik 63(1):87–116. https://hrcak.srce.hr/36920

Laharnar B (2018) Šilentabor od prazgodovine do srednjega veka [Šilentabor from Prehistory to the Middle Ages]. Interview by Milan Trobič. Nedeljska reportaža 3:26–11:51. https://prvi.rtvslo.si/podkast/nedeljska-reportaza/96/174567026

Laharnar B (2022) From Ocra to Albion: Notranjska between prehistory and antiquity. Narodni muzej Slovenije, Ljubljana

Laharnar B, Lozić E, Štular B (2019) A structured Iron Age landscape in the hinterland of Knežak, Slovenia. In: Cowley DC, Fernández-Götz M, Romankiewicz T, Wendling H (eds) Rural settlement: relating buildings, landscape, and people in the European Iron Age. Sidestone Press, Leiden, pp 263–271

Leusen V, Martijn P, Pizziolo G, Sarti L (eds) (2011) Hidden landscapes of Mediterranean Europe: cultural and methodological biases in pre- and protohistoric landscape studies. Archaeopress, Oxford

Lane FC (1973) Venice: a maritime republic. The Johns Hopkins University Press, Baltimore and London

Leban V (1950) Nanos: gospodarska povezava s sosedstvom [Nanos: les rapports économiques avec les régions voisines]. Geografski vestnik 22:100–137

Lewis SL, Maslin MA (2015) Defining the Anthropocene. Nature 519:171–180. https://doi.org/10.1038/nature14258

Luzzatto-Fegiz P (1929) La popolazione di Trieste (1875–1928). Officine grafiche de la Editoriale Libraria, Trieste

Malez M (1984) Povijest speleoloških istraživanja u Hrvatskoj [Geschichte der Höhlenforschungen in Kroatien]. In: Deveti jugoslavenski speleološki kongres. Savez speleologa Jugoslavije and Speleološko društvo Hrvatske, Zagreb, pp 73–102

Matičetov M (1948) O etnografiji in folklori zapadnih Slovencev [De l'ethnographie et du folklore des Slovènes occidentaux]. Slovenski etnograf 1:9–56

McNeill JR (1992) The mountains of the Mediterranean world: an environmental history. Cambridge University Press, Cambridge

McNeill JR (2003) Observations on the nature and culture of environmental history. History and Theory 42:5–43. https://doi.org/10.1046/j.1468-2303.2003.00255.x

McNeill JR (2011) The historiography of environmental history. In: Schneider A, Woolf D (eds) Historical writing since 1945, Vol. 5 of The Oxford history of historical writing. Oxford University Press, Oxford, pp 159–176

Mitchell D (2000) Cultural geography—a critical introduction. Blackwell, Oxford and Malden, MA

Mlacović D (2014) The communes of the northern Istria in the Late Middle Ages and wood. In: Štih P, Zwitter Ž (eds) Man, nature and environment between the northern Adriatic and the Eastern Alps in premodern times. Ljubljana University Press, Faculty of Arts, and Historical Association of Slovenia, Ljubljana, pp 86–101

Mlekuž Vrhovnik D, Fabec T (2024) The formation of the Kras landscape from prehistory to the Early Modern Period. In: Fuerst-Bjeliš B, Mrgić J, Petrić H, Zorn M, Zwitter Ž (eds) Environmental histories of the Dinaric Karst. Springer, Cham, pp 81–107. https://doi.org/10.1007/978-3-031-56089-7_4

Moritsch A (1969) Das nahe Triester Hinterland: Zur wirtschaftlichen und sozialen Entwicklung vom Beginn des 19. Jahrhunderts bis zur Gegenwart. Hermann Böhlau's Nachf, Wien, Köln, and Graz

Mrgić J (2013) Zemlja i ljudi: iz istorije životne sredine Zapadnog Balkana [Land and people: from the environmental history of the Western Balkans]. Equilibrium, Beograd

Mrgić J (2014) Rocks, waters and bushes—what did the Ragusan commune acquire from the Bosnian king in 1399? In: Štih P, Zwitter Ž (eds) Man, nature and environment between the northern Adriatic and the Eastern Alps in premodern times. Ljubljana University Press, Faculty of Arts, and Historical Association of Slovenia, Ljubljana, pp 52–73

Novak V (1970) Živinoreja [Die Viehzucht]. In: Blaznik P, Grafenauer B, Kos M, Vilfan S, Zwitter F, Žontar J (eds) Agrarno gospodarstvo. Vol. 1 of Gospodarska in družbena zgodovina Slovencev: Zgodovina agrarnih panog. Slovenska akademija znanosti in umetnosti and Državna založba Slovenije, Ljubljana, pp 343–394

Palladino I, Bidovec M (2008) Johann Weichard von Valvasor (1641–1693): Protagonist der Wissenschaftsrevolution der Frühen Neuzeit: Leben, Werk und Nachlass. Böhlau, Wien, Köln, and Weimar

Panjek A (2015) Paesaggio culturale e ambiente del Carso: L'uso delle risorse naturali in età moderna. Založba Univerze na Primorskem, Koper

Pavelić F, Petrić H, Radošević M (2024) Deforestation and reforestation attempts of the Military Frontier's coastal Dinaric Karst (1762–1881), Croatia. In: Fuerst-Bjeliš B, Mrgić J, Petrić H, Zorn M, Zwitter Ž (eds) Environmental histories of the Dinaric Karst. Springer, Cham, pp 169–201. https://doi.org/10.1007/978-3-031-56089-7_7

Perko F (2016) Od ogolelega do gozdnatega krasa: pogozdovanje krasa [From bare to forested karst: karst afforestation]. Zveza gozdarskih društev Slovenije–Gozdarska založba, and Jutro, Ljubljana

Pfister C, Wanner H (2021) Climate and society in Europe: the last thousand years. Haupt, Berne

Remec M (2024) Afforestation with non-native *Pinus nigra* in the karstic areas of southwestern Slovenia since the mid-nineteenth century: environmental implications. In: Fuerst-Bjeliš B, Mrgić J, Petrić H, Zorn M, Zwitter Ž (eds) Environmental histories of the Dinaric Karst. Springer, Cham, pp 203–237. https://doi.org/10.1007/978-3-031-56089-7_8

Rogić V (1976) Socijalno-geografski aspekt Dinarskog krša, dinarskog kulturnog areala i dinarskog brdsko-planinskog prostora [The socio-geographical aspect of Dinaric Karst, the Dinaric cultural area and the Dinaric mountainous space]. Geografski glasnik 38:253–269

Rože Kravanja A (2024) Water infrastructure in Kras in the 19th century between continuity and change. In: Fuerst-Bjeliš B, Mrgić J, Petrić H, Zorn M, Zwitter Ž (eds) Environmental histories of the Dinaric Karst. Springer, Cham, pp 295–337. https://doi.org/10.1007/978-3-031-56089-7_11

Schmitt OJ (2016) Introduction: the Ottoman conquest of the Balkans: research questions and interpretations. In: Schmitt OJ (ed) The Ottoman conquest of the Balkans: interpretations and research debates. Österreichische Akademie der Wissenschaften, Wien, pp 7–45

Schöller R (2010) Schafe. In: Jaeger F (ed) Enzyklopädie der Neuzeit. Vol. 11, Renaissance–Signatur. J. B. Metzler, Stuttgart and Weimar, pp 634–637

Smerdel I (1989) Ovčarstvo na Pivki: transhumanca od srede 19. do srede 20. stoletja ali trije "ovčarji": etnološka razprava [Sheep farming in the Pivka region: transhumance from the mid-nineteenth to mid-twentieth century, or three "shepherds": an ethnological treatise]. Lipa, Koper

Šilc U, Čarni A, Košir P, Marinšek A, Zelnik I (2008) Litter-raking forests in SE Slovenia and in Croatia. Hacquetia 7(1):71–88

Šumrada J (1986) Gospodarske posledice turških vpadov na Kras do tridesetih let 16. stoletja [Die wirtschaftlichen Folgen der türkischen Einfälle auf den Karst bis zu den dreißeiger Jahren des 16. Jahrhunderts]. In: Dolgan M, Dolinar D, Jakopin F, Pokorn D (eds) Družbena in kulturna podoba slovenske reformacije, 17–25. Slovenska akademija znanosti in umetnosti, Ljubljana

Tekić I, Fuerst-Bjeliš B, Durbešić A, Radeljak Kaufmann P, Cvitanović M (2024) Landscape change and fire risk in the Croatian Dinaric Karst—looking back and moving forward. In: Fuerst-Bjeliš B, Mrgić J, Petrić H, Zorn M, Zwitter Ž (eds) Environmental histories of the Dinaric Karst. Springer, Cham, pp 109–139. https://doi.org/10.1007/978-3-031-56089-7_5

Tratnik-Volasko M, Košir M (1995) Čarovnice: predstave, procesi, pregoni v evropskih in slovenskih deželah [Witches: notions, trials, and persecutions in Europe and in the Slovenian provinces]. Znanstveno in publicistično središče, Ljubljana

Tripathy-Lang A (2021) The difficulty of defining the Anthropocene. Eos 102. https://doi.org/10.1029/2021EO156556

Umek E (1957) Prispevki k zgodovini ovčereje na Krasu in v slovenski Istri [Beitrag zur Geschichte der Schaftzucht am Karst und im slowenischen Istrien]. Slovenski etnograf 10:71–76

Ureta S, Lekan T, von Hardenberg WG (2020) Baselining nature: an introduction. Nature Space 3(1): 3–19. https://doi.org/10.1177/2514848619898092

Vilfan S (1957a) K obdelavi polja v slovenski Istri: nekaj misli ob popotnih zapiskih [Einige Gedanken zur Feldbearbeitung im slowenischen Istrien]. Slovenski etnograf 10:61–70

Vilfan S (1957b) Podobe iz nekdanje živinoreje med Trstom in Slavnikom [Selected topics from the history of livestock farming between Trieste and Slavnik]. Kronika: časopis za slovensko krajevno zgodovino 5(2):69–87

Voje I (1978) The influence of the Ottoman Empire on Slovenian countries in the 15th and 16th centuries. In: César J (ed) Ottoman rule in Middle Europe and Balkan in the 16th and 17th centuries. Oriental Institute in Academia, Prague, pp 108–141

Voje I (1996) Slovenci pod pritiskom turškega nasilja [Die Slowenen unter dem Druck der türkischen Gewalt]. Znanstveni inštitut Filozofske fakultete, Ljubljana

Winiwarter V, Knoll M (2007) Umweltgeschichte: Eine Einführung. Böhlau, Köln, Weimar, Wien

Winiwarter V, Schmid M, Hohensinner S, Haidvogl G (2013) The environmental history of the Danube River Basin as an issue of long-term socio-ecological research. In: Singh SJ, Haberl H, Chertow M, Mirtl M, Schmid M (eds) Long-term socio-ecological research: studies in society-nature interactions across spatial and temporal scales. Springer, Dordrecht, pp 103–122. https://doi.org/10.1007/978-94-007-1177-8_5

Zalasiewicz J, Waters CN, Ellis EC, Head MJ, Vidas D, Steffen W, Thomas JA, Horn E, Summerhayes CP, Leinfelder R et al (2021) The Anthropocene: comparing its meaning in geology (chronostratigraphy) with conceptual approaches arising in other disciplines. Earth's Future 9(3):e2020EF001896. https://doi.org/10.1029/2020EF001896

Zorn M, Kumer P, Ferk M (2015) Od gozda do gozda ali kje je goli, kamniti Kras [From forest to forest or where is the bare rocky Karst?]. Kronika: časopis za slovenko krajevno zgodovino 63(3):561–574

Zwitter Ž (2014a) Od zgodnjega kapitalizma do konca prve svetovne vojne: razvoj cest in cestnega prometa v novem veku od začetka 16. stoletja do leta 1918 [From early capitalism to the end of the First World War: roads and road traffic developments in the Modern Period from the early 16th century to 1918]. In: Ficko G (ed) Ceste na Slovenskem skozi čas, Vol 1. DRC, Ljubljana, pp 141–237

Zwitter Ž (2014b) Okolje na Kranjskem v 17. stoletju po *Slavi vojvodine Kranjske* [Die Umwelt in Krain im 17. Jahrhundert in der *Die Ehre Deß Hertzogthums Crain*]. In: Weiss J (ed) Studia Valvasoriana: zbornik spremnih študij ob prvem integralnem prevodu Die Ehre Deß Hertzogthums Crain v slovenski jezik. Zavod Dežela Kranjska, Ljubljana, pp 611–702

Zwitter Ž (2015) Izbrani vidiki okoljske zgodovine kobilarne Lipica od ustanovitve do začetka tridesetih let 17. stoletja [Selected environmental history aspects of the Lipica stud farm from its foundation to the early 1630s]. Kronika: časopis za slovensko krajevno zgodovino 63(3):457–474

Zwitter Ž (2019) Tretji svetovni okoljskozgodovinski kongres (3rd world congress of environmental history), Florianópolis, Brazilija, julij 2019 [3rd world congress of environmental history, Florianópolis, Brazil, July 2019]. Econ and Ecohist: J Econ Hist Environ Hist 15:160–162

Zwitter Ž (2020) Eine Wissensgeschichte der Wiesen und Weiden im Südosten der Alpen im 16. und im frühen 17. Jahrhundert. Mitteilungen des Instituts für Österreichische Geschichtsforschung 128(1):49–82

Zwitter Ž, Rasran L (forthcoming) Historical biodiversity in the Alps: grassland agroecosystems in the last millennium. Austrian Academy of Sciences, Vienna

Žiga Zwitter Assistant professor of early modern history. He graduated in history and geography. In 2015 he obtained his Ph.D. in history. In 2016–2017 he was a postdoctoral research fellow at the Vienna's Institute of Social Ecology. His research has focused on environmental history. He has been leading the Croatia–Serbia–Slovenia region in the *European Society for Environmental History*, and has served as a vice-president of the *Historical Association of Slovenia.* He is the main author of the forthcoming monograph *Historical Biodiversity in the Alps: Species-Rich Grassland Agroecosystems in the Last Millennium.*

Borna Fuerst-Bjeliš Senior researcher and full professor. Her research interests include environmental history and landscape change, focusing spatially mainly on the Mediterranean karst region, including the development of research methods (historical cartography and GIS technology), borderlands, spatial perception and spatial (regional) identities. She is (co-)author/(co-)editor of books on topics of *Historical Geography of Croatia, Mediterranean Identities, and Globalization, Marginalization and Conflict*. She served as Vice-President of the *European Society for Environmental History* (2017–2019) and she is currently Chair of the *Commission on Marginalization, Globalization and Regional and Local Responses of the International Geographical Union* (2020–present).

Jelena Mrgić Associate professor. She teaches *Historical geography* and *Introduction to environmental history*. Her focus in research changed from historical geography of the mediaeval and early Ottoman Bosnia to preindustrial environmental history of the Western Balkans, changes of settlement patterns and cultural landscapes, wine and brandy production history in lieu of climate changes during the LIA, to emotional history as an aspect of studying environment, past and recent crises. She has been a representative of Serbia in the *European Society for Environmental History*

(ESEH) since 2010, now jointly represented in the Region of Slovenia-Croatia-Serbia. She has written three books, several book chapters in edited volumes and research articles in peer reviewed journals and is active in reviewing papers for research journals in environmental history.

Hrvoje Petrić Professor of history. His research deals with environmental history, economic history and Croatian early modern history. He has written several books and a series of articles (more than 20). He is the editor of "Ekonomska i ekohistorija" [*Economic and Environmental History*] Journal.

Matija Zorn Head of the Anton Melik Geographical Institute of the Research Centre of the Slovenian Academy of Sciences and Arts (Ljubljana, Slovenia), and a lecturer at the Faculty of Humanities of the University of Primorska (Koper, Slovenia). He is a geographer and a historian, PhD. His main research interest includes physical geography (especially geomorphology), natural hazards, land degradation, environmental history, historical geography and geographic information systems. Most recently he co-edited a book titled *The Geography of Slovenia: Small But Diverse* (Springer Nature). He is Chairman of the *Commission on Land Degradation and Desertification of the International Geographical Union.*